MOLECULAR BASIS OF SPECIFICITY IN NUCLEIC ACID-DRUG INTERACTIONS

THE JERUSALEM SYMPOSIA ON
QUANTUM CHEMISTRY AND BIOCHEMISTRY

Published by the Israel Academy of Sciences and Humanities,
distributed by Academic Press (N.Y.)

1. *The Physicochemical Aspects of Carcinogenesis* (October 1968)
2. *Quantum Aspects of Heterocyclic Compounds in Chemistry and Biochemistry* (April 1969)
3. *Aromaticity, Pseudo-Aromaticity, Antiaromaticity* (April 1970)
4. *The Purines: Theory and Experiment* (April 1971)
5. *The Conformation of Biological Molecules and Polymers* (April 1972)

Published by the Israel Academy of Sciences and Humanities,
distributed by D. Reidel Publishing Company (Dordrecht, Boston, Lancaster, and Tokyo)

6. *Chemical and Biochemical Reactivity* (April 1973)

Published and distributed by D. Reidel Publishing Company
(Dordrecht, Boston, Lancaster, and Tokyo)

7. *Molecular and Quantum Pharmacology* (March/April 1974)
8. *Environmental Effects on Molecular Structure and Properties* (April 1975)
9. *Metal-Ligand Interactions in Organic Chemistry and Biochemistry* (April 1976)
10. *Excited States in Organic Chemistry and Biochemistry* (March 1977)
11. *Nuclear Magnetic Resonance Spectroscopy in Molecular Biology* (April 1978)
12. *Catalysis in Chemistry and Biochemistry Theory and Experiment* (April 1979)
13. *Carcinogenesis: Fundamental Mechanisms and Environmental Effects* (April/May 1980)
14. *Intermolecular Forces* (April 1981)
15. *Intermolecular Dynamics* (Maart/April 1982)
16. *Nucleic Acids: The Vectors of Life* (May 1983)
17. *Dynamics on Surfaces* (April/May 1984)
18. *Interrelationship Among Aging, Cancer and Differentiation* (April/May 1985)
19. *Tunneling* (May 1986)
20. *Large Finite Systems* (May 1987)

Published and distributed by Kluwer Academic Publishers
(Dordrecht, Boston, London)

21. *Transport through Membranes: Carriers, Channels and Pumps* (May 1988)
22. *Perspectives in Photosynthesis* (May 1989)

VOLUME 23

MOLECULAR BASIS OF SPECIFICITY IN NUCLEIC ACID-DRUG INTERACTIONS

PROCEEDINGS OF THE TWENTY-THIRD JERUSALEM SYMPOSIUM ON QUANTUM CHEMISTRY AND BIOCHEMISTRY HELD IN JERUSALEM, ISRAEL, MAY 14–17, 1990

Edited by

BERNARD PULLMAN

*Institut de Biologie Physico-Chimique
(Fondation Edmond de Rothschild), Paris, France*

and

JOSHUA JORTNER

Department of Chemistry, University of Tel-Aviv, Israel

KLUWER ACADEMIC PUBLISHERS

DORDRECHT / BOSTON / LONDON

ISBN 0-7923-0897-2

o 4761595

CHEMISTRY

Published by Kluwer Academic Publishers,
P.O. Box 17, 3300 AA Dordrecht, The Netherlands.

Kluwer Academic Publishers incorporates
the publishing programmes of
D. Reidel, Martinus Nijhoff, Dr W. Junk and MTP Press.

Sold and distributed in the U.S.A. and Canada
by Kluwer Academic Publishers,
101 Philip Drive, Norwell, MA 02061, U.S.A.

In all other countries, sold and distributed
by Kluwer Academic Publishers Group,
P.O. Box 322, 3300 AH Dordrecht, The Netherlands.

Printed on acid-free paper

Printed in the Netherlands

PREFACE

One of the central problems in the study of the mechanism of DNA-ligand interactions is the existence and nature of sequence specificity with respect to the base pairs of DNA. The presence of such a specificity could be of particular significance because it might possibly mean the involvement of specific genes in the effectiveness of the different drugs. The elucidation of the factors responsible for the specificity could then be important for the development of compounds susceptible to contribute to the control of gene expression and also to the development of rationally conceived, improved new generations of effective and specific chemotherapeutic agents. Important recent achievements, experimental and theoretical, in the analysis of such sequence specificities open prospects for possible rapid progress in this field.

The 23rd Jerusalem Symposium was devoted to the exploration of these recent achievements in relation to many types of ligand, with special emphasis on antitumor drugs.

All major types of interaction, intercalation, groove binding, covalent linking, coordination, have been considered. So was also the effect of the interaction on the structure and properties of the nucleic acids and the relationship between the interaction and biological or pharmacological activities.

We feel that this Volume presents a relatively complete up-to-date account of the state of the art in this important field of research.

As the twenty two preceding ones this Symposium was held under the auspices of the Israel Academy of Sciences and Humanities and the Hebrew University of Jerusalem. It was sponsored by the Institut de Biologie Physico-Chimique, Fondation Edmond de Rothschild of Paris. We wish to express once again our gratitude to the Baron Edmond de Rothschild for his constant and generous support which makes this continuous endeavour possible.

We wish also to present our grateful appreciation to the Administrative Staff of the Israel Academy, and in particular to Mrs Avaigail Hyam, for the efficiency and excellency of the local arrangements.

Bernard PULLMAN
Joshua JORTNER.

TABLE OF CONTENTS

Mutual Conformational Adaptation of Both Ligand and Receptor in Antitumor Drug-DNA Complexes.

Andrew H.-J. Wang*, Yen-Chywan Liaw, Howard Robinson and Yi-Gui Gao

Department of Physiology and Biophysics
University of Illinois at Urbana-Champaign
Urbana, IL 61801

ABSTRACT. Many antitumor/anticancer drugs bind and interact with DNA double helix to exert their biological activities. The consequence of the binding process is that both the drug and the DNA molecule change their conformations to accommodate each other to optimize the binding interactions. Two series of drug-DNA complexes associated with intercalator and minor-groove binder, with their structures derived from the high resolution x-ray diffraction analysis, are used to illustrate this concept of mutual conformational adaptation between ligand and receptor. Anthracylcine drugs, including daunomycin, adriamycin and nogalamycin, intercalate between CpG base pairs using the aglycone chromophore with its elongated direction almost perpendicular to the C1'-C1' vector of the neighboring base pairs. Around the anthracycline intercalator, DNA stretches the two complementary backbones in a different manner to move the base pairs 6.8 Å apart. On the one side, the dC changes the ε/ζ combinatory torsion angles to ~[-100°/180°], while keeping the glycosyl χ angle near high *anti* range [-90°]. On the other side, all torsion angles are maintained close to those of B-DNA with the exception of the glycosyl χ angle changing to normal *anti* range [*ca.* -150°]. All sugar puckers are in the C2'-*endo* family. Daunomycin and adriamycin adjust the glycosyl ether linkage (between ring A and amino sugar) torsion angle such that the amino sugar fits better in the minor groove. In contrast, nogalamycin has a gentle bend in the long direction of the aglycone chromophore, bringing the aminoglucose and nogalose closer to each other. In the minor groove binding drug-DNA dodecamer complexes, there is a wide range of backbone torsion angles in the DNA molecules despite the uniform narrow minor groove width associated with the central AT sequences. Drug molecules (netropsin, distamycin and Hoechst 33258) exhibit sufficient flexibility and adjust their conformations to follow the contour surface of the right-handed B-DNA minor groove.

INTRODUCTION

DNA plays central roles in the biological activities of all living cells. The intricate and precise regulation of the expression of various genes encoded in the DNA nucleotide sequences by many regulatory proteins and enzymes have become an extremely critical issue to be unraveled in modern biology. Gene regulation is presumably determined by the DNA itself and its interactions with proteins. Therefore it is not surprising that many small molecular ligands

----

*To whom correspondence should be addressed.

1

B. Pullman and J. Jortner (eds.), Molecular Basis of Specificity in Nucleic Acid-Drug Interactions, 1–21.
© 1990 *Kluwer Academic Publishers. Printed in the Netherlands.*

interacting with DNA, including many antitumor drugs, have profound effects on the function of DNA. Indeed, it is now widely recognized that DNA double helix is the target molecule of many antitumor drugs. In fact, DNA molecules may be considered as the ultimate receptors for these drugs. Therefore the understanding of the molecular basis of functional roles of those antitumor drugs, e.g., their binding affinity and specificity (Wang, 1987) has become a topic of interest for many scientists in different disciplines. This is particularly important in that the correlation of the biological activities of the drugs with the manner in which the drug molecules bind the DNA double helix may provide new rationales to improve their chemotherapeutic properties.

An important aspect in the understanding of the interactions between these antitumor drugs and DNA (the receptor molecule) is the conformational polymorphism of the DNA molecules. Over the years, it has been assumed that the biologically relevant DNA structure *in vivo* is the right-handed double helical B-DNA. However, there has been significant advances in our understanding of the way in which DNA can adopt different conformations depending on its nucleotide sequence and many other extrinsic factors. For example, the left-handed Z-DNA double helix is favored by alternating C-G sequence (Wang, et al., 1979a). The interconversion between Z-DNA and the right-handed B-DNA is influenced by metal ions, ionic strength, supercoiling and Z-DNA binding proteins (Rich, et al. 1984; Wells and Harvey, 1987). Similarly, sequences with a string of guanines may have a propensity to adopt an A-DNA conformation (Wang, et al., 1982; Thomas & Wang, 1988). Another conformational state of DNA, namely bent DNA, has received a great deal of attention due to its potential role in gene regulation and nucleosome phasing (Widom, 1985). Bent DNA is apparently sequence directed and is strongly favored by oligo-$(dA)_n$ [n=4-6] stretches when they are appropriately and repeatedly spaced along DNA (Diekmann, 1987). More recently, other alternative DNA structures such as H-DNA in which triple stranded helix with both Watson-Crick and Hoogsteen base pairings coexist has been proposed (Lyamichev, et al., 1986). Several models have been suggested for the structure associated with the telomeric DNA sequences (Henderson, et al., 1987; Sen and Gilbert, 1988; Sundquist and Klug, 1989; Williamson, et al., 1989). These results underlie the importance of the full appreciation of DNA polymorphism and its implication for the binding mechanism of antitumor drugs.

Broadly speaking, there are five different types of DNA binding interactions with antitumor drugs, namely, *intercalation, non-covalent groove binding, covalent binding/cross-linking, DNA cleaving,* and *nucleoside-analog incorporation.* Questions related to the molecular consequence of the drug binding have been approached by solving the crystal structure of several complexes between antitumor drugs and DNA oligonucleotides at high resolution by x-ray diffraction. Several structures of these complexes have been successfully determined. These include the DNA minor groove binding antitumor drugs (e. g. netropsin, distamycin, Hoechst 33258) complexed to a series of related DNA dodecamers (Kopka, et al., 1985; Coll, et al., 1987, 1989; Carrondo, et al., 1989; Wang and Teng, 1990). A number of intercalator antitumor drug-DNA structures, including anthracyclines (daunomycin, adriamycin and nogalamycin) (Wang, et al., 1987; Liaw, et al., 1989; Moore, et al., 1989; Frederick, et al., 1990; Williams, et al., 1990a, b) and quinoxalines (triostin A and echinomycin) (Wang, et al., 1984, 1986; Ughetto, et al., 1985; Quigley, et al., 1986) have also been elucidated. The results of these structural analyses allow us to visualize the fine details of the ways in which the drug molecules interact with the base pairs and the backbones in the grooves of the double helix. They also provide a satisfactory correlation with results from solution studies.

We noticed that in all cases, both the drug and DNA molecules change their respective conformations in order to accommodate each other to achieve the optimal interactions. This concept of mutual conformational adaptation between the ligand and receptor molecules, observed in antitumor drugs and DNA complexes, may be a common one in other biological systems.

INTERCALATOR DRUGS

Intercalator antitumor drugs constitute an important class of compounds for cancer chemotherapy (Denny, 1989). Although it is generally believed that the ability of these compounds to insert their planar chromophores between DNA base pairs is a requirement for their biological activities, it is not sufficient for them to be effective as useful drugs, since many intercalators are not active anticancer agents. Other components of the molecules besides the intercalator chromophore are critical in determining whether they possess antitumor activity or not. Presumably these components contribute in making the compounds have different DNA binding affinity or DNA sequence specificity. Additionally, they may alter the ways in which proteins (e.g., polymerases or topoisomerases) interact with the drug-DNA complexes (Lown, 1988; D'Arpa and Liu, 1989). By solving the structure of the drug-DNA complexes, one can start to gain insights on the roles of various functional components in these compounds.

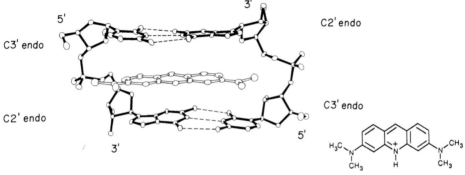

Figure 1. Three dimensional structure of the acridine orange-r(CpG) complex at 0.9 Å resolution. Notice the intercalator binds from the major groove direction.

There are several types of intercalators depending on their chemical structures. Many of the well-known intercalator anticancer drugs, such as daunomycin, are of the mono-intercalator type. The distortion in the DNA double helix associated with the binding of simple intercalators (e.g., ethidium, acridine orange) have been extensively studied by many techniques. For example, it has been shown that ethdium unwinds the double helix by -26° (Wang, 1976). This observation was corroborated by a series of crystal structure analyses of simple intercalators complexed to DNA and RNA dinucleoside monophosphates (for review, see Saenger, 1984). Figure 1 shows the structure of the acridine orange-r(CpG) complex (Wang, et al., 1979b) as an example of these complexes. Acridine orange binds, using only electrostatic, van der Waals interactions and minimization of exposed hydrophobic surface area, in the cavity created by stretching the backbone of the two CpG molecules. It is interesting to point out that the acridine orange molecule is oriented so that the bulky dimethyamino side chains are located in the major groove. This is in contrast to most of the DNA-binding small molecular ligands which bind mainly in the minor groove as discussed later.

A common observation in these simple intercalator complexes is that the DNA/RNA backbone has a C3'-*endo*-(5',3')-C2'-*endo* mixed sugar puckering conformation as shown in Figure 1. This was thought to be a plausible explanation for the nearest neighbor exclusion binding of intercalator molecules to DNA. However, this observation seems to be confined to the simple intercalator-dinucleotide complexes, which may have strong end effects. It is thus likely that DNA can have a wide range of possible ways to generate the intercalation cavity.

Anthracycline drugs: Daunomycin and Adriamycin

Daunomycin and adriamycin (Figure 2) are important anticancer drugs currently in clinical use (Crooke and Reich, 1980; Lown, 1988). These compounds possess an aglycone chromophore containing four fused rings and a positively charged amino sugar. They act by binding to DNA and inhibiting both DNA replication and transcription (Crooke and Reich, 1980; Lown, 1988). The detailed interactions between this family of drugs and DNA were elucidated recently by the high resolution crystal structure analysis of a series of daunomycin/adriamycin and DNA hexanucleotide complexes (Wang, et al., 1987; Moore, et al., 1989; Frederick, et al., 1990; Williams, et al., 1990b). The common features seen in these structures are represented by the daunomycin-d(CGTACG) complex in Figure 3. Two daunomycins are intercalated in the d(CpG) sequences at both ends of the distorted B-DNA double helix. The long axis of the daunomycin aglycone chromophore is oriented at nearly right angles to the long direction of the DNA base pairs with the ring D protruding into the major groove of the double helix and the cyclohexene ring A lying in the minor groove. The specificity of daunomycin-DNA interactions is provided by the hydrogen bonds between the hydroxyl group O9 of daunomycin and the N3 and N2 positions of the guanine base adjacent to the aglycone intercalating ring. These hydrogen bonds appear to be important for the biological activity of daunomycin and adriamycin, as it has been shown that anthracycline derivatives without the O9 group are not active. Additionally, the N2 amino group of G2 guanine base forms a hydrogen bond to the O7 atom of the glycosyl ether linkage.

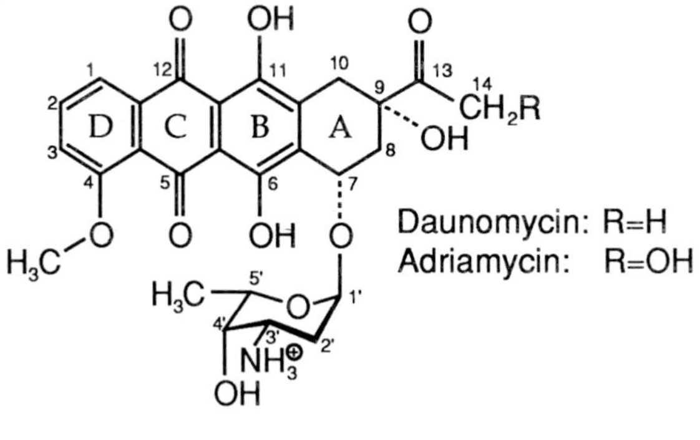

Figure 2. Molecular Formula of Daunomycin and Adriamycin.

Recently, we have determined and compared the structures of two other related complexes, daunomycin-d(CGATCG) and adriamycin-d(CGATCG), at high resolution (Frederick, et al., 1990). Most of the structural features are conserved in these three structures. However, there is an interesting sequence dependence of the binding of the amino sugar to the AT base pair outside the intercalation site. In the d(CGATCG) complexes, there are additional direct hydrogen bonds between the positively charged N3 amino group in the sugar and the O2 of both C11 and T10 residues of DNA. This suggests that daunomycin/adriamycin may bind to 5'-CGA sequence slightly better than to 5'-CGT sequence. These high resolution structures also provide reasonable explanation for the DNA base triplet specificity derived from solution (Chen et al., 1986; Chaires et al., 1987) and theoretical studies (Pullman, 1989).

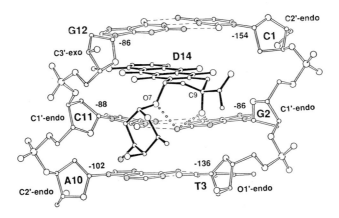

Figure 3. The three dimensional molecular structure of the daunomycin-d(CGTACG) complex. Some selective torsion angles are shown. The hydrogen bonds between daunomycin (O9 and O7) and DNA (guanine G2) are shown as small circles.

The DNA conformation in these complexes resembles that of B-DNA, except near the intercalation site. All sugar puckers are in the C2'-*endo* family as shown in Figure 3. The glycosyl torsion angles of four out of the six nucleotides surrounding the drug are in the high *anti* range near ~[-90°], except for the C1 [-154°] and T3 [-136°]. This combination of χ angles along with other conformational changes in the backbone, in particular ζ (becoming *trans* near -170°), produces the extended conformation to accommodate the large anthracycline drug. Surprisingly, the extension of the backbones is coupled with a small helix unwinding (-10° per drug). An interesting distortion in DNA conformation in the anthracycline complexes, not seen in other simple intercalator complexes, is the large buckle of the base pairs above and below the intercalator.

These structural analyses allow us to identify three major functional components of daunomycin and adriamycin: the aglycone intercalator (ring B-D), the anchoring function associated with ring A, and the amino sugar. Each component plays an important role in the biological activity of the drug. Aglycone, by intercalating into the base pairs, causes a distortion in the DNA double helix which may be recognized by enzymes (e.g., topoisomerase II). The amino sugar is essential as it lies in the minor groove with its functional hydroxyl and amino groups available for interaction with enzymes in a specific manner. The O9 hydroxyl provides key hydrogen bonds to DNA anchoring the drug firmly in the double helix. Finally, the configuration at the C7 position in ring A is important, as it joins the amino sugar to the aglycone with a right handed chirality such that the drug can position the amino sugar in the minor groove of a right-handed B-DNA double helix.

Anthracycline drugs: Nogalamycin and derivatives

Another interesting antitumor anthracycline antibiotic, nogalamycin (Figure 4), is active against a number of tumor cell lines (Bhuyan and Reusser, 1970; Wiley, 1979). It contains two sugar moieties (nogalose and aminoglucose) attached to rings A and D respectively. Many natural and semi-synthetic derivatives of nogalamycin have been studied to attempt to identify better agents for therapeutic purpose (Wiley, 1979). One of them, menogaril, is currently under phase II clinical trial (Adams, et al., 1989).

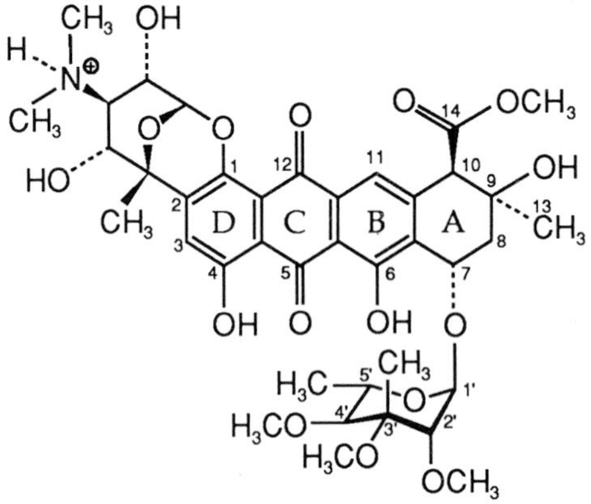

Figure 4. Molecular formula of nogalamycin. Two sugars are attached to the aglycone with nogalose at C7 and aminoglucose at the C1/C2 positions. The two hydroxyl groups in the aminosugar ring are labeled as O2G and O4G in the text.

Nogalamycin, with bulky sugars attached at both ends of the chromophore, poses an interesting question with respect to the ways in which it inserts itself between the base pairs. It has been suggested that the drug binds only to premelted DNA region (Fox and Waring, 1986). Based on model building, nogalamycin seems to bind DNA double helix with these two sugars lying separately in the major and minor grooves (Arora, 1983; Collier, et al., 1984). The nucleotide sequence specificity of nogalamycin has also been a matter of confusion (Fox and Waring, 1986). Recently the structure of nogalamycin complexed to DNA hexamers has been determined at high resolution and it has provided a great deal of information on the conformational flexibility in both nagalamycin and DNA (Liaw, et al., 1989; Williams, et al., 1990a). These results are discussed here using the 1.3 Å resolution structure of the 2:1 complex of nogalamycin-d[CGT(pS)ACG] as a representative case.

In the structure of the 2:1 complex of nogalamycin-d[CGT(pS)ACG], the two nogalamycin molecules are intercalated between the CpG steps at both ends of a distorted B-DNA double helix. This is illustrated in a detailed view of the interactions between nogalamycin and DNA in Figure 5. As in the structure of the daunomycin-DNA complex, the elongated aglycone chromophore (rings A-D) of nogalamycin penetrates the DNA double helix such that it is almost perpendicular to the C1'-C1' vectors of the two GC base pairs above and below the studied by NMR spectroscopy (Searle, et al., 1988). That work suggested that in the 2:1 complex two nogalamycin aglycone chromophores intercalated between the CpA (and its complement TpG) steps with the aminoglucose and nogalose lying in the major and minor grooves respectively. The structures of these two 2:1 complexes are shown schematically in Figure 6.

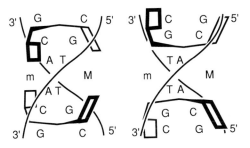

Figure 6. Schematic diagrams of the nogalamycin-d(CGTACG) and nogalamycin-d(GCATGC) complexes. Nogalamycin prefers to bind at CpN or NpG sites.

Based on these and other (including footprinting) data, we suggest that nogalamycin has a DNA sequence preference for NpG or CpN steps (Gao, et al., 1990). More specifically, the aglycone chromophore prefers to intercalate at the 5'-side of a guanine (between **NpG**), or at the 3'-side of a cytosine (between **CpN**) with the sugars facing toward the GC base pair. Taken together from the above discussion, it seems that nogalamycin prefers G-C sequence (e. g. CpG) embedded in a stretch of A-T sequences. It is interesting to note that DNA sequences in the promoter regions often possess such features, i.e., AT-rich sequence sprinkled with GC sequences.

Comparison of two anthracycline drugs

Overall Comparison. There are several similar features in the structures of the complexes between these two families of anthracylcine drugs with DNA. However, there are some interesting differences between them. Figure 7 compares the view of the two complexes from a direction perpendicular to the plane of the aglycone ring. The long dimension of both aglycones lies across the base pairs and reaching both grooves. The location of the aglycone ring relative to the base pairs in the nogalamycin complex is "pulled" toward the minor groove by about 2.0 Å, in comparison to that of the daunomycin-d(CGTACG) complex (Wang, et al., 1987; Liaw, et al., 1989). Ring D of nogalamycin is stacked underneath the N4 amino group of C1 residue. In the daunomycin/adriamycin and DNA complexes, the positively charged aminosugar lies in the minor groove of the distorted B-DNA double helix. In contrast, in the nogalamycin-DNA complex, the positively charged aminoglucose resides in the major groove and the neutral nogalose occupies the minor groove. This suggests that the location of the positive charge of the drug could be in either the major or the minor grooves and it is not an overriding factor in determining the binding specificity.

intercalator. The drug spans the two grooves of the helix with the nogalose in the minor groove and the aminoglucose in the major groove. The two sugars are on the same side of the flat aglycone chromophore, wrapping around the second (and the fourth) GC base pair and they both point toward the AT region in the middle of the helix.

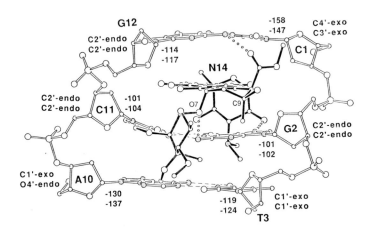

Figure 5. The three dimensional structure of the nogalamycin-d(CGTACG) complex. Hydrogen bonds are found in both major and minor grooves and they are shown as lines of small circles.

There are several direct hydrogen bonds between nogalamycin and DNA. In the major groove, the O2G hydroxyl of the aminoglucose is 2.85 Å (ave) from N7 of guanine G2, while O4G is 3.15 Å (ave) from N4 of cytosine C11. Indeed, the disposition of this pair of hydrogen bond donor (O2G) and acceptor (O4G) appears to have excellent complementarity with a single G-C base pair in the major groove. The dimethylamino group in the aminoglucose is bridged through a water molecule to the N6 of adenine A10 one base pair below. In the minor groove, nogalamycin receives two hydrogen bonds from guanines. The carbomethoxy group at C10 position in ring A of nogalamycin forms a hydrogen bond (ave. ~3.05 Å) with the N2 amino group of G12. In addition, the O7 atom of the glycosyl linkage is 3.20 Å from the N2 of G2 nucleotide. These hydrogen bonding interactions determine the sequence specificity of nogalamycin.

In order for the drug to intercalate the DNA, the double helix needs to open transiently with sufficient room for the bulky sugars to slide through between base pairs. This requirement should favor the A-T sequences, since they open up more readily. This binding process is expected to be slow, as has recently been shown (Fox and Waring, 1986). Our recent NMR study of the solution containing nogalamycin and unmodified d(CGTACG) duplex in a 2:1 ratio showed that the solution structure is substantially similar to the crystal structure (Robinson, et al., 1990). In addition, the 1:1 complex reveals that there are two forms of complexes (both 2:1 and 1:1) as well as the free DNA in slow equilibrium on the NMR time scale (Robinson, et al., 1990). Recently, the interactions between nogalamycin and another DNA hexamer d(GCATGC), which contains the putative preferred binding sites of nogalamycin, have been

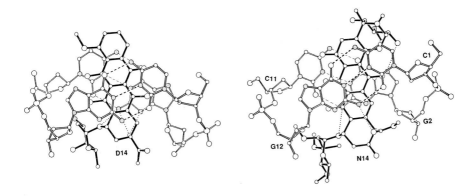

Figure 7. Comparison of the interactions surrounding the intercalation site in the daunomycin-d(CGTACG) and nogalamycin-d(CGTACG) complexes. The difference in the orientation of the aglycone in the CpG cavity is due primarily to the different hydrogen bonding interactions between the anthracycline drugs and DNA.

Conformational adjustment in drug. It is interesting to note that although the drug molecules (nogalamycin and daunomycin) in the complex have an overall conformation similar to those of the free drug due to the rigidity of the ring systems, they differ in an interesting way. The hemi-saturated ring A has some conformational flexibility. In both complexes, the C9 atom deviates from the mean plane with the largest displacement of 0.59 Å and 0.54 Å in daunomycin and nogalamycin respectively. This pucker of ring A juxtaposes various groups of ring A quite differently for the two drugs. For example, the O9 hydroxyl in daunomycin is positioned axially which enables it to form hydrogen bonds with guanine bases. In contrast, the O9 in nogalamycin is in the equatorial position pointing away from the DNA without interacting with DNA directly. Consequently, the two drugs have quite different DNA binding interactions.

The drug molecule can also use other features to fine tune its binding to DNA. For example, the torsion angles around the glycosyl ether linkage (C8-C7-O7-C1' and C7-O7-C1'-C2') in daunomycin can vary to various extents to fit the daunosamine sugar in the minor groove as seen in Figure 8 (top). Similarly, the nogalamycin in complexes differs in a subtle way from the free drug. For example, all the methoxy groups in nogalose (in chair conformation) are oriented with respect to one another in almost exactly the same manner as found in the free drug. However, when the molecules in the complex and in uncomplexed state are overlaid on one another by a least-square fitting of the ring A (Figure 8, bottom), the aglycone chromophore of the bound drug in the complex seems to bend down making the aminoglucose slightly closer to the nogalose. Presumably, this bending occurs so that (1) the pair of hydroxyl groups of the aminoglucose can form hydrogen bonds to the G-C base pair in the major groove, (2) the O7 and the carbonyl group of the carbomethoxy at the C10 position of the ring A can receive a hydrogen bond in the minor groove, and finally (3) maximum van der Waals contacts can occur in both grooves. Using semi-empirical and *ab initio* calculations, we estimate the energy cost for this bending in anthracyclines is not very great (unpublished results).

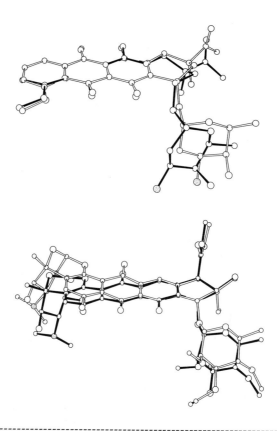

Figure 8. Comparison of free (open bond) and bound (filled bond) drugs. Top: daunomycin. The bound drug has rotation in two bonds in the glycosyl ether linkage. Bottom: nogalamycin. The bound nogalamycin bends gently along the long direction of the aglycone chromophore.

DNA conformation.   In all anthracycline-DNA complexes, significant rearrangements of the DNA backbone torsion angles have occurred around the drug molecule due to the binding of drug molecules to DNA. Figure 9 shows the comparison of the two d(CGTACG) backbones from the daunomycin and nogalamycin complexes. It can be seen that while they have similar overall conformations, they are significantly different in details. This is mainly due to the changes in the shape of the intercalation cavity (at CpG steps) in order to accommodate those two complicated anthracycline drugs. We summarize all the torsion angles in Figure 10 using conformational wheel diagrams. The largest deviations of these angles from the normal, uncomplexed B-DNA are associated with the C-G residues involved in the intercalation sites as expected. These include the ε of C5 and C11 (ranging from -85.0° to -90.2°) with the concomitant changes of ζ to -174.8° and -171.6° respectively, using nogalamycin complex as an example. The puckers of all the sugars remain in the C2'-*endo* type which has a pseudorotation angle P of 162°. No obvious pattern of the so-called C3'-*endo*-(5',3')-C2'-*endo* mixed sugar pucker (Tsai, et al., 1977; Wang, et al., 1978; see Figure 1) is seen.

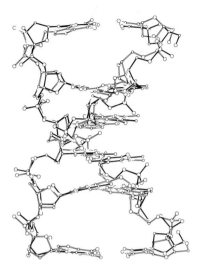

Figure 9. A least square fitting of the two d(CGTACG) hexamers derived from the daunomycin and nogalamycin complexes. The same hexamer adjusts its backbone differently to accommodate these two drugs. Some common features are seen. For example, base pairs surrounding the cavity have large buckles.

Other important distortions in these complexes involve the inner GC base pairs (G2-C11 and C5-G8) in the intercalated CpG steps which are significantly buckled (ranging from 15° to 25°), as they are in close contact with the sugars and they adjust to accommodate the bulky drug molecules. The terminal two G-C base pairs, which are located away from the drug sugars, are significantly less buckled (ranging from 6° to 10° for C7-G6 base pair).

One interesting feature is that the helical twist angle of the two base pairs across the intercalator in the anthracycline drug-DNA complexes (ave. 36° in daunomycin series and ave. 37° in nogalamycin series) is very close to that of B-DNA (36°), suggesting very little helix unwinding at the intercalating site. In fact, the helix unwinding occurs elsewhere in the d(CGTACG) hexamer helix with the rest of the steps have a combined unwinding angle of -12°, mostly in the GpT step next to the intercalator. Therefore, the overall unwinding angle of the DNA helix due to the intercalation of both daunomycin and nogalamycin is estimated to be about -10°, significantly lower than the value observed for the simple intercalator ethidium (-26°) (Wang, 1974). This low DNA unwinding angle may be a common feature of anthracycline drugs.

In summary, the recent structural determination of several anthracycline antibiotics complexed to DNA (Wang, 1987; Moore, et al., 1989; Liaw, et al., 1989; Frederick, et al., 1990;

Williams, et al., 1990a, b) allows some generalizations on the binding of those drugs to DNA to be made. First, the elongated aglycone chromophore intercalates between base pairs with its long direction almost perpendicular to the C1'-C1' vector of the base pairs above and below the intercalator ring. Second, specific hydrogen bonds between DNA and drug are formed using hydroxyls (as donor and acceptor) and ether or carbonyl group(s) (as acceptor only) of the drug molecule and the respective acceptor atoms (N7, N3 of purines and O2 of pyrimidines) and donors (N2 of guanine, N4 of cytosine and N6 of adenine) from the DNA. The DNA backbone oxygen atoms, including the negatively charged phosphate oxygens, are less frequently involved in the direct hydrogen bonding interactions. Third, water molecules or hydrated metal ions can further stabilize the complex by bridging functional groups. Fourth, van der Waals interactions between hydrophobic groups (e.g. methyls) and DNA in the grooves are important.

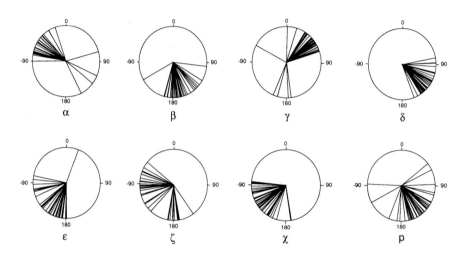

Figure 10. A summary of the DNA bonebone torsion angles (α to ζ and glycosyl χ, pseudorotation angle P) in DNA hexamers from six anthracycline-DNA complexes using the wheel diagram. Most of the angles are clustered in one region around the wheel. ζ is found to occupy in two regions, ~[-90°] and ~[-180°], the latter is associated with concommitant changes either in ε (to -90°) or in α (to 130°).

An important observation in the comparison of the structures of these anthracycline-DNA complexes is that both anthracycline (guest) and DNA (host) molecules have changed their respective conformations in order to form a tight complex. The change in nogalamycin is particularly surprising since the free drug seems to possess little flexibility. Instead, the bound drug bends gently along the long direction of the aglycone chromophore. This points out that when doing model building studies of drug-receptor binding interactions, the possible conformational flexibility in both drug and receptor have to be taken into consideration carefully. We may now use these general rules to predict and model the binding of different natural (e.g., aclacinomycin, steffimycin) and synthetic (e.g. mitoxantrones) intercalator compounds to DNA. More structural analyses like this would enable us to contemplate designing new compounds with different DNA binding properties.

MINOR-GROOVE BINDING DRUGS

A number of natural and synthetic compounds are known to bind to DNA double helix in a non-intercalative manner. Several well-known compounds of this type, such as the antitumor antibiotics netropsin and distamycin, and the synthetic Hoechst 33258, are shown in Figure 11. Recently, the molecular basis of their interactions with DNA have been studied intensively (for reviews, see Zimmer and Wahnert, 1986; Wang and Teng, 1990). The picture emerging from these studies is that this class of compounds binds in the narrow minor groove of the B-DNA double helix using a combination of interactions including hydrogen bonds, ionic charge attractions, as well as van der Waals interactions. Interestingly, they have a binding preference to stretches of AT-rich sequences with a gradation of binding affinity toward various A-T rich sequences (Zimmer and Wahnert, 1986). For example, netropsin binds tetranucleotides with the affinity in the order of AAAA, AATT > ATAT >> ACTG, CGCG. Such sequence binding microheterogeneity are now better understood based on the recent structural information available from several x-ray diffraction analysis of drug-DNA dodecamer complexes (for review, see Wang and Teng, 1990).

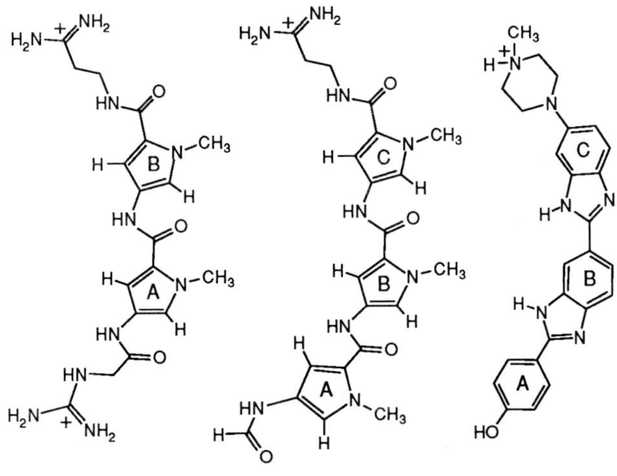

Netropsin   Distamycin A   Hoechst 33258

Figure 11. Molecular formula of netropsin, distamycin and Hoechst 33258.

The overall structure of the minor-groove binding drug-DNA complex is exemplified by the distamycin-d(CGCAAATTTGCG) structure shown with a diagram looking into the minor groove of the right-handed B-DNA double helix in Figure 12 (Coll, et al., 1987). The elongated, flat distamycin molecule (in van der Waals representation) binds in the narrow minor groove of the DNA double helix (in skeletal drawing) at the central AT region covering slightly over five base pairs. The drug molecule is sandwiched by the side walls of the minor groove which is made of the two anti-parallel sugar-phosphate backbones. Many atoms from DNA have close van der Waals contacts to both faces of the flat distamycin molecule.

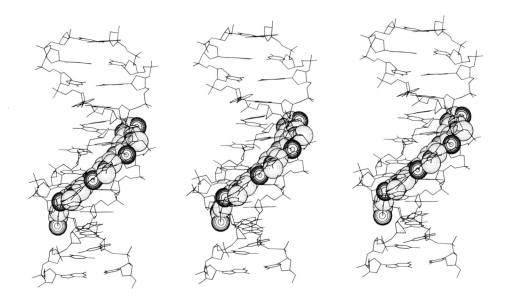

Figure 12. Stereoscopic diagram of the three dimensional structure of distamycin-d(CGCAAATTTGCG) complex in triplicate form. The two figures on the left are for the cross-eyed view and the two figures on the right are for the paralell-eyed view. The distamycin (in van der Waals diagram) binds in the central AAATTT region in which many AT base pairs are highly propeller-twisted.

The complex is stabilized by several intermolecular hydrogen bonds between the drug and the nucleophilic oxygen and nitrogen atoms of DNA at the floor of the minor groove. For example, the amide NH groups of distamycin form an array of bifurcated hydrogen bonds to DNA. The large number of bifurcated hydrogen bonds in the complex are possible due to the fact that the AT base pairs in the A_3T_3 sequence adopt a highly propeller twisted conformation such that the disposition of N3 and O2 atoms of the A_3T_3 sequence at the floor of the minor groove are optimized for such hydrogen bonding interactions.

The crescent-shaped netropsin drug hugs closely to the DNA double helix with the pyrrole HC5 and HC11 hydrogens approaching the HC2 hydrogens of adenine bases. For example, the HC5 atom of pyrrole A is 2.6 Å away from the HC2 atom of adenine A7 and the HC11 atom of pyrrole B is 2.6 Å from the HC2 atom of adenine A19. These close van der Waals contacts imply that there would be serious clashes between these drugs and guanine bases in the minor groove, if AT is replaced by GC base pairs. As suggested previously, the N2 amino group of a guanine base in the minor groove presents a major hindrance for the entry of the drugs into the minor groove, thereby providing the discrimination for a more favored binding toward AT base pairs (Kopka, et al., 1985).

The comparison of the structures of the complexes between the drugs and DNA oligonucleotides allows us to visualize the fine details of the manner in which a minor groove binding drug interacts with a B-DNA double helix. We may summarize the processes by which the drugs bind to the DNA double helix in the minor groove at the AT regions, as shown schematically in Figure 13. There are four major factors which are responsible for the AT binding preference for netropsin, distamycin and related compounds. First, all of them are positively charged molecules, thereby enhancing the initial non-specific attraction between the drug and DNA. In addition, it has been suggested that the deepest negative charge potential of a DNA double helix resides near the bottom of the minor groove at the AT-region (Zarkezewska, et al., 1983). The drug molecule may be driven into the vicinity of the AT region by such potential. The third factor is that the AT sequence may be associated with a higher tendency to adopt a narrow minor groove, which would provide many van der Waals stabilizing interactions with the sugar-phosphate side walls of the groove. The dipole interactions between the O4' atoms with the aromatic pyrrole π-electron clouds help immobilize the drug in place so that the amide NH group can form hydrogen bonds to N3 of adenine and O2 of thymine on the floor of the minor groove. Finally, the amino NH_2 group of guanine bases presents a prominent steric hindrance to the entry of the drug into the groove, hence establishing a definite AT binding preference.

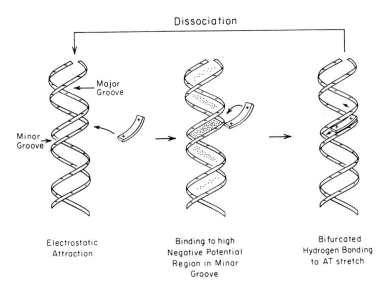

Figure 13. A highly schematic representation of the dynamic binding of minor groove binder to DNA.

Conformational changes in drugs

Netropsin and distamycin are N-methylpyrrole-containing antitumor antibiotics, while Hoechst 33258 is a synthetic dye. All of them have an elongated flat crescent-shaped geometry, ideal for the binding in the B-DNA minor groove. These molecules adjust their conformation and move along the minor groove seeking the optimal binding sites. They bind to different AT sequences with different affinities. For example, netropsin and distamycin form more hydrogen bonds with AATT and AAATTT sequences respectively than the netropsin with the alternating ATAT sequence (Kopka, et al., 1985; Coll, et al., 1987,1989). In the netropsin-d(CGCGAATTCGCG) and distamycin-d(CGCAAATTTGCG) complexes, the NH groups from the peptide bonds form a series of bifurcated (three-centered) hydrogen bonds to the N3 of adenine and O2 of thymine bases in the AT sequences. In the netropsin-d(CGCGATATCGCG) complex, no bifurcated hydrogen bond was seen. Figure 14A and B are two views of the composite drawing of the free and the bound netropsin molecules which show the difference in the dihedral angle between the two pyrrole rings and the flexibility of the end groups in these three molecules.

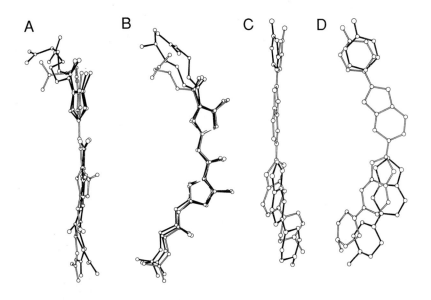

A   B   C   D

Figure 14. Comparion of the free and bound minor groove binding drugs. (A and B). Netropsin in free state (open bond) and bound state (filled bond). (C and D). Hoechst 33258 in bound state from two complexes with DNA d(CGCGAATTCGCG) (open bond) and d(CGCGATATCGCG) (filled bond).

Figure 14C and D compares two Hoechst 33258 molecules found in their complexes with DNA dodecamers determined in our laboratory. Even though the molecule is made of four rigid rings (two benzimidazoles, one phenol and one piperazine), the connections between the rings still have some flexibility to allow the molecule to follow the groove surface. They have similar, but distinct, binding modes toward the AT segments in those dodecamers. In the Hoechst 33258-d(CGCGAATTCGCG) complex (Teng, et al., 1988), the drug molecule binds at the AATT sites of the duplex. In contrast, the drug binds at another sequence (GATA) in the complex with d(CGCGATATCGCG) (Carrondo, et al., 1989). This suggests that there is a range of tetranucleotide sequences to which the Hoechst 33258 molecule binds almost equally well. This may be associated with the fact that the Hoechst compound has only two NH sites which it can use to form hydrogen bonds with the base pairs in the minor groove, resulting in a less stringent sequence specificity.

Influence of drugs on DNA conformation

The contributions of the binding affinity of those compounds to DNA derive not only from hydrogen bonding interactions, but also from the many van der Waals interactions between the sugar-phosphate backbone of DNA with the aromatic ring of the drug. The latter stabilizing contribution is due to the narrow minor groove in the AT region. DNA can achieve this narrow minor groove width using a wide range of torsion angles of the sugar phosphate backbones. This is reflected in the similar overall conformation of several DNA dodecamer duplexes despite the differences in their fine details. Figure 15 compares the DNA torsion angles of three drug-DNA complexes which have the conformation typical of the right-handed B-DNA. Despite of these great variations in their torsion angles, they all possess a prominently characteristic feature of a very distinct narrow minor groove at the central AT region. This is easily seen by comparing the minor groove width of all the known crystal structures of dodecamer duplexes diagrammatically shown in Figure 16. A consistent trend is obvious in these curves where they all reach the lowest values (~4 Å) near the 8/21 residue, except for the AATT dodecamer without a bound drug. This narrow minor groove is likely to be an intrinsic feature associated with AT-rich sequences. We have recently solved the crystal structure of a dodecamer duplex made of three separate strands [d(CGCGAAAACGCG) +d(CGCGTT)+d(TTCGCG)] (Aymami, et al., 1990) resulting nicked DNA duplex. The molecule

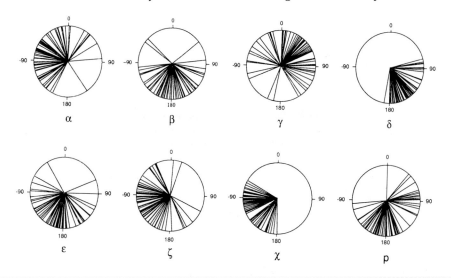

Figure 15. A summary of the DNA bonebone torsion angles in DNA dodecamers from three minor-groove binder-DNA complexes using the wheel diagram. Large distributions of all torsion angles are found, indicating the fluidity of DNA double helix.
----

crystallizes in a different crystal lattice, yet it still has the same distinctive narrow minor groove. Another example was seen in the complex of the phage 434 repressor protein with a 22-

mer containing O$_R$1 operator sequence, which is very rich in AT base pairs (Aggarwal, et al., 1988). The DNA molecule in this complex has a minor groove width ranging from ~9 Å in the central AT region to ~14 Å near the end of the helix.

In these drug-DNA complexes, large propeller twists (ω>20°) are associated with the A$_n$T$_n$ segments of the DNA dodecamers. In some structures, the high ω values of certain AT steps have resulted in some very interesting intramolecular hydrogen bonds. In those AT steps, a number of novel bifurcated hydrogen bonds in the major groove between the N6 amino group of an adenine to the carbonyl O4 groups of two adjacent thymines of the opposite strand are formed (Figure 12; Nelson, et al., 1987). Using this nucleotide structural motif incorporating bifurcated hydrogen bonds in the major groove, we were able to construct a modified B-DNA model of poly(dA)·poly(dT), which has a very narrow minor groove (Coll, et al., 1987; Aymami, et al., 1989). In contrast, the mean propeller twist angle for the alternating AT base pairs (e.g., ATAT in d[CGCGATATCGCG] complexes) is only 11°, a low value compared to those found in other dodecamers and about the same as that found in the GC stretch. Although it has been suggested previously that the narrow minor groove may be due to the high propeller twist of AT base pairs which can be stabilized by the spine of hydration or drug molecules (Wing, et al., 1980; Kopka, et al., 1985), the structures of dodecamer d(CGCGATATCGCG) complexed to netropsin and Hoechst 33258 have low averaged propeller twist angle in the AT region. This leads us to ponder what is the principal factor responsible for the narrow minor groove.

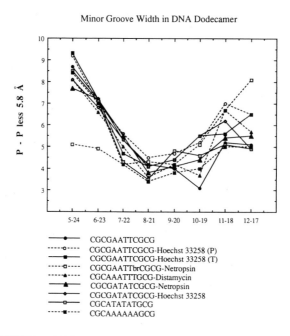

Figure 16. Diagram showing the narrow minor groove width in the AT region of seven DNA dodecamers.

Recently NMR (Patel, 1978; Pelton and Wemmer, 1988, 1989; Leupin, et al., 1986; Wang, et al., 1989) and chemical footprinting experiments (Dervan, 1986; Ward, et al., 1988) have demonstrated that the interaction of the minor groove binder to DNA is a highly dynamic process. This dynamic binding process reflects the fluidity of DNA molecule. Our results show that the role of a drug like netropsin extends beyond being merely a passive ligand just being attracted to the binding sites. Instead, much like an intercalator which extends the helix length by 3.4 Å, netropsin is capable of inducing the collapse of a "normal" minor groove into a "narrow" groove, suggesting the flexibility of DNA as shown in the netropsin-d(CGCGATATCGCG) (Coll et al., 1989) and Hoechst 33258-d(CGCGATATCGCG) (Carrondo, et al., 1989) structures. The ATAT sequence in the dodecamer d(CGCGATATCGCG) alone may have a low intrinsic propeller twist angle which may disfavor the formation of a narrow minor groove. The binding of netropsin to the ATAT segment will allow the sugar-phosphate backbones to come in close van der Waals contacts with netropsin to create a narrow groove, while still maintaining the low propeller twist in the AT base pairs. The readjustment of the backbone would cost slightly extra energy which may account partly for the low affinity of netropsin to alternating A-T sequence, in comparison to that for an oligo(A) sequence. This kind of small but significant conformational rearrangement undoubtedly can also be achieved by proteins interacting with DNA. Similarly, the minor groove can be expanded when two distamycin molecules are bound to one DNA double helix (e.g., d[CGCAAATTTGCG]) as shown by the recent NMR studies (Pelton and Wemmer, 1989).

SUMMARY AND CONCLUSIONS

High resolution x-ray diffraction analysis is the most powerful method to investigate the fine details of the molecular interactions between antitumor/anticancer drugs and their cellular receptor molecule, DNA. In this paper, we illustrate some of the recent crystallographic results on the interactions of two types of antitumor drugs, intercalators and minor groove binders, with DNA. From these studies, we start to understand the possible biological functions associated with different parts of the drug molecule. Some parts are essential for their binding to DNA either by intercalation, hydrogen bonding, van der Waals interactions, ionic charge interactions, or the combination of them. Other parts provide the necessary binding specificity. Finally, other parts may be important in thier interference with many important cellular enzymes such as polymerases or topoisomerases. Apparently, all drug molecules are capable of adjusting their conformation to place those various functional parts in the strategic places in DNA double helix to achieve the optimal binding affinity and specificity.

These structural studies also reinforce the observation associated with the fluidity of the DNA double helix. It should be pointed out that while B-DNA is presumed to be the predominant conformation *in vivo*, other stable conformations such as A-DNA or Z-DNA may have important roles in the biolocal functions of antitumor drugs. For example, chromomycin A_3 has recently been shown to bind to the wide and shallow minor groove of A-DNA double helix (Gao and Patel, 1989). It would not be surprising that there are other antitumor drugs which interact with A-DNA in a sequence-specific manner. Similarly, some compounds specifically interact with Z-DNA due to the unique physical and chemical properties of Z-DNA. The question of whether any of those compounds possesses antitumor activity remains to be explored.

In conclusion, the conformational interplay between the antitumor drug (ligand) and DNA (receptor) is an important concept which should be carefully considered in the rational design of any new drug directed toward receptor for which a high resolution structure has been determined.

ACKNOWLEDGEMENTS. This work was supported by NSF and NIH (A. H.-J. W). Y.-C. L. acknowledges the support from the Institute of Molecular Biology, Taiwan (ROC).

REFERENCES

Adams, W. J., McGovern, J. P., Dalm, E. A., Brewer, J. E. and Hosley, J. D. (1989) *Cancer Res.* **49**, 6328-6336.

Aggarwal, A. K., Rodgers, D. W., Drottar, M., Ptashne, M. and Harrison, S. C. (1988) *Science* **242**, 899-907.

Arora, S. K. (1983) *J. Amer. Chem. Soc.* **105**, 1328-1332.

Aymami, J., Coll, M., Frederick, C. A., Wang, A. H.-J. and Rich, A. (1989) *Nucleic Acids Res.* **8**, 3229-3245.

Aymami, J., Coll, M., van der Marel, G. A., van Boom, J. H., Wang, A. H.-J. and Rich, A. (1990) *Proc. Natl. Acad. Sci. USA* **87**, 2526-2530.

Bhuyan, B. K. and Reusser, F. (1970) *Cancer Res.* **30**, 984-989.

Carrondo, M., Coll, M., Aymami, J., Wang, A. H.-J., van der Marel, G. A., van Boom, J. H. and Rich, A. (1989) *Biochemistry* **28**, 7849-7859.

Coll, M., Frederick, C. A., Wang, A. H.-J. and Rich, A. (1987) *Proc. Natl. Acad. Sci. USA* **84**, 8385-8389.

Coll, M., Aymami, J., van der Marel, G. A., van Boom, J. H., Rich, A. and Wang, A. H.-J. (1989) *Biochemistry* **28**, 310-320.

Collier, D. A., Neidle, S. and Brown, J. R. (1984) *Biochem. Pharmacol.* **33**, 2877-2880.

Crooke, S. T. and Reich, S. D. (eds.) *Anthracyclines* (Academic Press, NY, 1980).

D'Arpa, P. and Liu, L. F. (1989) *Biochim. Biophy. Acta* **989**, 163-177.

Denny, W. A. (1989) *Anti-cancer Drug Design* **4**, 241-263.

Dervan, P. B. (1986) *Science* **232**, 464-471.

Diekmann, S. in *Nucleic Acids and Molecular Biology* Vol. 1 (eds Eckstein, F. & Lilley, D. M.) 138-156 (Springer, Berlin, 1987).

Fox, K. R. and Waring, M. J. (1986) *Biochemistry* **25**, 4349-4356.

Frederick, C. A., Williams, L. D., Ughetto, G., van der Marel, G. A., van Boom, J. H., Rich, A. and Wang, A. H.-J. (1990) *Biochemistry* **29**, 2538-2549.

Gao, X., and Patel, D. (1989) *Biochemistry* **28**, 751-762.

Gao, Y.-G., Liaw, Y.-C., Robinson, H., and Wang, A. H.-J. (1990) *Submitted for publication.*

Henderson, E., Hardin, C. C., Walk, S. K., Tinoco, I. and Blackburn, E. H. (1987) *Cell* **51**, 899-908.

Kopka, M. L., Yoon, C., Goodsell, D., Pjura, P. and Dickerson, R. E. (1985) *Proc. Natl. Acad. Sci. USA* **82**, 1376-1380.

Leupin, W., Chazin, W. J., Hyberts, S., Denny, W. A. and Wuthrich, K. (1986) *Biochemistry* **25**, 5902-5910.

Liaw, Y.-C., Gao, Y.-G., Robinson, H., van der Marel, G. A., van Boom, J. H., and Wang, A. H.-J. (1989) *Biochemistry* **28**, 9913-9918.

Lown, J. W. (ed.) *Anthracycline and Anthacenedione-based Anticancer Agents* (Elsevier, NY, 1988).

Lyamichev, V. I., Mirkin, S. M. and Frank-Kamenetskii, M. D. (1986) *J. Biomol. Struct. Dyn.* **3**, 667-669.

Moore, M. H., Hunter, W. N., Langlois d'Estaintot, B. and Kennard, O. (1989) *J. Mol. Biol.* **206**, 693-705.

Nelson, H. C. M., Finch, J., Luisi, B. F. and Klug, A. (1987) *Nature (London)* **330**, 221-226.

Patel, D. J. (1978) *Eur. J. Biochem.* **99**, 369-379.

Pelton, J. G. and Wemmer, D. E. (1988) *Biochemistry* **27**, 8088-8096.

Pelton, J. G. and Wemmer, D. E. (1989) *Proc. Nat. Acad. Sci. USA* **86**, 5723-5727.

Pullman, B. (1989) *Advances in Drug Design* **18**, 1-113.

Rich, A., Nordheim, A. and Wang, A. H.-J. (1984) *Ann. Rev. Biochem.* **53**, 791-846.

Robinson, H., Liaw, Y.-C., van der Marel, G. A., van Boom, J. H., and Wang, A. H.-J. (1990) *Submitted for publication.*

Saenger, W. *Principles of Nucleic Acid Structure* (Springer, Berlin, 1984).

Searle, M. S., Hall, J. G., Denny, W. A. and Wakelin, L. P. G. (1988) *Biochemistry* **27**, 4340-4349.

Sen, D. and Gilbert, W. (1988) *Nature (London)* **334**, 364-366.

Sundquist, W. I. and Klug, A. (1989) *Nature (London)* **342**, 825-829.

Teng, M.-K., Frederick, C. A., Usmann, N. and Wang, A. H.-J. (1988) *Nucleic Acids Res.* **16**, 2671-2690.

Thomas, Jr., G. J. and Wang, A. H.-J. in *Nucleic Acids and Molecular Biology* Vol. 2 (eds Eckstein, F. & Lilley, D. M.) 1-30 (Springer, Berlin, 1988).

Tsai, C.-C., Jain, S. C. and Sobell, H. M. (1977) *J. Mol. Biol.* **114**, 301-315.

Ughetto, G., Wang, A. H.-J., Quigley, G. J. and Rich, A. (1985) *Nucleic Acids Res.* **13**, 2305-2323.

Wang, A. H.-J., Quigley, G. J., Kolpak, F.J., Crawford, J. L., Van Boom, J. H., van der Marel, G. A. and Rich, A. (1979a) *Nature (London)* **282**, 680-686.

Wang, A. H.-J., Quigley, G. J. and Rich, A. (1979b) *Nucleic Acids Res.* **6**, 3879-3890.

Wang, A. H.-J., Fujii, S, van Boom, J. H. and Rich, A. (1982) *Proc. Nat. Acad. Sci. USA* **79**, 3968-3972.

Wang, A. H.-J., Ughetto, G., Quigley, G. J., Hakoshima, T., van der Marel, G. A., van Boom, J. H. and Rich, A. (1984) *Science*, **225**, 1115-1121.

Wang, A. H.-J., Ughetto, G., Quigley, G. J., and Rich, A. (1986) *J. Biomol. Struct. Dyn.* **4**, 319-342.

Wang, A. H.-J. in *Nucleic Acids and Molecular Biology* Vol. 1 (eds. Eckstein, F. & Lilley, D. M.) 32-54 (Springer, Berlin, 1987).

Wang, A. H.-J., Ughetto, G., Quigley, G. J. and Rich, A. (1987) *Biochemistry* **26**, 1152-1163.

Wang, A. H.-J., Cottens, S., Dervan, P. B., Yesinowski, J. P., van der Marel, G. A. and van Boom, J. H. (1989) *J. Biomol. Struct. Dyn.* **7**, 101-117.

Wang, A. H.-J. and Teng, M.-k. (1990) in *Crystallographic and Modeling Methods in Molecular Design*, Springer-Verlag, (in press).

Wang, J. C. (1974) *J. Mol. Biol.* **89**, 783-801.

Ward, B., Rehfuss, R., Goodisman, J. and Dabrowiak, J. C. (1988) *Biochemistry*, **27**, 1198-1204.

Wells, R. D. and Harvey, S. C. (eds) *Unusual DNA Structures*, (Springer-Verlag, NY,1987).

Widom, J. (1985) *BioEssays 2*, 11-14.

Wiley, P. F. (1979) *J. Natural Prod.* **42**, 569-582.

Williams, L. D., Egli, M., Gao, Q. Bash, P., van der Marel, G. A., van Boom, J. H., Rich, A. and Frederick, C. A.(1990a) *Proc. Natl. Acad. Sci. USA.* **87**, 2225-2229.

Williams, L. D., Egli, M., Ughetto, G., van der Marel, G. A., van Boom, J. H., Rich, A., Wang, A. H.-J. and Frederick, C. A.(1990b) *J. Mol. Biol.* (in press).

Williamson, J. R., Raghuraman, M. K. and Cech, T. R. (1989) *Cell* **59**, 871-880.

Wing, R., Drew, H., Takano, T., Broka, C., Tanaka, S., Itakura, K. and Dickerson, R. E. (1980) *Nature (London)* **287**, 755-758.

Zakrzewska, K., Lavery, R. and Pullman, B. (1983) *Nucleic Acids Res.* **11**, 8825-8839.

Zimmer, C. and Wahnert, U. (1986) *Prog. Biophys. Mol. Biol.* **47**, 31-112.

DNA DRUG INTERACTIONS STUDIED WITH POLARIZED LIGHT SPECTROSCOPY: THE DAPI CASE

B. Nordén, S. Eriksson, S. K. Kim, M. Kubista, R. Lyng and B. Åkerman
Department of Physical Chemistry
Chalmers University of Technology
S–412 96 Gothenburg
Sweden

Attention is focused on flow linear dichroism (FLD) measured with phase–modulation technique as a simplistic tool for detecting and characterizing DNA drug interactions. FLD combined with DNA–induced circular dichroism (CD), and in some instances fluorescence polarization anisotropy (FPA), has thus for a number of intercalators and non–intercalators provided important information about effective binding geometries and orientational dynamics under varying solution and binding conditions. The present study of a classical non–intercalator, DAPI (4',6–diamidino–2–phenylindole), offers several surprising features. FLD supports that DAPI binds in the minor groove of DNA and poly(dA–dT), but suggests that it may be intercalated in poly(dG–dC), albeit fluorescence quenching by iodide ion contradicts classical intercalation. The existence of two "binding sites" on DNA, one with an extremely low density as evidenced from CD, is surprisingly enough also found with the homopolymer poly(dA–dT). Three possible explanations for this observation are discussed: 1) the low–density site may be a "transient site" owing to DNA conformational dynamics, 2) the second site may reflect interacting DAPI monomers or 3) be an effect of a small allosteric conformational change in DNA. However, first at very high degrees of occupancy exciton DAPI–DAPI interaction indicates the presence of true dimers. Results from molecular mechanics calculations are in agreement with that different binding modes are preferred in poly(dG–dC) and poly(dA–dT) but give no indication as to the origin of two DAPI binding sites in the latter.

1. Introduction

1.1. BACKGROUND OF LINEAR DICHROISM SPECTROSCOPY

Anisotropy in flowing systems is a well–known phenomenon: it was discovered by Maxwell more than 100 years ago for viscous fluids which were found to show birefringence when subjected to shear in a so–called Couette cell, i.e. in the gap between two concentric cylinders rotating against each other. A similar device has been used occasionally to produce macroscopically aligned DNA in solution and for studying the flow linear dichroism (FLD) of the DNA bases in the ultraviolet region.[1-4] An early observation of the same normalized FLD magnitudes of planar heterocyclic dyes bound to DNA as for the intrinsic DNA chromophores, was taken as support for the binding mode known as intercalation.[5] Some 15 years ago, in our laboratory,[6] FLD, when measured sensitively with phase modulation

23

B. Pullman and J. Jortner (eds.), Molecular Basis of Specificity in Nucleic Acid-Drug Interactions, 23–41.
© 1990 *Kluwer Academic Publishers. Printed in the Netherlands.*

technique[4,7,8] on DNA solutions oriented in a high–performance Couette flow device, was found to be a simplistic, rather general way of detecting DNA interaction with chromophoric drugs.[3] For example, with purely anionic dyes the FLD in the dye absorption band is generally identically zero owing to the fact that the negative dye molecules in the solution stay far away from the negatively charged flow–oriented DNA double–helix. By way of contrast, all studied positively charged drugs have been found to display some kind of interaction with DNA, as evidenced by a non–vanishing FLD signal. Intercalators give a strongly negative signal corresponding to an orientation of their planes parallel to the DNA bases,[3,6,9,10] whereas groove–binders[11-13] generally give a strong positive FLD owing to an orientation more parallel to the helix. Weak electrostatic complexes frequently give weak FLD signals, with the FLD sign depending on the orientational bias.[14,15] Also cubic or randomly oriented drugs will display a minor FLD owing to a coupling with electronic transitions in the DNA bases that will not average to zero when integrated over all directions in space.[16]

In a study of DNA interaction with propeller shaped trigonal metal complexes, the two enantiomers Δ and Λ of Ru(1,10–phenanthroline)$_3^{2+}$ were found to exhibit varying affinity and enantioselectivity for DNA, poly(dA–dT) and poly(dG–dC).[15] Interestingly, FLD showed that they interact in terms of only two types of binding geometries (one for each kind of enantiomer) irrespective of DNA base composition and over a wide range of binding ratios and ionic conditions. This observation may indicate the presence of a basic interaction mechanism recognizing the chiral texture of the double helix, possibly through steric interaction upon binding in a groove.[14,15]

For both intercalators and groove–binders the circular dichroism (CD) induced in the drug chromophore transitions, due to the presence of the chiral DNA, can be an adjoint to FLD for further assignment of binding geometry. With rigid, planar polycyclic heteroaromatics such as acridine dyes the relatively weak induced CD is due to non–degenerate coupling with the $\pi\pi^*$ transitions of the DNA bases;[17,9] the dependence of sign and magnitude of the CD has in non–empirical calculations been correlated with different possible binding geometries with alternating GC and AT sequences.[18,19] With multifunctional oligoacridine ligands, FLD and induced CD have provided inference about potential multiintercalation and about angular orientation[10] and location[20] of the acridine moiety relative to the intercalation site.

With flexible ligands there is in addition a possibility of inherent chirality by preferred chiral conformations upon the DNA interaction. This is maybe one source of the CD in the case of DAPI considered here.

Finally, FLD measured in the DNA absorption region may also provide information about the DNA conformation through the local orientation of the DNA bases (tilt and twist): the reduced linear dichroism (see Results) is strongly negative and constant in the region 240 – 300 nm for the B and Z DNA forms where the bases are effectively perpendicular to the helix.[3,21-24] It is also negative but exhibits a pronounced wavelength dependence for the A form where the bases are tilted so that the different transitions in the base–plane form different angles to the helix axis.[21,22] For strand–separated DNA, where the bases owing to the flexibility of the DNA have a poor orientation, the dichroism is positive but very weak.[25] For a review on FLD and induced CD applied for studying structure and interactions of DNA, see Ref. 26.

1.2. BACKGROUND OF DAPI

DAPI is known to bind strongly and specifically to double–stranded DNA, preferentially to AT rich regions.[27-30]

Like ethidium, DAPI shows a large increase in fluorescence quantum yield upon binding to ds–DNA[31,32] which has made it frequently used in cytochemical work[32-38] and as a DNA marker in electrophoresis,[39,40] and fluorescence microscopy,[41,42] and in protein–DNA interaction studies.[43-45] DAPI is believed to be a minor–groove binder, similar to netropsin; evidence in support of this hypothesis includes FLD[13], fluorescence[46] and footprinting.[47] More recently a crystal structure[48] of its complex with a synthetic oligonucleotide verifies that it can bind in the minor groove to AT rich sequences. On the other hand there have been strong indications from viscometry and NMR[49] that DAPI may intercalate in alternating GC sequences.

From CD and fluorescence studies on natural calf thymus (CT) DNA a picture with two distinct DNA binding sites with high affinity for DAPI has emerged.[13,30,50] The first, stronger binding site is characterized by a very low binding site density ($n_1 = 0.02$) whereas the second binding site has a high saturation limit ($n_2 = 0.2$). In addition, at higher binding ratios ($R > 0.08$), there is evidence for a third spectroscopic species,[29,13] which has been assigned to a neighbour exciton interaction of an "accidental dimer" type.[9,13] As we shall see, however, this more or less established picture of two DNA binding sites for DAPI may be too simple an explanation of the spectral behaviour of this drug.

2. Results

CD and FLD spectra have been measured in the DAPI and DNA absorption regions for solutions with varying mixing ratios (R = drug/DNA phosphate) of DAPI with CT–DNA, poly(dA–dT) or poly(dG–dC). Fig. 1 shows how the CD in the 300 − 450 nm region of DAPI, induced by the presence of the chiral DNA, varies as a function of binding ratio. Up to R = 0.15 DAPI/DNA phosphate more than 99% of the amount of added DAPI can be considered as bound according to current stability constants at the present ionic strength.[30,49] Poly(dA–dT) and CT–DNA are found to behave in much the same ways. For both biopolymers the induced CD of DAPI at low R values shows a maximum at 335 nm with a shoulder at 365 nm. With increasing amount of associated DAPI, the CD induced per DAPI molecule is redistributed to give a CD maximum at 375 nm and a shoulder at 340 nm; an isodichroic point is observed at 351 nm for CT–DNA and at 341 nm for poly(dA–dT), indicating the presence of two spectroscopic species only. Whereas the 335 nm feature is essentially the same in the two DNAs, the 375 nm maximum for poly(dA–dT) increases to almost twice the magnitude

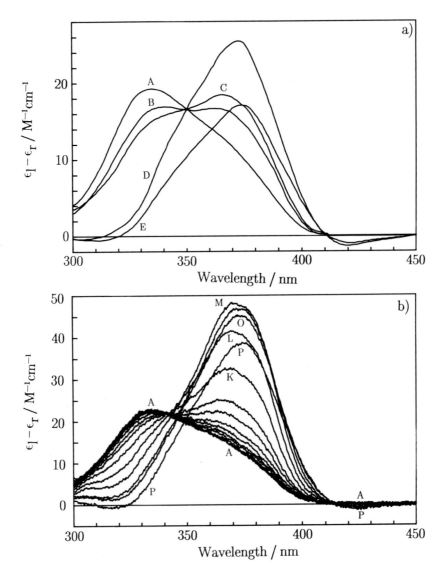

Figure 1: Circular dichroism (CD) spectra of DAPI–DNA complexes, normalized to unit DAPI concentration (**a–c**) or unit polynucleotide concentration (**d–e**). Mixing ratios drug–DNA phosphate (R) as indicated. Concentrations determined spectrophotometrically using $\epsilon_{340}=27000$ $M^{-1}cm^{-1}$ for DAPI in water, $\epsilon_{260}=6600$ $M^{-1}cm^{-1}$ for CT–DNA and poly(dA–dT), $\epsilon_{260}=8400 M^{-1}cm^{-1}$ for poly(dG–dC). All solutions were 0.25 mM EDTA, pH 7 and 10 mM (**a,b,d**) or 1mM (**c,e**) NaCl.
a) DAPI–CT–DNA, mixing ratios, A–E, 0.003, 0.03, 0.05, 0.2, 0.3.   **b)** DAPI–poly(dA–dT), mixing ratios, A–P : 0.0042, 0.0053, 0.0064, 0.0085, 0.013, 0.017, 0.023, 0.031, 0.047, 0.062, 0.093, 0.12, 0.15, 0.19, 0.23, 0.33.

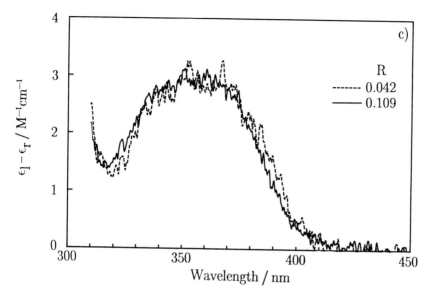

Figure 1: (continued) **c)** DAPI–poly(dG–dC).

observed with normal CT–DNA.

Above binding ratios of about 0.08 DAPI/DNA phosphate (0.15 with poly(dA–dT)) there appears a negative CD feature in the long–wavelength wing of the DAPI absorption band (at 410 − 430 nm), which will be referred to as "exciton CD". This effect, which can be referred to dimer interaction, is accompanied by a marked decrease and broadening of both the absorption and fluorescence spectra (see below).

The most striking result in Fig. 1 is the low apparent "binding site" density of the first bound species (approximately 0.02) in the DAPI binding to poly(dA–dT) as evidenced from the CD. In contrast to the heterogeneous CT–DNA, the homopolymer is not easily expected to provide any specific sites with such a low binding density.

Comparison of CD spectra for three different R values (Fig. 1d) normalized to equal nucleotide concentration, reveals an increase of the CD in the 260 nm region as compared to pure poly(dA–dT). The CD in the 260 nm and the 300–450 nm regions change differently with the binding ratio. This might imply that conformational changes occur when increasing the amount of DAPI bound to the polynucleotide. Regarding the strength of the DAPI CD in the 360 nm band, we cannot tell whether the change at 260 nm is due to a contribution from DAPI or to a change in the polynucleotide CD.

The CD spectrum of DAPI bound to poly(dG–dC) (Fig. 1c) shows a different behaviour: it is an order of magnitude weaker than for the other DNAs and follows essentially the shape of the absorption profile (cf Fig. 3c). The CD per bound DAPI is constant over a wide range of binding ratios indicating one type of binding site only. In contrast to the case of poly(dA–dT), a change in CD in the

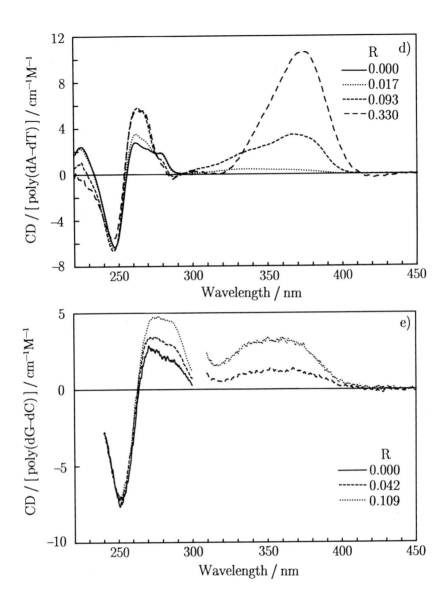

Figure 1: (continued) **d)** DAPI–poly(dA–dT). **e)** DAPI–poly(dG–dC).

260 nm region when DAPI binds to poly(dG–dC) scales exactly with the binding ratio and thus reveals no change of the intrinsic CD of poly(dG–dC). As will be inferred below, our spectroscopic results show that DAPI is oriented with its long–axis perpendicular to the poly(dG–dC) helix axis and that it may be partially intercalated in this polynucleotide.

Figure 2 shows the FLD spectra, both in the DAPI and the nucleotide absorption regions, for DAPI complexes with CT–DNA, poly(dA–dT) and poly(dG–dC). These results should be interpreted in terms of the *reduced linear dichroism* (LDr), which is the dimensionless ratio between the measured LD = A_z – A_y and the normal absorbance of the same sample at isotropic conditions (solution at rest). Here A_z denotes the absorbance measured with polarized light having the direction of polarization parallel to the flow direction (Z) and A_y the absorbance with the polarization perpendicular to the direction of flow. The light propagates radially through the Couette silica cylinders, the Y direction coinciding with the rotor axis of the device. For geometry and experimental details of the flow linear dichroism experiment, see Refs. 1–3, 8, 25.

The reduced linear dichroism is related to the local structure of the light–absorbing chromophore system and to the degree of macroscopic orientation of DNA as follows:[3]

$$LD^r = \frac{LD}{A_{iso}} = \frac{3}{2}\left(3\cos^2\alpha - 1\right) \cdot S \tag{1}$$

where α is the angle between the electronic transition moment responsible for the light absorption and the local DNA helix axis, and S ($0 \leq S \leq 1$) is an orientation function depending on the contour length and flexibility (persistence length) of the DNA as well as on the flow gradient and flow symmetry and on the viscosity of the solution. Hydrodynamic theory can be used for the determination of S.[2] For the present purpose a more reliable value is obtained from the LDr of the DNA chromophores as an internal standard:[3] assuming a retained B conformation structure of DNA (corresponding to an effective α equal to 86°),[21,22] and after appropriate correction of the LDr for any contribution from anisotropic absorption of the drug chromophore in the UV region, one has from Eq. 1:[21,22].

$$LD^r_{DNA}(260\ nm) = -1.47 \cdot S \tag{2}$$

which thus permits S be determined. Applying Eq. (1) to the LDr for a drug transition we get a corresponding angle α characterizing the drug orientation (a distribution owing to dynamic and static variations in α over the DNA must be recalled, though, and $\cos^2\alpha$ should strictly speaking be replaced by an ensemble average, $<\cos^2\alpha>$).

The conclusions from the FLD results in Fig. 2 may be summarized as follows:

1)   With poly(dA–dT), as with CT–DNA, the positive sign and the magnitude of LDr in the DAPI absorption region $330 - 390$ nm (which is known to arise from two essentialy long–axis polarized $\pi\pi^*$ transitions)[51] correlates with $\alpha = 43° \pm 5°$. This orientation of the DAPI long–axis relative to the DNA helix axis is consistent with a binding of DAPI in the

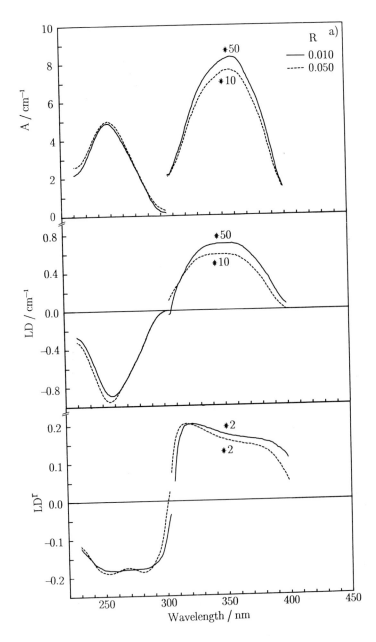

Figure 2a: Absorbance (A), linear dichroism (LD) and reduced linear dichroism (LD^r) spectra of the DAPI–CT–DNA complex at different mixing ratios, DNA concentration 0.72mM.

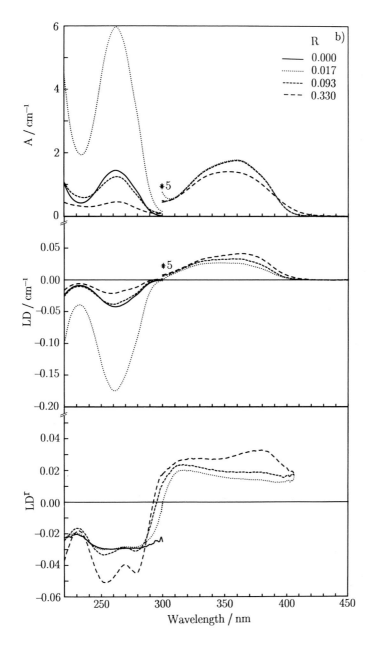

Figure 2b: Absorbance (A), linear dichroism (LD) and reduced linear dichroism (LD^r) spectra of the DAPI–poly(dA–dT) complex at different mixing ratios, DAPI concentration 14.7 μM.

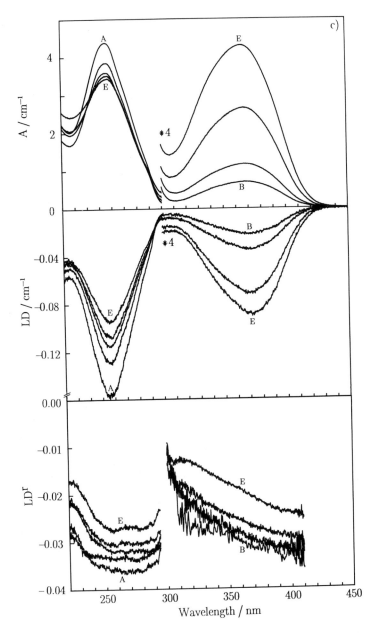

Figure 2c: Absorbance (A), linear dichroism (LD) and reduced linear dichroism (LD^r) spectra of the DAPI–poly(dG–dC) complex at different mixing ratios: A–E: 0, 0.023, 0.045, 0.110, 0.200. DNA concentration: 519, 453, 405, 365, 321 μM.

minor groove of DNA ($\alpha = 45° \pm 10°$ as from model building).* The variation in LD^r values for the two in–plane transitions suggests that DAPI binds with its plane intersecting deep into the groove in agreement with crystal work[48] (see Discussion).

2) For poly(dA–dT) approximately the same LD^r/S value, observed over a wide range of binding ratios, is consistent with a single geometric mode of binding. An increase in LD^r amplitude both in the drug and DNA absorption regions with higher binding ratios indicates that the structure of poly(dA–dT) becomes more rigid upon accommodating DAPI (S has increased by ca 50% at R = 0.3).

3) With poly(dG–dC), in contrast to poly(dA–dT) and DNA, the LD^r of DAPI is strongly negative, but decreases with increasing amount of bound drug. The same trend is seen in the absorption band of the polynucleotide showing that binding of DAPI decreases the degree of orientation of poly(dG–dC).

4) The LD^r/S spectrum of DAPI poly(dG–dC) is independent of R up to at least R = 0.11, again evidencing a single binding site. In the long wavelength region of the DAPI absorption band the LD is the same as that of the DNA bases, which means that the moment of this long axis polarized transition is effectively parallel to the plane of the bases ($\alpha = 86° \pm 5°$). Towards shorter wavelengths, LD^r decreases in magnitude and from the value at 320 nm one obtains $\alpha = 74°$ for the second transition moment[51] of DAPI.

Fig. 3 shows the absorption and emission profiles of DAPI upon association to poly(dA–dT). Like in the case of CT–DNA the DAPI–nucleotide interaction is associated with a red shift (20 nm) and a hypochromic change (ca 15%) of the 340 nm absorption band of DAPI. For poly(dA–dT) the hypochromicity is within $\pm 2\%$ constant up to about R = 0.1. Also, the shapes of the absorption and the emission profiles are the same at low and intermediate degrees of occupancy. The behaviour of DNA differs from that of poly(dA–dT) by showing continued decrease in absorption and emission, clearly evidencing binding heterogeneity in this case.

With poly(dG–dC) (Fig. 3c) the large hypochromicity (41%) and red shift (29 nm) suggest even stronger interactions with the DNA bases, in consistency with the proposed intercalation of DAPI in this polynucleotide.

At higher binding ratios all spectral quantities show clear influence from DAPI–DAPI interactions in the poly(dA–dT) and CT–DNA cases. In the CD spectrum a new (negative) band emerges at 420 nm and the isodichroic behaviour at 350 nm is lost by the strongly decreasing CD at shorter wavelengths. The absorption shows a further decrease at the maximum and an increasing wing of extra intensity at 400–420 nm. Also the fluorescence emission decreases and at very high binding ratios (R=0.3) a new band is found to build up at 530 nm, possibly due to excimer formation, (curve P in Fig. 3b).

---

*Binding in the major groove can be excluded on the basis of fluorescence experiments with DAPI bound to poly(dA–I⁵dU) and poly(dI–Br⁵dC) as shown by Härd et al.[46]

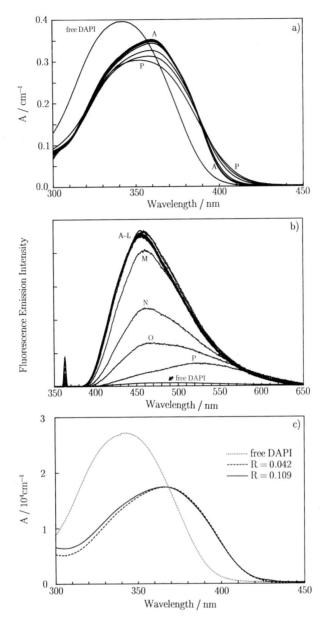

Figure 3 :   **a)** Absorbance (A) spectra of DAPI–poly(dA–dT) at different mixing ratios (see Fig 1b for an explanation). Spectra corrected for light scattering and absorption of the polynucleotide.   **b)** Fluorescence emission spectra of the DAPI–poly(dA–dT) complex in **a)**. Excitation wavelength 363 nm.   **c)** Absorbance of DAPI–poly(dG–dC) complexes at different mixing ratios. The spectra are normalized to unit DAPI concentration.

Figure 4 shows quenching experiments using iodide ion as a quencher. The quenching efficiency of iodide ion on the fluorescence of DAPI follows the order: free DAPI > DAPI–poly(dG–dC) > DAPI–poly(dA–dT) ~ DAPI–DNA. The results indicate that DAPI is more accessible to iodide ion in poly(dG–dC) than in poly(dA–dT) (see Discussion).

The results from the energy minimization computations (to be presented elsewhere) suggest, not surprisingly, that both minor–groove binding and intercalation can give, from an enthalpic point of view, stable DAPI complexes for both of the polynucleotides. A significantly less energy gain with groove–binding in poly(dG–dC), compared to intercalation, can explain the experimental finding suggesting a preferred intercalative type of binding with this polynucleotide.

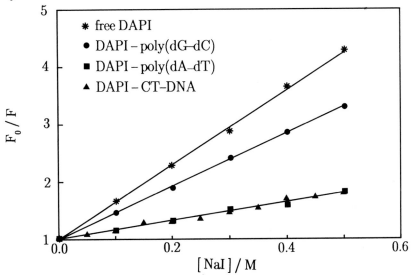

Figure 4 :   Stern–Volmer plot of quenching of DAPI fluorescence using iodide ion as a quencher (NaI). The ionic strength constant by adding NaCl to sodium concentration 0.5 M, DAPI concentration 0.91 μM.
(▲) DAPI–CT–DNA, R=0.008,   (■) DAPI–poly(dA–dT), R=0.007,
(●) DAPI–poly(dG–dC), R=0.007,   (*) DAPI/water.

3. Discussion

3.1. DAPI–POLY(dG–dC)

A DAPI orientation more or less parallel to the DNA baseplanes in poly(dG–dC) is evidenced from the strongly negative LD^r for DAPI. The LD^r of the first transition is in fact in accordance with the long axis being perfectly parallel to the baseplanes. This observation as well as the extreme red shift and strong hypochromicity suggests intercalation, in harmony with evidence from NMR, viscometry and binding data recently presented by Wilson and coworkers.[49] The induced CD in the $300 - 400$ nm absorption is also relatively weak ($\Delta\epsilon/\epsilon = + 1.9$ x 10^{-4}), as expected for a planar (achiral) chromophore, for which non–degenerate coupling with the chirally arranged DNA transitions is the main source of rotational strength.[9,17-19] The moment directions of the two DAPI transitions diverge from each other by less than $20°$,[51] so their $\Delta\epsilon/\epsilon$ values are theoretically expected to be nearly the same,[17] in accordance with the experimental CD spectrum having effectively the same shape as the absorption envelope.

That DAPI does not bind in the minor groove of poly(dG–dC) can be seen as an effect of the presence of the bulky amino group in 2–position on guanine.[48] The molecular mechanics results advocate an intercalative orientation where the DAPI long axis lies essentially along the long axis of the intercalation pocket. However, with this orientation, if the drug were located at the centre of the intercalation pocket, a negative induced CD would be expected according to calculations for a B form structure of poly(dG–dC).[18] The experimental, positive CD may therefore suggest that DAPI is only partially intercalated (for instance, the calculations predict positive CD in the minor groove).[18] Furthermore, the observation of a drop in the LD^r magnitude at the 330 nm transition (Fig. 2) shows that this transition, which is some 15–20° away from the DAPI long axis,[51] is not oriented perpendicular to the DNA helix axis ($\alpha = 74°$). Thus, whereas the perfectly perpendicular DAPI long axis and the strong hypochromicity and red shift observed in the absorption are in favour of DAPI being intercalated in poly(dG–dC), the behaviour of the induced CD and the LD^r of the 330 nm band, as well as the impaired polynucleotide orientation, indicate a severe deviation from a classical intercalative binding geometry.

Another strong indication against intercalation, in the classical sense, is the observation (Fig. 4) that DAPI in poly(dG–dC) is more easily accessible to quenching by iodide ion than when bound in poly(dA–dT). The quantitative interpretation of the Stern–Volmer plots in Fig. 4 has to consider the lifetimes of DAPI in the bound states. Our conclusion of greater iodide ion accessibility of DAPI when bound to poly(dG–dC) than to poly(dA–dT) is justified by the observation of very similar lifetimes of DAPI in these polynucleotides.[29,46,52]

3.2. DAPI–POLY(dA–dT)

The interaction of DAPI with poly(dA–dT) differs markedly from that with poly(dG–dC), both regarding binding angle and number of binding modes, as evidenced from the LD and CD results. The first important conclusion about a non–intercalative binding of DAPI to poly(dA–dT) comes from the positive LD at 370 nm which immediately implicates an angle less than 55° between the long axis of DAPI and the local helix axis of the homopolymer. Evaluation of the LD^r/S values for the two transitions at 335 and 370 nm reveals angles of about 43° and 46°, respectively. These angles are in excellent agreement with the angle between the DAPI plane and the helix axis of the Dickerson dodecamer observed in the crystal study of Larsen et al.[48] Considering the angle (15–20°) between the transition moments in the DAPI chromophore, the tiny difference between 43° and 46° is consistent with DAPI being bound in an essentially planar conformation with the molecular plane bisecting the minor groove, as indicated from the crystal results.[48]** Our spectroscopic results thus indicate a similar binding geometry of DAPI in its complex with poly(dA–dT) as the one reported in its crystal complex with the dodecamer CGCGAATTCGCG, where DAPI binds in the minor groove at the three central ATT bases.[48] It is also understandable how this mode of binding, with DAPI perfectly filling the groove, leads to an increased rigidity of the DNA structure as inferred from the markedly increased orientation degree S of the DAPI–poly(dA–dT) complex.

Whereas the LD[r], the absorption and the fluorescence spectra are essentially independent of binding ratio up to intermediate binding ratios, consistent with a single mode of DAPI binding, the shape of the CD spectrum changes considerably with R. Since CD can be anticipated to be highly sensitive to the orientation of the drug chromophore relative to its chiral host and to the conformation of the drug, these changes may reflect just minor alterations in the binding geometry as more DAPI is bound. The CD results suggest the presence of two "binding sites", the first one with an abundancy as low as only two DAPI molecules per five helical turns of poly(dA–dT). We do not at present understand the true origin of this apparent low–density site, which is indeed a remarkable finding in a homopolymer. We propose three possibilites that we intend to consider further:

1) the low density site is related to conformational dynamics of the polynucleotide that provides transient sites for which the drug has high affinity, or
2) the second site reflects interacting DAPI monomers, or
3) initial binding induces allosteric conformational changes that affect the binding of additional dyes.

The three alternatives differ in the responsible component for the two sites. In 1) the two sites are a property of the DNA only, in 2) the sites are entirely a property of DAPI, whereas in 3) they are caused by both DAPI and the DNA.

The transient site in alternative 1, might correspond to regions close to local DNA opening (breathing). The binding and dissociation kinetics of DAPI is extremely rapid (unpublished results) and DAPI could be able to locate these spots. The abundance of such openings is likely to be of the same order as the saturation level for the high affinity binding site. In alternative 2 the presence of one DAPI molecule affects the spectroscopic properties of other closely bound DAPI–molecules. Although close proximity between chromophores indeed is expected to affect the spectroscopic properties, clear indications of dimer formation is not observed at binding ratios below 0.12. Alternative 3 involves both DAPI and DNA and is therefore most flexible. When DAPI binds it may induce an allosteric conformational change in the DNA, altering the affinity for additional DAPI molecules.

When two or more DAPI molecules bind in the same region, a two–state situation similar to that of a dimer can be anticipated, but without necessarily

---

**Taking 43° to represent the angle of the DAPI plane to the helix axis, one has $(\cos^2 46°) = (\cos^2 43°)(\cos^2 \Phi)$ in accordance with an angle $\Phi = 18°$ between the two transitions (see eq. 1 in Ref. 25), in agreement with the moment determination in DAPI.[51]

involving any observable DAPI–DAPI interactions. The CD of DAPI may be due to a small twist around the internuclear bond between the indole and phenyl moieties, which in turn could be sensitively dependent on the pitch of the DNA. Simple fitting arguments suggest that DAPI in the minor groove of B–DNA may be preferentially twisted into a right–handed helical conformation, in agreement with the positive CD observed for the two long axis polarized transitions by analogy with the optical rotatory power of skewed dienes.[53] The dihedral angle is presumably small (no more than a few degrees according to the crystal structure)[48] but might be subject to large *relative* changes as response to even small alterations in the DNA structure.

Another possible mechanism of DNA–induced CD is by coupling with the DNA transitions.[17-19] A change in the orientation of the DAPI transition moments relative to the inducing moments of the bases, either due to a changed DAPI orientation or to a change in DNA helix angle, could thus influence the observed CD pattern.

At very high binding ratios (R>0.15) where exciton interactions appear, the positive LD^r shows a drop above 400 nm, indicating that the resulting transition moment of the low–energy exciton component of the interacting DAPI–DAPI pair has a resulting transition moment oriented more perpendicular than parallel relative to the DNA helix (the lower LD^r at $300 - 340$ nm can be ascribed to free DAPI, contributing in A_{iso} but not in LD). This could indicate that some DAPI molecules are intercalated at higher binding ratios and interacts "accidentally" with DAPI molecules in the minor groove. A head–tail interaction between two close–lying DAPI molecules in the groove is expected to give the opposite behaviour for the low energy coupling mode (positive LD), whereas a stacked type of dimer could explain the experimental observation by giving a perpendicularly oriented resulting moment. Recently Gao and Patel have reported that the antitumour drug chromomycin binds as a symmetrical, stacked dimer in the minor groove to a GC sequence; the crowding in the groove, though, forces the oligonucleotide to adopt an A–like conformation.[54]

3.3 DAPI–DNA

Earlier studies of DAPI–DNA interactions could be discussed in the light of the new findings for the interactions with the homopolynucleotides. In view of that natural DNA contains all four bases, in similar proportions, one could expect the occurrence of (at least) two structurally different complexes: minor groove adducts ($\alpha = 45°$) in AT sequences, and potential intercalators (α close to 90°) in GC sequences. The great resemblance of the LD as well as the CD features, and qualitative similarities of absorption and fluorescence too, between the DAPI complexes with CT–DNA and poly(dA–dT) clearly demonstrates that binding of DAPI to DNA is dominated by the groove–binding mode(s) observed with poly(dA–dT). It has been found from footprinting experiments that DAPI, like Hoechst 33258 and distamycin, displays a preferential binding to regions with at least four contiguous AT basepairs.[55,47] The fact that the observed average DAPI orientations, at low DAPI binding ratios, are approximately the same for DNA and poly(dA–dT), indicates that binding of GC type is insignificant in DNA. The similarity of the CD feature characteristic of the low–density binding sites in DNA and poly(dA–dT) also implies that one has to reconsider the nature of this site in DNA, as it is apparently not necessarily related to any sequence specific effect.

At high DAPI binding ratios clear differences can be noted between DNA and poly(dA–dT) in the CD titrations. First, the maximum CD in DNA amounts to only about half the amplitude observed in the case of poly(dA–dT) ("site 2"). Secondly, the exciton CD appears earlier in DNA: already at $R = 0.08$ compared to 0.12 in poly(dA–dT). Both of these observations are understandable since the number of AT–type binding sites must be less in DNA than in poly(dA–dT). The lower CD in DNA can be explained by heterogeneity. The CD of a molecule, for example owing to a chiral conformation, is generally reduced by a broadening of the conformational distribution (for the extreme case of an inversion–symmetric distribution, the CD vanishes). Correspondingly, the appearance of exciton CD (evidencing DAPI–DAPI interactions) at a lower binding ratio in DNA than in poly(dA–dT) can be seen as an effect of an earlier crowding of the available AT–containing regions, which are fewer in DNA than in the homopolymer.

Closer inspection of the absorption, fluorescence and LD^r spectra of DAPI bound to DNA shows, in contrast to the case of poly(dA–dT), certain variations also at low and intermediate binding ratios, evidencing a heterogenous binding distribution. For example, a decrease by some 10% in fluorescence intensity from $R = 0.01$ to 0.05[13] and a corresponding decrease in LD^r/S also by about 10%, might be explained by the presence of a minor amount of "GC type" of binding at the higher binding ratios. Anyhow, the GC type of DAPI binding is clearly infrequent on DNA.

4. Conclusions

Our results on the complexes of DAPI with the synthetic homopolymers poly(dA–dT) and poly(dG–dC) reveal several unexpected, potentially important features of DNA–drug interaction. We may summarize our conclusions as follows:

1. DAPI binds to poly(dA–dT) in the minor groove, at an angle of about 45° to the helix axis. The double–helix structure thereby gains rigidity.

2. While absorption, fluorescence and linear dichroism indicate a single binding geometry of the DAPI–poly(dA–dT) complex, up to a ratio of 0.15 drug/nucleotide, circular dichroism evidences two distinct (possibly only slightly different) binding conformations, one corresponding to a very low apparent "site density" of 0.02 drug/nucleotide.

3. A very similar spectroscopic behaviour of calf thymus DNA supports that DAPI also here binds in the minor groove, to AT–containing regions. The similar "site" behaviour in DNA and poly(dA–dT) indicates that the established low density site in DNA does not necessarily correspond to a specific sequence.

4. DAPI binds to poly(dG–dC) with a perpendicular orientation relative to the helix axis and shows strong optical interaction with the nucleic bases, which can indicate intercalation. A classical intercalation geometry is contrasted, though, by fluorescence quenching experiments indicating that DAPI is accessible to iodide ions, and by the LD^r vs wavelength

dependence suggesting that the molecular plane is tilted relative to the bases.

Our DAPI results, presented here, clearly demonstrate that properties such as the shape of a potential DNA ligand do not necessarily make it either a groove–binder or an intercalator, but that the mode of interaction may depend as well on the target sequence.

5. Acknowledgement

This research was supported by the Swedish Natural Science Research Council.

6. References

(1)   Wada, A., and Kozawa, S. (1964) *J. Polym. Sci.*, *Part A* 2, 853–864.
(2)   Wada, A. (1972) *Appl. Spectr. Rev.* 6 , 1–30.
(3)   Nordén, B. (1978) *Appl. Spectr. Rev.* 14, 157–248.
(4)   Schellman, J. A., and Jensen, H. P. (1987) *Chem. Revs.* 87, 1359–1399.
(5)   Lerman, L.S. (1963) *Proc. Natl. Acad. Sci.* 49, 94–102.
(6)   Nordén, B., and Tjerneld, F. (1976) *Biophys. Chem.* 4, 191–198.
(7)   Davidsson, Å., and Nordén, B. (1976) *Chem. Scr.* 9, 49–53.
(8)   Nordén, B., and Seth, S. (1985) *Appl. Spectrosc.* 39, 647–655.
(9)   Nordén, B., and Tjerneld, F. (1982) *Biopolymers* 21, 1713–1734.
(10)   Wirth, M., Buchardt, O., Koch, T., Nielsen, P. E., and Nordén, B. (1988) *J. Am. Chem. Soc.* 110, 932–939.
(11)   Tjerneld, F., and Nordén, B. (1977) *Chem. Phys. Letters* 50, 508–512.
(12)   Nordén, B., Tjerneld, F. and Palm, E. (1978) *Biophys. Chem.* 8, 1–15.
(13)   Kubista, M., Åkerman, B. and Nordén, B. (1987) *Biochemistry* 26, 4545–4553.
(14)   Tjerneld, F., and Nordén, B. (1976) *FEBS Letters* 67, 368–370.
(15)   Hiort, C., Nordén, B., and Rodger, A. (1990) *J. Am. Chem. Soc.* 112, 1971–1982.
(16)   Schipper, E., and Nordén, B. (1981) *Chem. Physics* 57, 365–376.
(17)   Schipper, E., Nordén, B., and Tjerneld, F. (1978) *Chem. Phys. Letters* 70, 17–21.
(18)   Lyng, R., Nordén, B., and Härd, T. (1987) *Biopolymers* 26, 1327–1345.
(19)   Kubista, M., Åkerman, B., and Nordén, B. (1988) *J. Phys. Chem.* 92, 2352–2356.
(20)   Fornasiero, D., Kurucsev, T., Lyng, R., and Nordén, B. (1989) *Croatica Chimica Acta* 62, 337–347.
(21)   Matsuoka, Y., and Nordén, B. (1982) *Biopolymers* 21, 2433–2452.
(22)   Matsuoka, Y., and Nordén, B. (1982) *Biopolymers* 22, 1731–1746.
(23)   Eriksson, M., Nordén, B., Lycksell, P.–O., Gräslund, A., and Jernström, B. (1985) *J. Chem. Soc. Chem. Commun.*, 1300–1302.
(24)   Johnson, W.C.Jr. (1988) in *Polarized Spectroscopy of Ordered Systems* (Eds. B. Samori and E. W. Thulstrup) Kluwer A P, Dordrecht 1988, pp 167–183.
(25)   Nordén, B., and Seth, S. (1979) *Biopolymers* 18, 2323–2339.

(26) Nordén, B., and Kubista, M. (1988) in *Polarized Spectroscopy of Ordered Systems* (Eds. B. Samori and E. W. Thulstrup) Kluwer A P, Dordrecht 1988, pp 133–165.

(27) Chandra, P., and Mildner, B. (1979) *Cell. Mol. Biol.* **25**, 137–146.

(28) Kapuscinski, J., and Szer, W. (1979) *Nucleic Acids Res.* **6**, 3519–3534.

(29) Cavatorta, P., Masotti, L., and Szabo, A. G. (1985) *Biophys. Chem.* **22**, 11–16.

(30) Manzini, G., Barcellona, M. L., Avitabile, M., and Quadrifoglio, F. (1983) *Nucleic Acids Res.* **11**, 8861–8876.

(31) Kapuscinski, J., and Skoczylas, B. (1977) *Anal. Biochem.* **83**, 252–257.

(32) Williamson, D. H., and Fennel, D. J. (1975) *Methods Cell Biol.* **12**, 335–351.

(33) Schweizer, D. (1976) *Exp. Cell Res.* **102**, 408–409.

(34) Langlois, R. G., Carrano, A. V., Stay, J. W., and Van Dilla, M. A. (1980) *Chromosoma* **77**, 229–251.

(35) Coleman, A. W., Maguire, M. J., and Coleman, J. R. (1981) *J. Histochem. Cytochem.* **29**, 959–968.

(36) Tijssen, J. P. F., Beekes, H. W., and van Steveninck, J. (1982) *Biochim. Biophys. Acta* **721**, 394–398.

(37) Lee, G. M., Thornthwaite, J. T., and Rasch, E. M. (1984) *Anal. Biochem.* **137**, 221–226.

(38) Brown, R. N., and Hitchcock, P. F. (1989) *Dev. Brain Res.* **50**, 123–128.

(39) Kapuscinski, J., and Yanagi, K. (1979) *Nucleic Acids Res.* **6**, 3535–3542.

(40) Naimski, P., Bierzynski, A., and Fikus, M. (1980) *Anal. Biochem.* **106**, 471–475.

(41) Yanagida, M., Miraoka, Y., and Katsura, I. (1980) *Cold Spring Harbor. Symp. Quant. Biol.* **47**, 177–187.

(42) Schwartz, D., and Koval, M. (1989) *Nature* **338**, 529–522.

(43) Kania, J., and Fanning, T. G. (1976) *Eur. J. Biochem.* **67**, 367–371.

(44) Stepien, E., Filutowicz, M., and Fikus, M. (1979) *Acta Biochim. Pol.* **26**, 29–38.

(45) Mazus, B., Falchuk, K. H., and Vallee, B. L. (1986) *Biochemistry* **25**, 2941–2945.

(46) Härd, T., Fan, P., and Kearns, D. R. (1990) *Photochem P.* **51**, 77–86.

(47) Jeppesen, C., and Nielsen, P. E. (1989) *Eur. J. Biochem.* **182**, 437–447.

(48) Larsen, T. A., Goodsell, D. S., Cascio, D., Grzeskowiak, K. and Dickerson, R. E.. (1989) *J. Biomol. Struct. Dyn.* **7**, 477–491.

(49) Wilson, W. D., Tanious, F. A., Barton, H. J., Strekowski, L. and Boykin, D.W. (1989) *J. Am. Chem. Soc.* **111**, 5008–5009.

(50) Kapuscinski, J., and Skozylas, B. (1978) *Nucleic Acids Res* ,**5**, 3775–3799.

(51) Kubista, M., Åkerman, B., and Albinsson, B. (1989) *J. Am. Chem. Soc.* **111**, 7031–7035.

(52) Barcellona, M.L. and Gratton, E. (1989) *Biochim. Biophys. Acta* **999**, 174–178.

(53) Moscowitz, A., Charney, E., Weiss, U., and Ziffer, H. (1961) *J. Am. Chem. Soc.* **83**, 4661–4670.

(54) Gao, X., and Patel, D. J. (1989) *Biochemistry* **28**, 751–762.

(55) Portugal, J., and Waring, M. (1988) *Biochim. Biophys. Acta* **949**, 158–168.

DRUG-DNA RECOGNITION: SEQUENCE-SPECIFICITY OF THE DNA MINOR GROOVE BINDER BERENIL

S NEIDLE, DG BROWN, TC JENKINS, CA LAUGHTON,
MR SANDERSON & JV SKELLY

Cancer Research Campaign Biomolecular Structure Unit
The Institute of Cancer Research
Sutton Surrey SM2 5NG United Kingdom

ABSTRACT. The crystal structure of a complex between the anti-trypanosomal drug berenil and the DNA sequence d(CGCGAATTCGCG)$_2$ is described. The drug is found in the minor groove, in the 5'-AAT region. Effects on various base-pair geometric parameters are described, and compared with those calculated from a molecular modelling study of berenil interacting with a long DNA sequence. Implications for the design of modified berenils with altered sequence selectivity are discussed.

1. Introduction

The non-covalent groove-binding interactions of various drugs with defined DNA sequences have been the focus of considerable interest over the past few years. The major goal of these studies is the eventual development of new ligands with enhanced affinities for defined DNA sequences that may have particular biological roles. It is axiomatic that the attainment of this objective requires a detailed picture of the molecular features of the drug-DNA interactions from X-ray crystallographic and NMR studies, together with reliable estimates of the relative contributions from hydrogen bonding, hydrophobic forces, electrostatic interactions and inherent sequence-specific DNA structural properties. In fact, comparative study of the structures of the complexes themselves, especially in comparison with the native structures, can indirectly provide information on the relative importance of these factors. Structural and molecular-modelling studies have been performed at the Institute of Cancer Research on one such drug, the anti-trypanosomal agent berenil and are discussed here in these terms. We also analyse the implications of this data for the rational design of berenil analogues with altered sequence selectivity.

43

B. Pullman and J. Jortner (eds.), Molecular Basis of Specificity in Nucleic Acid-Drug Interactions, 43–57.
© 1990 *Kluwer Academic Publishers. Printed in the Netherlands.*

1.1 BERENIL

Berenil (1,3-bis(4-amidinophenyl)triazene, Figure 1) is extensively used for the treatment of bovine trypanosomiasis, usually as the *N*-acetyl glycinate salt (Newton, 1975). It has mild cytotoxic and anti-viral properties (De Clercq and Dann, 1980).

Its binding to double-stranded DNA has long been implicated in these biological effects (Newton, 1975; Baguley, 1982), with preferential binding to AT regions being established (Zimmer and Wahnert, 1986). Interaction with kinetoplast DNA from cultured *Trypanosoma cruzi* cells (Bernard and Riou, 1980) is presumably at the phased A tracts that are present in this DNA. Further insight into the selectivity of action of berenil (and other anti-trypanocidal drugs) has been afforded by the recent finding that there is highly selective formation of DNA-protein cleavable complexes by

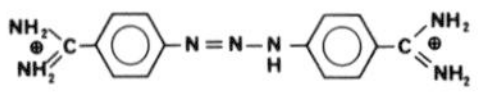

Figure 1. The chemical structure of berenil

kinetoplast DNA from these cells, and not by nuclear DNA (Shapiro and Englund, 1990). This linearisation of the DNA minicircles is believed to involve a specific mitochondrial DNA topoisomerase II.

The interaction of berenil with AT regions of DNA has been confirmed in detail by DNAase I footprinting studies on a 160-base pair *tyrT* fragment (Portugal and Waring, 1987), which have shown that the drug binds to both alternating AT and oligo A regions with at least three contiguous AT base pairs.

Several molecular-modelling studies have been reported on berenil-AT complexes (Gresh and Pullman, 1986; Pearl et al., 1987; Gago, Reynolds and Richards, 1988). All models concur in having the concave surface of the drug bound in the DNA minor groove, analogous to the arrangement observed in the crystal structure of a netropsin-dodecanucleotide complex (Kopka et al, 1985). The models also suggest that berenil spans two consecutive AT base pairs and hydrogen bond with the thymine bases via their O2 atoms.

Prior to detailed design of berenil analogues, we have determined the crystal structure of a berenil-oligonucleotide complex in order to provide experimental structural data on the geometry of binding.

2. Structural studies

Berenil was co-crystallised with the Dickerson-Drew dodecanucleotide duplex d(CGCGAATTCGCG)$_2$. The yellow crystals are isomorphous with the native structure, although surprisingly the unit-cell volume is 2.5% smaller. Intensity data were collected to a resolution of 2.5Å on an area detector and a rotating anode Xray source. The structure of the complex was solved by molecular replacement using the native structure as a starting-point. Difference and omit electron-density maps showed an extended continuous volume of density in the minor groove of the duplex which was unequivocally fitted to a berenil molecule. There was no evidence of disorder or multiple positions for the drug molecule, which refined satisfactorily. A total of 49 water molecules were also located, on the basis of good hydrogen-bonding geometry, and were included in the refinements. This relatively small number of water molecules reflects both our conservative criteria for acceptance and the inherent inability of diffraction data at this resolution to resolve more than a small percentage of the total water content. The final R factor is 0.177 for the 1759 reflections with intensities above the 2σ level. Details of the structure analysis are given in Brown et al (1990). Coordinates have been deposited in the Brookhaven Databank.

2.1 OVERALL FEATURES OF THE STRUCTURE

The structure shows berenil bound in the AATT region of the duplex (Figure 2). It is positioned slightly asymmetrically with respect to the pseudo (non-crystallographic) two-fold axis of the dodecamer, with the terminal protonated amidinium groups of the drug pointing into the groove.

The berenil molecule covers three base pairs with hydrogen bonds from the amidinium groups to adenines A5 and A18 on opposite strands. Thus the direct binding site is 5'-AAT. The hydrogen bond to A18 is a direct one, to N3, whereas that to A5 is indirect. It is mediated by a water molecule that is situated almost in the plane of A5 and is in hydrogen-bonding contact with an amidinium nitrogen atom of berenil (2.80Å), N3 of adenine A5 (3.10Å) and the O4' deoxyribose ring oxygen atom of A6 (2.82Å). Even small movements of the drug molecule so as to achieve perfect two-fold symmetry for the complex result in a loss of the good hydrogen-bonding geometry. Molecular graphics analysis of the structure has shown that the presence of the water molecule is essential for hydrogen-bonding interaction to take place between this end of the drug molecule and DNA, and thus to achieve good isohelicity of fit. The amidinium nitrogen ... adenine N3 distance is too long, by at least 0.75Å. This structural feature contrasts with that in models derived from canonical B-DNA, where it is possible to achieve direct berenil-DNA hydrogen bonding at both ends of the drug molecule (Brown et al.,

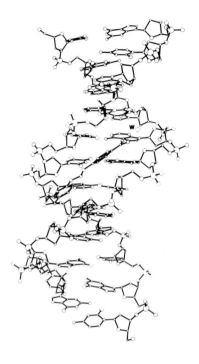

Figure 2. Overall view of the berenil-d(CGCGAATTCGCG)$_2$ complex

1990). Such molecular mechanics-minimised models do not have the characteristic sequence-dependent features of, for example, the narrow AATT minor groove seen in the crystal structures of both native and drug-complexed dodecamers.

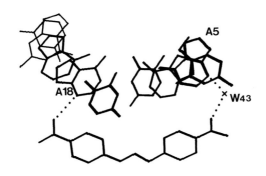

Figure 3. View of the berenil-d(CGCGAATTCGCG)$_2$ complex projected onto the plane of the drug molecule. Hydrogen bonds are shown as dashed lines.

This then raises the important questions of whether and to what extent the crystal structure(s) and their sequence-dependent features are affected by packing forces, and whether the bound water molecule is retained in solution. There is compelling enzymatic evidence in favour of sequence-controlled groove variability (Drew and Travers, 1985) that correlates with features observed in the dodecamer structure, suggesting that packing forces can be relatively unimportant, at least in the central region of a helix. Determination of more B-type oligomer crystal structures, of varying base-pair lengths, is clearly necessary. Binding of berenil to the 5'-AAT sequence in the dodecamer has recently received important verification from 2-D NMR studies in aqueous solution (Lane, Jenkins, Neidle and Brown, to be submitted); the fortuitously large number of aromatic protons on the drug molecule has enabled a suitably large number of inter-proton NOE distances to be unequivocally assigned. Such NMR studies cannot comment on the bridging water molecule seen in the crystal structure.

2.2 THE BOUND DRUG AND MINOR GROOVE WATER

The Dickerson-Drew dodecamer structure (Drew and Dickerson, 1981) includes well-ordered minor-groove bound water molecules, especially in the AATT region. These are necessarily displaced on binding a ligand such as berenil. The bridging water molecule in the structure of the complex does not correspond to any located in the native structure - this raises the question of whether the water is one that is included in the solvent shell of the drug in free solution before interaction, or whether it is a redistributed spine of hydration water.

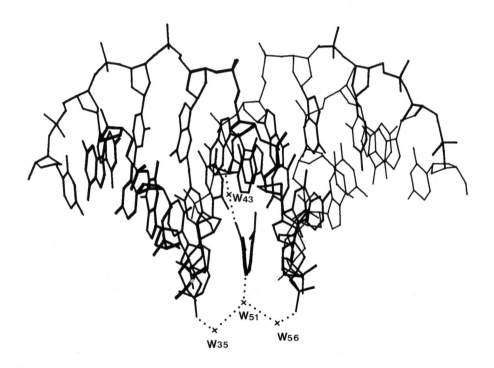

Figure 4. View of the berenil-d(CGCGAATTCGCG)$_2$ complex looking down the minor groove and showing the cluster of water molecules described in the text.

There is a well-ordered water arrangement at the mouth of the berenil complex that links the drug to phosphate groups on both strands (Figure 4). One nitrogen atom of the central triazene link of berenil is hydrogen-bonded to W51 (a distance of 3.01Å), suggesting that this nitrogen carries a hydrogen atom; high-resolution Xray studies on several berenil derivatives (Walton, Jenkins and Neidle, to be published), have shown that the triazene group has a localised hydrogen atom with asymmetric N-N bond lengths. Water W51 is linked by weak (ca 3.5Å long) hydrogen bonds to waters W35 and W56. These interact with phosphate oxygen atoms of thymine residues T19 and T8, with O(W) ... O distances of 2.95Å and 2.90Å respectively. There is also a loose arrangement from W51 via W42 to the phosphate on T21. The overall effect is of a clustering of water molecules around the relatively polar centre of the drug - this cluster is not observed in the native structure. There is no reason to consider it to be a sequence-related phenomenon.

2.3 EFFECT OF DRUG BINDING ON BASE-PAIR PARAMETERS

The effects of berenil on DNA structure have been examined in terms of the various base-pair parameters, as defined by the Cambridge convention. Geometries for the berenil-dodecamer complex have been compared with those available in the Brookhaven Database for the native structure (Drew and Dickerson, 1981) and its minor groove complexes with netropsin (Kopka et al., 1985) and Hoechst 33258 (Pjura, Grzeskowiak and Dickerson, 1987). Both of these drug molecules are somewhat larger than berenil, with netropsin spanning the whole of the AATT region and the Hoechst compound interacting with the ATTC sequence. The rise parameter (Figure 5) is barely affected by any of the drugs, with berenil producing the least effect, especially at its binding sequence. The largest changes are observed at the helix termini; this effect, which is common to all helical parameters, probably reflects small differences in packing and water environment that are accentuated at the termini. Slide of one base pair relative to its neighbour, again changes least for the berenil complex compared to the native structure. Changes in base-pair inclination with respect to the overall helix axis are minimal for all the complexes at the two central AT base pairs, whereas inclinations, especially on the 3' side, are markedly different for the three drugs. Inclination provides a rough measure of helix bending - it can be seen from Figure 5 that berenil barely affects the bending of the dodecamer structure whereas netropsin decreases it, by about 8° (Kopka et al., 1985). Values for helical twist along the length of the sequence are barely affected by the berenil binding, whereas Hoechst 33258 in particular appears to produce several significant changes.

The pattern of least distortion being induced by berenil is maintained by the propeller twist, roll and buckle parameters (Figure 5), although for all three, small but significant changes are apparent at the binding site itself. There is minimal change in propeller twist at and around the binding site, except at the 3' end AT base pair (an increase of 4°). This perturbation is needed in order to produce good adenine N3 ... berenil amidinium hydrogen bond geometry. There are changes in buckle, by up to 3°, in the ATT binding region and a sharp increase in roll, by 5° at the first AT step, that are also probably related to optimisation of the drug-DNA specific hydrogen bonding geometry.

These structural results indicate that propeller twist, roll and buckle are the most easily deformable base-pair parameters, whereas at least some of the others appear relatively invariant to minor-groove drugs. These conclusions require reinforcing with analyses of non-dodecamer complexes (to minimise crystal packing effects that may be particular to the dodecamer unit cell), as well as a study of the consequences of different refinement protocols on geometry.

Figure 5. Plots of various base-pair parameters for the native d(CGCGAATTCGCG)₂ structure and its complexes with berenil, netropsin and Hoechst 33258. (See facing pages).

3. Molecular modelling studies.

We have developed a procedure for partially-restrained molecular mechanics enthalpy minimisation which has enabled the sequence-dependence of berenil (and other non-intercalating ligands) to be studied for arbitrary sequences of considerable length (Laughton et al., 1990a,b). The computed interactions with a 60 base-pair sequence from the *tyrT* promotor (assigned a canonical B-DNA structure prior to minimisations) have been compared with those obtained by DNAase I and hydroxyl radical footprinting (in collaboration with KR Fox). The latter reagent in particular has provided a highly detailed picture of the sequence preferences shown by berenil, although this sequence does not represent more than a fraction of the potential binding-site sequences. In general, there is excellent agreement between observed and predicted binding sites. By contrast with some other DNA-binding drugs, berenil does not produce regions of enhanced cleavage to the footprinting reagents, suggesting that the drug does not induce changes in DNA structure distant from actual binding sites. These are all in AT-rich regions, towards their 3' ends. The total interaction energies showed

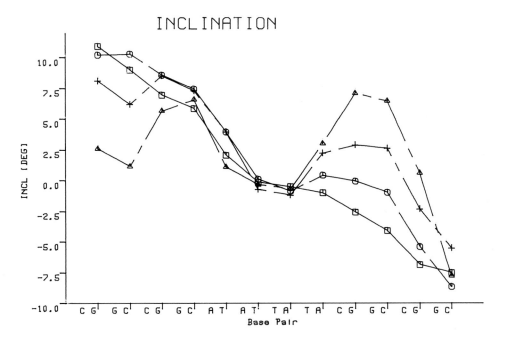

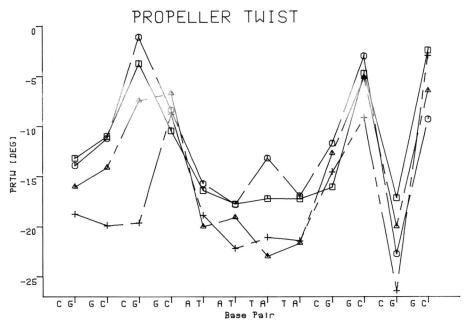

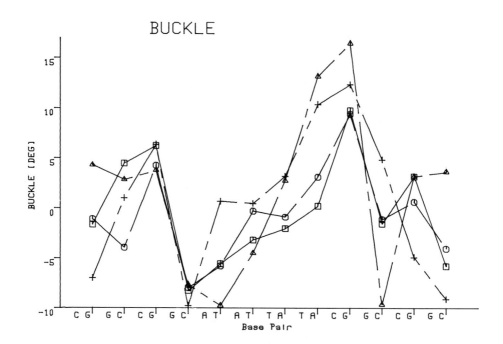

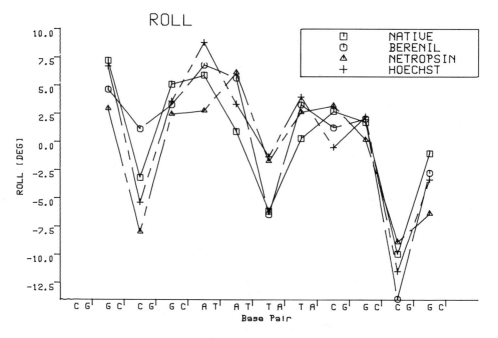

little sequence-dependence. On the other hand there is increased electrostatic interaction in AT sequences, coupled with lower DNA perturbation energies. Characteristic changes were also found in various helical parameters for the calculated binding sites, with roll, buckle, propeller twist and tilt showing the greatest changes. Buckle became more negative to the 5' side of the binding site and more positive to the 3' side, reducing steric clashes between the bases and the inward-facing hydrogen atoms of the berenil phenyl rings as well as facilitating base-berenil hydrogen-bonding. Deviations in propeller twist are symmetrical with respect to binding sites, enabling greater penetration of berenil into the minor groove. Roll becomes more negative on either side of the centre of a binding site, thus making the affected bases more closely coplanar. It appears that these changes do not differ greatly between 'good' (ie energetically favoured) and less good sites.

3.1 COMPARISON WITH XRAY STRUCTURAL DATA

There is a high degree of correspondence between the experimental and modelling findings of propeller twist, roll and buckle being most affected by berenil binding. The signs of the changes show less agreement, reflecting in part the fact that the modelled, minimised structures are considerably closer to standard B-form in the AT regions than is the AATT region in the crystal structure. Thus, distinct structural changes are required in order to achieve effective interaction with berenil. The inability of the modelling study to accommodate explicit solvent molecules, as found in the crystal, may be a further factor. Nonetheless, taken together, the experimental and theoretical studies strongly point to the inherently greater tendency of roll, propeller twist and buckle to respond to berenil binding in the DNA minor groove binding, compared to other helical and base-pair parameters.

4. Implications for the design of new sequence-specific agents.

The importance of DNA flexibility in the sequence-dependency of ligand binding is only now beginning to become clear, although further structural and modelling studies are still needed in order before detailed general rules analogous to the 'Calladine' ones can be defined with confidence. It appears from the modelling studies outlined above, that DNA structural deformation itself may be sequence-independent, whereas DNA flexibility, which is sequence-dependent, determines the extent to which deformations can be implemented. It is not clear as yet which modes of DNA distortion are the most important for optimal ligand binding, and which, if any, are ligand-independent.

It is now clear from the studies of Lown, Pullman and their colleagues that the strategy of reducing ligand charge in order to favour GC binding at the expense of AT binding is to some extent a well-founded one. Our results on the importance of perturbation energetic factors in the AT selectivity of berenil suggest that the charge strategy will only be effective if contributors to the binding energy other than the electrostatic one, also favour GC sites. The Van der Waals energy favours GC, although the perturbation energy does not. Thus, in the design of ligands that may recognise the N2 of guanine, the active avoidance of steric clash with N2 rather than positive hydrogen bonding with it, may be the more important factor. Clearly, a lack of DNA flexibility in GC sequences means that careful attention will have to be paid to the accurate matching of groove and ligand surfaces, whereas a much cruder match may suffice for AT recognition in view of the increased flexibility in these regions.

Groove width determines the accessibility of ligands to base-pair hydrogen bonds. The narrow width of minor-groove AT regions in B-DNA is well-documented, following its discovery in the Dickerson-Drew dodecamer crystal structure. It is clear that this width, coupled with the flexibility of AT sequences, is well-suited to ligands such as berenil and netropsin. The width is too narrow for a ligand to hydrogen-bond to both A and T in an individual base-pair. There is as yet little structural information on other, non-AT sequences in B-type oligomer crystal structures (we exclude the 5' and 3' ends of a sequence since packing and fraying effects are here at a maximum and thus tend to alter groove width). The recent crystal-structure analysis of a (mismatched) dodecamer with the central sequence being 5'-AGCT (Webster et al., 1990), has revealed a minor-groove width of 6-7Å (Figure 6). This suggests a further strategy for selective GC recognition, coupled with a down-weighting of AT binding. Ligands would be modified with appropriate, sterically bulky substituents such that they could not fit into an AT minor groove, but could enter a GC one. The relative lack of flexibility in the GC region implies that the molecular design would need to be accurate. The significantly wider minor groove in even the mixed-sequence AGCT also suggests that there is accessibility for two or even three hydrogen bonds to an individual GC base pair.

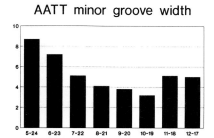

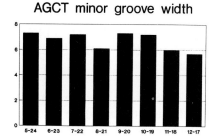

Figure 6. Minor-groove widths in (top) the d(GCGCAATTCGCG)$_2$ crystal structure (Drew and Dickerson, 1981), and (bottom) the d(CGCAAGCTGGCG)$_2$ crystal structure (Webster et al., 1990).

ACKNOWLEDGEMENTS

We are grateful to T Brown (Edinburgh), E Garman and DI Stuart (Oxford) for their contributions to the crystallographic aspects of the work reported in this paper. DB is a Cancer Research Campaign Research Student.

REFERENCES

Baguley, B.C. (1982) 'Nonintercalative DNA-binding antitumour compounds', Molecular Pharmacology, 43, 167-181.

Bernard, J and Riou, G.F. (1980) 'In vivo effects of intercalating and nonintercalating drugs on the tertiary structure of kinetoplast deoxyribonucleic acid', Biochemistry, 19, 4197-4201.

Brown, D.G., Sanderson, M.R., Skelly, J.V., Jenkins, T.C., Brown, T., Garman, E., Stuart, D.I. and Neidle, S. (1990) 'Crystal structure of a berenil-dodecanucleotide complex: the role of water in sequence-specific ligand binding', The EMBO Journal, 9, 1329-1334.

De Clercq, E. and Dann, O. (1980) 'Diaryl amidine derivatives as oncornaviral DNA polymerase inhibitors', J. Med. Chem., 23, 787-795.

Drew, H.R. and Dickerson, R.E. (1981) 'Structure of a B-DNA dodecamer. III. Geometry of hydration', J. Mol. Biol., 151, 535-556.

Drew, H.R. and Travers, A.A. (1984) 'DNA structural variations in the *E. Coli* tyrT promotor', Cell, 37, 491-502.

Gago, F., Reynolds, C.A. and Richards, W.G. (1989) 'The binding of nonintercalative drugs to alternating DNA sequences', Molecular Pharmacology, 35, 232-241.

Gresh, N. and Pullman, B. (1984) 'A theoretical study of the non-intercalative binding of berenil and stilbamidine to double-stranded (dA-dT)$_n$ oligomers', Molecular Pharmacology, 25, 452-455.

Kopka, M.L., Yoon., C., Goodsell, D., Pjura, P. and Dickerson, R.E. (1985) 'Binding of an antitumour drug to DNA. Netropsin and C-G-C-G-A-A-T-T-^{Br}C-G-C-G', J. Mol. Biol., 183, 553-563.

Laughton, C.A., Fox, K.R., Jenkins, T.C. and Neidle, S. (1990a) 'Interaction of berenil with the *tyrT* DNA sequence: a footprinting and molecular modelling study. Implications for the design of sequence-specific DNA recognition agents', Nucleic Acids Res., submitted.

Laughton, C.A., Jenkins, T.C., Fox, K.R. and Neidle, S. (1990b) 'Interaction of berenil with the *tyrT* DNA sequence: analysis of the effects of sequence and berenil binding on the conformation of the DNA in terms of base-pair helical parameters', Nucleic Acids Res., submitted.

Newton, B.A. (1975) 'Berenil: a trypanocide with selective activity against extranuclear DNA', in J.W. Corcoran and F.G. Hahn (eds.), Mechanism of Action of Antimicrobial and Antitumour Agents, Springer-Verlag, Berlin, pp. 34-47.

Pearl, L.H., Skelly, J.V., Hudson, B.D. and Neidle, S. (1987) 'The crystal structure of the DNA-binding drug berenil: molecular modelling studies of berenil-DNA complexes', Nucleic Acids Res., 15, 3469-3478.

Pjura, P.E., Grzeskowiak, K and Dickerson, R.E. (1987) 'Binding of Hoechst

33258 to the minor groove of B-DNA', J. Mol. Biol., 197, 257-271.

Portugal, J. and Waring, M.J. (1987) 'Comparison of binding sites in DNA for berenil, netropsin and distamycin. A footprinting study', European J. Biochem., 167, 281-289.

Shapiro, T.A. and Englund, P.T. (1990) 'Selective cleavage of kinetoplast DNA minicircles promoted by antitrypanosomal drugs', Proc. Natl. Acad. Sci. USA, 87, 950-954.

Webster, G.D., Sanderson, M.R., Skelly, J.V., Neidle, S., Swann, P.F., Li, B.F and Tickle, I.J. (1990) 'Crystal structure and sequence-dependent conformation of the A.G mis-paired oligonucleotide d(CGCAAGCTGGCG)', Proc. Natl. Acad. Sci. USA, in press.

Zimmer, C. and Wahnert, U. (1986) 'Nonintercalating DNA-binding ligands: specificity of the interaction and their use as tools in biophysical, biochemical and biological investigations of the genetic material', Prog. Biophys. Molec. Biol., 47, 31-112.

Binding of Minor Groove Ligands to Short DNA Segments: Berenil Complexed with d(GCAATTGC)$_2$ and d(GCTTAAGC)$_2$

Richard H. Shafer, Mitsuru Yoshida, Debra L. Banville and Sungho Hu

Department of Pharmaceutical Chemistry, School of Pharmacy,

University of California, San Francisco, California 94143

B. Pullman and J. Jortner (eds.), Molecular Basis of Specificity in Nucleic Acid-Drug Interactions, 59–65.

INTRODUCTION

Berenil is a trypanosidal drug which binds to double helical DNA. The binding of berenil to DNA increases the absorption maximum of the drug from 370 nm to 380 nm (Newton, 1967), but does not produce unwinding of closed circular DNA (Waring, 1970). This latter result indicates that it does not bind by intercalatiion. It is most likely that berenil binds in the minor groove of duplex DNA, as do netropsin and distamycin (Kopka et al., 1985a, 1985b). One berenil molecule has been reported to bind to 4 base pairs (Waring, 1970; Braithwaite & Baguley, 1980), with a preference for A-T sequences over G-C sequences has been observed (Braithwaite & Baguley, 1980; Portugal & Waring, 1987).

Structure of Berenil

Theoretical calculations of the berenil-DNA complex have been carried out by Gresh and Pullman (1984) and Pearl et al. (1987). Gresh and Pullman, reported that berenil interacts in the minor groove, where two hydrogen bonds are formed between the amidine groups of the drug and the oxygen atoms of the thymines in adjacent AT pairs. Pearl et al. (1987) used the coordinates determined from the crystal structure of berenil to model its interaction with DNA. Their model involved the same drug-DNA hydrogen bonds for the $d(AT)_4 \cdot d(AT)_4$ sequence as reported by Gresh and Pullman. In addition, they described the formation of hydrogen bonds between berenil and the N3 atoms of the diad-related central adenines for the $d(TA)_4 \cdot d(TA)_4$ oligomer.

In the studies described below, we have analyzed the interaction of berenil with $d(GCAATTGC)_2$ and $d(GCTTAAGC)_2$ using 1D and 2D ^1H NMR spectroscopy.

RESULTS

Both oligonucleotides studied, $d(GCAATTGC)_2$ and $d(GCTTAAGC)_2$ are self-complementary and have twofold symmetry. Detailed NMR studies have shown that these oligonucleotides belong to the family of B DNA conformations in solution (Yoshida et al., 1990; Hu et al., 1990).

Imino proton studies:

Berenil was added to a solution of $d(GCAATTGC)_2$ to examine the formation of the drug-oligonucleotide complex by 1-D ^1H NMR spectroscopy. In the absence of berenil, the imino proton region consists of four signals corresponding to the imino protons of the two guanines and thymines. During the titration the number of imino proton signals does not change, although their chemical shifts are altered significantly. Separate signals for drug-free DNA and drug-bound DNA are not detected, which indicates that the berenil-DNA complex is in fast exchange on the NMR time scale. Upon addition of berenil there was a large downfield shift of the internal thymine imino proton and a smaller upfield shift of the external thymine imino proton. This result is consistent with drug binding at the A-T base pair region of the oligonucleotide. No change in the imino protons of guanine were observed until the ratio of berenil to duplex exceeded 1:1.

Binding of berenil stabilized the duplex against heat denaturation. The signals of the non-terminal imino protons in the duplex alone broadened by $45^{\circ}C$ and were unobservable at $50^{\circ}C$ due to strand dissociation. At a 1:1 berenil to duplex stoichiometry, however, the imino proton signals did not appear to broaden significantly until $55^{\circ}C$, and they disappeared at $60^{\circ}C$. This result indicates that the central part of the duplex is protected from thermal denaturation by the drug.

Similar experiments with d(GCTTAAGC)$_2$ showed quite different results in that addition of berenil led to a substantial downfield shift of both thymine imino protons. Furthermore, there was significant broadening of these peaks while the guanine imino protons remained sharp. Temperature studies demonstrated only a slight stabilization of the imino protons against solvent exchange.

Two-dimensional ^1H NMR studies:

When exchange is fast on the time scale of the cross-relaxation rate, the sum of the bound and free magnetization decays through the equilibrium-concentration-weighted sum of the bound and free relaxation, and transferred nuclear Overhauser enhancements (TRNOE) can be observed (Landy & Rao, 1989). If the observed signals from two components are indistinguishable, only their sum can be measured. Since the binding constants of berenil for poly(dAdT)•poly(dAdT) and poly(dA)•poly(dT) are on the order of 10^4- 10^5 M^{-1} (unpublished data), it can be assumed that the equilibrium for berenil binding to the oligonucleotides lies far to the bound state at NMR concentrations. Under these conditions, i.e. with only bound duplex and berenil at a 1:1 stoichiometry, the relaxation rate equation is dependent only on the bound state rate matrix, and we can measure NOE effects as if we were looking at a stable complex without complications arising from exchange with the free state.

Assignments of the berenil-d(GCAATTGC)$_2$ complex were made by standard methods involving COSY and NOESY experiments at various mixing times (Yoshida et al., 1990). The duplex retains the B-conformation in the complex with berenil. In addition, the twofold symmetry of the oligonucleotide is retained in the berenil complex, where only one set of drug and DNA resonances can be observed.

The major changes in chemical shift upon complex formation between berenil and d(GCAATTGC)$_2$ are summarized in the table below:

SELECTED DNA CHEMICAL SHIFTS

	- berenil	+ berenil	change
Ai H2	7.67	8.08	0.41
Ti H6	7.20	6.96	-0.24
Te H6	7.35	7.05	-0.30
Ti H1'	5.95	5.54	-0.41
Te H1'	5.94	4.97	-0.97
Ti H2'	2.04	1.74	-0.30
Te H2'	2.14	1.82	-0.32
Ti H2"	2.60	2.24	-0.36
Te H2"	2.52	2.12	-0.40
Ti H3'	4.90	4.60	-0.30
Te H3'	4.96	4.66	-0.30
Ti H4'	4.27	2.77	-1.50
Te H4'	4.21	3.23	-0.98
Ti H5'/5"	4.40	3.90	-0.50
Ti H5'/5"	4.40	3.93	-0.47
Te H5'	4.16	3.90	-0.26
Te H5"	4.16	3.61	-0.55
Gi H5'	4.19	3.71	-0.48
Ce H5"	4.36	4.10	-0.26

Similarly, the following table describes the chemical shift changes observed for the ligand:

BERENIL CHEMICAL SHIFTS

		-DNA	+DNA	Shift
aromatic	H2	7.66	7.95	0.29
	H3	7.83	8.20	0.37
amidine		8.03	8.79	0.76
			8.87	0.84

A variety of berenil-d(GCAATTGC)$_2$ intermolecular contacts were observed in the NOESY experiments and the strongest of those are listed below. A NOESY experiment in 90 % H$_2$O/10 % D$_2$O at 250 msec mixing time using the 1331 pulse was performed to observe the exchangeable protons and shows contacts between the berenil amidine protons and the adenine H2 protons. Two separate signals were assigned to the berenil amidine protons: the one observed at higher field was assigned to the NH$_2$ group facing into the minor groove based on the stronger NOE contacts with the adenine H2 protons, and the other was assigned to the NH$_2$ group pointing outside of the groove. The difference in the chemical shift of the two amidine signals is 0.08 ppm. From this, we may estimate that any exchange between these two NH_2 groups must have a rate less than $90 \ s^{-1}$.

NOE Contacts between Berenil and d(GCAATTGC)$_2$ Protons

	H2	H3
Ae H2	+++	++++
Ai H2	++++	++++
Ai H1'	+++	+++
Ti H1'	++++	++++
Te H1'	+++	+++
Te H4'	+++	+
Gi H4'	+++	++++
Ti H5'/H5"		
/Te H5'	+++	+++
Te H5"	+++	++
Gi H5'		+++
Gi H5"	+++	+++

In contrast to berenil, the asymmetric minor groove binders, netropsin (Patel and Shapiro, 1986) and distamycin (Klevit et al., 1986) removed the twofold symmetry of the oligomers studied and revealed that these drugs undergo flip-flop motions between two binding sites. Evidence for this motion was obtained from proton exchange peaks between the two chemically inequivalent DNA strands. The symmetry of berenil and its complex with d(GCAATTGC)$_2$ makes it very difficult to determine unequivocally if a similar flip-flop motion occurs. The equivalence of the berenil protons H2 and H3 with H5 and H6 indicates that either the phenyl ring or the entire ligand does undergo a this type of motion.

Preliminary 2D NMR experiments with the complex formed between berenil and d(GCTTAAGC)2 indicate fewer but similar drug-DNA contacts as those observed with the other sequence.

Modeling of the complex of berenil with d(GCAATTGC)$_2$:

In order to obtain a model geometry for the berenil-DNA complex, we hand-docked berenil with d(GCAATTGC)$_2$ in the B-conformation according to the NOE contacts observed in our NOESY experiments. Two possible models could be developed to explain the drug-DNA contacts. Model I is characterized by hydrogen bonds between berenil amidine protons and O2 of the external thymines while Model II has hydrogen bonds between the amidine protons and N3 of the internal adenines. Model II is consistent with the berenil-d(TA)$_4$ • d(TA)$_4$ structure presented by Pearl et al. (1987), where the central sequence, 5'A-T3', is identical to ours. Since there is only a relatively small change in the orientation of berenil molecule between models I and II, the berenil-

DNA NOE contacts in general are very similar. One prominent difference is that the model I predicts a strong NOE contact between the amidine protons and H1' of the external thymidine and of the internal guanosine, whereas model II predicts a strong contact between the amidine protons and H1' of the internal adenosine. Unfortunately, the amidine-thymidine H1' cross-peak could not be detected in the 1331 pulse NOESY spectrum because the resonance of the thymidine H1' (4.97 ppm) lies very close to the H_2O signal. Since both the amidine-guanosine H1' and the amidine-adenosine H1' cross-peaks are clearly absent from this NOESY spectrum, the two models remain indistinguishable based on these predicted contacts.

A quantitative comparison of the two models was carried out with the CORMA method (COmplete Relaxation Matrix Analysis) described by Keepers and James (1984). Model II had consistently smaller RMS deviations between computed and experimental NOE peak volumes. The smaller RMS values for Model II at both mixing times of 150 and 250 ms suggests that berenil may preferentially interact with the oligonucleotide by hydrogen bonding its amidine protons with the N3 of the internal adenine.

CONCLUSIONS

This study represents the first attempt to determine the solution structure of both $d(GCAATTGC)_2$, $d(GCTTAAGC)_2$, and their complexes with berenil by NMR spectroscopy. Results described above for the oligonucleotides both alone and complexed with berenil demonstrate that the duplex is in a B-conformation. NOE experiments provide evidence for binding of the drug in the minor groove of the oligonucleotide at the central A-T sequences. Molecular models for this drug-DNA interaction that are consistent with our NMR data were generated with computer graphics and analyzed by CORMA.

ACKNOWLEDGMENTS

We are indebted to Dr. S. Neidle, Institute of Cancer Research, Sutton, Surrey, UK, for providing us with the coordinates of berenil for the computer graphics and to Prof. T.L. James for making his CORMA computer program available. It also is a pleasure to acknowledge the Computer Graphics Laboratory at UCSF, Prof. Robert Langridge, Director. This work was supported by grant CA27343 awarded by the National Cancer Institute, DHHS.

REFERENCES

Braithwaite, A. W., & Baguley, B. C.(1980) *Biochemistry 19*, 1101-1106.
Ferrin, T. E., & Langridge, R. (1980) *Computer Graphics 13*, 320.
Gresh, N., & Pullman, B. (1984) *Mol. Pharmacol. 25*, 452-458.
Hore, P. J. (1983) *J. Magn. Reson. 55*, 283-300.
Hu, S, James, T.L. and Shafer, R.H. (1990) in preparaton.
Keepers, J. and James, T.L. (1984) J. Magn. Res. **57**, 404-426.
Klevit, R. E., Wemmer, D. E., & Reid, B. R. (1986) *Biochemistry 25*, 3296-3303.

Kopka, M. L., Yoon, C., Goodsell, D., Pjura P., & Dickerson, R. E. (1985a) *J. Mol. Biol. 183*, 553-563.

Kopka, M. L., Yoon, C., Goodsell, D., Pjura P., & Dickerson, R. E. (1985b) *Proc. Natl. Acad. Sci. USA 82*, 1376-1380.

Landy, S. B., & Rao, B. D. N. (1989) *J. Magn. Reson. 81*, 371-377.

Newton,B. A. (1967) *Biochem. J. 105*, 50-51.

Patel, D. J., & Shapiro, L. (1986) *J. Biol. Chem. 261*, 1230-1240.

Pearl, L. H., Skelly, J. V., Hudson, B. D., & Neidle S. (1987) *Nucleic Acids Res. 15*, 3469-3478.

Portugal, J., & Waring M. J. (1987) *Eur. J. Biochem. 167*, 281-289.

Sanders, J. K. M., & Hunter, B. K. (1988) in *Modern NMR Spectroscopy*, pp 208-214, Oxford University Press, Oxford, New York, Toronto.

Waring, M. (1970) *J. Mol. Biol. 54*, 247-279.

Yoshida, M., Banville, D.L. and Shafer, R.H. (1990) Biochemsitry, in press.

The sequence specificity of damage caused by [125I]-labelled Hoechst 33258 and UV / iodoHoechst 33258 in intact cells and in cloned sequences of purified DNA which differ by a small number of base substitutions.

Vincent Murray
School of Biochemistry,
University of New South Wales,
PO Box 1, Kensington,
NSW 2033,
Australia.

ABSTRACT. An examination was made of the sequence specificity of several DNA damaging agents using DNA sequencing gels. These agents were [125I] Hoechst 33258, bleomycin, UV-iodoHoechst 33258, UV-bromodeoxyuridine, and UV-iododeoxyuridine. Cleavage was determined for human alpha DNA in three circumstances - cloned DNA, purified genomic human DNA, and DNA inside cells. The sequence specificity of [125I] Hoechst 33258 was compared in intact cells and in purified DNA. An analogous comparison was made for bleomycin. In general, similar results were obtained for DNA in both environments. The use of base substitutions enabled the ligand DNA binding sites to be investigated. In addition to the expected disruptions caused by base substitutions at the binding site, base substitutions in neighbouring sequences also had a significant effect.

1. Introduction

The use of DNA sequencing gels has enabled the sequence specificity of DNA damaging agents to be located to the exact base pair (bp). This type of data can be obtained for DNA in several environments :- cloned DNA, purified cellular DNA, and DNA in intact cells (see figure 1). By comparing DNA damage in various situations, the influence of different parameters can be investigated. The tandemly repeated human sequence called alpha DNA was used as target sequence for DNA damage in all three environments.

The crucial sequences for DNA ligand binding can be investigated using cloned DNA sequences that differ by a number of base substitutions. In this manner, the effect of a base substitution on the degree of damage can reveal the importance of that base for ligand binding. Again alpha DNA was employed as target DNA. The natural occurrence of random base substitutions in the 50,000 copies of alpha DNA per haploid genome, enabled, through cloning and DNA sequencing, the generation of a library of clones of DNA sequences that differ by a number of known base substitutions (Murray and Martin, 1987).

B. Pullman and J. Jortner (eds.), Molecular Basis of Specificity in Nucleic Acid-Drug Interactions, 67–73.
© 1990 *Kluwer Academic Publishers. Printed in the Netherlands.*

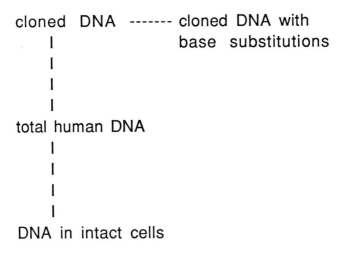

Figure 1

Several DNA damaging agents were employed. Bleomycin is glycosidic antibiotic that is successfully used in cancer chemotherapy (Crook and Bradner, 1977). It causes DNA single- and double-strand breaks and release of free base and base propenal. Its cytotoxic activity is generally assumed to be due to DNA damage.

Hoechst 33258 is a bis-benzimidazole that binds in the minor groove of DNA and preferentially binds to AT rich regions of DNA. Two X-ray crystallographic studies of Hoechst 33258 bound to the oligonucleotide CGCGAATTCGCG have shown that in one case it binds to the sequence ATTC (Pjura et al,1987) and in the other AATT (Teng et al,1988). Experiments with [^{125}I] Hoechst 33258 have shown that at least three consecutive AT bps are required for strong binding (Murray and Martin,1988b). In these experiments the decay of ^{125}I damages the DNA at the site of Hoechst 33258 binding.

On irradiation of iodoHoechst 33258 bound to DNA with UV light at 350nm, DNA cleavage occurred. This cleavage is thought to be mediated by a free radical mechanism (Martin et al,1990). The UV light splits the carbon-iodine bond and produces a carbon centred free radical on the ligand. This free radical then abstracts a hydrogen atom from DNA which eventually results in phosphodiester bond cleavage.

DNA which is substituted with bromodeoxyuridine (BrUdR) or iododeoxyuridine (IUdR) , becomes sensitive to UV light (Hutchinson and Kohlein,1980). A free radical is again central to the reaction mechanism through the splitting of the carbon-halogen bond by UV light. A uracilyl radical is produced which abstracts a hydrogen atom from the 5' deoxyribose and subsequently leads to phosphodiester

	cloned DNA	cloned DNA with base substitutions	total human DNA	DNA in intact cells
Bleomycin	+	+	+	+
[^{125}I] Hoechst 33258	+	+	+	+
UV \ BrUdR	+	+		
UV \ IUdR	+	+		
UV \ iodoHoechst 33258	+	+		

Table 1

bond cleavage.

2. Materials and Methods

An alpha DNA - M13 hybridisation system was used in the experiments with bleomycin and [^{125}I] Hoechst 33258 (Murray and Martin,1985c). This hybridisation system allows the separation of the DNA strands and also purification of alpha DNA from non- alpha DNA sequences.

An adaption of the Sanger DNA sequencing system (Murray and Martin,1989) was employed to examine the sequence specificity of DNA damage caused by UV-iodoHoechst 33258, UV-BrUdR and UV-IUdR. This system uses an oligonucleotide primer and single-stranded M13 to synthesise double-stranded DNA which is then cleaved by the DNA damaging agent. For both systems damage is located to the exact bp using DNA sequencing gels.

3. Results and Disscussion

3.1 DNA DAMAGE IN INTACT CELLS

Bleomycin damages DNA in intact cells and in purified DNA under appropriate conditions.The DNA sequence specificity of bleomycin damage in intact human cells was investigated at the base pair level using the alpha DNA-M13 hybridisation system and DNA sequencing gels (see table 1) (Murray and Martin,1985a). In intact cells the overall degree of damage was reduced relative to purified DNA (probably due to cell permeability, inactivation of bleomycin and/or nucleosome core regions protecting the DNA). In fact about 30-fold more bleomycin was necessary for whole cells to achieve the same degree of damage as for purified DNA.

For each strand, damage was quantified by densitometer analysis at each cleavage site and the relative intensities determined. No significant differences were observed in the position or relative extent of cleavage at each site. Bleomycin preferentially cleaved in the linker region of nucleosomes. Alpha DNA (which was used as the target DNA sequence in these experiments) is a tandem repeat but the nucleosomes are randomly phased over the same sequence. Hence there is a series of overlapping linker regions of DNA along alpha DNA. This results in a cleavage pattern for intact cells which is very similar to that of purified DNA.

The DNA sequence specificity of [^{125}I]-Hoechst 33258 damage in intact human cells was investigated (Murray and Martin,1988a). [^{125}I]-Hoechst 33258 causes exclusively double strand breaks (dsb). An M13 hybridisation system was developed that could detect dsb in intact cells. These experiments were the first determination of dsb damage to the exact base pair in intact mammalian cells. As for bleomycin the overall degree of damage was reduced in intact cells compared to purified DNA , and the position and relative extent of damage was similar. This implies that [^{125}I]-Hoechst 33258 is fully accessible to DNA even when DNA is complexed with nucleosomes and other DNA binding agents in chromatin.

2.2 THE EFFECT OF BASE SUBSTITUTIONS ON DNA DAMAGE

As mentioned in the Introduction, a library of alpha DNA clones that differed by a small number of base substitutions, was generated and sequenced.These clones were used to examine the effect of base substitutions on the degree of damage caused by the DNA damaging agents. Three types of effects were observed 1) base substitutions at the ligand binding site; 2) base substitutions in close proximity to the ligand binding site; 3) base substitutions two or more bps from the ligand binding site.

2.2.1 *Bleomycin* By use of a long sequence of DNA (more than 300 bps), the dinucleotides preferentially cleaved by bleomycin have been determined (Murray and Martin,1985b) - the dinucleotides GT and GC were cleaved on all occasions, GA most of the time, and other dinucleotides to lesser degrees. An important feature of the bleomycin data is that for a particular dinucleotide, the extent of cleavage can vary from one dinucleotide location to another (D'Andrea and Haseltine,1978; Kross et al,1982; Murray and Martin,1985b). Obviously neighbouring DNA sequences are modifying the response. Statistical analysis of the data also showed that the 5' base to the purine-pyrimidine dinucleotides was the most important factor in determining the degree of damage caused by bleomycin (see table 3) (Murray and Martin,1985b). The experiments with bleomycin and clones containing base substitutions (Murray et al,1988) confirmed the above results but in addition gave information on the modulation of the degree of DNA cleavage caused by base substitutions up to twelve bps away.

DNA is not a perfectly regular Watson-Crick double helix but significant microstructural variations in bond angles and other features occur (Dickerson and Drew,1981). These variations are apparently a response to the presence of the neighbouring DNA sequence. Thus base substitutions are expected to alter the microstructure of DNA . It is likely that DNA damaging agents are sensitive to microstructural variations. Thus DNA damaging agents can be used as probes of DNA microstructure.

Bleomycin is a relatively large ligand of approximate molecular weight 1500. It is thought to intercalate into DNA and derive its sequence specificity from this interaction. These properties of bleomycin would be expecteded to make it sensitive to variations in DNA microstructure.

2.2.2 *[125I]-Hoechst 33258* [125I]-Hoechst 33258 required at least three consecutive AT bps for strong binding and cleavage (Murray and Martin,1988b). Four consecutive AT bps gave rise to strong binding in almost all cases; but with three consecutive AT bps, strong binding occurred in the minority of cases - again in this latter case neighbouring DNA sequences are significantly affecting the degree of cleavage.

The data on the effect of base substitutions on [125I]-Hoechst 33258 damage, showed that runs of consecutive AT bps were the most important parameter that determined [125I]-Hoechst 33258 binding. However, other factors including nearest neighbour and long range interactions were also important.

The X-ray crystallographic data for Hoechst 33258 bound to the oligonucleotide CGCGAATTCGCG have shown that in one case it binds to the sequence ATTC (Pjura et al,1987) and in the other AATT (Teng et al,1988). One interpretation of this data is that there are at least two modes of binding of [125I]-Hoechst 33258 to DNA . The binding of [125I]-Hoechst 33258 to three AT bps could be analogous to the Pjura et al (1987) structure and four AT bps to the Teng et al (1988) structure.

2.2.3 *UV-bromodeoxyuridine and UV-iododeoxyuridine* Thymidine can be fully substituted with BrUdR or IUdR in the M13 synthesis reaction using the appropriate deoxynucleoside triphosphates. UV light causes the carbon-halogen bond to split and through a free radical mechanism, DNA strand cleavage occurs. DNA damage is only associated with BrUdR or IUdR incorporation (Murray and Martin,1989). BrUdR and IUdR damage occurred at the same locations at approximately the same intensity. (For the rest of this discussion BrUdR and IUdR will be considered to be the same.) At the sites of most intense cleavage a consensus sequence was found (see table 3). Similarly a consensus sequence was found for sites of no significant damage .

Experiments with clones containing base substitutions indicated that IUdR incorporation was necessary for cleavage, and neighbouring sequences (especially in the consensus sequence) were significantly affecting the degree of damage.

The reaction mechanism is thought to occur by abstraction of a hydrogen atom from the 2'-carbon of the 5'-deoxyribose by a uracilyl radical (Hutchinson and Kohlein,1980). It has been postulated (Murray and Martin,1989) that the distance

	nsensus sequence at y Strong damage sites	Consensus sequence at very Weak damage sites
Bleomycin	TGṪ TGĊ	TĠ ĊG
[^{125}I] Hoechst 33258	4(or more) A\Ts	GĊs
UV \ BrUdR	RCḂrBrG\Br	GḂrR
UV \ IUdR	RCİIG\I	GİR
UV \ iodoHoechst 33258	3-5 Ts	GĊs

N.B.   R is a purine G or A; Br is BrUdR; I is IUdR

Table2

between the uracilyl radical and the hydrogen atom on the 2'-carbon of the 5'-deoxyribose determines the extent of the reaction - with smaller distances giving more cleavage than larger distances. Obviously the most important parameter in this situation is the microstructure of DNA which is determined by the neighbouring DNA sequences.

2.2.4 *UV-iodoHoechst 33258*   The halogenated ligand iodoHoechst 33258, when bound to DNA, sensitises DNA to UV light (Martin et al,1990). As mentioned in the introduction, this reaction proceeds via a free radical centred on the Hoechst 33258. An unusual aspect of the reaction is that the cleavage site is always (for intense damage) at the 5'-end of the binding site. If both strands are examined, cleavage is at opposite ends of the binding site on each strand. Thus iodoHoechst 33258 can bind in both orientations but the free radical is closer to one strand than the other.

Whereas with [^{125}I]-Hoechst 33258 the majority of the strong damage sites had four or more consecutive AT bps, with UV-iodoHoechst 33258 the majority of strong cleavage sites had three consecutive AT bps. Runs of consecutive T bps were particularly favoured with UV-iodoHoechst 33258.

IodoHoechst 33258 was incorporated inside cells and made the cells very sensitive to killing by UV light. When the iodoHoechst 33258 is added to the medium to give a concentration of 4μM, irradiation resulted in 3-4 log cell kill at a UV dose that did not have a significant effect in the absence of iodoHoechst 33258.

3. Acknowledgements

I would like to thank my collaborators R. Martin, G. D'Cunha, M. Pardee, E. Kampouris, A. Haigh, D. Kelly, G. Hodgson, L. Tan and J. Matthews. This work was supported by the NHMRC, ACCV, ARGS and the Peter MacCallum Cancer Institute.

4. References

Crook,S.T. and Bradner,W.T. (1977) J. Med 7 333-428 .
D'Andrea,A.D and Haseltine,W.A. (1978) Proc. Natl. Acad. Sci. USA. 75 3608-3612.
Dickerson,R.E. and Drew,H.R. (1981) J. Mol. Biol. 149 761-786.
Hutchinson,F. and Kohlein,W. (1980) Prog. Mol. Subcell. 7 1-42.
Kross,J., Henner,W.D., Klecht,S.M. and Haseltine,W.A. (1982) Biochem. 21 4310-4318.
Martin, R.F., Murray, V., D'Cunha, G., Pardee, M., Kampouris, E., Haigh, A. and Hodgson, G.S. (1990) Int. J. Rad. Biol. (in press).
Murray, V. and Martin, R.F.(1985a) J.Biol. Chem. 260 10389-10391.
Murray, V. and Martin, R.F.(1985b) Nucl. Acid. Res. 13 1467-1481 .
Murray, V. and Martin, R.F.(1985c) Gene Anal. Tech. 2 95-99 .
Murray, V. and Martin, R.F.(1987) Gene 57 255-259 .
Murray, V. and Martin, R.F.(1988a) J. Mol. Biol. 201 437-442.
Murray, V. and Martin, R.F.(1988b) J. Mol. Biol. 203 63-73 .
Murray, V. and Martin, R.F.(1989) Nucl. Acids Res. 17 2675-2691 .
Murray, V., Tan, L., Matthews, J. and Martin, R.F. (1988) J. Biol. Chem. 263 12854-12859
Pjura,P.E., Grzeskowiak,K. and Dickerson,R.E. (1987) J. Mol. Biol. 197 257-271.
Teng,M., Usman,N., Frederick,C.A. and Wang,A.H-J. (1988) Nucl. Acids Res. 16 2671-2690.

Structure and Dynamics of a [1:1] Drug-DNA Complex: Analysis of 2D NMR Data Using Molecular Mechanics and Molecular Dynamics Calculations.

Ramaswamy H. Sarma[1], Mukti H. Sarma[1], Kimiko Umemoto[1], Goutam Gupta[2] and Angel E. Garcia[2]

[1] Institute of Biomolecular Stereodynamics,
Department of Chemistry,
State University of New York at Albany,
Albany, NY 12222.

[2] Theoretical Biology and Biophysics Group,
T-10, MS K710,
Los Alamos National Laboratory,
Los Alamos, NM 87545.

Abstract

1D/2D NMR studies are reported for a [1:1] complex of d(GA$_4$T$_4$C)$_2$ and Dst2 (an analogue of distamycin A). Full-Matrix NOESY Simulations, Molecular Mechanics and Molecular Dynamics Calculations are performed to analyze the NMR data. Results show that drug-DNA complex formation is driven by static features like H-bonding and steric interactions in the minor-groove of DNA. As a consequence of drug binding, a non-linear oscillatory mode is activated. In this mode the molecule samples equilibrium structural states of different degrees of bending. It is noted that these structures belong to three distinctly different energy wells that satisfy the same NMR data.

B. Pullman and J. Jortner (eds.), Molecular Basis of Specificity in Nucleic Acid-Drug Interactions, 75–93.

Introduction

Based upon physico-chemical studies [1-8], it is often concluded that static features like H-bonding and steric interactions in the minor-groove of DNA account for the sequence specificity of the minor-groove binding drugs. However, it is quite possible that some elements of specificity can be dynamic in nature. In other words, in presence of the drug, the target DNA molecule may exhibit characteristic motions at the site of binding and adjacent to it. This prompted us to analyze the dynamics of *Dst2*-d(GA$_4$T$_4$C)$_2$ [1:1] complex by combining 2D NMR and theoretical methods (*Dst2* is an analogue of distamycin A-see Fig. 1).

Methodology

NMR.   The details of NMR methods were previously reported [9]. All NMR experiments on *Dst2*-d(GA$_4$T$_4$C)$_2$ complex in which drug:DNA ratio was kept at 1:1 molar equivalent. Assignment of the exchangeable and non-exchangeable DNA and drug protons in the complex were derived by combining 1D/2D NMR data of the free DNA, the free drug and the drug-DNA complex in H$_2$O/D$_2$O [9]. The NOESY data of the *Dst2*-d(GA$_4$T$_4$C)$_2$ [1:1] complex for mixing times τ_m= 50, 100 and 150 msec, were used to carry out the Full-Matrix NOESY Simulations and associated R-factor test on several structures generated using constrained least-square method [10]. R-factor is defined as

$$\text{R-Factor} = (1/N) \; \frac{\Sigma^N_{ij} \; a^o_{ij} - a^c_{ij}}{\Sigma^N_{ij} \; a^o_{ij}} \qquad (1)$$

where

N= total number of measureable NOESY (i,j) peaks,

a^o_{ij} = observed NOESY intensity for the (i,j) peak

& a^c_{ij} = calculated NOESY intensity for the (i,j) peak.

For overlapping NOESY cross-peaks, it is hard to estimate the experimental value of the intensity due to a single pair-wise interaction. Therefore, for overlapping NOESY peaks, theoretical NOESY slices were constructed taking into consideration of the contributions of all (i,j) pairs in the region of overlap. Theoretical NOESY slices, so obtained, were then compared with the corresponding observed ones. The details of the methodology employed to treat the overlapping peaks were previously discussed [9,11].

Constrained Energy Minimization. NOESY data resulted in a set of about 30 conformationally similar DNA models that showed similar agreement with the NOESY data. From these DNA models about 60 independent average inter-proton distances were extracted as structural constraints required for agreement with the NMR data of the Dst2-d(GA$_4$T$_4$C)$_2$ [1:1] complex. A pseudo potential = K(d_{const} - d_{actual})^2 was added in the potential function with K= 10 Kcal/Mole/Å^2 for distances < 3Å and K=5 Kcal/Mole/Å^2 for distances > 3Å. The *Dst2*-d(GA$_4$T$_4$C)$_2$ [1:1] complex was energy minimized using the AMBER force field [12]. All atoms (i.e., heavy atoms and hydrogens) of Dst2 and d(GA$_4$T$_4$C)$_2$ were included in all calculations. Dst2 was docked at the center of the minor-groove of the target DNA duplex consistent with the NOESY data. The B-DNA model of Arnott and Hukins [13] was used as the starting DNA structure for energy minimization using NOESY constraints.

Constrained Molecular Dynamic Simulation. The NOESY constrained energy minimized structure of the Dst2-d(GA4T4C)2 [1:1] complex was used as the starting structure for the MD simulation. An 8-psec MD run was used for equilibriation. A 52-psec (after equilibriation) constrained MD simulation at constant energy (with T close to 300 K) was performed on the complex using the 60 NMR constraints with the associated pseudo-potential as mentioned above. The last 30-psec MD data were used to analyze the variations in the NMR constraints and other structural parameters. 104 structures at regular intervals of

0.5 psec were taken out of the 52-psec MD trajectory. Root-Mean-Square(RMS) deviations among all these structures were computed and presented in the form of a (104 X 104) matrix. Global translations and rotations were appropriately deducted from the RMS matrix such that the matrix reflected true conformational differences among these structures. Each of the 104 structures from the MD trajectory was fully energy minimized (to an average gradient of 0.01 Kcal/Mole/Å) with the NMR constraints. Structural variations among the 104 energy minimized structures were analyzed again by constructing another (104 X 104) RMS matrix.

Results

Dst2 is located at the central 5 basepairs of the Dst2-d(GA4T4C)2 [1:1] complex. The numbering scheme of the DNA duplex and Dst2 is shown in Figure 1. The presence of H-bonded amide protons of Dst2 were located within 9.3-8.6 ppm [9]. The length of Dst2 suggests that the drug can span about 5 consecutive A.T pairs in the minor-groove of a B-DNA duplex. From Fig. 1, it appears that Dst2 can bind to d(GA4T4C)2 in 5 different ways i.e., it can span

A2 - A3 - A4 -A5 -T6
T19-T18-T17-T16-A15 ,

A3 - A4 - A5 - T6 -T7
T18- T17-T16-A15-A14 or their symmetry
equivalents etc.

The following data [9] supported the notion that Dst2 is predominantly bound at the central 5 base-pairs i.e., the segment

A4 - A5 - T6 - T7 - T8
T17-T16-A15-A14-A13
or its symmetry equivalent
A14-A15-T16-T17-T18
T7 - T6 - A5 - A4 - A3   (See Fig. 1).

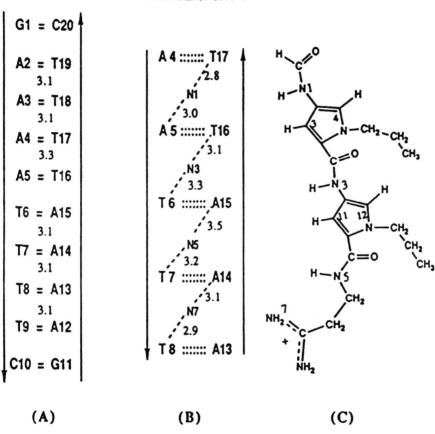

(A)   **(B)**   **(C)**

Figure 1.   Description of bifurcated H-bonds in the major-
and minor-groove of the constrained energy minimized
structure of the $Dst2$-d(GA$_4$T$_4$C)$_2$ [1:1] complex. (A)
Numbering scheme of the DNA duplex in the complex. The
array of bifurcated H-bonds along their lengths are shown.
These H-bonds are formed between N6(A) and O4(T) across
the strand between two neighboring A.T pairs. (B) The array
of H-bonds in the minor-groove where N1, N3, N5 and N7
are the proton donors and O2(T) and N3(A) are the
acceptors. (C) The numbering scheme of the Dst2 atoms.

(i) Strong NOEs (at τ_m= 50, 100 and 150 msec) are present between HC3/HC11 of Dst2 and H2(A5/A15) of DNA.

(ii) Weaker NOEs are observed between HC3/HC11 of Dst2 and H1' of T6/T16, T7/T17; these peaks begin to take shape at τ_m= 100 msec or higher.

(iii) Upon Dst2 binding H2 of A4/A14,A5/A15 are down field shifted and H1' of A5/A15, T6/T16, T7/T17 are high field shifted. Other protons show very little change upon drug binding.

(iv) There is no resonance doubling of DNA and Dst2 protons i.e., chemical shift values of H8 of A4 and A14 are equivalent and so are H6 of T6 and T16 and similarly for all other protons. This observation eliminates the possibility of any other binding mode present as a major population.

However, NMR data cannot rule out the possibility of minor conformers showing other than the central binding mode [Fig.1].

A Static Structure Obtained from the Constrained Energy Minimization. Full-Matrix NOESY Simulations were preformed on the DNA structures generated by a constrained least-square method and calculated NOESY intensities for proton pairs (i,j) were compared with the corresponding observed ones at τ_m=50, 100 and 150 msec. In this simulation, an isotropic correlation time τ_c of 3 nsec was used and Dst2 protons were not included in the relaxation matrix (inclusion of Dst2 protons did not significantly alter the NOESY pattern involving the DNA protons). Details of the methodology were previously reported [9,11].

NOESY Simulations resulted in a set of about 30 conformationally similar models that give similar agreement (i.e., a low R-Factor) with the observed NOESY data. From these 30 models, an estimate of average inter-proton distances can be obtained for various pair-wise interactions in the DNA; e.g., for T7 we obtain estimates of distances like

Figure 2. Calculated and observed NOESY (for τ_m=50 and 100 msec) are shown for several H8 and H6 protons through three panels A, B and C.

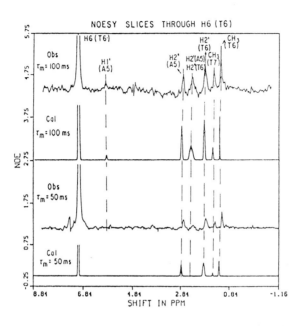

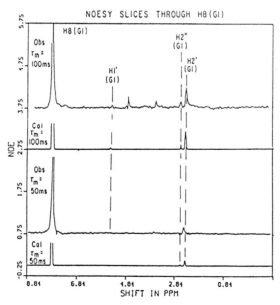

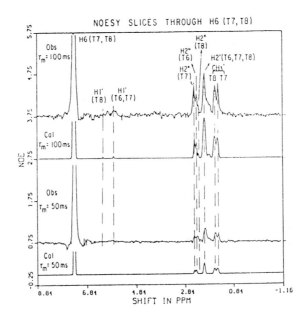

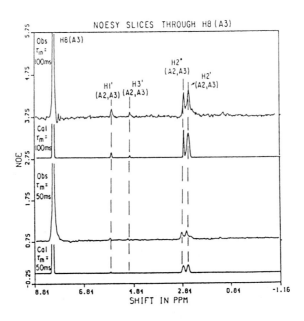

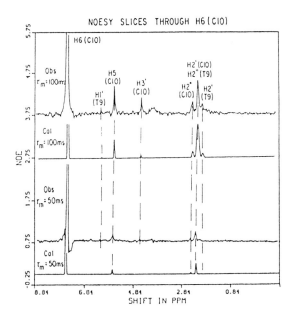

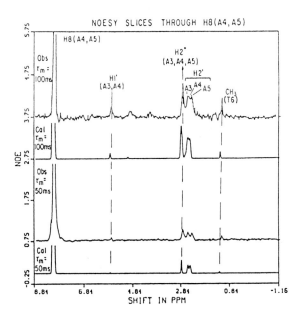

H6(T7)---H2'(T7), H6(T7)---H2"(T6), H6(T7)---H1'(T7),
H6(T7)---H1'(T6), H6(T7)---CH3(T8) etc. and similarly for
other residues. Estimates of inter-proton distance were
extracted for 60 pair-wise interactions by considering 30
NMR models. These 60 distances were taken as structural
constraints required for agreement with the NMR data. The
B-DNA model of Arnott and Hukins [13], instead of any one
of the NMR models, was chosen as the starting model of DNA
in the complex. This was done to ensure an unbiased search
of the conformational space during minimization. Indeed,
energy minimization led to a model of the $Dst2$-d(GA4T4C)$_2$
[1:1] complex that showed agreement with the NOESY data.
The energy minimized model have the following features:

(i) A.T pairs are propellar twisted leading to an array of
bifurcated H-bonds in the major-groove of the A/T tract
[Fig. 1].

(ii) The amide protons of Dst2 form a series of H-bonds in
the minor-groove of DNA [Fig. 1]. The location of the drug at
the center of the DNA molecule results in the following
drug-DNA contacts: H2(A5/A15)---HC3/HC11= 2.4 Å and
H1'(T6/T16,T7/T17)---HC3/HC11= 3.2-3.5 Å. This distance
profile mimics the observed drug-DNA NOESY pattern [9].

Figure 2 shows the observed and theoretically
constructed NOESY slices through H8/H6 protons for τ_m=50
and 100 msec. Theoretical NOESY slices are constructed
using the cartesian coordinates of the DNA and drug protons
in the energy minimized model of the $Dst2$-d(GA4T4C)$_2$ [1:1]
complex . Note that the energy minimized model reproduces
the observed NOESY pattern.

*An Ensemble of Structures derived from the 52-
psec Constrained MD Simulation.*   The structure
obtained after NOESY constrained energy minimization is
merely a stable static structure of the $Dst2$-d(GA4T4C)$_2$ [1:1]
complex. In reality, NMR data correspond to an ensemble of
structures. These structures originate due to the presence of
low frequency internal motions in the molecule and there is
a continuous interconversion among these structures within

the NMR time scale. In order to obtain an ensemble of structures, we carried out a 52-psec MD simulation (at constant energy with temperature T close to 300 K). The same set of 60 inter-proton distances were kept as structural constraints during the MD simulation and the energy minimized model of the $Dst2$-d(GA$_4$T$_4$C)$_2$ [1:1] complex was taken as the initial structure before temperature equilibriation. Hence, this enabled us to monitor the MD trajectory of the structures that agree with the NMR data.

We selected structures along the last 30-psec of the MD trajectory to analyze the NOESY constraints and other structural parameters. 3,000 structures (at time steps of 0.01 psec) were chosen from the trajectory. For the purpose of illustration, we only show here the result of our analysis describing the variations in the intra-nucleotide distances of the type H8/H6(i)---H2'(i) and the inter-nucleotide distances of the type H8/H6(i)---H2"(i-1). These distances are the shortest (hence the most reliable ones) estimated from the Full-Matrix NOESY Simulation [9]. It appears that H8(A$_i$)---H2'(A$_i$) for all A's show similar distance distribution profiles and so do H6(T$_i$)---H2'(T$_i$) for all T's. Therefore, for a given structure in the trajectory, we computed the distance H8(A$_i$)---H2'(A$_i$) averaged over all A's and similarly for H6(T$_i$)---H2'(T$_i$). Figure 3A shows the P(x) vs. x (P= % of structures with a given x) for H8(A$_i$)---H2'(A$_i$) [dotted line] and H6(T$_i$)---H2'(T$_i$) [solid line]. Primary NOESY pattern indicate that the H8(A$_i$)---H2'(A$_i$) distances are larger than the H6(T$_i$)---H2'(T$_i$) distances. The distribution in Fig. 3A is consistent the observed NOESY pattern. Figure 3B shows P(x) vs. x plots for three types of pairwise interactions i.e., H8(A$_i$)---H2"(A$_{i-1}$) averaged over all A's in a given structure of the trajectory [dotted line], H6(T$_i$)---H2"(T$_{i-1}$) averaged over all T's except T6/T16 [solid line] and H6(T6/T16)---H2"(A5/A15) [solid-dotted line]. Note that three distance distribution profiles are distinctly different; the same conclusion is also derived from the primary NOESY pattern. The primary NOESY pattern revealed a particularly interesting feature i.e., inside the A/T tract NOE for the intra-nucleotide H6(T$_i$)---H2'(T$_i$) interaction is stronger (by a factor of about 2) than the NOE

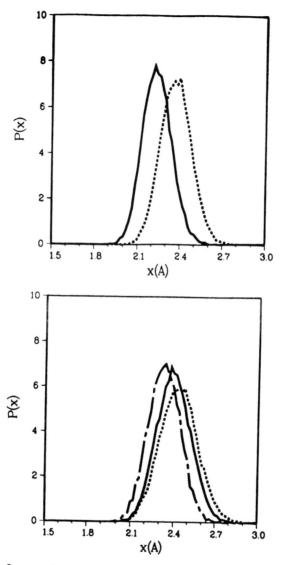

Figure 3.   Analysis of the short inter-proton distances using 3,000 snap-shots taken from the last 30-psec of the MD trajectory. (A) P(x) vs. x plot for the intra-nucleotide distances i.e., H8(A)---H2'(A) [dotted line] and H6(T)---H2'(T) [solid line]. (B) P(x) vs. x plot for inter-nucleotide distances i.e., H8(A_i)---H2"(A_{i-1}) [dotted line], H6(T_i)---H2"(T_{i-1}) except for the ApT sequence [solid line] and H6(T6/T16)---H2"(A5/A15) [solid-dotted line].

Table I

Average values and corresponding standard deviations (Å) of the short inter-proton distances in DNA as obtained from the 3,000 snap-shots along the last 30-psec of the MD trajectory.

Proton-Pair	Average Distance, $\underline{x}$	Standard Deviation, σ	Deviation from Gaussian, γ
1. H8(A)---H2'(A)	2.38	0.11	0.14
2. H6(T)---H2'(T)	2.25	0.10	0.25
3. H8(Ai)---H2"(Ai-1)	2.45	0.13	0.17
4. H6(Ti)---H2"(Ti-1) (except for ApT)	2.42	0.12	0.18
5. H6(T6/T16) ---H2"(A5/A15)	2.35	0.11	0.12

$\gamma = (1/\sigma^3) < (x - \underline{x})^3 >$ measures the deviation of $P(x)$ from a Gaussian distribution in Fig. 3.

for the inter-nucleotide $H6(T_i)$---$H6(T_{i-1})$ interactions. However, at the ApT sequence intra- and inter-nucleotide NOEs show similar NOE values. Comparison of Fig. 3A and Fig. 3B shows that the distribution plots for intra- and inter-nucleotide distances involving H6 at the ApT junction are similar in nature which is in agreement with the NOESY data [9]. Table I lists average inter-proton distances (and standard deviations) for the pair-wise interactions described in Fig. 3.

Three Energetically Stable But Distinctly Different Structures Isolated from the MD Trajectory. In summary, it is possible to construct an ensemble of structures from the MD trajectory that agree with the NOESY data. However, the question is : how different are these structures from each other? We address this question in this section.

104 Structures were chosen as representative snap-shots from the 52-psec MD trajectory; two neighboring snap-shots were separated by a time step of 0.5 psec. One RMS (104 X 104) matrix was constructed; each element (i,j) in the matrix gives the RMS deviations between two structures i and j in the MD trajectory. Global translation and rotations were appropriately deducted from from each element. Therefore, the RMS matrix reflects true conformational differences among the 104 structures. It is noted that values of the elements along a row or a column of the RMS matrix show a periodic variation i.e., at first the values gradually increase and then reaches a maximum and finally drops to a minimum. The period of flipping from one structure to another is approximately 5 psec.

Structures along the MD trajectory can either belong to the same energy minimum or to different ones. If they belong to the same energy minimum, upon minimization they should show a small RMS deviation (≤ 0.5 Å) among each other. If, however, the structures belong to different energy minima, upon minimization RMS deviations among the structures should still be fairly large (≥ 1.5Å). Therefore, NOESY constrained energy minimization was performed on all the 104 structures from the MD trajectory. The (104 X 104) RMS matrix of the 104 minimized also shows a similar periodic variation as observed in the previous RMS matrix before minimization i.e., along a row or a column, at first the RMS value gradually increases and then goes to a maximum and finally drops to a minimum. What is more interesting is that upon minimization three structures (designated as I, II and III) are isolated as three well defined minima that show quite large deviations ($\geq$ 1.5Å) among each other. Thus, in this way we are able to locate at least three distinctly different energy minima in the potential surface. And all these three minimized structures satisfy the NOESY data. Examination of the two (104 X 104) RMS matrices (i.e., before and after minimization) shows that transitions between I and III are very frequent (time period of 5 psec) while transitions between I and II are very rare (happens only once during 52-psec). Thus, emerges a simple picture of dynamics of the $Dst2$-d(GA$_4$T$_4$C)$_2$ [1:1] complex. The structures are located in three distinctly different energy wells and there is a continuous interconversion of the structures within one well and between structures in two different wells (I and III).

Table II lists the base and helix parameters for these structures calculated using the methodology of Soumpasis and Tung [14]. The values of the corresponding parameters for structures I, II and III are different even though it is not immediately obvious how exactly these differences translate into their 3-dimensional structures. However, differences among these structures are clearly visible from Figure 4 in which structures I, II and III are shown along their long-axis. Structures I, II and III clearly show three different states of bending in the DNA decamer duplex. In structure III, the segment

G1- A2 - A3
C20-T19-T18

seems to show a sharp kink with the rest of the molecule. In structure II, this kink is less abrupt while in structure I it is barely visible. Therefore, it appears that upon Dst2 binding, the region of the target DNA adjacent to the site of binding, can show a characteristic motion. Although, it is visually clear that bending may be a principle component of motion, a quantitative estimate of bending has not yet been done.

R. H. SARMA ET AL.

Table II

Base and Helix Parametrs of Structures I, II and III Calculated Using the Methodology of Soumpasis and Tung [14].

Base	Tilt	Helical Twist	Roll	Wedge	Xdis	Ydis	Helical Rise	Prop.	Buckle
I	-	-	-	-	-	-	-	20	-12
G II								14	-14
III								12	-22
A I	5	26	7	8	-1.0	-0.9	3.0	27	4
II	3	31	4	5	-1.2	-1.0	3.0	17	-4
III	2	32	5	6	-1.1	-0.7	3.1	18	-16
A I	0	31	-13	13	-0.9	-0.3	3.1	18	11
II	0	28	-5	5	-1.2	-0.2	3.0	16	1
III	1	32	2	2	-0.8	0.1	3.0	21	-8
A I	-5	35	-13	14	-0.9	0.2	3.1	12	6
II	-4	31	-4	6	-1.1	0.3	3.0	16	0
III	-3	30	3	4	-0.6	0.2	3.0	22	2
A I	-6	25	-5	8	-0.7	0.3	3.2	13	-2
II	-6	23	-1	7	-0.4	0.3	3.1	17	-2
III	-5	22	0	5	-0.2	-0.2	3.1	18	0
T I	-9	41	-4	10	-0.1	0.5	3.0	17	3
II	-9	40	-3	9	-0.1	0.6	3.0	20	9
III	-7	41	-3	8	0.0	0.3	3.1	16	8
T I	8	45	-3	8	-0.9	-0.4	2.9	20	3
II	8	39	0	8	-0.8	-0.4	3.0	17	7
III	5	45	-3	9	-0.8	-0.6	2.9	17	7
T I	6	31	-1	6	-0.8	0.0	3.1	19	3
II	6	35	-3	7	-0.8	0.0	3.0	19	3
III	5	31	-3	6	-0.9	0.1	3.1	15	3
T I	4	42	-2	4	-0.8	-0.2	3.0	20	9
II	4	42	-2	5	-0.8	-0.4	3.0	20	9
III	2	42	-5	5	-0.8	0.0	3.0	18	5
C I	-1	40	1	2	-1.2	-0.1	3.1	13	20
II	-1	41	1	1	-1.3	-0.1	3.1	13	20
III	-2	41	1	3	-1.2	0.1	3.1	14	17

I, II and III are the three minimized structures discussed in the text.

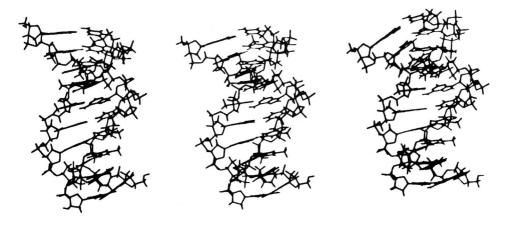

Figure 4.   Energy minimized structures I, II and III shown from left to right. All three structures show drug-DNA interactions in the minor-groove that are consistent with the NMR data. Although nucleotide conformations of all residues in structures I, II and III belong to an average BI geometry, there are small but finite differences among the three structures.   These small changes are sufficient to produce different bending states as exhibited by structures I, II and III. The RMS values for the three structures are : RMS(I,III)= 3.1 Å, RMS(I,II)= 2.2 Å and RMS(II,III)= 1.9 Å. The total potential energies of the three structures are within about 2 Kcal/Mole of each other (one mole is in terms of one mole of the drug-DNA complex).

Conclusion

In this article we show that an ensemble of structures can be constructed for the $Dst2$-d$(GA_4T_4C)_2$ [1:1] complex that agree with the NOESY data. Our studies also reveal that the dynamics of the system can be described by a simple picture in which structures are located in at least three well defined energy wells and there is a continuous interconversion of the structures within one well or between structures in two different wells. It is also apparent that upon drug binding, there is a characteristic motion at the site of binding and adjacent to it.

Acknowledgement

This work was supported by the U.S. Department of Energy, by grants from the National Institute of Health (GM29787) and by a contract from the National Foundation of Cancer Research. The high-field NMR experiments were performed at the Francis Bitter National Magnet Laboratory, MIT. The NMR facility is supported by the grant PR0095 from the Division of Research Resources of the NIH and by the National Science Foundation under contract C-670.

Authors wish to thank Dr. C. -S. Tung for providing us with the base and helix parameters in Table II and for many helpful discussions.

References

1. Zimmer, C. (1975), *Prog. Nucl. Acid Res. Mol. Biol., 15,* 285.
2. Dervan, P. B. (1986), *Science, 232,* 464.
3. Krowicki, Lee, M., Hartley, J. A., Ward, B., Kissinger, K., skorobogaty, A., Dabrowiak, J. C. and Lown, J. W. (1988), *in Structure & Expression, Vol 2: DNA its drug Complexes,* Eds. Sarma, R. H. and Sarma, M. H., Adenine Press, NY, pp 251.
4. Zimmer, C., Luck, G., Burckhardt, G. and Krowicki, K. and Lown, J. W. (1988), *in Structure & Expression, Vol 2: DNA its drug Complexes,* Eds. Sarma, R. H. and Sarma, M. H., Adenine Press, NY, pp 291.
5. Kopka, M. L., Yoon, C., goodsell, D. and Dickerson, R. E. (1985), in *Structure & Motion: Membranes, Nucleic Acids and Proteins, Eds., Clementi, E., Corongiu, G., Sarma, M. H. and Sarma, R. H.,* Adenine Press, NY, pp 461.
6. Klevit, R. E., Wemmer, D. E. and Reid, B. R. (1986), *Biochemistry, 25,* 3296.
7. Umemoto, K., Sarma, M. H., Gupta, G., Luo, J. and Sarma, R. H. (1990), J. Amer. Chem. Soc., in press.
8. Breslauer, K., Farrante, R. and Marky, L. a., Dervan, P. B. and Youngquist, R. S. (1985), *in Structure & Expression, Vol 2: DNA its drug Complexes,* Eds. Sarma, R. H. and Sarma, M. H., Adenine Press, NY, pp 273.
9. Sarma, M. H., Gupta, G., Garcia, A. E., Umemoto, K. and Sarma, R. H. (1990), *Biochemistry*, in press.
10. Gupta, G., Sarma, M. H. and Sarma, R. H. (1988), *Biochemistry, 27,* 7909.
11. Umemoto, K., Sarma, M. H., Gupta, G. and Sarma, R. H. (1990), *Biochemistry*, in press.
12. Weiner, S. J., Kollman, P. J., Case, D. A., Singh, U. C., ghio, C., Alagona, G., proteta, S. and Weiner, P. (1984), *J. Amer. chem. Soc., 106,* 765.
13. Arnott, S. and Hukins, D. W. l., (1972), *J. Mol. Biol., 81,* 93.
14. Soumpasis, D. M. and Tung, C. -S. (1988), *J. Biomol. Str. Dyn., 6,* 397.

DETERMINATION OF DISTAMYCIN-A BINDING MODES BY NMR

D.E.Wemmer, P.Fagan & J.G.Pelton
Department of Chemistry
University of California
Berkeley, California 94720   USA

ABSTRACT. The application of NMR to characterization of
the binding of distamycin A to DNA oligomers is discussed.
The orientational preference has been determined for single
distamycins binding to A-T rich sequences. In sites
containing more than the minimal four A-T pairs there is
evidence for a sliding transfer of the drug between
neighboring binding sites at a high rate. In sites
containing five and six A-T pairs there is a second binding
mode in which two distamycins bind side-by-side in the
minor groove, in close contact with one another in addition
to the walls of the groove. There also appears to be a
sliding transfer of drugs within the 2:1 complex when the
binding site is more than the minimal five A-T pairs.

INTRODUCTION

Distamycin-A is one of a large class of DNA binding drugs
which have substantial affinity for A-T rich sequences.
The compounds in this class all have a number of aromatic
rings, linked with an overall curvature which matches that
of the DNA grooves reasonably well.

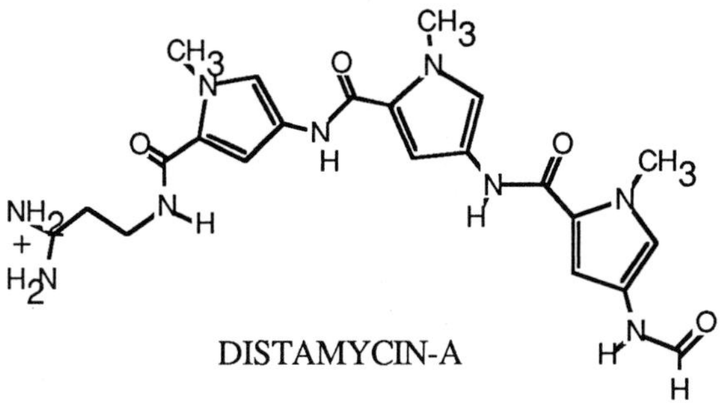

DISTAMYCIN-A

Early studies showed that intercalation does not occur, and
it was suggested from protection studies that instead these

95

B. Pullman and J. Jortner (eds.), Molecular Basis of Specificity in Nucleic Acid-Drug Interactions, 95–101.
© 1990 Kluwer Academic Publishers. Printed in the Netherlands.

compounds bound in the minor groove. This was confirmed in
NMR studies by Patel [1] for netropsin (a close relative of
distamycin) by showing that NOEs between the drug pyrole
HC3 and adenosine HC2 resonances could be observed.
Further information about sequence specificity of binding
was provided by affinity cleaving studies using distamycin
which was tagged with an EDTA(Fe) moiety, which generates
radicals at binding sites and leads to local cleavage of
the DNA [2]. These studies showed quite clearly that there
is discrimination among different A-T rich sequences, and
that there is an orientational preference at sites with
several A residues occuring on the same strand. These
studies, however, provide little information about the
details of the complexes formed. Further information at
the atomic level subsequently came from x-ray diffraction
analysis of cocrystals of netropsin and distamycin with
several different DNA oligomers, containing binding sites
of AATT [3], ATAT [4] and AAATTT [5]. In parallel with
these crystallographic studies, a number of NMR studies
have been carried out on this class of molecules.

IDENTIFICATION OF COMPLEXES

Our laboratory has concentrated on NMR studies distamycin-A
in complexes with a number of different sequences. This
work has provided some new insight into the binding
characteristics, and has lead to intentification of a new
class of complexes, which will be described. Although
Dervan and coworkers found that the preferred binding sites
for distamycin contained five AT base pairs, other evidence
suggested that four were sufficient, and our work began
with the DNA oligomer which Dickerson and coworkers had
analyzed in great detail: d(CGCGAATTCGCG)$_2$. 2D NMR,
particularly the NOESY experiment, was used to determine
contacts between the drug and DNA, with a number of
experiments being carried out to make semiquantitative
distance estimates. In this case the drug was found to
bind tightly (slow exchange on the NMR timescale) to the
central AATT site, with the pyrole C3 protons 1, 2 and 3
(numbered from the formyl end of the drug) in contact with
Adenosine C2 protons of A5, A6 and A18 respectively.
Molecular modelling of the complex suggested that the drug
fits snugly in the center of the minor groove, and that
many different interactions including electrostatics,
hydrogen bonding, and stacking all stabilize the complex[6].
Of course since the DNA is symmetric in this case, there is
no orientational preference for binding. When the
sequences AAAT and AAAA (with the complementary strand
present) were examined it was clear that there was only one

dominant form of complex, although binding in either
orientation should be possible. The section of a NOESY
experiment on the 1:1 complex of distamycin-A with AAAA is
shown below.

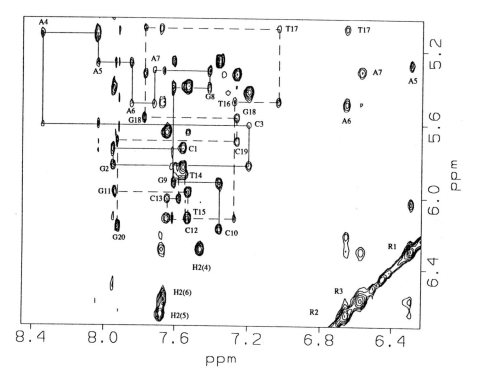

The sequential assignments of the two strands are
indicated. At the lower and right hand sections of this
region are seen NOEs connecting the drug HC3 resonances to
adenosine HC2 and sugar HC1' resonances respectively. From
these it is clear that the preferred orientation has the
formyl end of the drug in contact with the adenosine at the
5' end of the four As. The pattern of NOEs between the
drug and the sugars changes along the length of the drug,
and is somewhat different from that seen in AATT. This
indicates that the position of the drug relative to the DNA
is different. This could arise in part from a change in
shape of the DNA, poly A tracts are known to induce bending
in DNA oligomers. At the lower right of the spectrum there
are several weak cross peaks which have been determined to
arise from chemical exchange during the mixing time of the
NOESY. These peaks align with those from the drug HC3s of
the major form complex on one axis, but correspond to very
weak peaks in the spectrum on the other axis. These weak

peaks probably arise from binding of the drug in the
opposite orientation with a reduced affinity, not more than
about 5% of the major form. Further evidence for this
comes from sequences described below. With the sequence
AAAT, the predominant binding mode seems to be the same as
that seen with AAAA, the formyl end of the drug to the 5'
end of the A rich strand.

     Some preliminary studies have also been carried out
with an oligomer containing a central sequence TATA. In
this case as drug is titrated in the complex and free DNA
are in intermediate exchange (indicating a higher off rate
than for AATT), and it has not been possible to
characterize the complex in any detail.

COMPLEXES AT HIGHER DRUG:DNA RATIOS

     To go beyond the minimum binding site of four A-T base
pairs, the sequence AAATT was chosen. In this case as drug
was titrated in, two forms of complex were observed at low
drug:DNA ratios. Exchange/NOESY experiments carried out at
a ratio of 0.5:1 showed a number of different cross peaks,
from NOEs, exchange and a combination of the two.

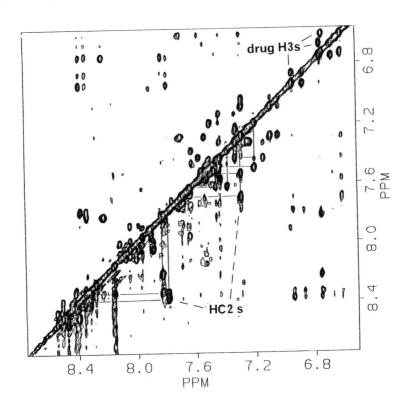

Although the spectra are complicated, it has been possible to sort out the origin of most of the peaks. Since free DNA resonances were identified in a NOESY of the DNA alone, it is straightforward to pick out the cross peaks representing exchange of the free DNA with the complex, leading to immediate assignment of some of the resonances in the complex. Additional exchange cross peaks were seen connecting drug HC3 resonances in the two forms of the complex. Finally, NOESY cross peaks were seen between the drug HC3s and A HC2s, which allowed us to identify the binding sites for the drug. We found that each HC3 seemed to be near to two A HC2s in each of the two forms of the complex. Bases on our experience with the other complexes, and on the crystallographic studies we believe that this arises from fast exchange of the drug between two neighboring binding sites, AAAT and AATT for the major form, and AATT and ATTT for the minor form (again using the convention that the ring 1 HC3 of the drug is opposite the first base listed, always written 5' to 3'). We do not see significant broadening of the resonances, so the exchange must be fast relative to chemical shift difference, requiring an estimated lifetime in either state of less than 1 ms. The basic features of binding are totally consistent with what was seen with sequences containing 4 A-T pairs, with slightly altered affinities for the same sequences. This is to say that there is a context effect, neighboring base pairs do modify the binding affinity for a particular sequence, as was seen with affinity cleaving studies. When an additional A-T pair is added, making AAATTT, then multiple NOEs are again seen at low drug/DNA ratios, but additional complications arise from another form of complex as discussed below. The data at low drug/DNA ratios are again suggestive of binding at multiple sites, with rapid sliding between them.

The titrations of both AAATT and AAATTT with distamycin showed that at near stochiometric ratios another form of complex was being populated. By continuing the titrations to higher ratios it was possible to show that in both cases the new complex contained two distamycins per DNA molecule. From a NOESY spectrum of the 2:1 complex with AAATT it was possible to identify all of the DNA resonances, and also resonances from the two distinct drugs. In addition to drug-DNA NOEs analogous to those seen before, there were also drug-drug NOEs. Taking these together it could be shown that the two drugs must occupy the minor groove of the DNA in a side-by-side manner. This requires that the groove be expanded by at least 3.5Å relative to the 1:1 complex to accomodate the second drug. In addition modelling of this complex showed that there are substantial changes in the contacts between the drug and the DNA [7]. When the 2:1 complex with AAATTT was examined in the same way it was found again that the drugs must be side-by-side, but that there must also be a sliding process present which changes the positions of the two drugs relative to one another [8]. In this case the process is somewhat slower than for a single drug sliding in the groove, leading to intermediate exchange for some of the protons at ambient temperature. This was somewhat surprising because the identical DNA sequence had been crystalized with distamycin to form a 1:1 complex with the drug bound at a single site. Reexamining the complex by NMR in the crystallization buffer indicated that the 2:1 complex was in fact the predominant form present. The differences between solution and crystal must arise from crystal packing constraints. A schematic drawing of the binding modes for the 2:1 complexes is given below:

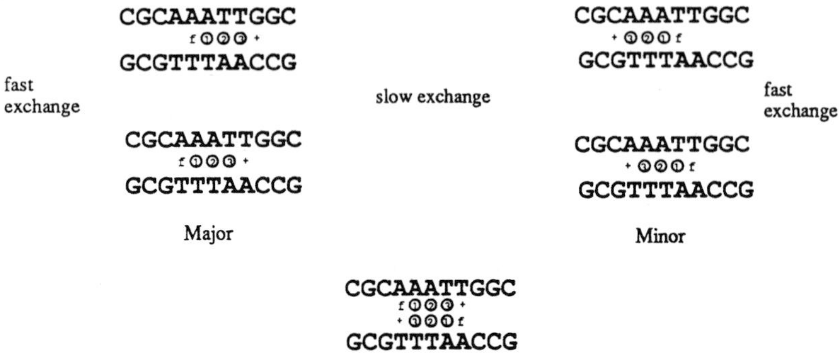

fast exchange

slow exchange

fast exchange

Major   Minor

"side by side" 2/1 complex

"side by side" 2/1 complexes which are rapidly interconverting

CONCLUSIONS

To summarize, NMR has been a powerful tool for analysis of the sequence specificity of binding of distamycin, for identification of new forms of complexes of distamycin and DNA. It has shown that high orientational specificity is present for some sequences, especially those with several sequential A residues on one strand. There is clear evidence for a sliding transfer of drug between neighboring binding sites when there are more than four A-T base pairs. At drug to DNA duplex ratios above 1:1 there are new complexes observed in which two distamycins bind in the A-T rich sequence in a side-by-side manor. This mode requires a minimum of five A-T pairs, when more are present binding in this mode is enhanced, and there is evidence for sliding transfer of drugs between different positions within the binding site. Overall the ability of NMR to analyse systems in which there are complicated equilibria present is unsurpassed, and makes it a powerful tool for the analysis of the structure and dynamics of minor groove binding drugs.

[1] Patel, D.J. (1982) 'Antibiotic-DNA interactions: Intermolecular nuclear Overhauser effects in the netropsin-d(CGCGAATTCGCG) complex in solution' Proc.Natl.Acad.Sci. USA 79, 6424-6428.

[2] Schultz, P.G. and Dervan, P.B. 'Distamycin and Penta-N-Methylpyrroloecarboxamide Binding Sites on Native DNA, A Comparison of Methidiumpropyl-EDTA-Fe(II) Footprinting and DNA Affinity Cleaving' (1984) J.Biomol.Struct.Dyn. 1, 1133-1147.

[3] Kopka, M.L., Yoon, C., Goodsell, D., Pjura, P. and Dickerson, R.E. 'The molecular origin of DNA-drug specificity in netropsin and distamycin' (1985) Proc.Natl.Acad.Sci. USA 82, 1376-1380.

[4] Coll, M., Aymami, J., van der Marel, J.A., van Boom, J.H., Rich, A. and Wang, A.H.-J. 'Molecular Structure of the Netropsin-d(CGCGATATCGCG) Complex: DNA Conformation in an Alternating AT Segment' (1989) Biochem. 28, 310-320.

[5] Coll, M., Frederick, C.A., Wang, A.H.-J. and Rich, A. 'A bifurcated hydrogen-bonded conformation in the d(A.T) base pairs of the DNA dodecamer d(CGCAAATTTGCG) and its complex with distamycin' (1987) Proc.Natl.Acad.Sci. USA 84, 8385-8389.

[6] Pelton, J.G. and Wemmer, D.E. 'Structural Modeling of the Distamycin A-d(CGCGAATTCGCG)2 Complex Using 2D NMR and Molecular Mechanics' (1988) Biochem. 27, 8088-8096.

[7] Pelton, J.G. and Wemmer, D.E. 'Structural characterization of a 2:1 distamycin A : d(CGCAAATTGGC) complex by two-dimension NMR' (1989) Proc.Natl.Acad.Sci. USA 86, 5723-5727.

[8] Pelton, J.G. and Wemmer, D.E. 'Binding Modes of Distamycin A with d(CGCAAATTTGCG)2 Determined by Two-Dimensional NMR' (1990) J.Am.Chem.Soc. 112, 1393-1399.

MOLECULAR MECHANISMS OF DNA SEQUENCE RECOGNITION BY GROOVE BINDING LIGANDS: BIOCHEMICAL AND BIOLOGICAL CONSEQUENCES

J. WILLIAM LOWN
Department of Chemistry
University of Alberta
Edmonton, Alberta, Canada
T6G 2G2

ABSTRACT. Progress in the development of DNA sequence selective minor groove binding agents is discussed. Among the factors contributing to the molecular recognition processes are: the presence and disposition of hydrogen bond accepting and donating groups, ligand shape, chirality, stereochemistry, flexibility, and charge. For longer ligands the critical feature is the phasing or spatial correspondence between repeat units in the ligand and the receptor. The application of these factors in the design and synthesis of novel agents which exhibit potent anticancer, and antiviral properties and inhibition of critical cellular enzymes including topoisomerases and, in the case of viruses, of reverse transcriptase is discussed. The emerging evidence of a relationship between sequence selectivity of the new agents and the biological responses they invoke is also described.

1. Introduction

There is considerable interest currently in the development of DNA sequence specific or selective agents for genetic targeting, *i.e.* the control of gene expression either for applications in diagnosis or ultimately in therapy (Stephenson and Zamecnik, 1978; Izant and Weintraub, 1984; Helené *et al.*, 1985; Holt *et al.*, 1986; Dervan, 1986; Miller *et al.*, 1986). DNA sequence specificity or selectivity has also recently become recognized as an important component of the efficacy of many cytotoxic agents including the pyrrolo(1,4)benzodiazepinone antibiotics (Hurley *et al.*, 1988), saframycins (Rao and Lown, 1990), CC-1065 (Reynolds *et al.*, 1985; Warpehoski and Hurley, 1988), caleachimicin (Lee *et al.*, 1987); bleomycin (Stubbe and Kozarich, 1987), netropsin and distamycin (Zimmer and Wahnert, 1986), several of which are of interest in the treatment of human malignancies.

Examples of cytotoxic agents isolated from natural sources exhibit DNA selectivities ranging from AT requirements (*e.g.* netropsin, distamycin and CC-1065) either in "binding" or in "bonding", to the GC preferences exhibited by *e.g.* saframycins, pyrrolo(1,4)benzodiazepinones and bleomycin.

The guiding principle behind efforts to develop DNA sequence specific agents is that greater biological response may be achieved for a given dose compared with a sequence neutral agent, thereby reducing toxic side effects. A related goal is the selective suppression of transcription from particular gene sequences (Stephenson and Zamecnik, 1978; Izant and Weintraub, 1984).

103

B. Pullman and J. Jortner (eds.), Molecular Basis of Specificity in Nucleic Acid-Drug Interactions, 103–122.
© 1990 *Kluwer Academic Publishers. Printed in the Netherlands.*

Conceptually there are a number of approaches to this problem, *e.g.* using β-oligonucleotides (Helené *et al.*, 1985; 1987) or their backbone modified counterparts (Miller *et al.*, 1985), which take advantage of the inherent Watson-Crick base pairing to target single strand sequences or with hybrid probes incorporating an intercalator. Another approach is to use the property of certain oligonucleotides to form triplex structures and thereby target double stranded nucleic acid sequences (Maher *et al.*, 1989).

DNA contains two channels of information, the major groove and the minor groove (Dickerson, 1983; Kopka *et al.*, 1985). In general the major groove is employed by control proteins (promoters, repressors) while the minor groove is used by certain polymerases and many xeno-biotics including antibiotics. The minor groove thus represents a vulnerable site of attack, in that it is normally unoccupied, and this is presumably the reason for the evolution of antibiotics to attack the DNA of competing organisms. Thus, although minor groove binders are, at first sight, less attractive as probes in that they target the less information-rich minor groove, they may nevertheless prove to have several complementary advantages with major groove ligands. Thus an alternative and complementary approach to the antisense oligonucleotide effort is to develop sequence specific probes based on naturally occurring DNA groove binding agents. We have reported such an approach based on the naturally occurring oligopeptide antibiotics netropsin and distamycin (Lown, 1988, 1989). Rational structural modification led to the development of lexitropsins, or information-reading agents, some of which are capable of recognizing unique sequences and which exhibit no memory for the preferred sequence of the parent antibiotic (Kissinger *et al.*, 1987). In an earlier review progress in understanding some of the factors contributing to the molecular recognition processes were described (Lown, 1988). The factors included: (i) the ability of certain hydrogen bond accepting heterocyclic moieties towards specific base pair recognition; (ii) the influence of ligand cationic charge in sequence selective binding; (iii) certain van der Waals contacts in 3'-terminal base pair recognition (Lee *et al.*, 1988a) as well as the importance of pharmacological factors such as the ready cellular uptake of the prototype lexitropsins and their subcellular distribution in living cancer cells with concentration in the nucleus (Bailly *et al.*, 1989).

In this review we will examine additional factors that have been uncovered that contribute to the molecular recognition processes between groove binding ligands and nucleic and receptors and that may consequently be incorporated into drug design. In addition we will describe some biochemical properties of the new agents developed and some of the biological implications of this program.

2. The Concept of DNA Base Site Acceptance

2.1. LIGANDS BEARING 2,5-DISUBSTITUTED FURANS

Our interest in the design of sequence specific DNA binding agents has led us to explore the molecular recognition properties of several hydrogen bond accepting heterocycles including imidazole, thiazole, and 1,2,4-triazoles (Lown, 1988, 1989). For example the bis-2,5-disub-stituted furan compound, *i.e.* formyl-furan-furan-CH_2CH_2-amidinium chloride was synthesized (Lee *et al.*, 1989c).

MPE-Fe(EDTA) footprinting of this monocationic bis-furan lexitropsin on a HindIII/EcoRI restriction fragment of pBR322 DNA revealed a series of four-base binding sites (all 5'→3') of

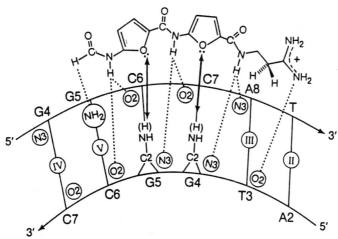

Figure 1. Molecular contacts deduced from NMR between bis-furan lexitropsin and preferred DNA sequence.

(primary), GCCA (Fig. 1), TGTA, TGAA, AAAT, ACAA, TTAT, and (secondary), CTAA, TCGT, TGTA, GTCA, and GGTT. Collectively the data support the inference of a GC recognizing capacity for a 2,5-substituted furan moiety within a lexitropsin.

3.   Concept of Base Site Avoidance

3.1. THIAZOLE-LEXITROPSINS

The structural aspect we then examined was the complementary one of base site avoidance *i.e.* whether a given heterocyclic moiety can exclude binding at a particular site. The corollary question then is whether this property can be incorporated into the design of agents capable of recognizing and binding to a unique sequence.

The synthesis of novel thiazole bearing oligopeptides related to the antibiotic distamycin was undertaken (Rao *et al.*, 1990a). The first group of compounds has the sulfur atom aligned inwards to the DNA minor groove (Fig. 2, **1-3**), whereas the second group of agents has the sulfur atom directed away from the DNA minor groove (Fig. 2, **4-5**).   All six compounds synthesized bind to double helical DNA with K_{app} values comparable with distamycin (Rao *et al.*, 1990b). The group of lexitropsins bearing nitrogen directed towards the DNA display comparable binding to poly(dA-dT) and to native DNAs, and complementary strand footprinting reveals their ability to accept and bind to mixed AT-GC sequences. The GC recognizing property plausibly arises from the hydrogen bonding between the thiazole nitrogen and G-2-NH$_2$ based on precedents. In contrast the group of lexitropsins bearing sulfur directed towards the floor of the

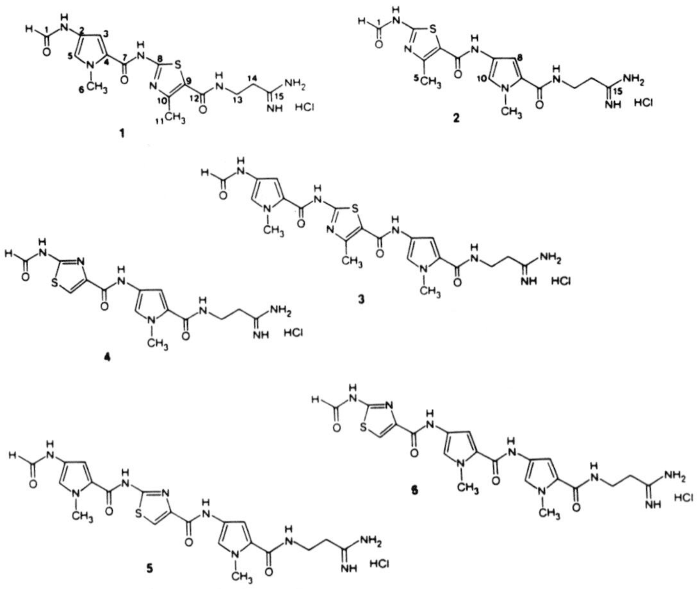

Figure 2. Structures of thiazole lexitropsins.

minor groove of DNA exhibit strict preference for AT sequences and are even more discriminating than distamycin. The latter agents, in common with the first group, bind firmly in the minor groove and with a binding site size of either 4 ± 1 or 5 ± 1 base pairs indicating intimate contact of all parts of the ligand. Therefore the property of GC site avoidance of these particular thiazole-lexitropsins is attributed to clash between the sterically more demanding sulfur and G-2-NH$_2$ groups.

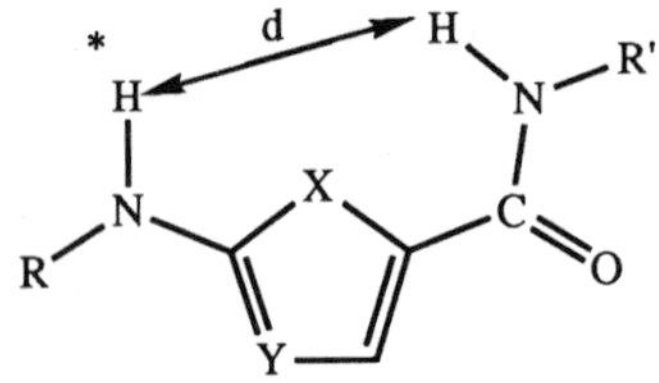

Figure 3. Repeat distance in lexitropsins.

Molecular mechanics calculations (Kumar *et al.*, 1989) showed when X = S; Y = N (as in the case of compounds **1-3**) the repeat distance between the amide hydrogens, d (Figure 3) is approximately 0.9 Å more than in the case of pyrrole or the isomeric thiazole (X = N; Y = S). Consequently the amide hydrogen (represented by * in Figure 3) is displaced from potential contact with the adjacent base resulting in failure of the amide hydrogen to participate in hydrogen bonding with N3 of adenine or O2 of thymine. NMR studies (Kumar *et al.*, 1990) on compound **2** with a decamer d[CGCAATTGCG]$_2$ and also the molecular mechanics calculations indicated that the steric clash between sulfur atom (van der Waals radius ~1.90 Å) and 2-NH$_2$ of guanine renders compounds **1-3** (where S is pointed towards the minor groove of DNA) strict AT sequence readers.

4. Effects of Ligand Chirality on DNA Binding

4.1. SEQUENCE SPECIFICITY OF ENANTIOMERIC ANTHELVENCINS

Thus far all the ligands examined have been achiral. However since the DNA receptor is chiral, the chirality of minor groove ligands may be anticipated to influence binding efficiency. This aspect was examined with the naturally occurring oligopeptide antibiotics anthelvencin and kikumycin.

The total synthesis of the two enantiomers of anthelvencin A [the naturally occurring isomer (4S)-(+)-**1a** and its enantiomer (4R)-(-)-**1b**] with enantiomeric excess of 80 ± 4 percent permitted the unambiguous assignment of absolute configuration of natural anthelvencin A as (4S)-(+)-**1a** (Lee *et al.*, 1988).

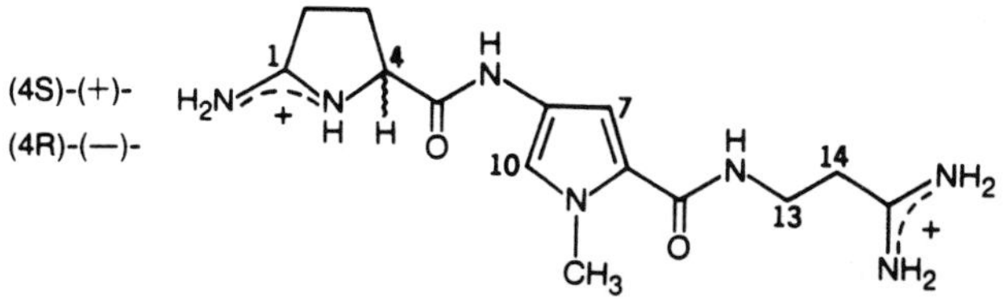

Figure 4. The structure, absolute configuration, and numbering system for dihydrokikumycin B.

The sequence specific binding of the antibiotic (4S)-(+)-dihydrokikumycin B and its (4R)(-) enantiomer, [(S)-I and (R)-I, respectively] to DNA were characterized by DNase I and MPE footprinting, calorimetry, UV spectroscopy, circular dichroism, and ^{1}H-NMR studies. Footprinting analyses showed that both enantiomers [(S)-I and (R)-I] bind to AT-rich regions of DNA (Lee *et al.*, 1989b).

^{1}H-NMR studies (ligand induced chemical shift changes and NOE differences) of the dihydro-kikumycins with d[CGCAATTGCG]$_2$ show unambiguously that the N to C termini of the ligands are bound to 5'-A$_5$T$_6$T$_7$-3' reading from left to right. From quantitative 1D-NOE studies, the AH2(5)-ligand H7 distance of complex A [(S)-I plus decamer (which is bound more strongly)] and complex B [(R)-I and decamer] are estimated to be 3.8 ± 0.3 Å and 4.9 ± 0.4 Å, respectively.

108   J. W. LOWN

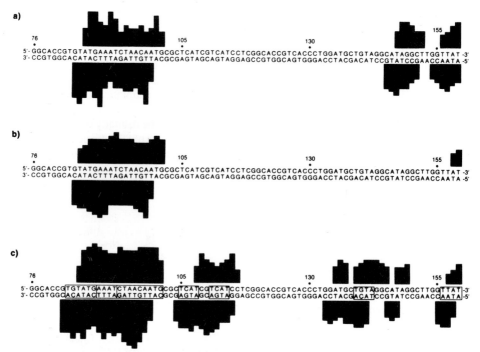

Figure 5. Histogram of MPE footprinting of the enantiomeric forms of dihydrokikumycin, (S)-**1** (a) and (R)-**1** (b), and netropsin (c) over bases 76 159 of pBR322 (r_t 0.78). Histogram height is proportional to the reduction of cleavage at each nucleotide relative to cleavage of unprotected DNA. Upper and lower footprints are from 5'- and 3'-end labeled DNA, respectively. Boxes in (c) indicate netropsin binding sites as determined by DNase I footprinting.

5. Phasing in the Binding of Ligands to Oligonucleotides

5.1. BIDENTATE AND MONODENTATE BINDING OF LIGAND-*BIS*-OLIGOPEPTIDES

One of the central problems in targeting relatively longer segments of DNA [15 to 16 base pairs defines a unique sequence in the human genome (Helené, 1987)] concerns the repeat distance of a nucleotide unit of DNA and the hydrogen bond and van der Waals contacts generated by oligo-peptidic-lexitropsins. This problem is sometimes referred to as the phasing problem (Kissinger *et al.*, 1990). In model studies involving idealized B-DNA coordinates it was demonstrated that as the netropsin-like ligand increases in length, the hydrogen bond contacts and van der Waals contacts between the ligand and DNA become seriously out of phase with the spacing between the nucleotide units of DNA. This phenomenon may explain the reduced binding affinities exhibited by certain poly-N-methylpyrrole peptides studied by Dervan and coworkers (Youngquist and Dervan, 1985).

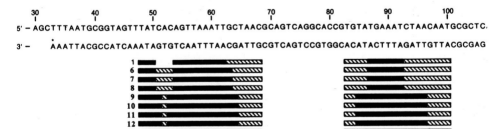

$$R_1—CO—(CH_2)_n—COR_1$$
$$R_2—CO—CH_2CH_2—CO—R_2$$
$$R_3—CO—CH_2CH_2—CO—R_3$$

6	$R_1—CO—R_1$
7	$R_1—COCH_2CO—R_1$
8	$R_1—CO(CH_2)_2CO—R_1$
9	$R_1—CO(CH_2)_3CO—R_1$
10	$R_1—CO(CH_2)_4CO—R_1$
11	$R_1—CO(CH_2)_5CO—R_1$
12	$R_1—CO(CH_2)_6CO—R_1$
13	$R_1—CO(CH_2)_7CO—R_1$
14	$R_1—CO(CH_2)_8CO—R_1$
15	$R_1—CO(CH_2)_9CO—R_1$
16	$R_1—CO(CH_2)_{10}CO—R_1$
17	$R_2—CO(CH_2)_2CO—R_2$
18	$R_3—CO(CH_2)_2CO—R_3$

Figure 6. Structures of linked lexitropsins.

An initial approach to this problem is to connect the netropsin moieties by a suitable tether. A series of tether-linked *bis*-netropsins have been synthesized (Fig. 6) in order to assess the phasing problem, which arises because of the lack of dimensional correspondence between oligopeptides and oligonucleotides on DNA binding characteristics (Lown *et al.*, 1989). The consequences of incorporating variable length, flexible and rigid tethers (polymethylene, Z and E ethylene, *m* and *p* phenylene) between the two netropsin-like moieties, on the DNA binding properties were assessed by DNase I footprinting (Kissinger *et al.*, 1990) (Fig. 7).

```
      30          40          50          60          70          80          90          100
      |           |           |           |           |           |           |           |
5' – AGCTTTAATGCGGTAGTTTATCACAGTTAAATTGCTAACGCAGTCAGGCACCGTGTATGAAATCTAACAATGCGCTC.

3' –  AAATTACGCCATCAAATAGTGTCAATTTAACGATTGCGTCAGTCCGTGGCACATACTTTAGATTGTTACGCGAG
```

Figure 7. Summary of areas of DNA protection by linked bis-netropsin of DNase I digestion.

The conformational freedom associated with two netropsins linked by a flexible methylene tether allows ligand binding in both a mono and bidentate fashion with bidentate binding requiring a minimum linker length of $[CH_2]_3$. For compounds possessing rigid tethers, for example *cis* and *trans* ethylene moieties, the *cis* geometry excludes bidentate ligation while the *trans* structure favors it (Fig. 8).

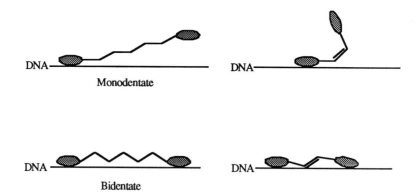

Figure 8. Mono and bidentate binding of bis-netropsins.

5.2. ANALYSIS OF PHASING BY FORCE FIELD CALCULATIONS

An examination of the repeat distance in oligopeptide ligands to permit accurate matching of recognition sites in the minor groove of the DNA receptor suggests different lengths are required according to the nature of the recognition process involved, *i.e.* hydrogen bond donation or

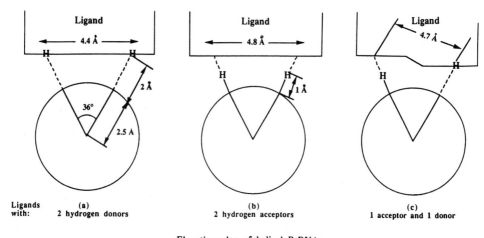

Figure 9. Considerations of phasing in ligand-DNA interactions.

acceptance. A preferred repeat distance of *ca.* 4.4 Å is predicted for ligands of the lexitropsin type which bear two hydrogen donors (Fig. 9). From this analysis one can now predict and interpret the properties of lexitropsins bearing different heterocyclic moieties (Kumar *et al.*, 1990).

One may see from the Table 1 that the repeat distance L for pyrrole at 4.7 Å is already too long and this explains why poly(N-methylpyrrole)lexitropsins both get out of phase and lose binding capacity.

Table 1. Dependence of repeat distance L in lexitropsins on heterocyclic moiety.

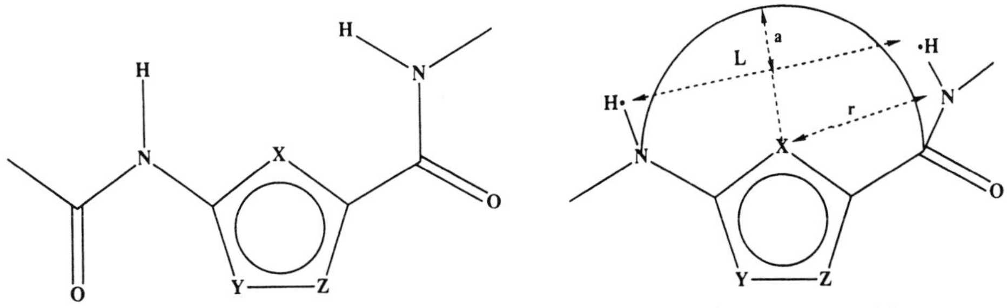

r = van de Waals radius of X

	X	Y	Z	r	a	L*
Pyrrole	CH	CH	N	(1.0+1.2)	1.2	4.7
Imidazole	N	CH	N	1.5	0.3	4.2
Thiazole	S	N	CH	1.8	1.5	5.5
Thiazole	N	S	CH	1.5	0.2	4.0
Triazole	N	N	N	1.5	0.3	4.2

*atoms in common plane

By contrast imidazole and 1,2,4-triazole moieties are predicted to be closer to the ideal repeat distance. The thiazole cases are of interest since the mode of interaction with the receptor is markedly dependent on the orientation. The isomer with N directed in towards the minor groove predicts an acceptable repeat distance of 4.0 Å whereas the isomeric structure (with S directed in to the groove) exhibits a repeat length of 5.5 Å clearly much too long to permit effective molecular recognition and binding when two or more such thiazole moieties are linked in a lexitropsin. As we have seen earlier the large van der Waals radius of the sulfur also effectively excludes GC recognition in the minor groove rendering such thiazole bearing ligands exclusively AT reading. The steric clash between the inward directed sulfur and minor groove sites also leads to intercalation.

It appeared possible to correct some of the deficiencies in repeat lengths of the component base recognizing heterocycles by linking units together with an appropriate tether.

5.2.1. *Properties of an Ideal Tether Unit.* Five different structural, stereochemical, conformational, and energetical aspects need to be examined for the alternative linkers. These are first described in terms of a hypothetical ideal linker (Rao *et al.*, 1990b) (Fig. 10).

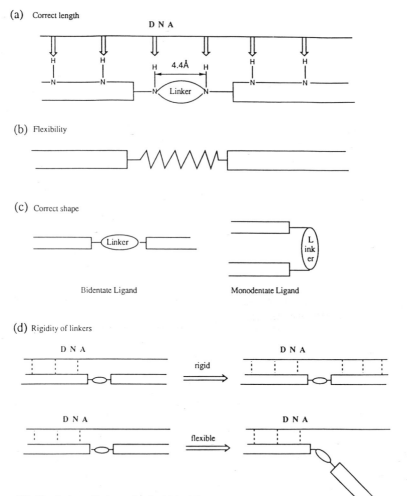

Figure 10. Depiction of a hypothetical ideal linker in terms of different structural properties.

5.2.2. Procedures for Molecular Mechanics Calculations. In screening the suitability of different liners the following procedure was adopted.

(a) The individual extended tether molecule (*e.g.* Fig. 6) was energy minimized using MM2 (Osawa and Musso, 1982) thereby obtaining the most stable conformation.

(b) The coordinates of this most stable conformation were fed into the program MacMOMO. The repeat length 1 was calculated as a function of the torsion angles $\zeta 1$ and $\zeta 2$ (Figure 11). The resulting data were then represented in a conformational plot.

(c) Those conformations with the correct repeat length 1 (4.4 ± 0.8 Å) were verified using molecular models. The head-to-head linked lexitropsin binding units must run in opposite directions and both of the hydrogen atoms on the amide groups must face in the same direction. Considerable variation of the repeat length can be seen in the conformation plot. The closer the

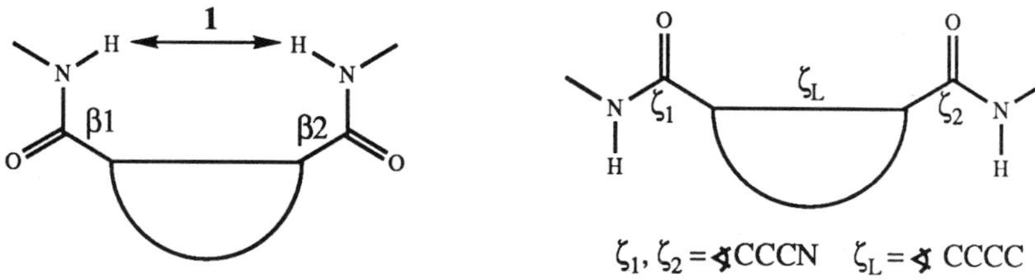

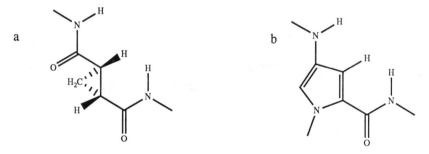

$$\zeta_1, \zeta_2 = \not{\zeta}\,\text{CCCN} \qquad \zeta_L = \not{\zeta}\,\text{CCCC}$$

Figure 11. Definition of dimensions and dihedral angles employed in MMX calculations.

contour lines (the steeper the slope) the better the tether can adjust the distance without changing the conformation of the ligand to any great extent.

(d) The energy of the conformations, which passed tests (a), (b), (c) were calculated by MM2. The resulting total strain energy was compared with that of the most stable conformation derived from (a) above.

The following alternative tethers were considered, cis and trans 1,2-disubstituted cycloalkanes from three to six membered rings, Z and E alkene and polymethylene linkers.

The force field analysis indicates that while none of the tested tethers appears to be ideal (Table 2) several of them have useful characteristics. The better tethers amongst those examined in this study are the trans-cyclopropane (Fig. 12), trans-cyclobutane, and trans-cyclopentane moieties, the fumaroic acid derivative and finally the succinate bridge. The preferred conformation of the trans-cyclopropane linked structure resembles that of N-methylpyrrole. In contrast none of the cis linked structures can adopt a suitable conformation.

Figure 12. Comparison of preferred conformation adopted by trans-cyclopropane compound with the N-methylpyrrole counterpart.

5.2.3. Trans Double Bond. All conformations in which $\zeta 2 = 180°-\zeta 1$ have the correct geometry for binding. The linker length is around 4.2 Å in all these conformations. MM2 calculation indicates a small difference (<1 kcal/M^{-1}) between the planar s-trans, s-cis conformation ($\zeta 1 = 180°$ and $0°$ respectively, $\zeta 2 = 0°$ and $180°$ respectively) and the most stable all s-trans compound. These parameters indicate a high probability for this compound to bind in a bidentate fashion to DNA. However the flexibility of the ligand is poor.

Table 2. Summary of Structural Characteristics of Alternative Linkers from Molecular Mechanics
Calculations as Potential

Compound	Shape/Length	Flexibility	Stability	Rigidity
cyclic compounds				
cis	d	—	—	—
n = 3				
n = 4	d	—	—	—
n = 5	c	b	b	b
n = 6	c	b	b	b
trans	b	b	b	b
n = 3				
n = 4	b	b	b	b
n = 5	b	b	b	b
n = 6	b	b	c	b
double bond				
cis	d	—	—	—
trans	a	c	a	b
polymethylene				
n = 0	d	—	—	—
n = 1	d	—	—	—
n = 2	a	b	a	c

a = very good
b = good
c = acceptable
d = unacceptable

The smallest linker length in an "allowed" conformation is about 4.2 Å. The relevant properties
of this linker for minor groove binding are very similar to those of the pyrrole moiety. Therefore
the binding of a compound bearing two N-methylpyrrole units (Net-2) joined by this tether should
be similar to that observed for Net-5. It has been noted that for longer ligands bearing
N-methylpyrrole units there will be a limit beyond which these oligo–(N-methyl–pyrrole)-carbox-
amides can no longer follow the right handed helical twist along the minor groove of B-DNA
(Youngqust and Dervan, 1985). One pyrrole unit is about 0.4 Å too long therefore a linker
joining two pyrrole units should have a linker length of about 3.6 Å (-8). Thus the *trans*-olefinic
link cannot completely compensate for the lack of dimensional correspondence between oligonucleotides and their complementary oligopeptides. However this tether should be ideal for
joining ligands which have the correct repeat length (for example 1,2,4-triazole and thiazole with
nitrogen directed in to the minor groove).

5.3. EXPERIMENTAL TEST OF MOLECULAR MECHANICS PREDICTIONS

5.3.1. Cycloalkane-linked Lexitropsins. The predictions made from the force field analysis were
tested experimentally.

The binding of a series of linked *bis*-oligopeptides, designed to examine the phasing problem of
the molecular recognition by longer ligands of DNA sequences, was examined by MPE
complementary strand footprinting on a EcoRI/HindIII restriction fragment of pBR322 DNA.

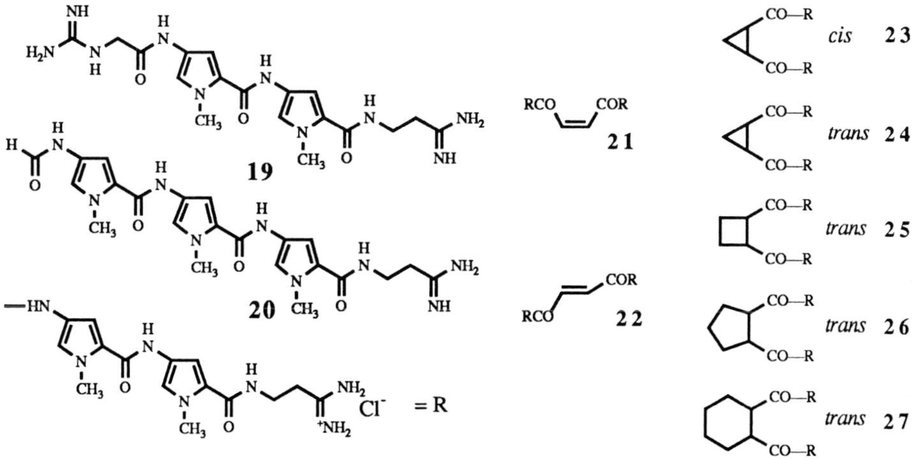

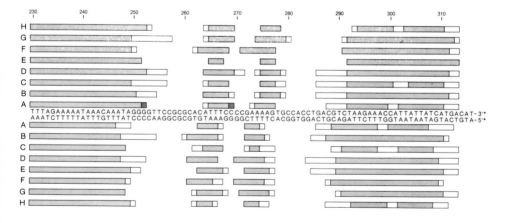

Figure 13. Structures of netropsin **19**, distamycin **20** and the new linked bis-oligopeptides.

Ligands bearing netropsin moieties linked by *trans* olefinic or *trans* 1,2-cycloalkane moieties (n = 3, 4, 5, 6) (Fig. 13) give evidence of bidentate binding in $(AT)_n$ rich sequences from footprinting at r = 0.16 (Fig. 14). By contrast those ligands linked by *cis* olefinic or *cis* 1,2-cyclopropane tethers exhibit monodentate binding. These results on the relative binding of *cis* and *trans* isomeric ligands are corroborated by detailed studies employing ultraviolet and circular dichroism spectroscopic studies on representative pairs of compounds which permit, *inter alia*, the determination of binding parameters, stoichiometry, and association constants (Rao *et al.*, 1990b).

Figure 14. Footprinting pattern of the new linked bis-netropsins and distamycin on the analyzed portion of the restriction fragment studied at lower (0.16) and higher (0.78) ratio. ☐ footprint at 0.78 ratio only (not observed at 0.16 ratio). ▧ footprint consistent both at 0.16 and 0.78 ratios. ▨ footprint at 0.16 ratio only (not observed at 0.78 ratio). A (**21**), B (**22**), C (**23**), D (**24**), E (**25**), F (**26**), G (**27**), and H (**20**).

The results suggest that the *trans* compound **24** binds to DNA in a bidentate fashion while its *cis* isomer **23** binds in a monodentate way (Fig. 15). If *trans* compound **24** binds to DNA in a true bidentate fashion (as in Fig. 15) one can expect a binding constant of this compound for poly(dA-dT) significantly greater than Dst which possesses three pyrrole rings and one cationic charge. Contrary to the prediction this compound has about six times lower binding constant than Dst. Probably this observed discrepancy in K_a value of compound **24** indicates that the rigid linker is not permitting both the *bis*-pyrrole arms of the compound to bind firmly in the minor groove of DNA at the same time. It is plausible to conclude that a sort of "dancing" interaction mechanism could be operating in the case of compound **24** with one arm sitting on DNA firmly at a given moment (as in Fig. 15). From the observed r_b and K_a values of compound **23** it is reasonable to assume that this compound binds to DNA with only one half of the molecule while the other half is positioned somewhat remote from the floor of the minor groove of DNA.

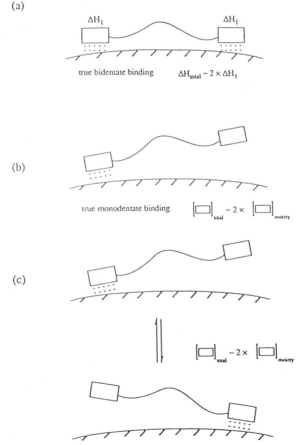

Figure 15. Depiction of alternative modes of binding of potentially bidentate ligands. (a) true bidentate binding, (b) true monodentate binding, (c) dancing mechanism or operational bidentate binding.

This analysis therefore indicates a dancing mode for **24** rather than true bidentate binding. A systematic analysis of thermodynamic data from microcalorimetry as well as detailed NMR experiments should permit a more reliable differentiation of the binding modes, and such studies are ongoing.

6. Pharmacological Properties of Lexitropsins

6.1. RELATIONSHIP OF CYTOTOXICITY OF EXTENDED POLYMETHYLENE-LINKED LEXITROPSINS TO THEIR MODE OF DNA BINDING

A group of oligopeptides have been synthesized which are structurally related to the natural antiviral antitumor antibiotics netropsin and distamycin bearing two such moieties linked by poly-methylene bridges. Cytostatic activity against both human and murine tumor cell lines (Table 3) and their *in vitro* activity against a range of viruses were examined. Enhanced antiviral activity was obtained against vaccinia virus. As a result of the introduction of the polymethylene linkers

Table 3. Inhibitory Effects of Linked Oligopeptides on the Proliferation of Murine Leukemia (P388) Cells.

Compound	n^b	ID_{50},[1] µg/mL	Compound	n^b	ID_{50},[a] µg/mL
20 (distamycin)		>20	12	6	0.70
6	0	8.0	13	7	0.75
7	1	0.91	14	8	0.98
8	2	1.75	15	9	18.0
9	3	6.4	16	10	4.8
10	5	4.4	17	2	>20
11	5	4.4	18	2	>20

[a]50% inhibitory dose measured by using 25000 cells/well and a 72-h incubation at 37°C with doses of 20, 10, 5, 1, and 0.5 µg/mL. [b]Number of CH_2 units in the linker.

Table 4. Inhibitory Effects of the Linked Oligopeptides on the Proliferation of Murine Leukemia (L1210), Murine Mammary Carcinoma (FM3A), Human B Lymphoblast (Raji), and Human T Lymphoblast (Molt/4F) Cells

Compound	n^b	ID_{50},[a] µg/mL			
		L1210	FM3A	Raji	Molt/4F
19 (netropsin)		245±92	321±18	139±63	
20 (distamycin)		27±4.7	31±2.4	24±3.7	
6	0	>100	>100	>100	>100
7	1	28.5±9.7	5.87±2.23	3.39±0.69	2.85±0.64
8	2	>100	47.2±27.8	26.4±2.8	33.8±0.9
9	3	>100	>100	>100	>100
10	4	24.3±8.9	57.8±32.7	13.3±5.6	5.62±0.60
11	5	>100	≥100	11.1±3.2	5.74±1.85
12	6	3.34±0.27	3.15±0.72	2.14±0.80	1.74±0.28
13	7	10.8±6.6	32.0±9.7	4.36±0.89	3.30±0.99
14	8	4.21±1.52	22.1±11.1	3.24±0.44	2.97±0.49
15	2	>100	>100	>100	>100
16	2	>100	4.29±1.12	41.4±8.4	2.85±0.64

[a]50% inhibitory dose. [b]Number of CH_2 units in linker.

$[(CH_2)_n$, with $n = 1, 2,$ and $6\text{-}8)]$, both the antitumor and antivaccinia virus activity was markedly enhanced, relative to that of the parent compounds netropsin and distamycin. The biological activity of these agents is discussed both in terms of their structural differences and in relation to their minor groove binding to duplex DNA (Tables 3, 4) (Lown *et al.*, 1989).

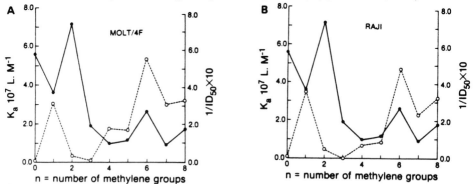

Figure 16. Correlations between DNA binding constants of linked oligopeptides (K_a, solid line) and observed inhibitory properties, expressed as reciprocal of ID_{50} values (dotted lines) for (A) human tumor cell line Molt/4F, (B) human tumor cell line Raji.

The results permit identification of some of the structural parameters that contribute to enhanced activity. First, binding to DNA of members of this homologous series is comparable ($\sim 10^7$ M^{-1}) both within the series and to that of the parents netropsin or distamycin, so this alone cannot explain the up to 30 fold increase in antileukemic potency (**12a** *vs* **20**). Similarly, differences in sequence selectivity are unlikely to be a dominant cause given the strict AT binding preference of al N-methylpyrrole-based oligopeptides. The strict AT sequence preference was also found to be an important determinant of biological activity in other novel classes of oligopeptide agents (Krowicki *et al.*, 1988). A more likely factor is the increased lipophilicity promoting cellular uptake as a result of introduction of the polymethylene linkers. Nevertheless, the rise and fall of both tumor cell cytotoxic and antiviral activity with $-(CH_2)_n$ reflecting maximal activity in the regions $n = 1, 2$ and $6\text{-}8$ (Fig. 16) suggests a subtle interplay of enhanced cellular uptake and phase-dependent binding. The latter aspect, which is beyond the scope of the present study will require careful quantitative footprinting and thermodynamic evaluation of ligand binding. A serendipitous finding is that introduction of the $-(CH_2)_n$ linker enhances preferential activity against human tumor cells. The introduction of the linker also appears to confer selective antivaccinia activity.

6.2. INHIBITION OF MOLONEY LEUKEMIA VIRUS REVERSE TRANSCRIPTASE

The inhibition of Moloney murine leukemia virus (MLV)-associated reverse transcriptase (RNA and DNA-directed DNA polymerase) activity by the linked oligopeptides was examined (Lown *et al.*, 1989). In hibition was observed for the homologous series (n = 1 to 10) in the range of 9.1 to 72.5 µg/mL. The greatest extent of inhibition was observed in the following cases (order of decreasing activity): **23a>22a>21a>19a>15a>20a>14a>16a>17a>18a**. A correspondence is observed between the length of the linker and the inhibitory activity against the reverse transcriptase (order of decreasing activity): **23a>22a>21a>19a>15a>20a>14a>16a>17a**

>**18a**. Again a correspondence is observed between the length of the linker and the inhibitory activity (Figure 17).

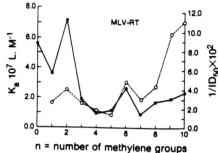

Figure 17. Correlation between DNA binding constants of linked oligopeptides (K_a, solid line) and observed inhibitory properties expressed in reciprocal ID_{50} values against Moloney leukemia virus reverse transcriptase.

7. Conclusions and Prospects

The concept of the lexitropsin, or information reading molecule derived by rational structural modification of natural DNA sequence selective agents, has proven to be both instructive and useful. It has proven instructive in that it has revealed several of the structural, stereochemical, conformational, electrostatic, and phasing factors that contribute to the molecular recognition processes that determine sequence selectivity of groove specific agents. The concept has also proven to be useful in that it has led directly to the development of several agents which exhibit significant biochemical properties including ready cellular uptake, concentration in the nucleus, as well as marked inhibitory properties against key enzymes including topoisomerases I and II and reverse transcriptase. The intriguing finding that inhibitory action against these critical cellular targets can, in the cases of individual classes of ligands, be related to their selectivity for the template has significant potential. Thus such lexitropsins may, on the one hand serve as probes of the mechanism of action of such enzymes, particularly with regard to the role of template selectivity, but may also provide new types of potent inhibitors of these same enzymes. A bonus of these studies has been the observation of significant anticancer and antiviral (including HIV) potency in the case of several of the lexitropsins. The implication that these latter biological properties may also be related to the sequence selectivity of the lexitropsins is the next challenge to be faced.

Acknowledgment

This research was supported by grants (to J.W.L.) from the Natural Sciences and Engineering Research Council of Canada and the Medical Research Council of Canada.

References

Bailly, C., Catteau, J-P., Henichart, J-P., Reszka, K., Shea, R.G., Krowicki, K., and Lown, J.W. (1989) "Subcellular distribution of a nitroxide spin-labeled netropsin in living KB cells: E.P.R. and sequence specificity studies", *Biochem. Pharmacol.* **38**, 1625-1630.

Dickerson, R.E. (1983) "The DNA helix and how it is read", *Scientific American* **249**, 94-111.

Dickerson, R.E. (1983) "Base sequence, helix structure and intrinsic information readout", in B. Pullman and J. Jortner (eds.), Nucleic Acids: The Vectors of Life, D. Reidel, Jerusalem, pp. 1-15.

Helené, C. (1987) Lecture on antisense probes in the "International Symposium on DNA as a Target for Chemotherapeutic Action", Seillac, France, October 11-17.

Helené, C., Monteray-Garestier, T., Saison, T., Takasugi, M., Toulme, J.J., Asseline, V., Lancelot, G., Maurizot, J.C., Toulme, F., and Thuong, N.T. (1985) "Oligodeoxynucleotides covalently linked to intercalating agents: A new class of gene regulatory substances", *Biochimie* **67**, 777-783.

Holt, J.T., Venkat, G.T., Moulton, A.D., and Nienhuis, A.W. (1986) "Inducible production of c-fos antisense RNA inhibits 3T3 cell poliferation", *Proc. Natl. Acad. Sci., U.S.A.* **83**, 4794-4798.

Hurley, L-H., Reck, T., Thruston, D.E., Langley, D.R., Holden, K.G., Hertzberg, R.P., Hoover, J.R.E., Gallagher, G., Fancette, L.F., Mong, S.M., and Johnson, R.K. (1988) "Pyrrolo-[1,4]-benzodiazepine antitumor antibiotics: Relationship of DNA alkylation and sequence specificity to the biological activity of natural and synthetic compounds", *Chem. Res. in Toxicol.* **1**, 258-268.

Izant, J.G. and Weintraub, H. (1984) "Inhibition of thymidine kinase gene expression by antisense RNA: A molecular approach to genetic analysis", *Cell* **36**, 1007-1015.

Kissinger, K.L., Krowicki, K., Dabrowiak, J-C., and Lown, J.W. (1987) "Monocationic imidazole lexitropsins that display enhanced GC sequence dependent DNA binding", *Biochemistry* **26**, 5590-5595.

Kopka, M.L., Yoon, C., Godsell, D., Pjura, P., and Dickerson, R.E. (1985) "The molecular origin of DNA-drug specificity in netropsin and distamycin", *Proc. Natl. Acad. Sci., U.S.A.* **82**, 1376-1380.

Krowicki, K., Balzarini, J., De Clercq, E., Newman, R.A., and Lown, J.W. (1988) "Novel DNA groove binding alkylators: Design, synthesis and biological evaluation", *J. Med. Chem.* **31**, 341-345.

Kumar, S., Jaseja, M., Zimmermann, J., Yadigiri, B., Pon, R.T., and Lown, J.W. (1990) "Molecular recognition and binding of a GC site-avoiding thiazole-lexitropsin to d[CGCAATTGCG]$_2$: ^{1}H-NMR evidence for thiazole intercalation", *J. Biomol. Struct. Dyn.*, submitted.

Kumar, S., Yadagiri, B., Pon, R.T., and Lown, J.W. (1990) "Sequence specific molecular recognition and binding by a Hoechst 33258 analog to the decadeoxyribonucleotide d[CATGGCCATG]$_2$: Structural and dynamic aspects deduced from high field ^{1}H-NMR studies", *J. Biomol. Struct. Dyn.*, submitted.

Lee, M.D., Dunne, T.S., Chang, C.C., Ellestad, G.A., Siegel, M.M., Morton, G.O., McGahren, W.J., and Borders, D.B. (1987) "Calichemicins, a novel family of antitumor antibiotics. Chemistry and structure of calcichemicin γ_1", *J. Am. Chem. Soc.* **109**, 3466-3468.

Lee, M. and Lown, J.W. (1987) "Synthesis of (4S) and (4R)-methyl-2-amino-1-pyrroline-5-carboxylates and their application to the preparation of (4S)(+) and (4R)(-) dihydrokikumycin", *J. Org. Chem.* **52**, 5717-5721.

Lee, M., Krowicki, K., Hartley, J.A., Pon, R.T., and Lown, J.W. (1988a) "Molecular recognition between oligopeptides and nucleic acids: Influence of van der Waals contacts in determining the 3'-terminus of DNA sequences read by monocationic lexitropsins", *J. Am. Chem. Soc.* **110**, 3641-3649.

Lee, M., Coulter, D.M., and Lown, J.W. (1988b) "Total synthesis and absolute configuration of the antibiotic oligopeptide (4S)(+) anthelvencin A and its (4R)(-) enantiomer", *J. Org. Chem.* **53**, 1855-1859.

Lee, M., Shea, R.G., Hartley, J.A., Kissinger, K., Pon, R.T., Vesnaver, G., Breslauer, K.J., Dabrowiak, J-C., and Lown, J.W. (1989a) "Molecular recognition between oligopeptides and nucleic acids: Sequence specific binding of the naturally occurring antibiotic (4S)(+) anthelvencin A and its (4R)(-) enantiomer to DNA deduced from ^{1}H-NMR, footprinting and thermodynamic data", *J. Am. Chem. Soc.* **111**, 345-354.

Lee, M., Shea, R.G., Hartley, J.A., Lown, J.W., Kissinger, K., Dabrowiak, J-C., Vesnaver, G., Breslauer, K.J., and Pon, R.T. (1989b) "Molecular recognition between oligopeptides and nucleic acids: Sequence specific binding of (4S)(+) and (4R)(-)-dihydrokikumycin B to DNA deduced from ^{1}H-NMR, footprinting studies and thermodynamic data", *J. Molec. Recogn.* **2**, 6-17.

Lee, M., Krowicki, K., Shea, R.G., Pon, R.T., and Lown, J.W. (1989c) "Molecular recognition between oligopeptides and nucleic acids: Specificity of binding of a monocationic bis-furan lexitropsin to DNA deduced from footprinting and ^{1}H-NMR studies", *J. Molec. Recogn.* **2**, 84-93.

Lown, J.W. (1988) "Lexitropsins: Rational design of DNA sequence reading agents as novel anti-cancer agents and potential cellular probes", *Anti-Cancer Drug Design* **3**, 25-40.

Lown, J.W. (1989) "Synthetic chemistry of naturally occurring oligopeptide antibiotics and related lexitropsins", *Org. Prep. and Proced. Int.* **21(1)**, 1-46.

Lown, J.W., Krowicki, K., Balzarini, J., Newman, R.A., and De Clercq, E. (1989) "Novel linked antiviral and antitumor agents related to netropsin and distamycin: Synthesis and biological evaluation" *J. Med. Chem.* **32**, 2368-2375.

Lown, J.W. and Beerman, T.A. (1990), unpublished results.

Maher, L.J., Wold, B., and Dervan, P.B. (1989) "Inhibition of DNA binding proteins by oligo-nucleotide-directed triple helix formation", *Science* **245**, 725-730.

Miller, P.S., Agris, C.H., Aureliani, L., Blake, K.R., Murakami, A., Reddy, M.P., Spitz, S.A., and Ts'o, P.O.P. (1986) "Control of ribonucleic acid function by oligonucleoside methyl phosphonates", *Biochimie* **67**, 769-776.

Osawa, E. and Musso, H. (1982) "Application of Molecular Mechanics Calculations" in N.L. Allinger, E.L. Eliel, S.H. Wilen (eds.), Organic Chemistry in Topics in Stereochemistry, Wiley, New York, p. 117.

Rao, K.E. and Lown, J.W. (1990) "Mode of action of saframycin antitumor antibiotaics: Sequence selectivities in the covalent binding of saframycins A and S to DNA", *Chem. Res. in Toxicol.* **3**, in press.

Rao, K.E., Krowicki, K., Balzarini, J., De Clercq, E., Newman, R.A., and Lown, J.W. (1990a) "Novel linked antiviral and antitumor agents related to netropsin 2: Synthesis and biological evaluation", *J. Med. Chem.*, submitted.

Rao, K.E., Zimmermann, J., and Lown, J.W. (1990b) "Sequence selective DNA binding by linked bis-N-methylpyrrole dipeptides: An analysis by MPE footprinting and force field calculations", *J. Org. Chem.*, submitted.

Rao, K.E., Bathini, Y., and Lown, J.W. (1990c) "Synthesis of novel thiazole containing DNA minor groove binding oligopeptides related to the antibiotic distamycin", *J. Org. Chem.* **55**, 728-737.

Rao, K.E., Shea, R.G., Yadagiri, B., and Lown, J.W. (1990d) "Molecular recognition between oligopeptides and nucleic acids: DNA sequence specificity and binding properties of thiazole-lexitropsins incorporating the concepts of base site acceptance and avoidance", *Anti-Cancer Drug Design*, in press.

Reynolds, V.L., Molineux, L.J., Kaplan, D., Swenson, D.H., and Hurley, L.H. (1985) "Reaction of the antitumor antibiotic CC-1065 with DNA. Location of the site of structurally induced strand breakage and analysis of DNA sequence specificity", *Biochemistry* **24**, 6228-6237.

Stephenson, M.L. and Zamecnik, D.C. (1978) "Inhibition of Rous sarcoma viral RNA translation by a specific oligodeoxyribonucleotide", *Proc. Natl. Acad. Sci., U.S.A.* **75**, 285-288.

Stubbe, J. and Kozarich, J.W. (1987) "Mechanisms of bleomycin-induced DNA degradation", *Chem. Revs.* **87**, 1107-1136.

Warpehoski, M.A. and Hurley, L.H. (1988) "Sequence selectivity of DNA covalent modification", *Chem. Res. in Toxicol.* **1**, 315-333.

Youngquist, R.S. and Dervan, P.B. (1985) "Sequence specific recognition of B-DNA by oligo(N-methylpyrrolecarboxamide)", *Proc. Natl. Acad. Sci., U.S.A.* **82**, 2565-2569.

Zimmer, C. and Wahnert, U. (1986) "Non-intercalating DNA-binding ligands: Specificity of the interaction and their use as tools in biophysical, biochemical and biological investigations of the genetic material", *Prog. Biophys. Mol. Biol.* **47**, 31-112.

DAUNOMYCIN BINDING TO DNA: FROM THE MACROSCOPIC TO THE MICROSCOPIC

Jonathan B. Chaires
Department of Biochemistry
The University of Mississippi Medical Center
Jackson, MS 39216-4505
U. S. A.

ABSTRACT. Results from footprinting titrations experiments that probe the sequence preference of daunomycin binding to DNA are described. The triplet sequences 5'(A/T)CG and 5'(A/T)GC are identified as the most preferred daunomycin binding sites within the 165 bp *tyr T* DNA fragment, where the notation (A/T) means that either A or T may occupy the position. Preferential binding is also observed to the triplet sequences 5'(A/T)C(A/T), but less frequently than to the triplets containing contiguous GC base pairs. Tentative estimates for the microscopic binding constants for daunomycin binding to these preferred sites indicate that their magnitude is at least ten times larger than the average macroscopic binding constant for binding of daunomycin to natural DNA of heterogeneous sequence. Spectrofluorometric titration and stopped-flow kinetic experiments identify a previously undocumented class of high affinity daunomycin binding sites in calf thymus DNA. These sites presumably corresponds to the preferred sites identified by footprinting experiments.

1. INTRODUCTION

The anthracycline antibiotics daunomycin (daunorubicin) and adriamycin (doxorubicin) are among the most potent compounds in current clinical use in cancer chemotherapy. Compelling experimental evidence suggests that the cellular target of these compounds is DNA (1,2), although the exact DNA site that is crucial for drug action remains unknown. Daunomycin and Adriamycin are potent inhibitors of topoisomerase II, which may be a primary intracellular target for anthracycline action (3). The physical chemistry of anthracycline - DNA interactions has been extensively studied, and has been recently reviewed (4, 5). High resolution X-ray crystallographic studies have provided the structure of daunomycin bound to the oligonucleotides 5'd(CGATCG)$_2$ (6) and 5'd(CGTACG)$_2$ (7) to 1.2 and 1.5 angstroms, respectively.

Whether or not daunomycin binds preferentially to particular DNA sequences has been a particularly vexing question. Early attempts to address this problem produced experimental results that were both confusing and contradictory (reviewed in [5,8]). The first coherent

123

B. Pullman and J. Jortner (eds.), Molecular Basis of Specificity in Nucleic Acid-Drug Interactions, 123–136.
© 1990 *Kluwer Academic Publishers. Printed in the Netherlands.*

proposal concerning the preferred anthracycline binding sites in DNA
emerged from the theoretical studies of the Pullman laboratory (8-10).
Daunomycin was found, in these studies, to bind preferentially to a
triplet sequence consisting of contiguous GC base pairs, flanked by an
AT base pair.   This entirely novel proposal could explain much of the
earlier confusion in the interpretation of the experimental results.

Reported here are recent equilibrium binding, footprinting titration,
and kinetic experiments that further explore the issue of the sequence
preference of daunomycin binding to DNA in solution.   The combined
results from these approaches provide a more refined view of the
sequence preference of daunomycin.

2. SUMMARY OF THE THERMODYNAMICS OF THE DAUNOMYCIN-DNA INTERACTION.

Six laboratories have now independently determined the equilibrium
binding constant for the interaction of daunomycin with calf thymus DNA
under identical solution conditions (pH 7.0, 200 mM Na^+), with
excellent agreement among the individual determinations (reviewed in
[5]).   The average values obtained for the neighbor exclusion binding
parameters are $K_0 = 6.7 (\pm 0.9) \times 10^5$ M^{-1} and n = 3.25 ± 0.4 bp.   The most
reliable binding parameters for adriamycin binding to calf thymus DNA,
under comparable solutions conditions, are $K_0 = 2.2 \times 10^6$ M^{-1} and n = 3.0
bp (11). K_0 in these cases refers to drug binding to an isolated site
along the DNA lattice.   Because of the sequence heterogeneity of the
calf thymus DNA, however, K_0 is a macroscopic quantity, representing
the weighted average of the binding constants for the interaction of
daunomycin with all possible sites of varying sequence along the DNA
lattice.

From the temperature dependence of the binding constant, the van't Hoff
enthalpy may be determined, and the binding free energy may then be
dissected into its enthalpic and entropic components (5,12,13).   The
thermodynamic profile obtained for daunomycin binding to calf thymus
DNA is (in 0.2 M NaCl, pH 7.0, 20°C):

$$\Delta G_0 = -7.8 \text{ kcal-mol}^{-1}$$

$$\Delta H_0 = -12.8 \text{ kcal-mol}^{-1}$$

$$\Delta S_0 = -17.0 \text{ cal-K}^{-1}\text{-mol}^{-1}$$

Daunomycin binding to DNA is thus strongly energetically favored, and
is driven by the large, negative enthalpy.

The positively charged amino group on the daunosamine moiety of
daunomycin imparts a salt dependency to the binding constant.   A number
of laboratories have experimentally determined that the salt dependency
of daunomycin binding to DNA is described by the relation

$(\delta\ln K / \delta\ln[Na^+]) = -0.88$ (5), a value in almost exact accord with the predictions of Record's theory (14). This value may be used to calculate a so-called thermodynamic binding constant, K_t^0, of $2.3 \times 10^5 M^{-1}$. This value is the binding constant corrected for the ionic contributions of the charged daunosamine. It is independent of Na^+ concentration. The large magnitude of K_t^0 indicates that the daunomycin-DNA complex is stabilized by extensive nonionic interactions, such as hydrogen bonding or van der Waals interactions. From the thermodynamic point of view, it is important to stress that the salt dependency arises from a nonspecific, general electrostatic effect in which the positive charge on the drug molecule replaces a condensed Na^+ ion upon binding. No specific molecular interactions between the charged daunosamine moiety and DNA phosphates need be invoked to explain the salt dependence of the binding constant.

The magnitude of the binding constant for the daunomycin-DNA interaction is strongly dependent on the DNA base composition, and increases with increasing GC content (reviewed in [5,15]). Both direct equilibrium binding studies using natural DNA samples of varying GC content and the method of competition dialysis were used to quantitatively study the GC preference of daunomycin (15,17). The dependency of its binding to DNA may be quantitatively described by the relation $K_{app} = (3.1 \times 10^6)f^2$, where f is the fractional GC content. The f^2 dependency suggests a minimal requirement for adjacent GC base pairs at the preferred daunomycin binding site (15). More complicated models involving triplet sites can also account for the experimentally observed base composition dependency of daunomycin binding (15,17), but we state here the simplest and most conservative model that describes the data. Preferential daunomycin binding to GC base pairs was independently demonstrated by circular dichroism spectroscopy (16). Notably, the dependency of the daunomycin binding constant on GC content is greater than that which is observed for actinomycin, which has been regarded as the paradigm for a GC specific binding antibiotic (15).

Studies utilizing synthetic deoxypolynucleotides yield data in apparent contradiction to the preferential binding of daunomycin to GC base pairs described above. Results from several laboratories now agree that daunomycin binds preferentially to polynucleotides of alternating purine-pyrimidine sequences (e.g.,poly[d(G-C)]; poly [d(A-T)]) compared to their nonalternating sequence counterparts (e.g.,poly dA-poly dT; poly dG-poly dC) (reviewed in [5]). However, daunomycin binding to poly [d(A-T)] is more affine than to poly[d(G-C)]. This apparent anomaly (when compared to the clear preference of daunomycin for GC rich natural DNA samples) may be explained by the computations of the Pullman laboratory.

3. THEORETICAL STUDIES OF DAUNOMYCIN SEQUENCE SELECTIVITY

The theoretical investigations of Chen, Gresh and Pullman (8-10)
provided key insights toward the identification of preferred DNA
binding sequences for the anthracycline antibiotics. Their first,
surprising, result was that a triplet sequence was required to specify
the preferred site. This was the first, and remains the only, case in
which greater than a dinucleotide sequence appears to be recognized by
an intercalator (8). Given that the site size of daunomycin, as
measured by equilibrium binding studies, is 3 bp (5), the requirement
for a triplet sequence is both reasonable and attractive. The second
important finding from the theoretical calculations was that the most
preferred daunomycin binding sites were of mixed base composition, with
drug intercalation between contiguous GC base pairs, but with an AT
base pair found at the 5' end of the triplet. The order of preference
of triplet binding sites for daunomycin was calculated to be (from most
to least preferred):

$$5'TCG \geq 5'ACG > 5'TCA \geq 5'ATA > 5'GCG \geq 5'GTA$$

The major molecular determinants of this order of sequence preference
are the 9-hydroxyl group and the daunosamine moieties of the
anthracycline. The former constituent participates in two hydrogen
bonds with the central GC base pair, but could only form one if an AT
bp were to occupy that position. The preference for an AT bp at the 5'
end is more subtle. If a GC base pair were to occupy that position,
substantial steric hindrance results from the clash between the guanine
NH_2 and the daunosamine moiety. Such hindrance is absent when an AT
base pair is present, resulting in an energetically more favorable
environment for the daunosamine within the minor groove. The most
general statement concerning the sequence preference of daunomycin to
emerge from the calculations was that "daunomycin should show the
highest preference to intercalate immediately upstream from a guanine
base, provided that there is an adenine or thymine immediately
downstream.... Depending upon the nature of the base immediately
upstream from G, there could be variations in the degree of binding."
(9). Subsequent experimental work has verified this prediction.

4. DNASE I FOOTPRINTING OF DAUNOMYCIN BINDING SITES

Deoxyribonuclease I (DNase I) footprinting experiments from this
laboratory provided experimental support for the notion of preferential
binding of daunomycin to a triplet sequence as proposed in the
theoretical studies (17). Sixteen protected sequences were identified
from three different DNA fragments. A common triplet motif of
contiguous GC base pairs flanked by an AT bp at the 5' side could be
found in all but two of these protected sites, a result in excellent
agreement with the predicted specificity from the theoretical studies.
Additional equilibrium binding studies and an examination of the effect
of daunomycin on the activity of the restriction endonucleases PvuI and
EcoRI provided strong additional experimental evidence in support of

the putative triplet recognition site.

However, DNase I footprinting results from a second laboratory appeared which were interpreted to indicate that the dinucleotide 5'CA was the preferred daunomycin binding site (19,20). The origin of the apparent discrepancy between results from our laboratory and theirs was not clear, and might arise from either differences in interpretation or in the primary data itself.

In order to reconcile the problem, higher resolution footprinting data were needed, which we obtained using a footprinting titration procedure, similar to that described in [20]. These titration experiments, in which the concentration dependency of protection of individual phosphate backbone positions is examined as a function of 0 to 3.0 μM daunomycin, afford single nucleotide resolution of preferred binding sites, and are a substantial improvement over previous attempts to map daunomycin binding sites. Full details of the experimental procedures and descriptions of the primary data are in press (21).

TABLE 1.   Most Strongly Protected Sites in the
tyr T DNA Fragment.
_____

Position	Sequence	A_f/A_0	$K''/10^6 M$
67	5'CTT^TAC	0.12	12.4
70	5'TAC^AGC	0.18	24.4
59	5'AGC^TAA	0.18	8.2
119	5'CGA^GGC	0.18	4.0
95	5'TGC^GCC	0.20	3.9
36	5'ACG^CAA	0.27	7.3
38	5'GCA^ACC	0.27	4.2
100	5'CCC^GCT	0.27	4.2
64	5'ACA^CTT	0.27	5.8

A_f/A_0 is the ratio of the product peak area at the end point of titration with daunomycin to the area of a control with no added daunomycin. A mean value of A_f/A_0 = 0.39±0.16 was found for all protected sites. K" is the preliminary estimate for the single site binding constant.
_____

By the footprinting titration procedure, we have now obtained single nucleotide resolution at 65 positions within the 165 bp try T DNA fragment. We observe no effect of added daunomycin at 21 positions, a fact that indicates clearly that daunomycin does indeed discriminate against certain potential binding sites. Four positions show enhanced cleavage in the presence of daunomycin. Forty positions show a

reduction in cleavage upon addition of daunomycin. The latter class
are thus protected from cleavage by daunomycin, and each protected site
was found to show a unique concentration dependency. The most strongly
protected sites, as judged by the magnitude of the reduced product peak
area at the end point of the daunomycin titration are listed in Table
1. Inspection of Table 1 reveals that these strongly protected sites
most commonly occur at the ends of a triplet sequence containing an AT
base pair flanked by contiguous GC base pairs. The GC base pairs are
in all cases in either an alternating purine/pyrimidine or
pyrimidine/purine arrangement. We identify these as the preferred
daunomycin binding sites. Such sites were predicted to be the most
energetically favored by the calculations of Chen, Gresh and Pullman
(9), and such sequences are the ones to which daunomycin is found bound
in the published x-ray diffraction studies (6,7). We interpret the
location of these sites as follows. DNase I covers approximately 10 bp
on one side of the DNA helix upon binding (22). The molecular
interactions that stabilize the enzyme-DNA complex occur in the minor
groove, and involve the 4 bp to the 5' side (with reference to the
strand that is cut) of the actual cleavage site. Ligand binding that
disrupts these minor groove interactions may register as "protected"
sites that are displaced in the 3' direction relative to the actual
physical location of the ligand. The ligand need not block the actual
cleavage site. All of the sites listed in Table I may be located
within the protected sequences identified in our previous lower
resolution footprinting study (17), but we now are able to pinpoint the
most strongly protected sites with greater precision. The results of
Table 1 suggest that the sequence 5'(A/T)CG or 5'(A/T)GC are preferred
daunomycin binding sites. Of the forty protected sites we have
identified, 20 are at the ends of triplets composed of contiguous GC bp
flanked at the 5' end by an AT bp. In 12 of these 20 cases, the GC
base pairs are in an alternating purine/pyrimidine or pyrimidine/purine
configuration. In the twenty-one sites at which we have single bond
resolution, but at which we observe no effect of added daunomycin, the
putative triplet recognition sequence occurs only 3 times. We note in
Table 1 two additional strongly protected sites, both occurring, at
first glance, at the end of the triplet 5'CTT. The one at position 64
is also at the end of the triplet 5'ACA, and the other, at position 67,
is only one bp away from the end of the same triplet, and also near the
strongly protected 5'AGC sequence at position 70. Protection at these
sites most probably arises from preferential binding to the triplet
5'ACA, which was third in the order of preference established by the
calculations of Chen, Gresh, and Pullman (8).

The footprinting titration procedure allows for the preliminary
estimation of single site, microscopic binding constants at the
locations observed to be protected. For an independent, noninteracting
site, the observed product peak area, A, at free drug concentration C
is given by the relation

$$A = \frac{(A_0 + A_\infty)KC}{1+KC} \qquad\qquad [1]$$

the sites detected in footprinting experiments.

New results from stopped-flow kinetic studies also support the existence of a class of preferred binding sites in natural DNA samples. Daunomycin binding to calf thymus DNA is kinetically complex, with at least three relaxation times required to fit the time course for either drug association or dissociation. The most reasonable explanation for this kinetic complexity is that it arises from daunomycin binding to different classes of sites. A detailed review of all published kinetic studies on the daunomycin-DNA interaction, and of the most likely kinetic mechanism governing the interaction may be found in [5]. Table 2 lists the dissociation constants recently determined for daunomycin initially bound to natural DNA samples of varying GC content and selected deoxypolynucleotides.

TABLE 2. First-Order Rate Constants for the Dissociation of Daunomycin from Natural DNA Samples and Selected Deoxypolynucleotides

DNA Sample	k_1	k_2	k_3
C. perfringens	1.4±0.1	4.5±0.5	17.7±3.1
Calf thymus	1.2±0.1	3.3±0.4	11.5±1.3
M. lysodeikticus	1.5±0.1	3.6±0.3	12.7±1.7
Poly [d(G-C)]	--	4.1±0.2	13.0±1.3
Poly [d(A-T)]	--	4.6±0.2	10.4±1.4
Poly [d(A-C)] [d(T-G)]	1.3±0.1	4.3±0.2	21.3±4.8

Data were obtained by the SDS dissociation method using a Dionex stopped-flow system equipped with an OLIS data acquisition system. Daunomycin dissociation was monitored by absorbance measurements at 480 nm. Kinetic data were analyzed as described in Chaires *et al* (1985) Biochemistry *24*, 260. Each rate is the average of at least five determinations. No concentration dependency was observed for k_1 or k_2 over the range 0 - 0.3 mol DM/mol bp.

Natural DNA sample all are characterized by three observable dissociation rates. These rates, as inspection of Table 2 shows, are essentially equal in the three samples examined. All natural DNA samples have one slow dissociation rate near 1 s^{-1}. In contrast, the synthetic deoxypolynucleotides poly [d(G-C)] and poly [d(A-T)] are characterized by only two dissociation rates, with the ≈ 1 s^{-1} rate observed in natural DNA samples absent. However, the full kinetic complexity is again seen in the synthetic deoxypolynucleotide poly [d(A-C)]-poly [d(G-T)], and the slow rate near 1 s^{-1} is again present. The latter polynucleotide contains a triplet 5'ACA sequence that is among those we rank a preferred daunomycin binding sites. The slow rate observed in this polynucleotide and that observed in natural DNA samples may thus most reasonably be assigned to daunomycin dissociation from its most preferred sites. Such sites are, then, kinetically as well as energetically distinct. Sadly, results from polynucleotides containing the 5'ACG motif are as yet unavailable. Such a sequence should exhibit complex and slow dissociation kinetics.

6. SUMMARY AND CONCLUSIONS.

The footprinting titration results described here show that daunomycin binds preferentially to certain sites along the DNA lattice. These sites are triplet sequences with the general sequence 5'AGX, where A and G represent AT and GC base pairs, respectively, and X indicates that any base pair is allowed. The most preferred site within this framework, from our experimental data, contains contiguous GC base pairs in an alternating purine/pyrimidine or pyrimidine/purine configuration. These results are fully compatible with the theoretical studies of Chen, Gresh and Pullman (8-10), and with the structures observed in recent x-ray crystallographic studies (6,7). Tentative estimates of the microscopic binding constants to these preferred sites indicate that their magnitude is at least one order of magnitude greater than the average macroscopic binding constant for daunomycin binding to natural DNA as determined by spectrophotometric titration. New spectrofluorometric titration data, however, provides evidence for a heretofore unobserved class of high affinity binding sites for anthracycline antibiotics in calf thymus DNA. Stopped-flow kinetic studies further support the existence of such sites. These sites presumably correspond to the preferred triplets identified by footprinting titration experiments.

7. POSSIBLE BIOCHEMICAL IMPLICATIONS.

The magnitude of the estimates for the microscopic binding constant for the interaction of daunomycin with its preferred binding sites ($\geq 10^7$M^{-1}) is approaching that of typical binding constants observed for specific protein-DNA interactions. Daunomycin may therefore be an effective competitor for the binding of particular proteins to specific DNA sites during gene expression. Indeed, our laboratory has shown the daunomycin can inhibit the highly specific interaction of restriction endonuclease PvuI with its recognition site(17). The recent results of Bartkowiak *et al* (24) are of particular interest in this light. Their experiments showed that adriamycin selectively displaced a unique set of nuclear proteins from their DNA binding sites within the nucleus.

Ethidium, in contrast, had no such effect. It is possible that the anthracycline antibiotics act by inhibiting protein-DNA interactions crucial for gene expression.

Another possible target for the anthracycline antibiotics is topoisomerase II. Results from two separate laboratories have now established that DNA binding and intercalation are necessary for the inhibition of topoisomerase II by anthracycline antibiotics (25,26). A careful analysis of the anthracycline-induced topoisomerase II cleavage sites in SV40 DNA show that the four most prominent sites (observed at 0.125 μM antibiotic concentration) are all adjacent to triplet sequences with either the 5'ACA or 5'ACG motif. Such triplets are the very ones identified here as preferred daunomycin binding sites. An additional previous study (27) has identified twelve topoisomerase II cleavage sites in the human c-fos proto-oncogene. All of those sites are immediately adjacent to triplets sequences to which daunomycin would preferentially bind. Daunomycin and topoisomerase II thus appear to bind to identical sequences, providing a suggestive link between preferential antibiotic sequence binding and the inhibition of the enzyme.

ACKNOWLEDGEMENTS. The work described here was supported by U. S. Public Health Service Grant CA35635. Travel support from the National Science Foundation (INT-8521004 & INT-8822374) and from the Alexander von Humboldt Foundation is gratefully acknowledged. Julio Herrera and Michael J. Waring stimulated much of the work described here.

REFERENCES

1. Valentini, L., Nicolella, V., Vannini, E., Menozzi, M., Penco, S, & Arcamone, F. (1985) *Il Farmaco 40*, 377 - 397.
2. Gigli, M., Doglia, S. M., Millot, J. M., Valentini, L. & Manfait, M.(1988) *Biocim. Biophys. Acta 950*, 13 - 20.
3. D'Arpa, P. & Liu, L. (1989) *Biocim. Biophys. Acta 989*, 163 - 177.
4. Fritzsche, H. & Berg, H. (1987) *Gazetta Chimica Italiana 117*, 331 - 352.
5. Chaires, J. B. (1990) *Biophys. Chem. in press*
6. Wang, A. H.-J., Ughetto, G., Quigley, G. J., & Rich, A. (1987) *Biochemistry 26*, 1152 - 1163.
7. Moore, M. H., Junter, W. N., Langlois d'Estaintot, B. & Kennard, O. (1989) *J. Mol. Biol. 206*, 693 - 705.
8. Pullman, B. (1989) *Advances in Drug Design 18*, 1 - 113.
9. Chen, K.-X., Gresh, N. & Pullman, B. (1985) *J. Biomol. Struc. & Dynam.3*, 445 - 466.
10. Chen, K.-X., Gresh, N. & Pullman, B. (1986) *Nuc. Acids Res. 14*, 2251 - 2267.
11. Graves, D. E. & Krugh, T. R. (1983) *Biochemistry 22*, 3941-3947.
12. Chaires, J. B., Dattagupta, N. & Crothers, D. M. (1982) *Biochemistry 21*, 3933 - 3940.
13. Chaires, J. B. (1985) *Biopolymers 24*, 403 - 419.

14. Record, M. T. Jr., Anderson, C. F. & Lohman, T. M. (1978) *Q. Rev. Biophys. 11*, 103 - 178.
15. Chaires, J. B. (1990) *in Advances in DNA Sequnce Specific Agents, Vol. 1* (L. H. Hurley, ed.), Jai Press, *in press*
16. Jones, M. B., Hollstein, U., & Allen, F. S. (1987) *Biopolymers 26*, 121 - 136.
17. Chaires, J. B., Fox, K. R., Herrera, J. E. Britt, M. & Waring, M. J. (1987) *Biochemistry 26*, 82227 - 8236.
18. Skorobogaty, A., White, R. J., Phillips, D. R., & Reiss, J. A. (1988) *FEBS Lett. 227*, 103 - 106.
19. Skorobogaty, A., White, R. J., Phillips, D. R. & Reiss, J. A. (1988) *Drug Design Deliv. 3*, 125 - 152.
20. Dabrowiak, J. C. & Goodisman, J. (1989) *in* Chemistry & Physics of DNA-Ligand Interactions (N. R. Kallenbach, ed.) Adenine Press, Schenectady, N. Y.
21. Chaires, J. B., Herrera, J. E. & Waring, M. J. (1990) *Biochemistry 29*, xxx - yyy.
22. Suck, D. & Oefner, C. (1986) *Nature 321*, 620 - 625.
23. Weber, G. (1965) *in Molecular Biophysics* (B. Pullman & M. Weissbluth, *eds.*) Academic Press, New York, pp.369 - 396.
24. Bartkowiak, J., Kapuscinski, J., Melamed, M. R. & Darzynkiewicz, Z. (1989) *Proc. Nat. Acad. Sci. USA 86*, 5151 - 5154.
25. Bodley, A. L., Liu, L. F.,Israel, M., Giuliani, F. C., Silber, R., Kirschenbaum, S. & Potmesil, M. (1989) *Cancer Res. 49*, 5969-5978.
26. Capranico, G., Zunino, F., Kohn, K. W. & Pommier, Y. (1990) *Biochemistry 29*, 562 - 569.
27. Darby, M. K., Herrera, R. E., Vosberg, H.-P. & Nordheim, A. (1986) *EMBO J. 5*, 2257 - 2265.

IN VITRO TRANSCRIPTION ANALYSIS OF THE SEQUENCE SPECIFICITY OF REVERSIBLE AND IRREVERSIBLE COMPLEXES OF ADRIAMYCIN WITH DNA

D.R. PHILLIPS, C. CULLINANE, H. TRIST and R.J. WHITE
Department of Biochemistry
La Trobe University
Bundoora   Victoria   3083
Australia

ABSTRACT. An *in vitro* transcription assay has been used to show that the preferred reversible (intercalative) binding sites for daunomycin and Adriamycin is at CpA sequences. This sequence alone is not sufficient as a determinant of preferential binding since there was little or no binding at some CpA sites. The role of flanking sequences was examined by a quantitative analysis of 12 highest affinity binding sites in a total of 260 bp, and revealed that 9 were at CpA sequences, with the preferred sequence being 5'-TCA. Exposure of an initiated transcription complex to Adriamycin under *in vitro* transcription conditions for several hours resulted in drug-induced transcriptional blockages at G of GpC sequences of the non-template strand. These blockages were stable over 3 h of elongation, and exhibited the properties of an irreversible adduct at that site. Bidirectional transcription footprinting revealed that the region occupied by the apparent adduct was a maximum of 2 bp comprising the GpC sequence. The implication of these reversible and irreversible complexes with DNA have been discussed with respect to future development of anthracycline derivatives based on these different interactions with DNA.

1. INTRODUCTION

Adriamycin and daunomycin (Fig. 1) have been in routine clinical use as an anticancer agent for two decades (Crooke and Reich, 1980; Schwartz, 1983; Lown, 1988). Full realisation of their antineoplastic potential is limited by an associated cardiotoxicity (Gianni *et al.*, 1982) and there have been extensive efforts to develop more active and/or less cardiotoxic derivatives to combat this problem (Acton *et al.*, 1984; Myers *et al.*, 1987; Grandi *et al.*, 1988) as evidenced by the fact that some 600 derivatives have now been isolated or synthesized (Crooke and Reich, 1980; Brown, 1983; Schwartz, 1983; Arcamone and Penco, 1988), and over 500 have been submitted to the NCI for screening (Myers *et al.*, 1988; Sinha, 1989). In order to place the development of new derivatives of these anthracyclines on a more rational basis, there have

137

B. Pullman and J. Jortner (eds.), Molecular Basis of Specificity in Nucleic Acid-Drug Interactions, 137–155.
© 1990 *Kluwer Academic Publishers. Printed in the Netherlands.*

been widespread and concerted efforts to delineate the mode of action of
these drugs.

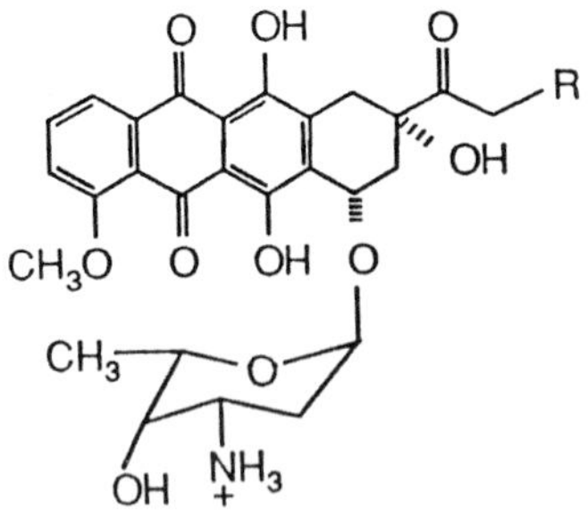

R = H   daunomycin

R = OH   Adriamycin

Figure 1.   Structure of daunomycin and Adriamycin.

Although the mode of action of Adriamycin is still debated, this
complex topic has been the subject of a recent comprehensive review
(Myers et al., 1988) which concludes that there are three major
mechanisms: the effects of membrane binding; impairment of enzymes such
as topoisomerase II; the effect of products of reductive activation.
The latter two processes may involve DNA, and there is now a large body
of evidence which suggests that anthracyclines act dominantly at the DNA
level (Schwartz, 1983; Valentini et al., 1985).
    Two extreme forms have been identified for the binding of Adriamycin
to DNA - a reversible, intercalation mode (Gianni et al., 1982; Neidle
and Sanderson, 1983; Myers et al., 1988; Ughetto, 1988) as well as an
irreversible interaction involving the formation of covalent adducts
(Sinha and Chignell, 1979; Sinha, 1980; Sinha and Sik, 1980; Sinha and
Gregory, 1981; Sinha et al., 1984; Myers et al., 1988).   If the
mechanism of action of the anthracyclines is dependent upon DNA binding
in some manner, then the logical design of potentially more active
derivatives will be greatly assisted by a knowledge of the major DNA
receptor site(s), and the mode of interaction at these sites.   For this
reason, there have been many attempts to elucidate the DNA sequence
specificity of the anthracyclines.

1.1.   Reversible Binding

Physico-chemical studies of the interaction of anthracyclines have led
to conflicting results of DNA sequence selectivity.   Early binding
studies with synthetic polydeoxynucleotides showed a preference of
Adriamycin for poly (dG-dC) (Tsou and Yip, 1976) and this has been
supported by a preference for GpC sites indicated by fluorescence
quenching (Du Vernay et al., 1979), Raman (Manfait et al., 1982) and
circular dichroism studies (Jones et al., 1987).   In contrast, the
structurally similar anthracycline daunomycin, exhibited some

selectivity for poly (dA-dT) compared to other synthetic polydeoxynucleotides (Phillips et al., 1978; Graves and Krugh, 1983), while competition dialysis experiments indicated a slight preference for G·C sites (Chaires et al., 1982). Stopped-flow SDS-induced sequestration studies of the dissociation of daunomycin from synthetic polydeoxynucleotides revealed little preference between ApT, GpG, CpG or ApA sequences (Grant and Phillips, 1979).

More definitive data on the sequence specificity of daunomycin and Adriamycin has emerged from DNase I footprinting studies of Chaires et al. (1987), with 7 distinct protected regions being observed in a 160 bp DNA fragment. Each of these footprints contained one or more triplets containing adjacent GC base pairs flanked by an AT base pair, suggesting that those sequences were preferred binding sites for daunomycin.

Theoretical calculations of daunomycin with tetrameric duplexes have suggested that the favoured intercalation site is at TpA or CpA sequences (Newlin et al., 1984), while ab initio self consistent field computations show that base-pair triplets are required to define the binding site for anthracyclines (Chen et al., 1985, 1986; Pullman, 1988). Similar interaction energies have been identified for Adriamycin at ACG, ATA and TCA sequences (Chen et al., 1986) where the intercalation site is between the 2 bp on the 3' end.

1.2. Irreversible Binding

Reduction of Adriamycin and other anthracyclines has been achieved by both chemical activation (Sinha and Chignell, 1980; Sinha and Gregory, 1981) and by a variety of enzymatic methods (Pan and Bachur, 1980; Sinha et al., 1984; Fisher et al., 1983, 1985). The one electron reduction facilitated by these processes results in the formation of a semiquinone intermediate which may be involved further in a number of reactions (Kappus, 1986). Under anaerobic conditions the semiquinone undergoes further reduction, accompanied by cleavage of the sugar residue to form the quinone methide (Moore, 1977). This species binds covalently to both electrophiles (Kleyer and Koch, 1983) and nucleophiles (Ramakrishnan and Fisher, 1983), and is known to form adducts with DNA (Sinha and Chignell, 1979; Sinha, 1980; Sinha and Gregory, 1981).

Reductive activation of Adriamycin with sodium borohydride results in the formation of base selective adducts with polydeoxynucleotides - maximal adducts were observed with poly dG (one adduct per 3 nucleotides) with lesser amounts for poly dC, poly dA and poly dT (one adduct per 17, 50 and 62 nucleotides respectively)·(Sinha, 1980). With calf thymus DNA, adduct ratios of one per 12-15 nucleotides were detected with Adriamycin, but only one per 135 nucleotides with daunomycin. Microsomal activation of Adriamycin and daunomycin has also been shown to produce reactive species which result in the formation of adducts with DNA, with maximal yields of one adduct per 125 or 280 nucleotides for Adriamycin and daunomycin respectively (Sinha and Gregory, 1981). On the basis of these data, and other ESR studies, it appears that the adducts produced by enzymatic activation of the anthracyclines are similar to those derived from chemical activation. Evidence for the role of such adducts *in vivo* comes from the detection of [^{14}C]-labelled Adriamycin as covalent adducts, predominantly to DNA, after intraperitoneal injection of the drug into rats (Sinha and Sik, 1980), and from the presence of Adriamycin and daunomycin-induced crosslinks in HeLa cells (Konopa, 1983).

1.3. *In vitro* Transcription

An *in vitro* transcription assay has recently been developed to study the sequence selectivity, relative affinity and dissociation kinetics of drug-DNA interactions (Phillips and Crothers, 1986; Crothers *et al.*, 1987; White and Phillips, 1988, 1989a; Phillips *et al.*, 1990a,b). This technique relies upon progression of RNA polymerase along the DNA template, until halted by an occupied drug site - the length of RNA transcript therefore defines the location of the drug site, and the amount of blocked transcript is a measure of relative drug affinity for that site.   We have utilised this technique to define the highest occupancy intercalation sites of Adriamycin and daunomycin, as well as the role of flanking sequences.   We also present the sequence specificity of more permanent Adriamycin-induced blockage sites, formed after extensive exposure of Adriamycin to DNA under *in vitro* transcription conditions.

2.   MATERIALS AND METHODS

2.1. Materials

All materials were as described by Trist and Phillips (1989) (reversible complexes) and Cullinane and Phillips (1990) (irreversible complexes).

2.2. Unidirectional Transcription Assay

The transcription assay relies upon the formation of a stable initiation complex with the RNA of all transcripts synchronized to a known length. When using the UV5 promoter this has been achieved by forcing transcription to begin selectively from one start site with high levels of GpA (complementary to the -1 and +1 positions) and allowing transcription to proceed up to the nucleotide (CTP) absent from the transcription mixture (Phillips and Crothers, 1986; White and Phillips, 1988).   This yields RNA comprising mainly 10-mers from the UV5 promoter as well as small amounts of 17 and 23-mers (Figure 2).   The 10-mer has proved to be sufficiently stable for all drug-DNA studies to date, but an 11-mer would be even more stable (Straney and Crothers, 1985), and can be formed using dGGA in the initiation mixture (Trist and Phillips, 1989).   The conditions employed for initiation of the λP_L and Tet^R promoters have been described by Trist and Phillips (1989).

The binary DNA/enzyme complex was formed by incubating one of the restriction fragments (50 nM) containing the desired promoter, with *E.coli* RNA polymerase (100 nM for 15 min at 37°C in transcription buffer (40 mM Tris.HCl (pH 8.0), 100 mM KCl, 3 mM MgCl$_2$, 0.1 mM EDTA, 10 mM

DDT, 125 µg/ml BSA and 1.6 u/µl RNase inhibitor). Heparin was then added (final concentration of 400 µg/ml) and the mixture incubated for 5 min at the appropriate temperature. The stable initiated complex was formed upon addition of the required dinucleotide, together with 3 nucleotides, one of which was labelled with ^{32}P. This mixture was incubated at 37°C for 10 min.

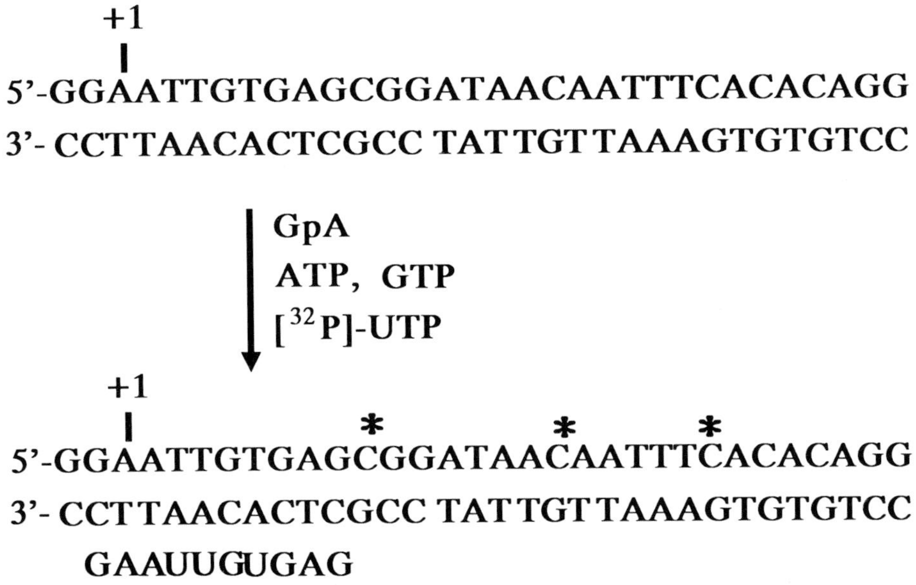

+1

5'-GGAATTGTGAGCGGATAACAATTTCACACAGG
3'- CCTTAACACTCGCC TATTGTTAAAGTGTGTCC

GpA
ATP, GTP
[^{32}P]-UTP

+1

5'-GGAATTGTGAGCGGATAACAATTTCACACAGG
3'- CCTTAACACTCGCC TATTGTTAAAGTGTGTCC
    GAAUUGUGAG

Figure 2. Synchronization of initiation. A stable initiation complex comprising a RNA 10-mer results from initiation of the *lac* UV5 promoter in the absence of CTP. A lesser amount of 17-mer and 23-mer are also produced. Radiolabel is incorporated into 3 residues of the 10-mer.

The initiated transcription complex was then equilibrated at the appropriate temperature with sub-saturating amounts of drug for defined periods of time. To minimise the extent of natural pausing, elongation was carried out with all 4 nucleotides at a final concentration of 2.5 mM, and at high salt levels (0.4 M) (White and Phillips, 1988). A summary of the procedure is shown in Figure 3.

2.3. Bidirectional Transcription Footprinting

A 315 bp DNA fragment containing both the UV5 and N25 promoters (counterdirected) was used for bidirectional transcription footprinting (White and Phillips, 1989b). In this process, one promoter was selectively initiated, to yield transcripts from one direction at a time. A combination of the two sets of transcripts reveal a series of non-overlapping regions which are the transcription footprints, and has been described by White and Phillips (1989b) and Phillips *et al.* (1990).

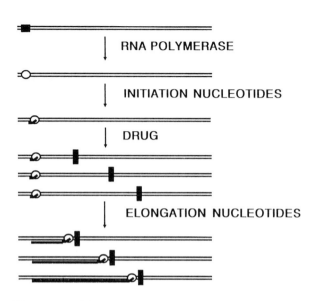

Figure 3. Diagrammatic representation of major steps in the transcription assay.

2.4   Gel Electrophoresis

Electrophoresis, autoradiography and densitometry were all carried out as described previously (White and Phillips, 1988; Trist and Phillips, 1989).

3.   RESULTS

3.1.   Reversible Adriamycin-DNA Complexes

3.1.1.   *5'-CpA Sequence Specificity*.   The initiated 497 bp UV5 complex was equilibrated at 10°C with daunomycin or Adriamycin for 15 min prior to elongation with all 4 nucleotides.   A time course for this elongation process is shown in Figure 4 (Skorobogaty *et al.*, 1988).   Transcription proceeds up to drug occupied sites, where it is halted for a period determined by the residence time of the drug at that site.

Quantitation of the 2 min lanes for daunomycin and Adriamycin yields the mole fraction of each blocked transcript, and this measure of the relative occupancy of drug at each site (Phillips *et al.*, 1990) is summarised in Figure 5.   Sites where drug-induced pausing coincides with natural pausing have been ignored, except where clear drug induced enhancement was evident compared to the control lane.   The drug binding site is taken to be immediately downstream of the blocked transcript, based on other studies with well documented intercalators which have

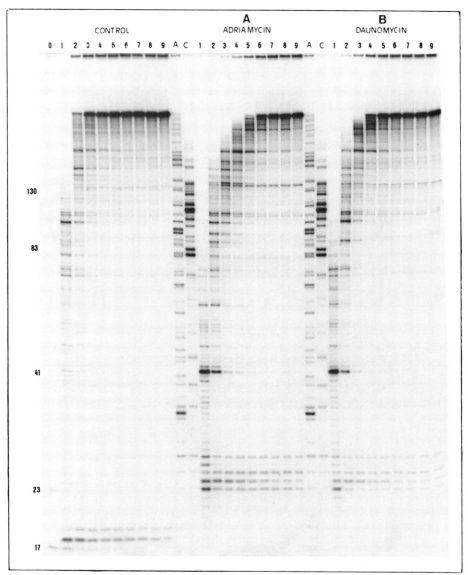

Figure 4. Sequencing gel autoradiogram of transcription blockage by daunomycin and Adriamycin (30 µM) at 10°C, 0.1 ionic strength, using a 497 bp *lac* UV containing DNA-fragment. Lane 0 represents the initiated transcription complex prior to the addition of nucleotides or drugs; A and C are sequencing lanes. Lanes 1 to 9 represent a time course (samples taken at 0.5, 2, 5, 7, 10, 15, 20, 30 and 60 min respectively) of elongation of the initiated complex which had been preincubated with either daunomycin, Adriamycin or merely additional buffer. From Skorobogaty *et al.* (1989).

shown that *E.coli* polymerase transcribes up to the base pairs forming part of an intercalation complex (Phillips and Crothers, 1986; White and Phillips, 1988b; Phillips *et al.*, 1990b).

The majority of all blocked transcripts induced by daunomycin and Adriamycin are immediately prior to a 5'-CA sequence (Figure 5). These results therefore suggest that the preferred binding site for these anthracyclines is at CpA sequences, and this has been further supported by transcriptional blockages induced by a bis-intercalating bis-daunomycin (Skorobogaty *et al.*, 1988). Other high occupancy sites are also apparent for the anthracyclines in Figure 5, suggesting that other sequences are only marginally less preferred than CpA.

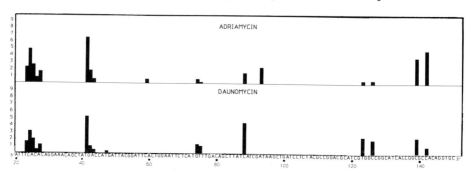

Figure 5.   Transcription detected drug block sites. The data were obtained from densitometric analysis of the 2 min elongations for daunomycin and Adriamycin at 10°C. The transcribed sequence is shown from the nucleotides +20 to +150 downstream of the GpA-initiated UV5 promoter. The ordinate represents the percentage blockage normalized against the total radioactivity incorporated into transcripts beyond 10-mers. The effects of natural pausing were subtracted from the apparent drug-induced blockage. Drug induced block sites were ignored if they did not exceed natural pausing by 15%. From Skorobogaty *et al.* (1989).

3.1.2.   *Role of Flanking Sequences*. Since some CpA sequences in Figure 4 were not associated with blocked transcripts, it appears that the apparent CpA intercalation site alone is not a sufficient determinant for preferential binding of these anthracyclines. It was therefore reasoned that in order to establish which flanking sequences were important as additional contributions to the sequence selectivity of Adriamycin, it was necessary to accumulate a data base of those sequences where there was high drug occupancy (as exhibited by a high mole fraction of blocked transcript), together with a data base of these sequences where there was little or no drug occupancy. This data was acquired using sequences probed by *E.coli* RNA polymerase after exposure to Adriamycin under identical conditions. Elongation of the initiated, Adriamycin treated complexes was derived from three promoter-containing DNA fragments: 1928 bp containing UV5, 1132 bp containing λP_L and 377 bp containing Tet^R (Trist and Phillips, 1989).

The relative drug occupancy at each site, as described above, does not allow for the statistical probability that even at low drug loadings, the RNA polymerase may not reach downstream drug sites on the DNA if the polymerase is blocked at earlier drug sites. This factor results in a progressive under-estimation of drug occupancy at downstream sites but was taken into account with a statistical factor, together with a factor to allow for read-through past each occupied site, as described previously (Trist and Phillips, 1989).

Twelve high occupancy sites were identified from 260 bp probed by this procedure, 9 of which were at 5'-CA sequences (Table 1), the other 3 being at ACGA, CGTA and ACCC (Trist and Phillips, 1989), where the position of the blocked transcript is denoted with a bar. Again, the preferred sequence is seen to be at CpA sites, with a reduced preference for other sequences.

TABLE 1.   Nucleotide sequence flanking all high occupancy CA binding sites, and the consensus 5' and 3' flanking sequences, where a modest probability of a defined bp is shown as a.t, and a high probability as G.C. From Trist and Phillips (1989).

Promoter	Flanking Sequences	
	5'	3'
UV5	ATTT	CACA
	TGGT	TAGC
	GATT	CTGG
λP_L	TCTT	GGGC
	CATT	AAGC
	CACA	CCCC
Tet^R	ATTT	TACA
	CGCT	TCGT
	CCGG	TAAC
consensus sequence	--tT CA aaG	
		...
		ttC

The 9 high occupancy sites in Table 1 exhibit a distinct preference (7 out of 9) for T at the 5' position to the CpA site, with a marginal suggestion of some preference for other flanking sequences. Overall, the unmistakable conclusion is that 5'-TCA is the preferred site for Adriamycin, (where the intercalation site is between the C and A base pairs), with additional flanking of A·T base pairs to both the 5' and 3' ends being somewhat preferred.

A converse analysis of all low occupancy CpA sites revealed 11 such locations. Three of these contained CpA palindromes, and since low occupancy at these sites may have been due to negative cooperativity, these regions were precluded from further analysis - the remaining eight sites revealed a consensus sequence for unfavourable binding of Adriamycin of 5'(G·C)CA, with a strong indication of additional G·C base pairs flanking both the 5'- and 3'- ends of this sequence (Table 2).

Table 2.   Nucleotide sequences flanking all low occupancy CA sites, and the concensus 5' and 3' flanking sequences. Adjacent G.C base pairs have been underlined, high probability bp shown as G.C and modest probability bp as g.c. From Trist and Phillips (1989).

Promoter	Flanking Sequence
λP$_L$	G<u>CG</u>T CA <u>CC</u>TT
	AAA<u>G CA</u> GAA<u>GG</u>C
Tet^R	CA<u>GG CA</u> <u>CC</u>GT
	A<u>GCG CA</u> TTGA
	CAT<u>C CA</u> <u>GGG</u>T
	ACA<u>G CA</u> TCCA<u>GGG</u>
	<u>GCC</u>TA CA <u>GC</u>AT
	TA<u>AC CA</u> A<u>GCC</u>
consensus sequence	G CA gG
	· ··
	C cC

3.2. Apparent Irreversible Adriamycin-DNA Complexes

3.2.1. *5'-GpC sequence specificity.*   Incubation of Adriamycin for several hours with the initiated 497 bp UV5 DNA fragment in transcription buffer resulted in transcriptional blockages which did not allow read-through of the RNA polymerase, even after 15 min of elongation time (Phillips *et al.*, 1989).   These drug-induced irreversible blockages were enhanced dramatically by the presence of Fe(II) or Fe(III) ions, but not by other physiological divalent metal ions (Phillips *et al.*, 1989).   Optimal conditions have now been established for the formation of these irreversible blockages (10 µM Adriamycin, 75 µM Fe(III), 24 h reaction time), and resulted in a 12-15 fold enhancement of the transcriptional blockages, compared to the absence of added Fe(III) ions (Figure 6) (Cullinane and Phillips, 1990).

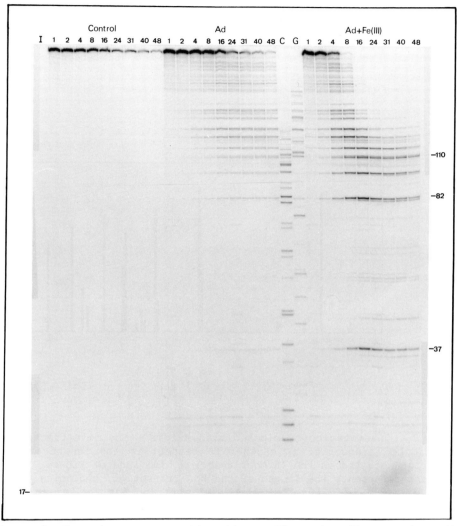

Figure 6. Dependence on reaction time. The autoradiogram shows the dependence of transcriptional blockages on the reaction time (37°C) of Adriamycin (10 μM) with the 497 bp *lac* UV5 containing DNA fragment in the absence (central 9 lanes entitled "Ad") and presence of 75 μM Fe(III) (right hand 9 lanes entitled "Ad + Fe(III)"), for reaction times of 1-48 hours. Control reactions in the absence of both Adriamycin and Fe(III) ions were also left for the corresponding times of 1-48 hours. Lane "I" is the initiated complex prior to elongation. All other samples were subjected to elongation conditions for 5 min. Lanes C and G are sequencing lanes using 3'-0-methoxy CTP and 3'-0-methoxy GTP respectively. From Cullinane and Phillips (1990).

The background blockages in the absence of added Fe(III) is assumed to
be due to the effect of trace amounts of Fe(III) in the transcription
buffer.   Nine high occupancy sites were identified, all of which
corresponded to transcription up to G of 5'-GpC sequences of the
non-template strand (Figure 7), most of which were stable for up to 3 h
of elongation.

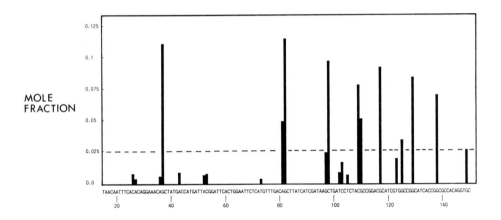

Figure 7.   Quantitation of blocked transcripts.   The mole-fraction of
each blocked transcript is shown with respect to the length of each
transcript detected in the 16 hour reaction time lane of Figure 1.   The
numbering represents the length of the RNA transcript, beginning from G
of GpA in each chain (where A is the +1 position for the start of the
transcriptional message from the UV5 promoter).   The dashed line is an
arbitrary level to define high occupancy sites as above a mole fraction
of 0.025.   From Cullinane and Phillips (1990).

3.2.2.   *Bidirectional transcription footprinting*.   The UV5/N25
bidirectional transcription footprinting procedure of White and Phillips
(1989b) was applied to the irreversible blockages induced by Adriamycin,
and 4 blockages were detected from each promoter (Figure 8), all of
which were at G of GpC sequences of the non-template strand.   Three of
the sites were detected by RNA polymerase from both promoters, and
revealed a physical site of blockage being restricted to the 2 bp GpC
element (Figure 8).

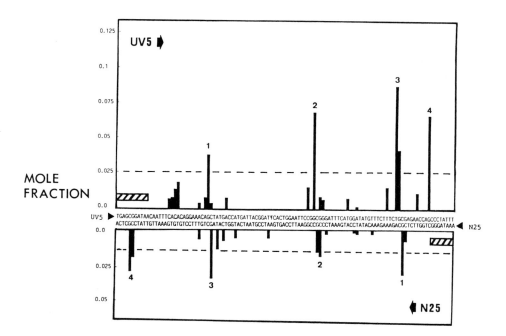

Figure 8. Relative intensity of bidirectional blockages. The mole-fraction of blocked transcripts were quantitated by densitometry for reaction of the initiated complexes for 24 hours with 10 μM Adriamycin and 75 μM Fe(III). The mole-fraction of each blocked transcript is shown with respect to the length of the transcript, as shown by the non-template DNA strand for transcription proceeding from each promoter. The dashed lines at arbitrary levels define high occupancy sites as above a mole fraction of 0.025 (UV5 promoter) or 0.0125 (N25 promoter). The hatched areas indicate the region occupied by each RNA polymerase, as defined by the length of RNA in the initiated transcription complex. From White and Phillips (1988).

4.   DISCUSSION

4.1. Reversible Complexes

The *in vitro* transcription data has revealed a preference for the intercalation of daunomycin and Adriamycin between the 2 bp comprising CpA sequences, with the highest affinity site for Adriamycin being at 5'-TCA sequences. This conclusion conflicts with that of Chaires *et al.* (1987) where daunomycin footprints were most readily explained in terms of a three bp site comprising two adjacent G·C base pairs, flanked by an A·T bp. However, it should be noted that the broad footprints resulting from such DNase I studies can also be largely explained by a preference of daunomycin for CpA sequences, since there are 11 such sites associated with the 7 regions of inhibition of DNase I activity (Figure 2 of Chaires et al., 1987). Furthermore, within the 7 broad footprints, there are 14 discrete, smaller footprints, 11 of which are associated with CpA sequences, and of these 11 high affinity sites, 5 are at 5'-TCA sequences.

When comparing the 2 experimental approaches employed to elucidate the sequence specificity of daunomycin and Adriamycin (i.e. in vitro transcription and DNase I footprinting) it is important to note several points:

(i)   these two approaches are the most definitive techniques currently available to yield the sequence specificity of DNA-binding drugs at a variety of individual sites

(ii)   both sets of data can be interpreted as arising from a specificity for CpA sequences, with an indication of the most preferred sequence being 5'-TCA

(iii)   there is only a marginal thermodynamic preference for 5'-TCA sequences, as both sets of data reveal high affinity binding at other, less frequently observed sequences

(iv)   it is easier to interpret the in vitro transcription data because the RNA polymerase reads right up to a drug occupied site, and reveals a narrow transcriptional blockage, compared to the much broader DNase I footprints.

The preference of the anthracyclines for 5'-TCA sequences has been predicted by the theoretical studies of Chen et al. (1986), where 3 major features were identified - a requirement for a 5'T·A base pair adjacent to the intercalation site (G at this position gives rise to large repulsions between the G-NH$_2$ and the daunosamine ammonium group), a preference for C at 5'side of the intercalation site (giving rise to favourable interactions with the 9-OH of daunomycin), and the need for 3bp to define the selectivity of the interaction. The consensus sequence of 5'-TCA for high affinity binding by Adriamycin validates all three of these theoretical predictions - a requirement for T, 5' to the intercalation site has been identified, a requirement for C as a 5' element of the intercalation site has been confirmed, and 3 bp have emerged as the dominant element of the consensus sequence.

The low occupancy CA sites have a consensus sequence (Table 2) which is fully consistent with that observed for high occupancy CA sites. In addition, these flanking sequences have a high distribution of G·C base pairs on both sides of the CpA site - since this feature is known to be associated with rigidity of the DNA (Hogan and Austin, 1987), it appears likely that the lack of flexibility adjacent to the intercalation site precludes the formation of good stabilising contacts possible in more flexible sequences.

4.2. Irreversible Complexes

Most of the transcriptional blockages induced by Adriamycin were exceedingly long-lived, with no evidence of read-through past these sites, even after 3 h of elongation time. Such terminated transcripts could arise from either a double-strand break in the DNA at that point, or from a halt to the elongation process by the presence of a long-lived complex on the DNA at that site. There is no evidence of DNA degradation (Cullinane and Phillips, 1989), and it must therefore be assumed that some species derived from Adriamycin is associated with the DNA, and exhibits extreme sequence selectivity. Given the stability of these blockages, it must be concluded that they behave as essentially irreversible, covalent complexes comparable to those detected with mitomycin C under similar transcription conditions (Phillips et al., 1989).

The most likely model for the formation of Adriamycin-dependent irreversible transcriptional blockages involves the reduction of Adriamycin to a quinone methide (Sinha, 1980). Since the amount of blockage is dependent on the level of Fe(III) ions, and Adriamycin exhibits a high affinity for Fe(III) ions, it strongly suggests that an Adriamycin-Fe(III) complex is involved. Reduction of the drug therefore probably occurs by DTT, mediated by Fe(III). The N-7 of guanine is the most nucleophilic and most accessible centre in DNA, and would therefore be the most likely site for alkylation by a quinone methide (Pullman and Pullman, 1981) - any such adducts would therefore result in transcription blockages preferentially at G-residues, as observed. The reason for the observed specificity of transcriptional blockages at GpC residues, rather than at all isolated guanine residues, is not yet clear. Additional intriguing questions are whether these apparent adducts are the same as those formed in vitro by $NaBH_4$ or enzymatic reduction of Adriamycin (Sinha, 1980; Sinha and Gregory, 1981) and how they relate to the cross-links formed by metabolic activation of Adriamycin in vivo (Konopa, 1980).

4.3. Biological Implications

Although there is a good body of evidence to show that the anthracyclines act in some way at the DNA level (Schwartz et al., 1983; Valentini et al., 1985; Myers et al., 1988), there is no consensus as to the exact or major mode of action. Indeed, the reasons why it has been so difficult to elucidate the dominant role of the anthracyclines have recently been reviewed (Myers et al., 1988). One of the most fundamental of these reasons is the very low free concentrations of drug which result in effective tumour cell kill - the plasma levels for prolonged infusion of Adriamycin in man are only 10^{-8} to 10^{-7} M (Legha et al., 1982), conditions difficult to simulate in vitro to enable definitive, comparable studies to be performed. If the process of intercalation of anthracycline into DNA is a critical factor, then at these low free drug levels, only the highest affinity DNA binding sites would be occupied, given the anthracycline-DNA intrinsic association constant of approximately 10^6 M^{-1} (Neidle and Sanderson, 1983). It is therefore of considerable interest to establish the single, highest affinity binding site, since this permits further analysis of possible effects of anthracyclines at those sites (e.g. are 5'-TCA sites predominant in promoter or enhancer regions?). In addition, future drug design can now be based on 5'-TCA as the dominant binding site, and this will be of particular use to those attempting to further refine three-dimensional structural features of the anthracycline-DNA complex using X-ray diffraction (Wang et al., 1987; Ughetto, 1988) or 2D NMR studies involving synthetic oligonucleotides.

A more likely role of anthracyclines at the DNA level arises from their apparent ability to form adducts with DNA, thereby completely terminating the process of transcription and DNA replication at those sites. If this process is based on a bimolecular reaction between DNA and an activated form of the anthracycline, such a process will continue, even at the low free drug levels noted above for clinical conditions in man (Legha et al., 1982). The specificity for GpC sequences suggests a possible crosslink, based on the formation of an adduct at each G residue - although no such crosslinks have yet been detected under these transcription conditions, a pseudo-crosslink has been suggested to account for these results (Cullinane and Phillips, 1990). This concept is attractive because one objection to the possible role of alkylation as a mechanism of action of the anthracyclines has been that such adducts deriving from the quinone methide would be monofunctional, in contrast to the bifunctional crosslinking observed for most alkylating agents which are effective as anticancer drugs (Myers et al., 1988). If the irreversible apparent adducts prove to be a decisive mode of action, then it will be critical to establish the mechanism of formation and chemical nature of the adduct, with a view to exploiting these features in the search for more effective derivatives of anthracyclines.

5.   REFERENCES

Acton, E.M., Mosher, C.W. and Gruber, J.M. (1982) 'Approaches to more effective anthracyclines by analog synthesis and evaluation', in H.S. El Khadem (ed.), Anthracycline Antibiotics, Academic Press, New York, pp. 119-139.

Arcamone, F. and Penco, S. (1988) 'Synthesis of new doxorubicin analogs' in J.W. Lown (ed.), Anthracycline and Anthracenedione-based Anticancer Agents, Elsevier, Amsterdam, pp. 1-54.

Brown, J.R. (1983) 'New natural semisynthetic and synthetic anthracycline drugs' in S. Neidle and M.J. Waring (eds.), Molecular Aspect of Anticancer Drug Action, MacMillan, New York, pp. 57-92.

Chaires, J.B., Dattagupta, N. and Crothers, D.M. (1982) 'Studies on the interaction of anthracycline antibiotics with deoxyribonucleic acid: Equilibrium binding studies on the interaction of daunomycin with DNA', Biochemistry 21, 3933-3940.

Chaires, J.B., Fox, K.R., Herrara, J.E., Britt, M. and Waring, M.J. (1987) 'Site and sequence specificity of daunomycin-DNA interaction', Biochemistry 26, 8227-8236.

Chen, K., Gresh, N. and Pullman, B. (1985) 'A theoretical investigation on the sequence selective binding of daunomycin to double stranded polynucleotides', Molec. Pharmacol. 30, 279-286.

Chen, K-X., Gresh, N. and Pullman, B. (1986) 'A theoretical investigation on the sequence selective binding of adriamycin to double-stranded polynucleotides', Nucleic Acids Res. 14, 2251-2267.

Crooke, S.T. and Reich, S.D. (1980) Anthracyclines: Current Status and New Developments, Academic Press, New York.

Crothers, D.M., Straney, D.C. and Phillips, D.R. (1987) 'Effects of antitumor drugs on transcription' in C. Chagas and B. Pullman (eds.), Molecular Mechanisms of Carcinogenic and Antitumour Activity, Adenine Press, New York, pp. 403-424.

Cullinane, C. and Phillips, D.R. (1990) 'Detection of irreversible adriamycin-DNA adducts by *in vitro* transcription: GpC specificity and dependence on Fe(II) ions', Biochemistry, in press.

Du Vernay, V.H., Pachter, J.A. and Crooke, S.K. (1979) 'DNA binding studies of several new anthracycline antitumor antibiotics: sequence preferences and structure activity relationship of marcellomycin and its analogues as compared to adriamycin', Biochemistry 18, 4024-4030.

Fisher, J., Ramakrishnan, K. and Becvar, J.E. (1983) 'Direct enzyme-catalyzed reduction of anthracyclines by reduced nicotinamide adenine dinucleotide', Biochemistry 22, 1347-1355.

Fisher, J., Abdella, B.R.J. and McLane, K.E. (1985) 'Anthracycline antibotic reduction by spinach ferredoxin-NADPH$^+$ reductase and ferrodoxin', Biochemistry 24, 3562-3571.

Gianni, L., Corden, B.J. and Myers, C.E. (1982) 'The biochemical basis of anthracycline toxicity and antitumour activity', Rev. Biochem. Toxicol. 5, 1-82.

Grandi, M., Giuliani, F.C., Verhoef, V. and Filppi, J. (1988) 'Screening of anthracycline analogs' in J.W. Lown (ed.), Anthracycline and Anthracenedione-based Anticancer Agents, Elsevier, Amsterdam, pp. 571-598.

Grant, M. and Phillips, D.R. (1979) 'Dissociation of polydeoxynucleotide-daunomycin complexes', Molec. Pharmacol. 16, 357-360.

Hogan, M.E. and Austin, R.H. (1987) 'Importance of DNA stiffness in protein-DNA binding specificity', Nature 329, 253-266.

Jones, M.B., Hollstein, U. and Allen, F.S. (1987) 'Site specificity of binding of antitumor antibiotics to DNA', Biopolymers 26, 121-135.

Kappus, H. (1986) 'Overview of enzyme systems involved in bioreduction of drugs and in redox cycling', Biochem. Pharmacol. 35, 1-6.

Kleyer, D.L. and Koch, T.H. (1983) 'Electrophilic trapping of the tautomer of 7-deoxydaunomycinone: A possible mechanism for covalent binding to DNA', J. Am. Chem. Soc. 105, 5154-5155.

Konopa, J. (1983) 'Adriamycin and daunomycin induce interstrand DNA crosslinks in HeLa S3 cells', Biochem. Biophys. Res. Commun. 111, 819-826.

Legha, S.S., Benjamin, R.S., Mackay, B., Ewer, M., Wallace, S., Valdivieso, M., Rasmussen, S.L., Blumenschien, G.R. and Freireich (1982) 'Reduction of doxorubicin cardiotoxicity by prolonged continuous intravenous infection', Ann. Int. Med. 96, 133-139.

Lown, J.W. (1988) Anthracycline and Anthracenedione-Based Anticancer Agents, Elsevier, Amsterdam.

Manfait, M., Alix, A.J.P., Jeannesson, P., Jardillier, J.C. and Theophanides, T. (1982) 'Interaction of adriamycin with DNA as studied by resonance Raman spectroscopy', Nucleic Acids Res. 10, 3803-3816.

Moore, H.W. (1977) 'Bioactivation as a model for drug design bioreductive alkylation', Science 197, 527-532.

Myers, C.E., Mimnaugh, E.G., Grace, C.Y. and Sinha, B.K. (1988) 'Biochemical mechanisms of tumour cell kill by the anthracyclines' in J.W. Lown (ed.), Anthracycline and Anthracenedione-based Anticancer Agents, Elservier, Amsterdam, pp. 528-570.

Myers, C.E., Muindi, J.R.F., Zweier, J. and Sinha, B.K. (1987) '5-Iminodaunomycin: an anthracycline with unique properties', J. Biol. Chem., 11571-11577.

Neidle, S. and Sanderson, M.R. (1983) 'The interaction of daunomycin and adriamycin with nucleic acids', in S. Neidle and M.J. Waring (eds.), Molecular Aspects of Anti-cancer Drug Design, MacMillan Press, London, pp. 35-55.

Newlin, D.D., Miller, M.J. and Pilch, D.F. (1984) 'Interactions of molecules with nucleic acids, VII. Interaction and T.A. specificity of daunomycin in DNA', Biopolymers 23, 139-158.

Pan, S.S. and Bachus, N.R. (1980) 'Xanthine oxidase catalysed reduction of anthracycline antibiotics and free radical formation', Mol. Pharmacol. 17, 955-99.

Phillips, D.R. and Crothers, D.M. (1986) 'Kinetics and sequence specificity of drug-DNA interactions: An in vitro transcription analysis', Biochemistry, 25, 7355-7362.

Phillips, D.R., Di Marco, A. and Zunino, F. (1978) 'The interaction of daunomycin with polydeoxynucleotides', Eur. J. Biochem. 85, 487-492.

Phillips, D.R., White, R.J. and Cullinane, C. (1989) 'DNA sequence specific adducts of Adriamycin and mitomycin C', FEBS Letters 246, 233-240.

Phillips, D.R., White, R.J., Dean, D.D. and Crothers, D.M. (1990a) 'Monte-Carlo simulation of multi-site echinomycin-DNA interactions detected by in vitro transcription analysis', Biochemistry, in press.

Phillips, D.R., White, R.J., Trist, H., Cullinane, C., Dean, D. and Crothers, D.M. (1990b) 'New insight into drug-DNA interactions at individual drug sites probed by RNA polymerase during active transcription of the DNA', Anti-Cancer Drug Design 5, 117-125.

Pullman, B. (1988) 'Binding affinities and sequence selectivity in the interaction of antitumor anthracyclines and anthracenediones with double stranded polynucleotides and DNA', in J.W. Lown (ed.), Anthracycline and Anthracenedione-based Anticancer Agents, Elsevier, Amsterdam, pp. 372-400.

Pullman, A. and Pullman, B. (1981) 'Molecular electrostatic potential of the nucleic acids', Quart. Rev. Biophysics 14, 289-380.

Ramakrishnan, K. and Fisher, J. (1983) 'Nucleophilic trapping of 7,11-dideoxyanthracyclinone quinone methides', J. Am. Chem. Soc. 105, 7187-7188.

Schwartz, H.S. (1983) 'Mechanisms of selective cytotoxicity of adriamycin, daunomycin and related anthracyclines', in S. Neidle and M.J. Waring (eds.) Molecular Aspects of Anti-cancer Drug Design, MacMillan Press, London, pp. 93-125.

Sinha, B.K. (1980) 'Binding specificity of activated anthracycline anticancer agents to nucleic acids', Chem.-Biol. Interact. 30, 66-77.

Sinha, B.K. (1989) 'Free radicals in anticancer drug pharmacology', Chem.-Biol. Interactions 69, 293-317.

Sinha, B.K. and Chignell, C.F. (1979) 'Binding mode of chemically activated semiquinone free radicals from quinone anticancer agents to DNA', Chem.-Biol. Interact. 28, 301-308.

Sinha, B.K. and Gregory, J.L. (1981) 'Role of one electron and two electron reduction products of Adriamycin and daunomycin in DNA binding', Biochem. Pharmacol. 30, 2626-2629.

Sinha, B.K. and Sik, R.H. (1980) 'Binding of adriamycin to cellular macromolecules *in vivo*', Biochem. Pharmacol. 29, 1867-1868.

Sinha, B.K., Trush, M.A., Kennedy, K.A. and Mimnaugh, E.G. (1984) 'Enzymatic activation and binding of adriamycin to nuclear DNA', Cancer Res. 44, 2892-2896.

Skorobogaty, A., White, R.J., Phillips, D.R. and Reiss, J.A. (1989) 'Elucidation of the DNA sequence preferences of daunomycin', Drug Design Delivery 3, 125-152.

Straney, D.C. and Crothers, D.M. (1985) 'Intermediates in transcription initiation from the *E.coli lac* UV5 promoter', Cell 43, 449.

Trist, H. and Phillips, D.R. (1989) *In vitro* transcription analysis of the role of DNA flanking sequences on adriamycin sequence selectivity', Nucleic Acids Res. 17, 3673-3688.

Tsou, K.C. and Yip, K.F. (1976) 'Effect of deoxyribonuclease on adriamycin-polynucleotide complexes', Cancer Res. 36, 3367-3373.

Ughetto, G. (1988) 'X-ray diffration analysis of anthracycline-oligonucleotide complexes', in J.W. Lown (ed.), Anthracycline and Anthracenedione-based Anticancer Agents, Elsevier, Amsterdam, pp. 295-334.

Valentini, L., Nicolella, V., Vannini, E., Menozzi, M., Penco, S. and Arcamone, F. (1985) 'Association of anthracycline derivatives with DNA: A fluorescence study', Il Farmaco Ed. Sci. 40, 377-389.

Wang, A.H.J., Ughetto, G., Quigley, G.J. and Rich, A. (1987) 'Interactions between an anthracycline antibiotic with DNA: Molecular structure of daunomycin complexed to d(CpGpTpApCpG) at 1.2 A resolution', Biochemistry 26, 1152-1163.

White, R.J. and Phillips, D.R. (1988) 'Transcriptional analysis of multi-site drug-DNA dissociation kinetics: Delayed termination of transcription by actinomycin D', Biochemistry 27, 9122-9132.

White, R.J. and Phillips, D.R. (1989a) 'Drug-DNA dissociation kinetics: *In vitro* transcription and SDS sequestration', Biochem. Pharmacol. 38, 331-334.

White, R.J. and Phillips, D.R. (1989b) 'Bidirectional transcription footprinting', Biochemistry 28, 2659-2669.

3' and 5' sides of the binding region the slopes are large and positive and decrease in magnitude away from the site. Although the errors are large, the slopes associated with Fe-MPE cleavage of the 139-mer in the presence of ActD also clearly show the location of the two drug binding sites, as shown in Figure 4. However, unlike for DNase I, the magnitudes of the positive slopes outside of drug binding sites are approximately the same. In the case of Fe-MPE, the slope data may also indicate weaker drug binding sites near positions ~52 and ~80 of the fragment. Since MnT4MPyP is specific for AT-rich regions of DNA [11], the cleavage sites for this probe on the 139-mer are few in number. The manganese complex is able to detect drug binding at positions 62-65 (negative slope at site 62), but the complex exhibits enhanced cleavage (positive slopes)

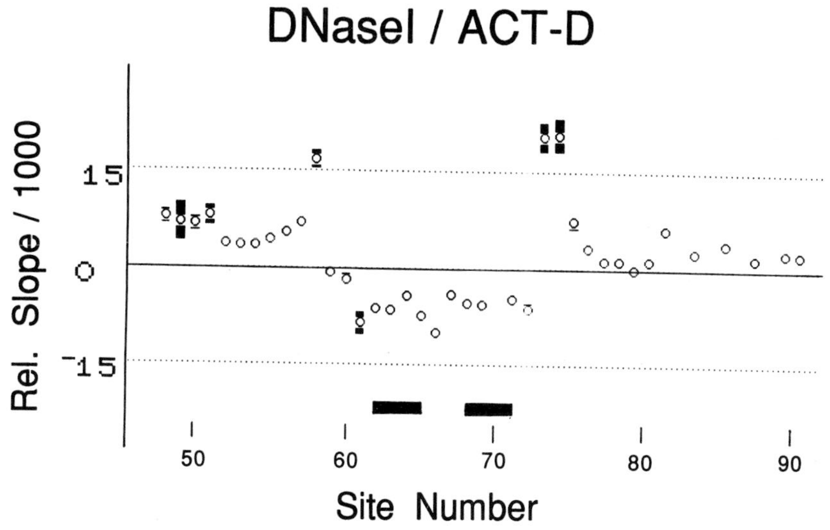

Figure 3. The relative slopes and their errors (vertical bars) of footprinting plots of the DNase I/ActD/139-mer experiment. The two strong ActD binding sites in the region, 48-90, of the fragment are indicated as rectangles.

at most other sites on the 139-mer, as shown in Figure 5. The cleavage rate appears to be significantly higher at site 60, located at the edge of an actinomycin D binding site, than at most of the other sites analyzed. This also seems to occur at site 95, which is located to the 5' side of the strong actinomycin D binding sites at positions 100-105 of the fragment.

Discussion

CALCULATION OF BINDING CONSTANTS FROM FOOTPRINTING DATA

The data used to measure binding constants from footprinting experiments consist of a series of plots showing the intensity of each oligonucleotide on the autoradiogram as a function of drug concentration (footprinting plots). Initially, these plots are inspected for their shapes and are grouped into those which exhibit inhibitions, binding

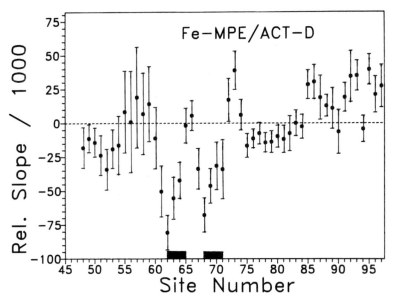

Figure 4. The relative slopes and their errors (vertical bars) of footprinting plots of the Fe-MPE/ActD/139-mer experiment. The two strong ActD binding sites in the region, 48-97, of the fragment are indicated as rectangles.

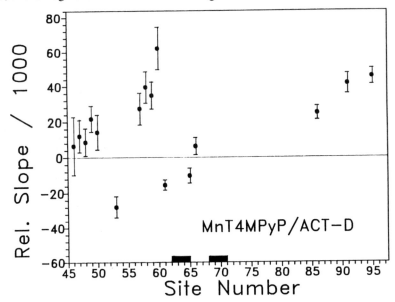

Figure 5. The relative slopes and their errors (vertical bars) of footprinting plots of the MnT4MPyP/ActD/139-mer experiment. The two strong ActD binding sites in the region, 46-95, of the fragment are indicated as rectangles.

sites, drug binding at one site would exclude drug binding to the second site on the same 139-mer fragment. Earlier NMR studies [3] have demonstrated that this sequence can in fact bind two actinomycin D molecules, but the interaction is anticooperative with the binding constant of the second drug being 1/20 that of the first. Interestingly, the binding constants derived for these two sites are significantly lower than those for the same sequence when it is imbedded in a short segment of DNA [12]. This and earlier studies on the antiviral agent netropsin [9] suggest that DNA length may influence the magnitude of the binding constant: short oligonucleotides return higher binding constants than does polymeric DNA.

The non-GC site at positions 124-127, CGTC, has a higher than expected binding constant, Table 1. Recent work by Snyder *et al.* [13] has demonstrated that, when this sequence is part of a two-fold symmetric duplex, two actinomycin D molecules can bind to the duplex in a cooperative manner, with a binding constant of $\sim 10^7 M^{-1}$. Since actinomycin D bound to this sequence possesses anomalous optical effects, it may be that the binding mechanism is different from that utilized at GC sites. Although this sequence, CGTC, occurs at 109-112 of the 139-mer, the binding constant of actinomycin D toward that site is significantly lower than for the site 124-127. Thus, flanking sequences and/or anticooperativity effects due to the adjacent GC sites in the region 100-105 may influence drug binding at this nonclassical site.

The restriction fragment possesses a number of GC sites which are part of the triplet sequence, GGC. In every case these sites have binding constants which are significantly lower than the GC sites found in Table 1. Although the reason for this is presently unknown, weak binding to GGC was previously observed in footprinting studies [7] and has been recently reported to occur with azido actinomycin D, a derivative which can photocleave DNA [14]. A possible exception is the GGC at 162-160 on the non coding strand which appears to have a high binding constant, Table 1.

CLEAVAGE RATE ENHANCEMENTS

Enhancements in cleavage rates away from drug binding sites are often observed in footprinting experiments. As was earlier reported by Ward *et al.* [7], DNase I exhibits rate enhancements immediately adjacent to drug sites which are larger than those expected from a simple redistribution effect (mass action) associated with the enzyme. If drug binding causes a change in DNA structure, and/or modifies the degree of ion and water binding, the free energy of binding of the cleavage agent would likely also be changed. Since the probe must bind in order to cleave DNA, these changes would be manifested as rate increases or decreases at the affected sites. In an effort to determine if different cleavage agents could detect such changes for actinomycin D binding to DNA, we studied the DNA cleavage patterns of Fe-MPE and MnT4MPyP in the presence of the drug. To uncover possible drug-induced alterations in cleavage rates, the early part (low drug concentration region) of each footprinting plot was fitted to a linear function. The slopes of these plots along with their standard deviations are shown in Figures 3-5. DNase I, which is sensitive to groove width and DNA flexibility [15], shows anomalous rate increases immediately adjacent to the actinomycin binding sites located at 62-65 and 68-71 of the fragment. Due to the lower quality of the data for Fe-MPE, the errors in slope are large and it is difficult to determine if unusual alterations in cleavage rate occur adjacent to these two actinomycin D binding sites (Figure 4). This probe binds to DNA via intercalation and cleaves through a Fenton type process involving hydroxyl radical [16]. Intercalation of the drug between the base pairs of DNA would be expected to produce a local distortion in the polymer, but it is not evident that this can be detected by Fe-MPE. The negative sites, and those displaying intensity enhancements, sites on the polymer where no binding is taking place. The plots for binding sites are grouped according to which loading events on the fragment they are measuring. This takes into account the cleavage geometry of DNase I, which cannot cleave ~3 nucleotides to the 3' side of the drug site [9]. In the case of overlapping sites, e.g. positions 100-105 of the fragment, the group of plots show all of the competing loading events in the overlapped region. As earlier described [7], the enhancements observed in footprinting experiments are mainly due to redistribution of the cleavage agent. Since redistribution affects footprinting plots for binding sites as well, it is important to include a number of sites on the fragment which monitor only redistribution in the calculations. Because actinomycin D may induce structural changes in DNA which in turn alter DNase I cleavage rates, the enhancement sites chosen were far from drug binding sites and representative of the average level of enhancement taking place over the entire fragment.

After accounting for the mass balance relationships which exist in the system, including equilibria with the carrier (calf thymus) DNA, footprinting plots can be calculated, given binding constants toward sites on the fragment, an average drug-site concentration on the carrier and the associated drug-binding constant toward the carrier. To derive binding constants from footprinting data, it is necessary to search for the set of binding constants which leads to calculated footprinting plots agreeing as well as possible with those experimentally determined, as measured by equation 1

$$D = \Sigma \ (I_{ij} - \tilde{I}_{ij})^2. \tag{1}$$

Here, I_{ij} is the intensity for cutting at site i at the jth drug concentration and $\tilde{I}_{ij}$ is the calculated value. For a given site concentration, the binding constants toward the fragment and the carrier DNA are found by minimizing equation 1 using a Simplex searching method. Eventually, when convergence is obtained, one finds a simplex such that every new step increases D rather than decreasing it. The error in equilibrium constants determined in this way can be estimated by examining the effect of a change in that equilibrium constant on D. It is usually found that a change of a few percent in any one equilibrium constant changes D by 10% or more, so that equilibrium constants are reported to two significant figures, Table 1. The calculated values of the binding constants toward the radiolabeled fragment depend on the value used for the site concentration on the carrier DNA. This quantity is not measured, but derived from the data on the concentration leading to the smallest D. Our model does not consider weak sites on the carrier; their inclusion could change the derived binding constants on the fragment.

BINDING CONSTANTS OF ACTINOMYCIN D TOWARD THE 139-MER

As mentioned above, the strongest actinomycin D binding site on the restriction fragment has the sequence TGCT at positions 136-139 of the polymer. Although this sequence also occurs at positions 62-65 of the fragment, its binding constant there is less than half that of the higher-numbered site. The reason for this is not clear, but possibly relates to the closeness of the second site to the strong binding site at positions 68-71 of the fragment (Figure 2). The occupancy of the latter site by ActD may have a negative cooperative effect on the binding of drug at 62-65; this does not occur for the isolated site at 136-139. The site at 100-103, TGCG, possesses a binding constant which is half that of the adjacent, overlapped, site at 102-105, CGCT. Both sites were assumed to bind drug in an independent fashion but, due to the proximity of the two slopes at positions ~52 and ~80 of the fragment may be due to weak actinomycin binding sites which are partially occupied over the drug concentration range used in the slope analysis.

The DNA cleaving metalloporphyrin, MnT4MPyP, binds in the minor groove of DNA [6, 11]. A variety of studies have shown that this cleavage agent is sensitive to DNA melting and readily cleaves at AT-rich sites. It will also cleave at high-melting sequences containing GC if these sites are placed in a premelted state [11]. In the presence of actinomycin D the slope analysis indicates that the cleavage rate of this agent increases at the 5' edge of the actinomycin D binding site, Figure 5. Specifically, the porphyrin cleavage site at position 60 on the fragment exhibits an anomalous cleavage rate increase when the actinomycin D site at 62-65 is occupied by drug. Although additional work is warranted, it may be that drug binding causes melting or premelting of sites adjacent to the actual intercalation position and that this effect is sensed by the porphyrin. A cleavage rate above that expected from a simple redistribution process is also found for position 95 of the restriction fragment, Figure 5. This may be caused by drug binding in the region 100-105, which contains two strong ActD sites.

In summary, when the strong sites on the 139-mer are occupied by actinomycin D, the enzyme DNase I and the metalloporphyrin, MnT4MPyP, exhibit anomalous rate increases adjacent to two drug binding sites. Based on the DNA structural features recognized by these two agents, enhanced cleavage may be due to alterations in groove width and/or decreases in the melting point of the site. Further work in progress involving other DNA modifying agents will attempt to elucidate the mechanism of this interesting drug-induced process.

Acknowledgment

We wish to thank the American Cancer Society, Grant #NP681, for support of this research. We are also thankful to Ms. Julie Neri for typing this manuscript.

References

1.   Gale, E.F., Cundliffe, E., Reynolds, P.E., Richmond, M.H. and Warnig, M.J. (1981) The Molecular Basis of Antibiotic Action, John Wiley & Sons, New York.
2.   Zhou, N., James, T.L. and Shafer, R.H. (1989) 'Binding of Actinomycin D to [d(ATCGAT)]$_2$: NMR Evidence of Multiple Complexes', Biochemistry, 28, 5231-5239.
3.   Scott, E.V., Jones, R.L., Banville, D.L., Zon, G., Marzilli, L.G. and Wilson, W.D. (1988), '^{1}H and ^{31}P NMR Investigations of Actinomycin D Binding Selectivity with Oligodeoxyribonucleotides Containing Multiple Adjacent d(GC) Sites', J. Am. Chem. Soc., 27, 915-923.
4.   Dabrowiak, J.C. (1983) 'Sequence Specificity of Drug-DNA Interaction, Life Sci., 32, 2915-2931.
5.   Rehfuss, R. (1988) 'Quantitative Footprinting Analysis of Actinomycin D Binding to DNA', Ph.D. Dissertation, Syracuse University, Syracuse, New York.

6.   Ward, B., Skorobogaty, A. and Dabrowiak, J.C. (1986) 'DNA Cleavage Specificity of a Group of Cationic Metalloporphyrins', Biochemistry, 25, 6875-6883.
7.   Ward, B., Rehfuss, R., Goodisman, J. and Dabrowiak, J.C. (1988) 'Rate Enhancements in the DNase I Footprinting Experiment', Nucleic Acids Res., 16, 1359-1369.
8.   Dabrowiak, J.C., Skorobogaty, A., Rich, N., Vary, C.P.H., and Vournakis, J.N. (1986) 'Computer Assisted Microdensitometric Analysis of Footprinting Autoradiographic Data', Nucleic Acids Res., 14, 489-499.
9.   Dabrowiak, J.C. and Goodisman, J. (1989) 'Quantitative Footprinting Analysis of Drug-DNA Interactions' Kallenbach, N. (ed.) Chemistry and Physics of Ligand-DNA Interactions, Adenine Press, Guilderland, N.Y., pp. 143-174.
10. Goodisman, J. and Dabrowiak, J.C. (1990) 'Quantitative Aspects of DNase I Footprinting', Hurley, L.H. (ed.) Advances in DNA Sequence Specific Agents, JAI Press, in press.
11. Raner, G., Goodisman, J. and Dabrowiak, J.C. (1989) 'Porphyrins as Probes of DNA Structure and Drug-DNA Interactions', Tullius (ed.) Metal-DNA Chemistry, ACS Symposium Series, 401, 74-89.
12. Rehfuss, R., Goodisman, J. and Dabrowiak, J.C. (1990) 'Quantitative Footprinting Analysis. Binding to a Single Site', Biochemistry, 29, 777-781.
13. Synder, J.G., Hartman, N.G., D'Estantoit, B.L., Kennard, O., Remeta, D.P. and Breslauer, K.J. (1989) 'Binding of Actinomycin D to DNA: Evidence for a Nonclassical High-Affinity Binding Mode That Does Not Require GpC Sites', Proc. Natl. Acad. Sci. USA, 86, 3968-3972.
14. Rill, R.L. Marsch, G.A. and Groves, D.E. (1989) '7-Azido-Actinomycin D: A Photo-Affinity Probe of the Sequence Specificity of DNA Binding by Actinomycin D', J. Biomol. Struct. Dynamics, 7, 591-605.
15. Suck, D., Lahm, A. and Oefner, C. (1988) 'Nicked DNA Bound to DNase I: Refined 2Å-Structure of a DNase I-Octanucleotide Complex', Nature (London), 332, 465-469.
16. Dervan, P.B. (1986) 'Design of Sequence Specific DNA-Binding Molecules', Science, 232, 464-471.

STRUCTURAL REQUIREMENTS FOR DNA TOPOISOMERASE II INHIBITION BY ANTHRACYCLINES

G. CAPRANICO and F. ZUNINO
Division of Experimental Oncology B
Istituto Nazionale per lo Studio e la Cura dei Tumori
20133 Milan, Italy

ABSTRACT. The nuclear enzyme DNA topoisomerase II has been proposed to be the primary cellular target of several antitumor drugs including anthracyclines. These agents stabilize a DNA-topoisomerase II complex wherein the DNA strands are broken and covalently linked to the enzyme. Studies using the alkaline elution technique showed that the extent as well as the persistance of protein-associated DNA breaks in drug-treated cells are equally important for the cell killing effects of anthracyclines. In a recent study on the sequence selectivity of doxorubicin-induced DNA cleavage, a ternary complex has been proposed to take place between topoisomerase II, DNA and doxorubicin, in order to explain the specific action of doxorubicin on topoisomerase II activity. Since anthracyclines are also able to inhibit several other DNA-dependent enzymes, it is proposed that the strong DNA binding in the intercalation process may account for the non-specific action of anthracyclines on DNA functions. Structure-activity relationships of the induction of topoisomerase II-mediated DNA cleavage by anthracyclines suggest that the mode of drug-DNA interaction is more critical than the strength of drug binding to DNA. It is likely that the mode of interaction critically alters drug interaction with topoisomerase II in the ternary complex. Such studies indicate critical structural requirements in the anthracycline molecule and suggest new drug design approaches.

1. INTRODUCTION

The anthracycline antibiotics, which include doxorubicin and daunorubicin, possess a significant therapeutic efficacy against several human malignancies. Due to the central role of doxorubicin in combination chemotherapy of a number of hematologic and solid tumors, several studies have focused on the mechanisms of cytotoxic action of anthracyclines. These drugs are among the best characterized DNA intercalators. Despite intensive study, the molecular mechanism by which anthracyclines exert antitumor and cytotoxic activity has continued to be the subject of debates over the past years because these agents are known to interfere with various cellular processes. Recent studies have emphasized the critical role of interference of DNA intercalators with the breakage-rejoining reaction of DNA topoisomerase II in their cytotoxic action [1-3]. This nuclear enzyme has

B. Pullman and J. Jortner (eds.), Molecular Basis of Specificity in Nucleic Acid-Drug Interactions, 167-176.
© 1990 Kluwer Academic Publishers. Printed in the Netherlands.

been recognized as an important multidrug target. DNA topoisomerase II is an essential nuclear enzyme that, together with DNA toposiomerase I, regulates the topological state of DNA during replication, transcription and other DNA processes [4, 5].

Anthracyclines exert two actions upon topoisomerase II DNA cleavage activity: 1) they stimulate the enzyme-mediated DNA cleavage at low drug concentrations and 2) globally abolish the DNA cleavage at higher drug concentrations [1, 6, 7]. The first effect is due to the drug interference with the DNA breakage-reunion activity of topoisomerase II, by freezing a covalent enzyme-DNA complex. SDS treatment of such complexes results in DNA cleavage and the covalent linking of the enzyme to the 5′ end of the broken DNA duplex [2-5]. Anthracyclines share this effect with epipodophyllotoxins, amsacrine, actinomycin D and ellipticines [2, 3]. The second effect is most likely due to the strong DNA binding affinity of the compound, since epipodophyllotoxins and amsacrine do not suppress DNA cleavage [2, 3]. Other strong DNA binders such as ethidium bromide, which does not stimulate topoisomerase II DNA cleavage, and ellipticines do suppress DNA cleavage at sufficiently high drug concentrations [1-3].

The cytotoxic effects of topoisomerase II-targeted drugs are associated with the induction of topoisomerase II-mediated DNA cleavage [2, 3, 7]. However, the ability of anthracyclines to exert two opposite effects upon topoisomerase II-mediated DNA cleavage resulted in an apparent absence of a correlation between cytotoxicity and extent of DNA cleavage, since highly cytotoxic drug concentrations did not induce DNA cleavage in the presence of purified topoisomerase II [1]. Filter elution studies also showed that short exposure of tumor cells to highly cytotoxic concentrations of some anthracycline derivatives produced much lower DNA break levels than those induced by less cytotoxic drug concentrations [6, 8]. Thus, whereas a correlation between cytotoxicity and DNA cleavage, determined immediately after drug treatment, was demonstrated for acridines and epipodophyllotoxins [2, 3], an absence of such a correlation was reported for anthracyclines [6].

A simple correlation between cell killing and cellular DNA cleavage measured at a fixed time point (i.e., after drug exposure) is not appropriate in the case of anthracyclines and other strong DNA binders. Since anthracyclines are able to induce and to suppress topoisomerase II-mediated DNA cleavage [1, 8] and the global kinetics of drug effects on cellular DNA is highly dependent on cellular pharmacokinetics [8, 10], not only the level, but also the persistance of DNA cleavage is critical for lethal events [7, 10, 11]. Indeed, by considering the persistence of cellular DNA breaks following drug removal from the culture medium, a correlation does exist between protein-associated DNA breaks and cytotoxicity [11, 12]. A recent study on anthracycline-induced cleavage of SV40 DNA in the presence of purified topoisomerase II further indicated that the cytotoxic activity of a number of anthracycline analogs with very different DNA binding constants was well correlated with the induction of SV40 DNA cleavage [13]. Taken together, these studies have provided evidence for a causal relation between cytotoxic potency of antitumor anthracyclines and their interference with topoisomerase II.

2. MOLECULAR MODELS FOR THE DOUBLE ACTION OF ANTHRACYCLINES UPON MAMMALIAN DNA TOPOISOMERASE II

Since the discovery that topoisomerase II is the target of clinically useful anti-cancer drugs, it has been questioned that drug intercalation into DNA is the molecular mechanism for the induction of topoisomerase II-mediated DNA cleavage by the drugs, since epipodophyllotoxins do not intercalate into DNA. Doubts on the role of drug binding into DNA also arise from the mechanism of action of quinolones and camptothecin, which interfere with the prokaryotic topoisomerase II, DNA gyrase, and the eukaryotic topoisomerase I, respectively, in a way similar to the mammalian topoisomerase II poisons. These drugs induce topoisomerase-mediated DNA cleavage and, like epipodophyllotoxins, do not intercalate into DNA [3-5]. Although one could conceive the intercalating agents as having a molecular mechanism different from that of the non-intercalating agents, the close similarity of the drug effect on topoisomerase II suggests a common molecular mechanism.

The relation between DNA binding of seven chromophore-modified anthracyclines and topoisomerase II-mediated cellular DNA cleavage was evaluated in P388 leukemia cells [6]. In this study, no quantitative correlation was observed between the DNA binding affinity of the drug and the formation of protein-associated DNA breaks. This finding was confirmed on a larger number of anthracycline derivatives by using a purified murine topoisomerase II [13]. It was suggested that the ability of the drug to intercalate is required, since non-intercalating derivatives lack significant antitumor activity, but is not sufficient for the induction of topoisomerase II DNA cleavage [13]. This suggestion is an accord with previous structure-activity studies, which indicated that drug intercalation is a necessary but not sufficient condition for optimal antitumor activity of anthracycline antibiotics [7, 14-16], and also of other classes of DNA intercalating antitumor compounds [17, 18], and which prompted the search for "minimal" DNA intercalating agents [17, 19].

Therefore, even intercalation of the anthracycline molecule into DNA by itself cannot account completely for the induction of topoisomerase II-mediated DNA cleavage.

2.1 A Highly Specific Interaction of Doxorubicin with the DNA-Topoisomerase II Complex

A more specific interaction between the anthracycline molecule and the DNA-topoisomerase II complex has been very recently proposed to take place and to be responsible for drug induction of topoisomerase II-mediated DNA cleavage [20]. In this study, the sequence specificity of doxorubicin-induced topoisomerase II-mediated DNA cleavage was investigated in SV40 DNA. Two types of topoisomerase II-mediated DNA cleavage sites were defined, depending on the effect of doxorubicin: 1) topoisomerase II-mediated DNA cleavage sites that were never enhanced by doxorubicin, and 2) topoisomerase II-mediated DNA cleavage sites that were stimulated by doxorubicin. This highly sequence-selective action of doxorubicin on topoisomerase II could be explained by an absolute requirement for an adenine base to the 3' end of the broken DNA strand in order for doxorubicin to stabilize the cleavable complex [20]. Analogous types of specific nucleotide requirements have been shown to occur locally at the DNA cleavage site induced by topoisomerase I in the presence of camptothecin [21].

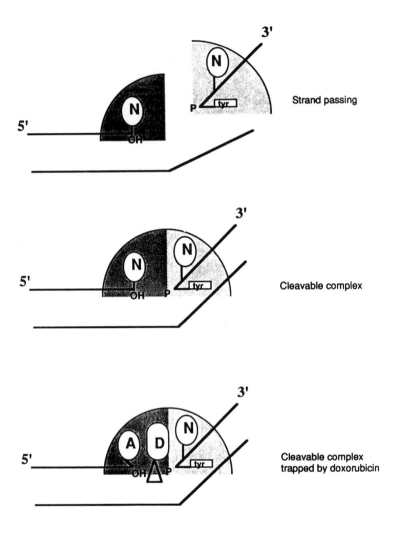

Figure 1. A proposed molecular model of the ternary complex between DNA-anthracycline-topoisomerase II. D, doxorubicin; A, adenine; N, any nucleotide.

In the case of doxorubicin, the absolute requirement for an adenine base at the break site suggests that this adenine is involved in a specific molecular interaction with a doxorubicin molecule. One could hypothesize that the drug molecule is situated at the DNA break site or very close to it and thus in close proximity to the active site of the enzyme. Doxorubicin might be situated in a hydrophobic pocket created by the interactions of the nucleotides of the broken DNA strand with non-polar amino acid residues of the enzyme (Fig. 1). Since all the mammalian topoisomerase II poisons are either DNA intercalators or have a planar ring system, a drug molecule could be inserted in this pocket in an intercalation-like manner. In such a case, the drug molecule might be positioned in a way to obstruct directly the strand break rejoining process.

2.2 A Role for Anthracycline Intercalation into DNA

The DNA binding affinity constant of anthracyclines was shown to be better correlated with the suppression, rather than the induction, of topoisomerase II-mediated DNA cleavage [13]. 4-Demethyl-6-0-methyl doxorubicin and 5-iminodaunorubicin, which have a lower DNA binding affinity constant than doxorubicin, did not suppress DNA cleavage at concentrations at which doxorubicin and other strong DNA binder analogues did [13]. Moreover, doxorubicin markedly reduced m-AMSA- and VM26-induced DNA cleavage at low drug concentrations and completely abolished such cleavage at higher concentrations [13]. Therefore, these findings support the idea that intercalation into DNA of anthracycline molecules is responsible for the suppressive effect of topoisomerase II-mediated DNA cleavage. This is consistent with previous observations that other strong DNA intercalators (ethidium bromide) suppress topoisomerase II-mediated DNA cleavage, whereas non-intercalators (epipodophyllotoxins) or weak DNA intercalators (m-AMSA) do not show this effect [1, 22, 23].

Anthracyclines are known to inhibit the activity of a number of other DNA-dependent enzymes, including DNA and RNA polymerases [14, 24, 25] and DNA ligase [26]. Therefore, it is possible that such compounds exert two effects at the DNA level: a) a specific action on topoisomerase II with the stabilization of the enzyme-DNA cleavable complex, and b) a non-specific inhibition of DNA functions. The latter effect is likely to be the consequence of distortions of the DNA helix following drug intercalation into adjacent base pairs of the double helix, thus interfering with template functions of DNA. Such an effect on DNA and RNA synthesis, which was presumably dependent on binding affinity, was not necessarily correlated with cytotoxic activity. Indeed, no correlation was found between inhibition of nucleic acid synthesis and cytotoxic potency of 4-demethoxy derivatives of daunorubicin [27].

3. STRUCTURAL REQUIREMENTS OF THE ANTHRACYCLINE MOLECULE FOR THE INDUCTION OF TOPOISOMERASE II-MEDIATED DNA CLEAVAGE

The definition of the relations between drug structure and the induction of topoisomerase II-mediated DNA cleavage by anthracyclines should consider that the overall extent of DNA cleavage induced by such agents is the result of a balance between the stimulative and suppressive effects on topoisomerase II-mediated DNA cleavage. In principle, an increased induction of DNA cleavage by an anthracycline analog might be due either to a reduced suppressive effect or to an enhanced stimulative effect of the cleavage. Therefore, a proper comparison of different anthracyclines should be made for derivatives with similar cleavage-suppressing activity. Since suppression activity is likely due to tight DNA binding of the drug in intercalation, various derivatives which have been studied for DNA cleavage in cultured cells or in in vitro systems are gathered in three groups according to their apparent DNA binding constants (Table 1).

TABLE 1. DNA cleavage activity and DNA binding constant (Kapp) of anthracycline derivatives[1]

Anthracyclines derivatives	Kapp $(M^{-1})^2$	Relative DNA cleavage	
		whole cells	Isolated DNA + purified topoisomerase II
Doxorubicin (DX)	4.8×10^6	+	+
4'-epi-DX	2.2×10^6	ND	+
9-deoxy-DX	1.6×10^6	−	ND
4-demethyl-6-deoxy-DX	1.6×10^6	+++	ND
4'-deoxy-4'-Iodo-DX	6.4×10^6	++	ND
Daunorubicin (DA)	4.8×10^6	ND	+
4-demethoxy-DA	2.4×10^6	+++	+++
4-demethoxy-11-deoxy-4'-epi-DA	3.1×10^5	+++	ND
4-demethyl-6-0-methyl-DX	2.0×10^5	++++	+
11-deoxy-DA	5.7×10^5	−	−
5-iminoDA	ND	ND	−/++
β-anomer-DX	1.5×10^4	ND	−
6-0-methyl-DA	3.5×10^4	ND	−

[1]DNA cleavage semiquantitatively evaluated from refs. 6, 8, 11, 13.
[2]Binding constant determined under the same ionic strength (0.1 M NaCl).
ND = not determined.

Among the strong DNA binder derivatives (1st group in Table 1), doxorubicin and daunorubicin have very similar DNA binding constants, which in agreement with experimental evidence [1, 13] suggests that the two drugs suppress DNA cleavage to the same extent. Since both drugs induce similar DNA cleavage levels in the presence of purified topoisomerase II (Table 1), the OH group in position 14 apparently does not affect the induction of topoisomerase II-DNA cleavable complexes. Removal of the C-4 methoxy group of the D ring greatly enhances the drug interference with topoisomerase II, whereas removing the hydroxyl group in C-9 has the opposite effect. These results suggest that substituents at positions C-4 and C-9 play a critical role in modulation of drug interaction with the enzyme and DNA. The OH group in position C-9 might be involved in a particularly important interaction (hydrogen bond) with an adjacent acceptor group. 4-Demethyl-6-deoxydoxorubicin is more effective than doxorubicin, again suggesting that the substituents at the C-4 position greatly influence drug activity. Epimerization at the 4' position has no significant effect on DNA cleavage activity, and a bulk substituent, like an iodine atom, in position 4 of the sugar does not decrease but rather slightly increases the cleavage activity of the drug, suggesting that no steric restrictions are present in the sugar moiety.

In the second group (Table 1), an enhanced activity of an analog compared to doxorubicin in inducing DNA cleavage could be due in principle to a decreased suppressive effect of the cleavage with respect to doxorubicin. Indeed, a decreased suppressive effect of DNA cleavage has been reported for anthracyclines with a lower DNA affinity than doxorubicin and daunorubicin, such as 5-iminodaunorubicin and 4-demethyl-6-0-methyl- doxorubicin [13]. Thus, although 5-iminodaunorubicin is more active than the parent drugs at high drug concentrations, no definitive conclusion can be drawn from the effect of the C-5 modification on cleavage induction activity. The effects of 11-deoxydaunorubicin indicate that removal of the C-11 OH group decreases drug induction of topoisomerase II-mediated DNA cleavage. Comparison between 11-deoxydaunorubicin and 4-demethoxy-11-deoxy-4'epidaunorubicin again shows that removal of the methoxy group in position C-4 increases drug induction of the cleavage, since 4'epimerization has no effect (see above). These observations suggest that the OH group in position C-11 may be involved in a molecular interaction which increases drug cleavage activity only when a bulk substituent is present in the C-4 position but which is dispensable when the methoxy group is absent. Finally, moving the methoxy group from position C-4 to position C-6 (4-demethyl-6-0-methyl-doxorubicin) results in a cleavage activity similar to doxorubicin in in vitro systems. However, the derivative has been shown to induce much more DNA single-strand breaks than doxorubicin and the 4-demethoxydaunorubicin derivatives in cultured P388 cells [6]. Although, the discrepancy between results in cultured cells and in in vitro systems remains unexplained, the derivative might have a different molecular interaction with the DNA-enzyme complex, compared to doxorubicin, which results in a production of relatively more single-strand breaks than doxorubicin. This may be the outcome of a different DNA-binding geometry at the intercalation site following modifications at the C-4 and C-6 positions [28].

The third group of derivatives (Table 1) includes two analogs that have very low DNA affinity constants and are completely inactive in inducing DNA breaks in the presence of purified topoisomerase II [13]. The inactivity of 6-0-methyldaunorubicin indicates that the presence of two methoxy groups in the chromophore completely abolishes the drug is ability to intercalate and to form a ternary complex with DNA and enzyme. In addition, the inactivity of the β-anomer of doxorubicin indicates that a precise orientation of the sugar relative to the chromophore is required for drug induction of the cleavage, although minor changes at the 4' position of the sugar are less critical. Recently, 3'-N-substituted analogs were shown to be inactive or weakly active in inducing topoisomerase II DNA cleavage [29]. Since the weakly active analogs had a lower DNA binding affinity than doxorubicin and accordingly the drug suppressive effect of the cleavage was greatly reduced

[29], modifications at the 3' position were unfavorable for drug induction of topoisomerase II-mediated DNA cleavage.

4. CONCLUSIONS

An excellent correlation between DNA binding affinity and relative antitumor and cytotoxic activity was found in several series of anthracycline derivatives modified in the amino sugar moiety [30, 31]. Structural modifications in the amino sugar causing a substantial reduction of DNA binding ability are also associated with a loss of activity, thus suggesting the critical importance of DNA binding in the biologic properties of anthracyclines. However, no precise correlation was found between DNA binding affinity and cytotoxic activity in chromophore-modified anthracyclines [6], in contrast to observations on other sugar-modified series. These results would be rationalized if it is assumed that changes in the chromophore (and in the mode of DNA interaction) alter drug interaction with topoisomerase II in the ternary complex and therefore the stability of the cleavable complex. Drug interaction with enzyme is proposed as a more specific effect than inhibition of template function of DNA. The importance of peculiar features of cleavable complexes produced by anthracyclines as a consequence of this "specific" interference with topoisomerase II activity (i.e. sequence selectivity of DNA cleavage, persistence of DNA lesions) remains unclear. Insight into the molecular mechanism of drug-topoisomerase II-DNA interaction is expected to provide new and rational bases for improved drug design approaches.

5. REFERENCES

1. Tewey, K.M., Rowe, T.C., Yang, L., Halligan, B.D. and Liu, L.F. (1984) 'Adriamycin-induced DNA damage mediated by mammalian DNA topoisomerase II', Science 226, 466-468.
2. Pommier, Y. and Koh, K.W. (1989) 'Topoisomerase II inhibition by antitumor intercalators and demethylepipodophyllotoxins', in R.I. Glazer (ed.), Developments in Cancer Chemotherapy, CRC Press, Boca Raton, pp. 175-195.
3. Liu, L.F. (1989) 'DNA topoisomerase poisons as antitumor drugs', Annu. Rev. Biochem. 58, 351-375.
4. Wang, J.C. (1985) 'DNA topoisomerases', Annu. Rev. Biochem. 54, 665-667.
5. Gellert, M. (1981) 'DNA topoisomerases', Annu. Rev. Biochem. 50, 979.
6. Capranico, G., Soranzo, C. and Zunino, F. (1986) 'Single-strand DNA breaks induced by chromophore-modified anthracyclines in P388 leukemia cells', Cancer Res. 46, 5499-5503.
7. Zunino, F. and Capranico G. (1990) 'DNA topoisomerase II as the primary target of antitumor anthracyclines', in preparation.
8. Capranico, G., Riva, A., Tinelli, S., Dasdia, T. and Zunino, F. (1987) 'Markedly reduced levels of anthracycline-induced DNA strand breaks in resistant P388 leukemia cells and isolated nuclei', Cancer Res. 47, 3752-3756.

9. Capranico, G., Tinelli, S. and Zunino, F. (1987) 'Kinetics of resealing of 4-demethoxydaunorubicin (dmDR)-induced DNA breaks in P388 leukemia cells sensitive and resistant to anthracyclines', in Progress in Antimicrobial and Anticancer Chemotherapy. Proceedings of the 15th International Congress of Chemotherapy, pp. 96-98.

10. Capranico, G., Tinelli, S. and Zunino, F. (1989) 'Formation, resealing and persistence of DNA breaks produced by 4-demethoxydaunorubicin in sensitive and resistant P388 leukemia cells', Chem. Biol. Interact. 72, 113-123.

11. Capranico, G., De Isabella, P., Penco, S., Tinelli, S. and Zunino, F. (1989) 'Role of DNA breakage in cytotoxicity of doxorubicin, 9-deoxydoxorubicin, and 4-demethyl-6-deoxydoxorubicin in murine leukemia P388 cells', Cancer Res. 49, 2022-2027.

12. Binaschi, M., Capranico, G., De Isabella, P., Mariani, M., Supino, R., Tinelli, S. and Zunino, F. (1990) 'Comparison of DNA cleavage induced by etoposide and doxorubicin in two human small cell lung cancer lines with different sensitivity to topoisomerase II inhibitors', Int. J. Cancer 45, 347-352.

13. Capranico, G., Zunino, F., Kohn, K.W. and Pommier, Y. (1990) 'Sequence selective topoisomerase II inhibition by anthracycline derivatives in SV40 DNA: relationship with DNA binding affinity and cytotoxicity', Biochemistry 29, 562-569.

14. Zunino, F., Di Marco, A., Zaccara, A. and Gambetta, R.A. (1980) 'The interaction of daunorubicin and doxorubicin with DNA and chromatin', Biochem. Biophys. Acta 607, 206-214.

15. Zunino, F., Casazza, A.M., Pratesi, G., Formelli, F. and Di Marco, A. (1981) 'Effect of methylation of aglycone hydroxyl groups on the biological and biochemical properties of daunorubicin', Biochem. Pharmacol. 30, 1856-1858.

16. Zunino, F., Barbieri, B., Bellini, O., Casazza, A.M., Geroni, C., Giuliani, F., Ciana, A., Manzini, G. and Quadrifoglio, F. (1986) 'Biochemical and biological activity of the anthracycline analog, 4-demethyl-6-0-methyl-doxorubicin', Invest. New Drugs 4, 17-23.

17. Denny, W.A. (1989) 'DNA-intercalating ligands as anti-cancer drugs: prospects for future design', Anti-Cancer Drug Design 4, 241-263.

18. Lepecq, J.-B., Dat-Xuong, N., Gosse, C. and Paoletti, C. (1974) 'A new antitumoral agent: 9-hydroxyellipticine. Possibility of a rational design of anticancerous drugs in the series of DNA intercalating drugs', Proc. Natl. Acad. Sci. U.S.A. 71, 5078-5082.

19. Denny, W.A., Atwell, G.J. and Baguley, B.C. (1987) 'Minimal DNA-intercalating agents as anti-tumor drugs: 2-styrylquinoline analogues of amsacrine', Anti-Cancer Drug Design 2, 263-270.

20. Capranico, G., Kohn, K.W. and Pommier, Y. (1990) 'Local sequence requirements for DNA cleavage by mammalian topoisomerase II in the presence of doxorubicin', in preparation.

21. Kjeldsen, E., Mollerup, S., Thomsen, B., Bonven, B.J., Bolund, L. and Westergaard, O. (1988). 'Sequence-dependent effect of camptothecin on human topoisomerase I DNA cleavage', J. Mol. Biol. 202, 333-342.

22. Nelson, E.M., Tewey, K.M. and Liu, L.F. (1984) 'Mechanism of antitumor drug action: poisoning of mammalian DNA topoisomerase II on DNA by 4'-(9-acridinylamino)-methanesulfon-m-anisidide', Proc. Natl. Acad. Sci. U.S.A. 81, 1361-1365.

23. Chen, G.L., Yang, L., Rowe, T.C., Halligan, B.D., Tewey, K.M., and Liu, L.F. (1984) 'Nonintercalative antitumor drugs interfere with the breakage-reunion reaction of mammalian DNA topoisomerase II', J. Biol. Chem. 259, 13560-13566.
24. Zunino, F., Gambetta, R.A., Di Marco, A., Velcich, A., Zaccara, A., Quadrifoglio, F. and Crescenzi, V. (1977) 'The interaction of adriamycin and its β-anomer with DNA', Biochim. Biophys. Acta 476, 38-46.
25. Zunino, F., Di Marco, A. and Zaccara, A. (1979) 'Molecular structural effects involved in the interaction of anthracyclines with DNA', Chem. Biol. Interact. 24, 217-225.
26. Montecucco, A., Pedrali-Noy, G., Spadari, D., Zanolin, E. and Ciarrocchi, G. (1988) 'DNA unwinding and inhibition of T4 DNA ligase by anthracyclines', Nucleic Acids Res. 16, 3907-3918.
27. Supino, R., Necco, A., Dasdia, T., Casazza, A.M. and Di Marco, A. (1977) 'Relationship between effects on nucleic acid synthesis in cell cultures and cytotoxicity of 4-demethoxy derivatives of daunorubicin and adriamycin', Cancer Res. 37, 4523-4528.
28. Quadrifolgio, F., Ciana, A., Manzini, G., Zaccara, A. and Zunino, F. (1982) 'Influence of some chromophore substituents on the intercalation of anthracycline antibiotics into DNA', Int. J. Biol. Macromol. 4, 413-418.
29. Bodley, A., Liu, L.F., Israel, M., Seshadri, R., Koseki, Y., Giuliani, F.C., Kirschenbaum, S., Silber, R. and Potmesil, M. (1989) 'DNA topoisomerase II-mediated interaction of doxorubicin and daunorubicin congeners with DNA', Cancer Res. 49, 5969-5978.
30. Di Marco, A., Casazza, A.M., Dasdia, T., Necco, A., Pratesi, G., Rivolta, P., Velcich, A., Zaccara, A. and Zunino, F. (1977) 'Changes of activity of daunorubicin, adriamycin and stereoisomers following the introduction of removal of hydroxyl groups in the amino sugar moiety', Chem.-Biol. Interact. 19, 291-302.
31. Bargiotti, A., Casazza, A.M., Cassinelli, G., Di Marco, A., Penco, S., Pratesi, G., Supino, R., Zaccara, A., Zunino, F. and Arcamone, F. (1983) 'Synthesis, biological and biochemical properties of new anthracyclines modified in the aminosugar moiety', Cancer Chemother. Pharmacol. 10, 84-89.

ACKNOWLEDGEMENTS

We wish to thank Drs. K.W. Kohn and Y. Pommier for many helpful discussions and for their hospitality during the stay of one of us at the National Institutes of Health. We thank L. Zanesi and B. Johnston for editorial assistance.
This work was supported in part by the Consiglio Nazionale delle Ricerche (Progetto Finalizzato "Oncologia") and in part by a grant from the Italian Ministero della Sanita'.

THERMODYNAMIC STUDIES OF AMSACRINE ANTITUMOR AGENTS WITH NUCLEIC ACIDS

D.E. GRAVES AND R.M. WADKINS
Department of Chemistry
University of Mississippi
University, Mississippi 38677

ABSTRACT.   The equilibrium binding of several anilinoacridine analogs are compared over a wide range of ionic strengths and temperatures. Although o-AMSA binds DNA with a higher affinity than m-AMSA it is not effective as an antitumor agent. Both m-AMSA and o-AMSA bind DNA in an intercalative manner. In an effort to gain insight into the physical chemical properties associated with these compounds and to correlate these properties with antitumor activity, an in-depth investigation into the thermodynamic parameters of these compounds and structurally related anilinoacridine analogs was performed. These studies demonstrate that substituient type and placement on the aniline ring of the anilinoacridines influence both the affinity of these drugs in binding to DNA and dictate whether the DNA binding is an enthalpy or entropy driven process. These data demonstrate that the antitumor agent m-AMSA interacts with DNA via an enthalpy driven process. In contrast, the structurally similar but biologically inactive o-AMSA binds to DNA through an apparent entropy driven process. The differences in thermodynamic mechanisms of binding between the two isomers along with molecular modeling studies reveal that the electronic and/or steric factors resulting from the positioning of the methoxy substituient group on the anilino ring directs the DNA binding properties through orientation of the methanesulfonamido group at the 1' position of the aniline ring. The orientation of this substituient group may result in favorable contacts through hydrogen bonding with neighboring base-pairs and ultimately influence the biological effectiveness as an antitumor agent.

Introduction

Amsacrine ((4'-(9-acridinylamino)methanesulfon-m-anisidide) or m-AMSA is a 9-anilinoacridine derivative of the parent 9-aminoacridine. Synthesized in 1974 by Bruce Cain and coworkers (Cain et al, 1975), this compound was found to have potent clinical activity against a variety of solid cell tumors and leukemias (McCredie, 1985). The interaction of m-AMSA with nucleic acids has been extensively studied (Waring, 1976, Wilson et al., 1981, and Denny and Wakelin, 1986). Both m-AMSA and its biologically inactive structural isomer, o-AMSA, (structures shown in Figure 1) have been demonstrated to bind DNA via intercalation

177

B. Pullman and J. Jortner (eds.), Molecular Basis of Specificity in Nucleic Acid-Drug Interactions, 177–189.
© 1990 Kluwer Academic Publishers. Printed in the Netherlands.

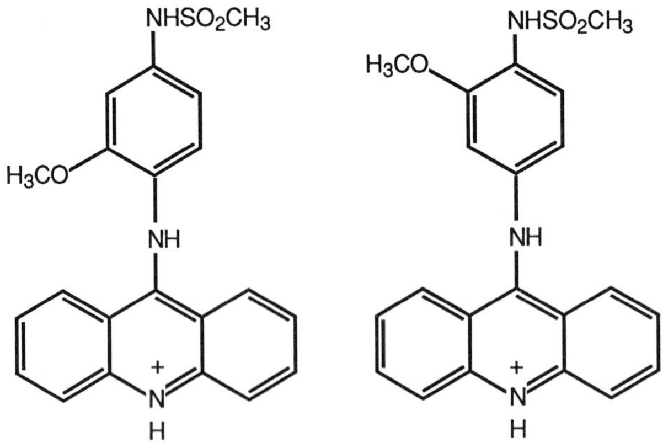

Figure 1. Chemical structure of the potent antitumor agent, m-AMSA and its biologically inactive isomer, o-AMSA.

of their planar acridine chromophore between adjacent base pairs of the DNA. However, the DNA binding affinities of these anilinoacridine compounds appear to be inversely related to their biological effectiveness as antitumor agents, with m-AMSA exhibiting the lowest affinity for binding DNA and o-AMSA characterized by a DNA binding affinity approximately 5 times greater than that of m-AMSA (Wadkins and Graves, 1989). This lack of correlation between DNA binding affinities and biological activities provide an intriguing paradox when compared to other antitumor agents such as the anthracyclines adriamycin and daunomycin. Slight changes in the drug structure via alteration of the position of the methoxy substituient (2' versus 3' placement for the o-AMSA and m-AMSA, respectively) result in dramatic differences in both the biophysical properties associated with the interactions of these compounds with DNA and in their effectiveness as antitumor agents. A related drug, AMSA, lacks the methoxy group at either the 2' or 3' position. However, this compound demonstrates effectiveness as an antitumor agent requiring a dosage 5-7 times that of m-AMSA to elicit a similar biological response (Cain et al., 1975). In contrast, this compound binds DNA with the highest affinity of the anilinoacridine analogs (Wilson et al., 1981).

The equilibrium binding of these anilinoacridine compounds has been investigated in an effort to gain insight into the molecular mechanism(s) responsible for their antitumor effectiveness. Viscosity measurements on the interactions of these compounds with DNA reveal unwinding angles of 20.5°, 20.6°, and 20.9° for m-AMSA, o-AMSA, and AMSA, respectively (Waring, 1976). The similarity of these values suggests that the DNA complexes formed by all three compounds are sterically equivalent. A physical manifestation of drug-DNA interactions that is often linked to antitumor activity is the kinetics of ligand binding. The potent antitumor antibiotic actinomycin D has a long residence time on the DNA. It has often been suggested that this residence time results in blockage of the replication fork during the cellular reproductive process, and that this blockage results in cell death (Waring, 1981). This relationship between residence time on the DNA and antitumor activity has also been observed with large number of other DNA intercalating drugs such as daunomycin, the ellipticines, bisantrene, mitonafide, and the anthracenediones (Feigon et al., 1984). An investigation of the dissociation kinetics of *m*-AMSA, *o*-AMSA and AMSA as well as 9-aminoacridine from calf thymus DNA has been performed (Denny and Wakelin, 1986). The kinetics profiles for AMSA and *o*-AMSA were virtually identical, and were very similar to that of 9-aminoacridine. In contrast, *m*-AMSA showed a much faster dissociation rate than the other compounds studied. Thus, the drug which demonstrates the highest potency as an antitumor agent also has the shortest residence time on the DNA, in direct contrast to the other intercalating antitumor drugs listed above. Thus, from these studies no apparent trends are evident which conclusively link DNA binding with antitumor activity.

Recently, a number of clinically important anticancer drugs have been proposed to exert their biological activity via affecting the function of topoisomerase II in mammalian cells. Both groove binders such as epipodophyllotoxins (etoposide and tenoposide) and intercalators and intercalating agents including m-AMSA, ellipticine, and adriamycin, have been shown to exert these topoisomerase II inhibition effects. The molecular mechanism(s) responsible for the neoplastic activities of *m*-AMSA are thought to arise as a result of topoisomerase II inhibition. Studies by Pommier and coworkers (1985a) and Robinson and Osheroff (1990) have shown *m*-AMSA to induce the formation of protein associated single strand breaks in nuclear DNA probably via a ternary complex formed between the m-AMSA, DNA and topoisomerase II. Although the specific mechanism remains unknown, it is thought that the presence of the m-AMSA alters the cleavable complex such that the strand passing activity or the religation step of the reaction is inhibited, resulting in the enhanced production of single and double strand breaks (D'Arpa and Liu 1989). Similar studies have demonstrated o-AMSA to be ineffective in generating DNA cleavage products (Pommier et al., 1985b, 1987). Although the site of drug action has yet to be determined, the interactions of AMSA and m-AMSA with DNA play a critical role in inhibiting topoisomerase II activity while the structurally similar o-AMSA is ineffective in eliciting this inhibitory response.

In an effort to resolve the distinctions between the DNA binding of m-AMSA and o-AMSA which may influence biological activities, a detailed analysis of the thermodynamic binding properties of these and selected anilinoacridines was performed. Within the scope of these studies, we wished to determine the thermodynamic properties associated with the interactions of these anilinoacridine analogs with DNA and compare discernable trends with modifications to the substituient type and position on the aniline ring. These studies provide evidence that the thermodynamic properties associated with the interactions of these drugs with DNA may be correlated to antitumor effectiveness. Molecular modeling studies provide information concerning possible structural properties associated with the interactions of these compounds with DNA and provide insight into a more rational approach to drug design and development.

Experimental

MATERIALS

Drug Preparations.   m-AMSA (NSC-249992) was obtained from the National Cancer Institute. The parent compound, 9-aminoacridine was purchased from Sigma Chemical Co. and used without further purification. The AMSA , o-AMSA, 3'-methoxy, 2'-methoxy, and N-phenyl analogs (structures shown in Figure 2) were synthesized according methods reported by Denny and co-workers (1982) and used as their hydrochloride salts. Authenticities of these compounds were confirmed by elemental analyses, and NMR spectroscopy (Bruker AC-300).

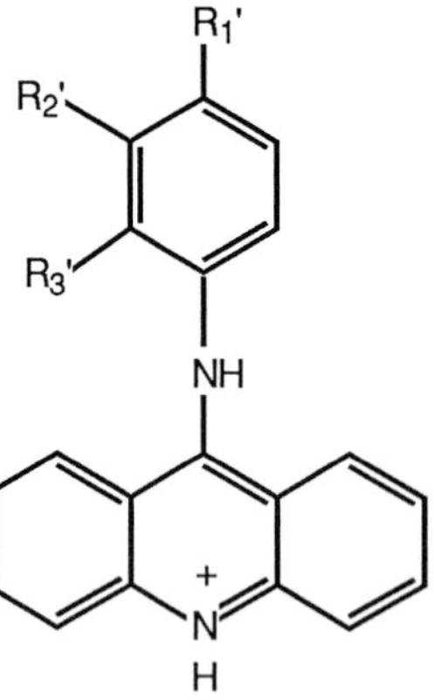

Compound	R1'	R2'	R3'
N-Phenyl	H	H	H
2'-Methoxy	H	-CH₃O	H
3'-Methoxy	H	H	-CH₃O
AMSA	-NHSO₂CH₃	H	H
m-AMSA	-NHSO₂CH₃	H	-CH₃O
o-AMSA	-NHSO₂CH₃	-CH₃O	H

Figure 2. Structure of the anilinoacridine analogs. Substituients to the aniline ring were added at the 1', 2', and 3' positions.

The compounds were dessicated and stored in the dark at -5°C until ready to be used. Due to their low aqueous solubility, dimethylsulfoxide (DMSO) was used to dissolve the compounds into concentrated stock solutions (~0.5 mg ml/ml). The DMSO-drug solution was then diluted 1to 10 in 0.01 M sodium phosphate (pH 7.0), 0.001 M disodium EDTA, and 0.1 M sodium chloride and filtered through a 0.22μ syringe filter (Millipore). The pKa values for all compounds were determined by visible spectroscopic methods. At pH 7.0, the m-AMSA was shown to be 89% protonated in contrast to the o-AMSA which was only 78% protonated.

DNA Preparations. Calf thymus DNA (sodium salt, Type I) was purchased from Sigma Chemical Co. and purified as described by Chaires et al. (1982). Briefly, the DNA was sheared by sonification for 30 minutes at 5°C in the presence of bubbling N₂. Afterwards, the DNA was subjected to T₁ RNase and Proteinase K (Boehringer Mannheim) digestions. The solutions were then repeatedly extracted with a 1:1 mixture of chloroform and phenol and the precipitated with cold absolute ethanol. The DNA pellet was the dissolved in sodium phosphate buffer, pH 7.0, 0.001 M disodium EDTA, and the desired concentration of sodium chloride. The DNA solutions were dialysed against the appropriate buffer and filtered through 0.45μ syringe filters (Millipore) prior to use. The concentrations of the DNA solutions are stated in terms of base pairs (bp) using the molar absorptivity of ε_{260nm} of 13,200 M^{-1}cm^{-1}.

METHODS

DNA Binding Studies. The optical titrations were performed using a Cary 2290 UV/visible spectrophotometer (Varian) equipped with a Lauda RC-6 circulating water bath. Temperatures were monitored by immersion of the thermistor probe directly into the sample cell. In an effort to maximize sensitivity, quartz cells of 10-cm pathlengths were used for these DNA binding studies. The drug's absorption (ranging from 0.2 - 0.3 as measured in the 10 cm cells) was read directly from the digital display of the spectrophotometer operating in the statistical mode. DNA binding isotherms for each of the AMSA derivatives were obtained by titrating measured quantities of a stock drug solution into a known volume of a calf thymus DNA and monitoring the resulting change in the absorption spectrum of the drug. DNA binding data are presented as Scatchard plots and quantitated using the neighbor exclusion models of McGhee and von Hippel (1974).

The examination of the DNA binding isotherms of the 3'-methoxy and N-phenyl compounds as a function of temperature resulted in relatively small changes in the slopes of the Scatchard plots, thus limiting the accuracy of the determination of the binding enthalpies using the classical van't Hoff methodology. To circumvent these problems, the method outlined by Chaires (1985) and Shimer et al. (1988) was used. Data were obtained by mixing a known concentration of drug and DNA and monitoring the absorption of the drug. The temperature of the drug-DNA solution was then adjusted as necessary and the change in absorption as a function of temperature was measured. A linear least squares analysis of a plot of the ln K versus $1/T$ (K) was used to determine the enthalpy from the van't Hoff relationship and the entropy of binding determined from the Gibbs free energy and enthalpy at 20°C.

Results and Discussion

EQUILIBRIUM BINDING

The anilinoacridine analogs shown in Figure 2 provide an excellent system for developing insight into the effects of substituient modification(s) via type and position on the DNA binding properties of intercalating drugs. The strategy employed for these studies was to construct a series of structurally similar acridine analogs starting with the parent 9-aminoacridine, and building up to the biologically active m-AMSA. The compounds listed in Figure 2 have been previously reported, making the synthetic procedures straightforward (Cain et al, 1975). The design of these compounds was developed such that changes in the equilibrium and thermodynamic binding properties could be directly correlated to changes in the drug structure and placement of the corresponding substituient groups.

Single Substituient Effects. Equilibrium binding studies reveal that upon addition of the N-phenyl ring to the parent 9-aminoacridine (N-phenyl analog shown in Figure 2) a dramatic decrease in the DNA binding affinity, relative to that of the parent 9-aminoacridine, is observed. The binding constant is shown to decrease by a factor of 4 in comparison to the equilibrium binding of the parent 9-aminoacridine to DNA as shown in Table I. This decrease may arise from the additional steric factors imposed by the bulky phenyl ring in the minor groove influencing the degree of insertion of the acridine chromophore into the intercalation site. Such a steric influence was predicted by the molecular modeling studies of the m-AMSA-d(TACGTA)$_2$ complex by Neidle and coworkers (Abraham et al., 1988).

Addition of the methoxy substituient group to the 2' position on the anilinoacridine molecule resulted in no change in the DNA binding affinity from that of the N-phenyl analog. Thus, at the 2' position the methoxy group exerts no steric or electronic effects that result in changes to the DNA binding affinity other than those imposed by the N-phenyl ring substituient. In contrast, movement of the methoxy group from the 2' to the 3' position on the N-phenyl ring results in a marked change, decreasing the equilibrium binding of both the 3'-methoxy and m-AMSA analogs by a factor of two, as compared to the DNA binding affinity of both the N-phenyl and 2'-methoxy anilinoacridine analogs. Clearly, the positioning of the methoxy group (2' or 3' sites) provides a pivotal role in influencing the DNA binding properties of these anilinoacridine analogs.

Addition of the methanesulfonamido moiety to the 1' position on the N-phenyl ring (AMSA) results in an increase in the binding affinity of this analog as compared to the N-phenyl and 2'- and 3'- methoxy analogs. Although the DNA binding affinity of AMSA is lower than that of the parent 9-aminoacridine analog, it is more than double that of the N-phenyl and 2'- and 3'-methoxy anilinoacridine analogs. The contribution of the 1'-methanesulfonamido group towards enhances binding may be through its ability to provide additional contacts within the minor groove of the DNA, as proposed by Denny et al (1983).

Table I. Binding Constants for Amsacrine Analogs at 0.1 M NaCl.

Compound[a]	K_{int}[b] (M^{-1})	n[c]
9-aminoacridine	1.4×10^5	3.2
N-phenyl	3.9×10^4	1.9
2'-methoxy	4.1×10^4	2.2
3'-methoxy	2.5×10^4	1.9
AMSA	8.8×10^4	2.8
o-AMSA	5.4×10^4	3.0
m-AMSA	1.8×10^4	3.8

[a]See Figure 2. [b]From fit to McGhee and von Hippel cooperative ligand equation (McGhee & von Hippel, 1974) (corrected for state of protonation). [c]Binding site size in terms of number of base pairs per bound drug.

Multisubstituient Effects. Placement of the methoxy group at the 2' and 3' positions of the 1'-methanesulfonamido analogs (i.e., AMSA to o-AMSA and m-AMSA, respectively) results in a dramatic effect to both the DNA binding affinities and on the antitumor activities of these anilinoacridine analogs. If the methoxy group is placed at the 2' position to form the 1'-methanesulfonamide-2'-methoxy analog (o-AMSA) a slight decrease in the DNA binding affinity as compared to AMSA is observed. In contrast, placement of the methoxy group at the 3' position, meta to the 1'-methanesulfonamido group (m-AMSA) results in a substantial decrease in the DNA binding affinity of this anilinoacridine analog. In fact, the binding constant determined for m-AMSA is the lowest of all of the anilinoacridine analogs examined. Thus, even though addition of the 1'-methanesulfonamido group results in an enhancement of the DNA binding affinity, placement of the methoxy group at the 3' position overrides this enhancement and results in the lowest DNA binding affinity. In addition, correlation of the biological activity with DNA binding affinity is further clouded by the observation that the most potent antitumor agent, m-AMSA binds to DNA with the lowest affinity, while AMSA, which also exhibits antitumor activity binds to DNA with the highest affinity of the anilinoacridines studied. Interestingly, o-AMSA which demonstrates no antitumor activity lies between m-AMSA and AMSA in its affinity for binding DNA.

THERMODYNAMICS

Equilibrium binding of the anilinoacridine analogs with DNA revealed that substituient modification and position plays a pivotal role in influencing the binding affinities of these compounds. However, the substituient effects on DNA binding properties did not provide a correlation with biological activities. Thus, our studies were extended to include an investigation into the thermodynamic properties associated with the interactions of these compounds with DNA. These studies consisted of van't Hoff analyses which provided the enthalpy values presented in Table II. The $\Delta G°$ values for the interactions of the anilinoacridine analogs as well as the parent 9-aminoacridine were determined from the equilibrium constants described in Table I. The values of $\Delta G°$ for the interactions of the anilinoacridine compounds range from -5.6 to -6.5 kcal/mol. The parent 9-aminoacridine has a much higher binding constant and is reflected in a $\Delta G°$ value of -7.0 kcal/mol.

Binding Enthalpies. As observed in Table II, the parent 9-aminoacridine exhibits the largest $\Delta H°$ value of -9.2 kcal/mol. Addition of a phenyl group to the N9 position of the acridine results in a substantial decrease in the $\Delta H°$ of binding. The three AMSA compounds, AMSA, o-AMSA and m-AMSA (all having the 1' methanesulfonamido moiety) exhibited similar values for $\Delta H°$ ranging from -5.2 to -7.6 kcal/mol. In contrast, the N-phenyl, 2'-methoxy, and 3'-methoxy anilinoacridine analogs had significantly lower $\Delta H°$ values, -2.7,-3.7, and -1.4 kcal/mol, respectively. These values are approximately half the magnitude of those observed for the compounds containing the methanesulfonamido moiety. This difference in the enthalpy values may provide significant information concerning the role of the methanesulfonamido group in the interactions of these compounds with DNA. Upon adding a phenyl ring to the exocyclic amino group on 9-aminoacridine, the enthalpy of binding is greatly reduced to about -3 kcal/mol. Assuming that the binding properties of the N-phenyl derivative are otherwise similar to that for 9-aminoacridine, this suggests that the acridine chromophore is no longer able to penetrate as deeply into the DNA duplex as the parent 9-aminoacridine or is less stacked with the adjacent base pairs of the intercalation site. Recent molecular modeling experiments by Neidle and coworkers (Neidle et al., 1988; Abraham et al. 1988) arrived at similar conclusions for the decamer duplex d(GATACGATAC)2, generated from the crystal structure of the proflavine-d(CpG) complex (Shieh et al., 1980). Their results showed the proflavine to be oriented in the intercalation site between the GC base pairs such that there was considerable overlap between the acridine ring and the cytosine base on the 3' side. The CG base pair on the 5' side has little overlap with the acridine ring, but the guanine O6 on this strand is stacked almost precisely vertically above the protonated ring nitrogen of the drug. In comparison, m-AMSA was shown to stack asymmetrically with respect to the bases of the two strands due to the van der Waals repulsions resulting from the N-phenyl ring, thus the acridine ring of m-AMSA shows less overlap with the bases in the intercalation site.

Data presented here suggests that addition of the methoxy group at the 3'-position further reduces the ability of the drugs to intercalate. The change in enthalpy between binding of the 3'-methoxy derivative as compared to the N-phenyl and the m-AMSA as compared to AMSA is approximately 1 kcal/mol less favorable due to the addition of the 3'-methoxy group.

For both AMSA and m-AMSA the enthalpies for binding are around 3 kcal/mol more favorable than for the N-phenyl and 3'-methoxy derivatives, suggesting a definite interaction between the methanesulfonamido group and the DNA. The 2'-methoxy group of o-AMSA may disallow necessary contacts between drug and DNA resulting in the observed binding enthalpy of -3.0 kcal/mol.

An interesting feature of the thermodynamic profiles is the near identical manner in which the N-phenyl, 2'-methoxy and o-AMSA derivatives bind to DNA and their similar thermodynamic profiles. This suggests that the 2'-methoxy group does not interact with DNA itself, and may, in the case of o-AMSA, prevent a contact between the DNA and the methanesulfonamido moiety. This latter point is illustrated by comparing the binding profiles of

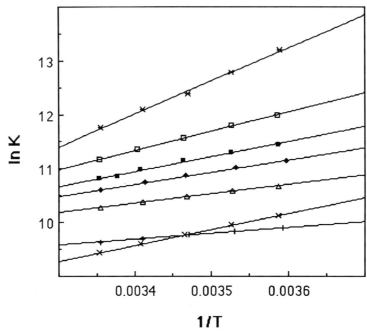

Figure 3. Sample van't Hoff plots for the anilinoacridine analog-DNA interactions. Solid lines represent linear least squares fits to the data. The plots are shown for 9-aminoacridine (✳), AMSA (□), m-AMSA (✗), o-AMSA (■), N-phenyl (Δ), 2'-methoxy (◆) and 3'-methoxy (✚) analogs. Binding constants were determined from both complete titrations as well as by increasing the temperature of a sample containing a fixed ratio of drug/DNA (bp) as described in the Methods section.

AMSA, m-AMSA and o-AMSA. Both antitumor agents, m-AMSA and AMSA, bearing the 1'-methanesulfamido group exhibit similar enthalpies of binding of -5.2 to -5.9 kcal/mol. In contrast, the binding enthalpy of the biologically inactive o-AMSA is only -3.0 kcal/mol, indicating that an additional contact with the DNA may be crucial in manifesting the antitumor activity of these compounds.

Entropy of Binding. Using the $\Delta G°$ and $\Delta H°$ values, the binding entropy was calculated for each of the analogs. The $\Delta S°$ values for four of the compounds, the N-phenyl, 2' and 3'-methoxy, and o-AMSA, exhibit large positive $\Delta S°$ values, ranging from 8.4 to 15 cal/mol K. Interestingly, the two biologically active compounds, AMSA and m-AMSA exhibit much small entropy components of 1 to 2 cal/mol K. These data reveal a correlation of binding entropy with biological activity. Both drugs which are effective antitumor agents (AMSA and m-AMSA) exhibit very small positive binding entropies while the structurally similar anilinoacridine analogs which are ineffective as antitumor drugs are characterized by larger entropy values.

Classical intercalators such as proflavine, 9-aminoacridine, and ethidium bromide are all characterized by negative entropies of -5.0, -6.4, and -5.4 e.u., respectively, indicating that their interaction with DNA results in the formation of a rigid complex resulting in the loss of rotational and translational freedom. Large positive entropies which characterize the binding of the 2'-methoxy, 3'-methoxy, N-phenyl, and o-AMSA analogs to DNA may occur from several sources. Partial insertion (rather than full insertion) of the acridine chromophore between adjacent base pairs of the DNA would result in greater freedom of motion for both the acridine and the N-phenyl rings providing the large entropy value that is observed.

Table II. Thermodynamic Data for Anilinoacridine Analogs at 20°C.

Compound	$\Delta G^{\circ a}$ (kcal/mol)	$\Delta H^{\circ\ b}$ (kcal/mol)^c	ΔS° cal/mol K)
N-phenyl	-6.2	-2.7	+11.9
2'-methoxy	-6.2	-3.7	+08.4
3'-methoxy	-5.9	-1.4	+15.4
AMSA	-6.6	-5.9	+02.4
m-AMSA	-5.7	-5.2	+01.7
o-AMSA	-6.3	-3.0	+11.2

^{a}Calculated from K_{int} in Table I. ^{b}Obtained from the slope of the van't Hoff plots shown in Figure 3. ^{c}The entropy was determined from the relationship $\Delta S^{\circ} = -(\Delta G^{\circ} - \Delta H^{\circ})/T$.

The added enthalpic contribution to the binding from the 1'-methanesulfonamido substituient in m-AMSA and AMSA requires that this group maintain a relatively limited number of conformation upon complex formation such that the methanesulfonamido moiety might interact with a site on DNA. Our entropic data support this theory, as both AMSA and m-AMSA are characterized by small entropy values. In contrast, o-AMSA exhibits entropy and enthalpy values similar to those observed for the biologically inactive N-phenyl, 2'-methoxy, and 3'-methoxy anilinoacridine analogs suggesting that the DNA binding modes and geometries of the DNA complexes are similar.

MOLECULAR MODELING

Standard computational techniques were used in an attempt to gain insight into the interactions of m-AMSA and the structurally similar anilinoacridine analogs with DNA. In thermodynamic studies described earlier, the DNA binding of both compounds which exhibit antitumor activity (i.e., m-AMSA and AMSA) was shown to be enthalpy driven. In contrast, the interaction of the structurally similar, but biologically inactive o-AMSA (and the other anilinoacridine analogs) were demonstrated to be entropy driven. Molecular modeling methods were used to explore the conformational modes available to these compounds and correlate these structures with their thermodynamic binding data described earlier.

Modeling of the Drug-DNA Complex. The sequence d(GCGC)$_2$ was constructed using the molecular editor (MOLEDT) subroutine of Insight/Discover (Biosym Technologies, Inc.). This structure is based on the B-DNA crystal structure of the Dickerson dodecamer (Dickerson et al., 1982) and was subjected to minimization using conjugate gradients and the AMBER force field until the RMS deviation was less than 0.001 kcal/$\mathring{A}^2$. A dielectric constant of 78.5 (water at 25°C) and no solvent molecules were included in the computations.

An intercalation site within the central C-G step of the d(GCGC)$_2$ was built by docking a proflavine within this site. Not only did this experiment provide an intercalation site within the d(CGCG)$_2$ tetramer but also served as a suitable control for comparison of the computer generated drug-DNA structure with the crystallographic structure of the d(CpG)$_2$-proflavine complex (Shieh et al., 1980). After docking of the drug, the complex was minimized to an RMS deviation of less than 0.001 kcal/$\mathring{A}^2$. Excellent agreement was observed between the minimized intercalation complex drug orientation and that of the proflavine-d(CpG)$_2$ crystal structure.

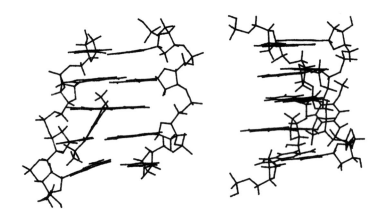

Figure 4. Structure of the m-AMSA-d(GCGC)$_2$ complex from the minor groove (left view) and the side (right view). Minimized structure was obtained as described in the text using the AMBER force field. Note the relative positions of the sulfonamido oxygens and guanine-NH$_2$ of the lower flanking guanine-cytosine base pair adjacent to the intercalation site.

Anilinoacridine-d(GCGC)$_2$ Complexes. Starting structures for the anilinoacridine-DNA complex minimizations were obtained by superimposing the acridine rings of the anilinoacridine compounds on the acridine ring of the proflavine in the proflavine-d(GCGC)$_2$ complex followed by selective removal of the proflavine. Final structures from these complexes determined after minimizing to an RMS deviation of 0.05 kcal/$\mathring{A}^2$.

The minimized drug-DNA structures comparing the o-AMSA and m-AMSA-DNA complexes are shown in Figures 4 and 5. Both drug-DNA complexes exhibit similar characteristics as far as base and backbone positioning are concerned with the drugs positioned such that their phenyl ring resides in the minor groove and acridine ring intercalated into the DNA helix.

These geometries are considerably different than those observed for the proflavine-DNA complex in terms of their base-base and base-drug orientations. The major structural perturbation that is observed is due to the proximity of the phenyl ring of the drug with respect to the lower cytosine within the intercalation site. Intercalation of the drug results in a forward displacement of the cytosine toward the major groove. A tilting effect of this nature has been previously observed in the crystal structure of ethidium-r(CpG)$_2$ complex which the phenyl group of ethidium induces a

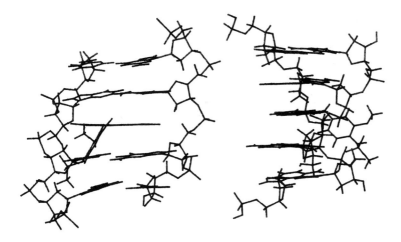

Figure 6.   Minimized structure of the o-AMSA-d(GCGC)$_2$ complex from the minor groove view (left) and the side view (right). The structure was obtained as described in the text using the AMBER force field. Note that the positions of the sulfonamido oxygens with respect to the adjacent guanine-NH2, a distance of ~ 4Å.

similar perturbation (Jain et al., 1977).   Unwinding angles of the anilinoacridine-oligomer complexes determined from the computer generated structures were found to be slightly lower than experimentally determined values (15 versus 20, respectively) (Waring, 1976; Denny and Wakelin, 1986). However, longer oligomers of varied sequences may be required to fully explore the unwinding angles induced by these drugs.

These structures demonstrate unique differences between m-AMSA and its biologically inactive conformer, o-AMSA, which may be correlated with the thermodynamic properties discussed earlier.   Analyses of the minimized structure for m-AMSA-d(CGCG)$_2$ complex in **Figure 4** reveals the potential for for formation an additional contact with DNA via hydrogen bonding between one of the oxygens of the sulfonamido group and the guanine 2-amino located in the minor groove. For *m*-AMSA, the distance between the sulfonamido oxygen and the amino hydrogen is 2.26 Å. In contrast, the distance between the sulfonamide oxygen of o-AMSA and the guanine-2-amino is 4.01 Å as demonstrated in Figure 5.   Examination of the structural orientation of the 2'- and 3'-methoxy moiety with respect to the methanesulfonamido group provides key insight into this hydrogen bonding potential. Intramolecular hydrogen bonding between the methoxy oxygen and sulfonamido hydrogen (2.3 Å) for the o-AMSA   (shown in Figure 6) may prevent the methanesulfonamido oxygens from assuming an orientation conducive for hydrogen bonding with the guanine-2-amino. This type of intramolecular hydrogen bonding has been previously observed for *o*-anisidine and *o*-acetanisidide (Sahini and Telea, 1978; Dyall and Kemp, 1966). In the case of m-AMSA the methoxy group located at the 3' position is in favorable proximity to form a hydrogen bond with the N9 amino group (2.3 Å), thus facilitating the orientation of the N-phenyl ring   such that the sulfonamido group is pointed towards the guanine 2-amino (Figure 6).

Although this particular intramolecular hydrogen bonding does not occur with AMSA, the methanesulfonamido group of this analog is not restricted (due to the 2'-methoxy hydrogen bond as in the case of o-AMSA) and also has the capability of forming an additional contact with the DNA via the methanesulfonamido group.

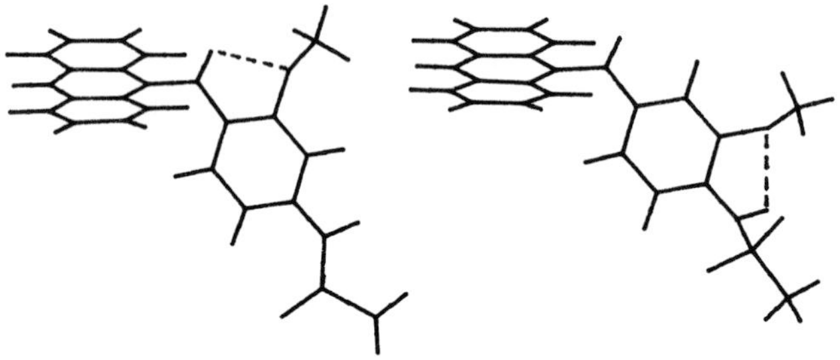

Figure 4. Minimized structures of m-AMSA (left) and o-AMSA (right) illustrating the potential for intramolecular hydrogen bonding within the drug molecules. Structures were obtained by "undocking" the drugs from their respective DNA complexes shown in Figures 4 and 5.

The hydrogen bonding between *m*-AMSA with the guanine 2-amino results in significant structural modifications at the intercalation site. As observed with the N-phenyl and *o*-AMSA analogs, the cytosine of the lower base pair is displaced into the helix axis toward the major groove, resulting in an overwinding of the C-G step. However, additional hydrogen bonding with the DNA forces the phenyl ring of the drug closer to the DNA helix, resulting in a more extensive perturbation of the cytosine. This structural perturbation (and stabilization) may be significant for expression of the antitumor activity of m-AMSA. Interestingly, the potential for formation of this hydrogen bond would indicate that the interaction of m-AMSA with DNA to be base specificity requiring that a G·C base pair flank the intercalation site. Currently, efforts are underway to determine the influence of flanking sequences on the sequence specificities of m-AMSA-DNA interactions.

Acknowledgements

This research was supported by NCI research grant CA-41474 from the National Cancer Institute.

References

Abraham, Z.H.L., Agbandje, M., Neidle S., Acheson, R.M. (1988) J. Biomol. Struct. Dyn. 6, 471-488.
Cain. B.F., Atwell, G.J. and Denny, W.A. (1975) J. Med. Chem. 18, 1110-1117.
Chaires, J.B., Dattagupta, N and Crothers, D.M. (1982) Biochemistry 21, 3933-3940.
Chaires, J.B. (1985) Biopolymers 24, 403-419.
D'Arpa, P. and Liu, L.F. (1989) Biochim. Biophys. Acta 989, 163-177.

Denny, W.A., Cain, B.F., Atwell, G.J., Hansch, C., Panthananickal, A., and Leo, A. (1982) J. Med. Chem. 25, 276-315.

Denny, W.A., Atwell, G.J. and Baugley, B.C. (1983) J. Med. Chem. 26, 1625-1630.

Denny, W.A. and Wakelin, L.P.G. (1986) Cancer Research 46, 1717-1721.

Dickerson, R.E., Drew, H.R., Conner, B.N., Wing, R.M., Fratini, A.V., and Kopka, M.L. (1982) Science 216, 475-485.

Dyall, L. K. and Kemp, J.E. (1966) Spectrochimica Acta 22, 483-493.

Feigon, J., Denny, W.A., Leupin, W., and Kearnes, D.R. (1984) J. Med. Chem. 27, 450-465.

Jain, S.C., Tsai, C.C. and Sobell, H.M. (1977) J. Mol. Biol. 114, 317-331.

McCridie, K.B. (1985) European Journal of Cancer 21, 1-3.

McGhee, J.D. and von Hippel, P.H. (1974) J. Mol. Biol. 86, 469-489.

Neidle, S., Pearl, L, Herzyk, P. and Berman, H.M. (1988) Nucleic Acids Research 16, 8999-9016.

Pommier, Y., Minford, J.K., Schwartz, R.E., Zwelling, L.A. and Kohn, K.W. (1985a) Biochemistry 24, 6410-6416.

Pommier, Y., Schwartz, R.E., Zwelling, L.A. and Kohn, K.W. (1985b) Biochemistry 6406-6410.

Pommier, Y., Covey, J.M., Kerrigan, D., Markovits, J. and Kohn, K.W. (1987) Nucleic Acids Research 15, 6713-6731.

Robinson, M.J. and Osheroff, N. (1990) Biochemistry 29, 2411-2515.

Sahini. V.E. and Telea, L. (1978) Revue Roumaine de Chemie 23, 483-487.

Shieh, H.-S., Berman, H.M., Dabrow, M., and Neidle, S. (1980) Nucleic Acids Research 8, 85-97.

Shimer, G.H. Jr., Wolfe, A.R. and Meehan, T. (1988) Biochemistry 27, 7960-7966.

Wadkins, R.M. and Graves, D.E. (1989) Nucleic Acids Research 17, 9933-9946.

Waring, M.J. (1976) European Journal of Cancer 12, 995-1001.

Waring, M.J. (1981) Annual Review of Biochemistry 50, 159-192.

Wilson, W.R., Baugley, B.C., Wakelin, L.P.G. and Waring, M.J. (1981) Mol. Pharmacology 20, 404-414.

KINETIC AND EQUILIBRIUM BINDING STUDIES OF A SERIES OF INTERCALATING AGENTS THAT BIND BY THREADING A SIDECHAIN THROUGH THE DNA HELIX

LAURENCE P.G. WAKELIN and WILLIAM A. DENNY
St. Luke's Institute of Cancer Research, Highfield Road, Rathgar, Dublin 6, Ireland and Cancer Research Laboratory, University of Auckland School of Medicine, Private Bag, Auckland, New Zealand.

We have investigated the DNA binding properties of a series of amsacrine derivatives substituted in the 4 position with carboxamide sidechains of varying structure. Our principal objective is to develop asymmetrically-substituted intercalating ligands that bind by positioning a sidechain in each groove of DNA. Compounds of this type have the potential for development as carriers for the groove-specific delivery of functionalised groups to DNA and as template inhibitors of transcription. Moreover, since these agents are obliged to thread a sidechain through the helical stack during the intercalation process, they bind with slow dissociation kinetics but with only moderate DNA affinities. This is a desirable combination of features from the perspective of antitumour drug design since a long residence time on DNA is likely to promote cytotoxic activity for a variety of chemical and biological reasons, while a low-to-moderate affinity for DNA will facilitate good tissue distribution and hence help promote solid tumour activity.

Viscometric titrations using circular DNA, UV/vis absorption spectroscopy and measurements of ligand effects on the chemical shift values of DNA imino protons confirm that the compounds studied bind by intercalation. Equilibrium measurements using synthetic polynucleotides and the ethidium displacement method show that the carboxamide sidechain imparts a GC basepair selectivity to the ligands. The affinity of compounds with positively charged carboxamide sidechains is unchanged or lowered compared to the (non-threading) parent ligand lacking the anilino function of amsacrine whereas the affinity of amsacrine-4-carboxamides with neutral sidechains increases in proportion to the bulk of the substituent. Stopped-flow measurements reveal the association kinetics to be complex and suggest that when the acridine chromophore is unsubstituted or the substituent is small the ligand can intercalate without requiring disruption of basepairing. However, when the threading group reaches a critical size the intercalation mechanism changes to one where transient opening of the basepairs is required to allow insertion. For these compounds the association rate is largely determined by the rate of basepair breathing and depends little on the structure of the sidechain. By contrast, dissociation rates are very variable and strongly dependent on sidechain structure, implying that the kinetic stability of the complex is determined in large part by specific interactions between functional groups on the sidechain and DNA. However, both association and dissociation rates depend on nucleotide sequence, each process being slower at GC than AT basepairs. Based on the results of these studies we propose a general model for the DNA complex of the amsacrine-4-carboxamides in which the anilino ring lies in the DNA major groove and the 4-carboxamide sidechain makes specific bonding interactions with component atoms of the bases in the minor groove.

191

B. Pullman and J. Jortner (eds.), Molecular Basis of Specificity in Nucleic Acid-Drug Interactions, 191–206.
© 1990 *Kluwer Academic Publishers. Printed in the Netherlands.*

Introduction

Practically all anticancer drugs kill cells by interfering with the biochemical functions of DNA and many clinically useful compounds do so by binding directly to DNA itself [1,2]. Amongst this latter category one finds that agents bind by a wide variety of mechanisms including reversible intercalation and groove binding (e.g., the anthracylines, actinomycin, amsacrine, mitoxantrone, chromomycin, mithramycin), irreversible alkylation (e.g., the nitrogen mustards, nitosoureas, cyclophosphamide, mitomycin), pseudo-irreversible metallation (e.g., the transition metal complex cisplatin) and free radical attack (e.g., bleomycin, neocarzinostatin) [1-3]. Whilst it is difficult to be certain exactly what feature of the DNA-drug complex ultimately leads to cell death, it is increasingly clear that most of the intercalating agents used in the clinic are cytotoxic by virtue of their capacity to poison the enzyme topoisomerase II [4]. The groove binding ligands are generally template inhibitors of transcription and it now appears that they too may also inhibit topoisomerase activity [1,4,5]. Alkylating agents cause interstrand crosslinks that are believed to be lethal for reasons associated with the difficulty of replicating segments of the genome in which the two strands of the parental DNA duplex are covalently welded together [3,6]. Cisplatin forms both inter- and intrastrand crosslinks and it remains uncertain which of these represents the cytotoxic lesion [7]. However, given that the intrastrand crosslinks are by far the more dominant species attention is increasingly focussing on their biochemical properties. In the case of the antibiotics that fragment DNA sugar rings by mechanisms involving activated oxygen species it is intuitively obvious that depolymerising DNA in this way will prove lethal to the dividing cell [1,3]. The reader will find that aspects of all of these topics are dealt with elsewhere in this volume and is referred thereto for relevant in-depth discussions.

An area of current activity in the search for new DNA-binding antitumour agents centres on the notion that antitumour efficacy might be enhanced in ligands that comprise features of more than one therapeutically useful compound in a single entity. One expression of this approach leads to attempts to improve the activity of agents of known clinical utility by combining their functional groups with a DNA-binding component that will efficiently target the reactive species to DNA. This concept is exemplified by the development of intercalating agents bearing alkylating and platinating groups [8-10]. In a similar vein, there is much interest in exploring the DNA-binding and pharmacological properties of various classes of alkylating minor groove binding ligand where the lead is taken from the example of antibiotics like anthramycin, CC-1065 and netropsin [1,11-13]. Clearly, if one wishes to investigate systematically the biological effects of targetting reactive species to DNA, it would be advantageous to have available a generalised carrier system that proffers the functional groups to DNA in a stereochemically controlled manner. It is this problem we address here, where we present a preliminary characterisation of the DNA binding properties of a class of agent intended to have the capacity to deliver chemically reactive groups selectively to the DNA minor and major groove.

Our design strategy was inspired by the example of the naturally-occurring antitumour antibiotic nogalamycin. The principal structural feature of nogalamycin that attracted our attention was that it possesses an anthracycline chromophore substituted with two dissimilar sidechains (one hydrophobic and one positively charged) that molecular models show must necessarily come to lie in separate DNA grooves in an intercalated complex [14]. Studies with closed circular DNA by Waring [1] confirmed nogalamycin as an intercalating agent and calculations by Neidle and associates [15] demonstrated that the sidechains could, in principle, be accommodated in either DNA groove. The issue of which sugar residue of nogalamycin lies in which DNA groove was resolved, at least for one nucleotide sequence, by our NMR study of the (nogalamycin)$_2$/d(GCATGC)$_2$ complex which revealed the hydrophobic nogalose sugar to be located in the minor groove and the positively-charged bicyclic sugar to be in the major groove [16]. This study led us to propose that components of the bicyclic sugar of nogalamycin make hydrogen bonding and electrostatic interactions with the O_6 and N_7 atoms of the guanine base in the 5'-CpA (5'-TpG) intercalation site [16]. These general features of the complex have been elegantly confirmed by the crystallographic studies of Wang and colleagues for nogalamycin bound at the sequence 5'-CpG described in reference [17] and elsewhere in this volume.

Nogalamycin also presented itself to us as a novel paradigm for drug design in a mechanisic context. The sidechains of nogalamycin are too bulky to insert through the DNA helical stack without disruption of basepairing which has interesting consequences for the thermodynamic and kinetic characteristics of the threading mechanism by which the antibiotic comes to be intercalated. The most significant issue here is that although nogalamycin intercalates into DNA several orders of magnitude more slowly than simple ligands, it dissociates even yet more slowly with dissociation time constants of thousands of seconds [18,19]. This results in a ligand with extremely slow dissociation kinetics but with only a moderate affinity for DNA (about 10^6 M^{-1}) [20]. This is a very desirable combination of features from the perspective of the design of DNA-binding antitumour drugs since a long residence time on DNA is likely to promote cytotoxic activity for a variety of chemical and biological reasons, while a low-to-moderate affinity for DNA will, in general, ensure good tissue distribution and hence help promote solid tumour activity [9].

Bearing in mind the structural and mechanistic features of the nogalamycin-DNA complex described above, we designed a series of compounds comprising a positively charged intercalating acridine chromophore, a bulky neutral anilino sidechain and a flexible carboxamide substitutent which is either positively charged or neutral. The two sidechains are attached at the acridine 9- and 4- positions so as to lie in opposite grooves of DNA when the acridine ring intercalates with maximum overlap with the basepairs. In this way the agents synthesised (the amsacrine-4-carboxamides, see compounds 4 to 11, Figure 1 and Table 1 for structures) embody the essential features of nogalamycin, in so much as they have an intercalating chromophore assymetrically substituted with a bulky hydrophobic sidechain and a positively charged 'hook' for making bonding interactions with the DNA, but they are not structural analogues, rather they are structural mimics. Here we present data which show that these compounds do indeed intercalate by a threading mechanism requiring the transient disruption of basepairing and we also demonstrate an 'uncoupling' of dissociation rates and binding affinity. We propose a model for the complex of the amsacrine-4-carboxamides in which the anilino ring lies in the DNA major groove and the carboxamide sidechain makes bonding interactions with components of the minor groove.

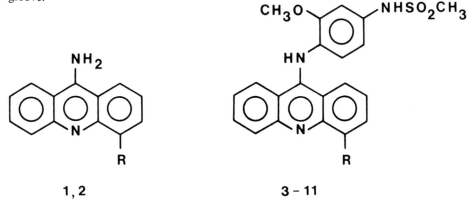

Figure 1. Formulae of the ligands studied. See Table 1 for the definition of the sidechain structures, R.

Materials and Methods

MATERIALS

The synthesis and characterisation of the amsacrine-4-carboxamides has been reported previously [21]. Covalently closed-circular plasmid pNZ 116 DNA was a gift from Dr. H.E.D. Lane, Department of Cell Biology, University of Auckland. Calf thymus DNA (type 1) was purchased from Sigma Chemical Co., Missouri, and poly (dG-dC).poly (dG-dC), poly (dA-dT).poly (dA-dT) and poly (dI-dC).poly (dI-dC) from Boehringer-Mannheim. The buffer used, designated 0.01 SHE, contained 2 mM HEPES, 10 μM EDTA and 9.4 mM NaCl dissolved in ultrapure water. It was adjusted to pH 7.0 at 20^0 with NaOH, the resultant ionic strength being 0.01. Solvents of higher ionic strength were obtained by supplementing buffer solutions with solid NaCl. The molecular weight of the calf thymus DNA was reduced to approximately 1 x 10^6 by sonicating solutions containing 2 mg ml^{-1} of DNA in 0.2 SHE buffer at 0^0 for 10 min on a Branson 150-watt sonicator, followed by dialysis into 0.1 SHE. All ligands were dissolved in 0.01 SHE buffer at a concentration of 1 mM and stored at -20^0.

EQUILIBRIUM AND SPECTROSCOPIC MEASUREMENTS

Association constants for binding to poly (dA-dT).poly (dA-dT) and poly (dG-dC).poly (dG-dC) were measured fluorimetrically by using the ethidium displacement method as previously described in 0.01 SHE buffer and are corrected for quenching effects [22]. The capacity of the ligands to remove and reverse the supercoiling of covalently closed circular pNZ 116 DNA was assessed viscometrically in 0.01 SHE buffer at 25^0 as previously described [23], and helix unwinding angles calculated by comparing equivalence binding ratios with that found for ethidium, taking the unwinding angle of the latter to be 26^0 [24]. Molar extinction coefficients at the wavelengths of maximum absorption in the visible spectrum were determined for the compounds free in solution, bound to calf thymus DNA, and when sequestered into sodium dodecyl sulphate (SDS) micelles using a Cary 219 spectrophotometer. All spectra were measured at a ligand concentration of 50 μM in 0.1 SHE buffer, DNA- and micelle-bound spectra being determined in the presence of 1 mM DNA and 10 mM sodium dodecyl sulphate (monomer concentration) respectively.

KINETIC MEASUREMENTS

These were performed using a Dionex D110 stopped-flow spectrophotometer coupled to a 64-kilobyte Apple II microcomputer as previously described [23,25,26]. Software has been developed to permit collection, storage, editing and analysis of up to 500 points per dataset, with a wide choice of sampling frequency [25]. The fastest rate of data collection is 0.1 ms per point. The spectrophotometer was fitted with a 20 mm light path optical cuvette and was operated in transmittance mode. For standard measurements of dissociation rates a solution of ligand-DNA complex containing 400 μM DNA (in nucleotide pairs) and 20 μM ligand in 0.1 SHE buffer was mixed with an equal volume of sodium dodecyl sulphate (monomer concentration 20 mM) in the same buffer at 20^0. In the association kinetics studies the mixing solutions contained 20 μM ligand and 400 μM DNA in 0.1 SHE buffer at 20^0. The spectrometer was operated with a time constant of 1 ms and the optical bandwidth set to 3 nm.

Table 1. Structures and spectral characteristics for binding to calf thymus DNA.

Compound	R	Free ligand[a] λ_{max}	$\varepsilon \times 10^{-3}$	DNA-bound ligand[b] λ_{max}	$\varepsilon \times 10^{-3}$	Micelle-bound ligand[c] λ_{max}	$\varepsilon \times 10^{-3}$
1	H	399	10.1	406	5.0	402	10.2
2	$CONH(CH_2)_2N(CH_3)_2$	407	9.4	421	4.9	412	9.4
3	H	434	12.1	439	8.8	437	12.2
4	$CONHCH_3$	438	11.6	457	7.4	449	12.6
5	$CONH(CH_2)_2OH$	441	10.4	457	7.9	450	12.2
6	$CONH(CH_2)_3OH$	439	10.9	456	8.3	450	12.8
7	$CONHCH_2CH(OH)CH_2OH$	425	10.4	455	7.4	448	11.4
8	$CONH(CH_2)_2N(CH_3)_2$	442	12.9	462	8.8	449	13.9
9	$CONH(CH_2)_2NH(CH_2)_2OH$	441	11.6	460	8.2	450	12.8
10	$CONH(CH_2)_3N(CH_3)_2$	441	11.6	458	8.5	449	13.3
11	$CONHCH_2CONH(CH_2)_2NH(CH_2)_2OH$	440	12.3	460	8.8	450	14.3

[a] 50 μM ligand in 0.10 SHE buffer.
[b] 50 μM ligand plus 1 mM calf thymus DNA (basepairs) in 0.10 SHE buffer.
[c] 50 μM ligand plus 1 mM calf thymus DNA (basepairs) plus 10 mM SDS (monomer concentration) in 0.10 SHE buffer.

Results

EVIDENCE FOR INTERCALATION

Three lines of evidence indicate that the amsacrine-4-carboxamides studied bind by an intercalative mechanism, namely; the nature of the perturbation to their visible absorption spectra on binding, their capacity to reverse the supercoiling of closed circular DNA and their ability to induce up-field shifts in the NMR chemical shift values of DNA imino proton resonances. As shown in Table 1, the absorption spectrum of each of the amsacrine carboxamides undergoes a bathochromic and hypochromic shift on binding, and the magnitudes of these effects are typical of those generally found for intercalating acridines (e.g. compounds **1** to **3** and refs [1,26-28]). Interestingly, the wavelengths of maximum absorption of compounds **4** to **11**, which differ simply in the structure of the carboxamide sidechain, range over 17 nm when free in solution but only over 7 nm when bound. Given that the visible absorption spectrum will reflect, in part, the degree of conjugation between the acridine ring and the carboxamide group, this may indicate that the carboxamide sidechains adopt a more restricted geometry with respect to the acridine ring in the DNA complexes than they do when the ligands are freely tumbling in solution. That the absorption spectrum of the amsacrine carboxamides is indeed sensitive to environment and the orientation of the sidechain, is made clear by the spectra obtained in sodium dodecyl sulphate micelles where all the compounds absorb at the same wavelength (within experimental error, Table 1).

Table 2. Binding constants and intercalation parameters for the compounds of Table 1.

Compound	Unwinding angle[a] (deg)	Imino ^{1}H Shift[b] (ppm)	Binding constant[c] $(K(o) \times 10^{-5} M^{-1})$	
			(dA-dT)	(dG-dC)
1	17	1.4	12	14
2	16	-	220	500
3	21	1.1	2.9	2.6
4	18	1.5	3.5	15
5	-	-	8.9	24
6	-	-	9.5	81
7	15	-	25	370
8	25	1.0	250	660
9	-	-	220	490
10	21	1.0	23	40
11	21	1.1	280	660

[a] Association constants for binding to poly (dA-dT). poly (dA-dT) and poly (dG-dC).poly (dG-dC) in 0.01 SHE buffer measured by the ethidium displacement method.
[b] Helix unwinding angle measured viscometrically using pNZ 116 DNA in 0.01 SHE buffer.
[c] Maximum shift, in parts per million, of the centre of mass of the GC imino proton envelopes. Data from [31].

Table 2 records the helix unwinding angles of representative compounds measured by comparing their capacity, with that of ethidium, to remove and reverse the supercoiling of plasmid DNA by viscometric titration. The values found for 9-aminoacridine (**1**), the 9-

aminoacridine-4-carboxamide **2**, and amsacrine (**3**) are indistinguishable from those previously reported [29,30]. The amsacrine carboxamides investigated reverse DNA supercoiling in the same manner as their individual constituent components (compounds **2** and **3**), with unwinding angles generally typical of acridines [1,29,30]. Compound **8**, however, stands out as having a larger unwinding effect, to an extent comparable with that of ethidium in fact, which suggests that the structure of the DNA-complex of this ligand may be somewhat different from the structure of the complexes of it's congeners.

Within a series of DNA-binding compounds such as this, where each member has an identical chromophore, a comparative measure of the extent of chromophore-basepair overlap and orientation can be obtained by NMR studies of ligand-induced perturbations to the chemical shift values of basepair imino protons. Such measurements have been reported [31] for several of the compounds studied here and the results are included in Table 2. Binding of compounds **4**, **8**, **10** and **11** induces the imino protons of GC basepairs to resonate about 1 ppm further up-field than in drug-free DNA which is the maximum shift calculated for both 9-aminoacridine (**10**) and amsacrine (**3**) when fully intercalated [32]. Thus, it appears that the amsacrine-4-carboxamides are intercalated with the major axis of their acridine rings oriented parallel to the major axes of the basepairs.

AFFINITY MEASUREMENTS

The affinities of compounds **1** to **11** for the synthetic DNAs poly (dA-dT).poly (dA-dT) and poly (dG-dC).poly (dG-dC) are shown in Table 2. 9-aminoacridine and amsacrine bind without selectivity to these polymers, the bulky anilino group reducing the binding constant of the latter by a factor of about 4. Adding the positively charged carboxamide sidechain to 9-aminoacridine, to give compound **2**, imparts a moderate GC-selectivity to the ligand and enhances affinity to the extent predicted by polyelectrolyte theory for binding of a di-cation [33]. The amsacrine-4-carboxamides **4** to **7** have uncharged sidechains and their affinities for the GC-containing polymer rise steadily, compared to that of amsacrine, in proportion to their steric bulk. By the time we reach the propane diol substituent (**7**) the binding constant is 143-fold greater than that of amsacrine. This trend is reflected in the poly (dA-dT) data as well, but to a lesser extent, so that the maximum enhancement is now only 9-fold. As a result, compound 7 has the greatest basepair selectivity and binds to poly (dG-dC) 15 times more tightly than to poly (dA-dT). Compounds **8** to **11** are di-cationic and, with the exception of **10**, bind to these two DNAs with affinities practically indistinguishable from those of ligand **2**. Thus, for compounds **8**, **9** and **11**, it is as though the bulky anilino sidechain makes no contribution to the thermodynamic stability of the DNA complex. Compound **10** appears to be somewhat of an enigma since it binds an order of magnitude less tightly than it's dictionic homologues and an order of magnitude more tightly than amsacrine.

ASSOCIATION KINETICS

The kinetics of association of the compounds to calf thymus DNA were studied by using stopped-flow spectrophotometry at analytical wavelengths close to the visible absorption maximum of the free ligand (Table 1), where the absorbance change on binding is greatest. Standard measurements were made at a molar ligand-to-basepair input ratio of 0.05 (i.e. one ligand molecule per two turns of the duplex) and a final DNA concentration of 200 µM, the data being presented in Table 3. The absorbance versus time plots were deconvoluted by a computerised curve-stripping procedure [25] with data from at least six independent kinetic runs being analysed separately and the results averaged. To ascertain the proportion of the association process being monitored in the kinetic measurements, the absorbance change observed kinetically was compared with the total absorbance change measured at equilibrium. The rates of association of 9-aminoacridine and amsacrine were too fast to measure and only a small fraction of the binding of the 9-aminoacridine-4-carboxamide 2 was discernible. By contrast, for the amsacrine carboxamides, one half to 90% of the equilibrium absorbance change was detectable in the stopped-flow time range. With the exception of compound 4 the kinetic profiles of the remaining ligands were basically similar, there being three resolvable exponential processes with time constants averaging 5, 25 and 90 msec respectively. The kinetically-determined absorbance changes are approximately equally distributed amongst the three processes for all the di-cationic amsacrine carboxamides and for the most bulky of the monocationic ligands (7). For the ligands with ethanol and propanol-containing sidechains (5 and 6), however, the middle process dominates and accounts for two-thirds of the total amplitude. Compound 4, bearing the methylcarboxamide sidechain, binds substantially more quickly than it's congeners with only two processes of equal amplitude being detectable. The rates of association of all the compounds were measured over a three-fold range in DNA concentration, keeping the ligand to basepair ratio constant at 0.05, and were found not to vary (data not shown). We deduce from this that the kinetic events detected are first-order processes associated with the insertion of the intercalating chromophore into the helical stack, and that the bimolecular association rates for these ligands are too fast to measure by this experimental approach.

TABLE 3. Kinetics of association for the binding of compounds of Table 1 to calf thymus DNA

Compound	Time constants[a] (ms)			Relative amplitudes[b]			F[c]	$1/\tau_{av}$[d] (s^{-1})
	τ_1	τ_2	τ_3	A1	A2	A3		
1	Too fast to measure							
2	4	19	-	52	48	-	7	89
3	Too fast to measure							
4	2	7	-	52	48	-	54	230
5	4	31	100	14	64	22	90	24
6	4	32	95	9	66	25	85	22
7	8	40	110	20	40	40	77	16
8	5	24	110	41	43	16	62	33
9	3	18	66	30	40	30	69	35
10	3	15	66	31	39	30	79	37
11	5	25	93	31	33	36	66	23

[a] Time constants describing the exponential processes associated with the binding of the compounds to calf thymus DNA in 0.1 SHE buffer at 20^0. Final [DNA] = 200 μM (basepairs) and final [ligand] = 10 μM.
[b] Amplitudes of the processes characterised by the corresponding time constants, expressed as a percentage of the sum of the amplitudes evaluated kinetically.
[c] The fraction of the observed equilibrium absorbance change associated with binding accounted for by the kinetic analysis.
[d] Reciprocal harmonic mean time constant defined as $1/\tau_{av} = 1/\Sigma(A_i/100 \times \tau_i)$

We also investigated the effect of increasing the ligand to basepair binding ratio on the rates of intercalation for the di-cationic amsacrine carboxamides 8, 9 and 11 (Table 4). We found there was no change in the number of resolvable exponentials or their relative amplitudes, but that the magnitudes of the time constants increase linearly with binding ratio. Thus, it appears that as the DNA lattice becomes more saturated the intercalation rate slows down.

TABLE 4. Variation of association rates with binding ratio for compounds **8**, **9** and **11**

Compound	R^e	Time constants[a] (ms)			Relative amplitudes[b]			F^c	$1/\tau_{av}{}^d$ (s^{-1})
		τ_1	τ_2	τ_3	A1	A2	A3		
8	0.05	5	24	110	41	43	16	62	33
8	0.10	8	35	130	41	46	13	72	28
8	0.20	12	53	300	39	48	13	60	14
9	0.05	3	18	66	30	40	30	69	35
9	0.10	7	36	130	40	49	11	68	29
9	0.20	16	60	300	44	48	8	60	17
11	0.05	5	25	93	31	33	36	66	23
11	0.10	8	37	100	33	30	37	74	20
11	0.20	10	71	220	39	48	13	65	15

[a-d] See footnotes to Table 3.
[e] Final ligand to DNA basepair binding ratio.

Finally, we probed the effect of basepair composition on intercalation rates for the amsacrine carboxamide **8**, by studying it's interaction with a variety of synthetic polynucleotides (Table 5). The kinetic spectrum for binding to poly (dG-dC) is very similar to that seen with calf thymus DNA, there being, once again, three resolvable processes. Although the amplitudes and time constants are somewhat different for calf thymus DNA and poly (dG-dC), the harmonic mean time constants are practically indistinguishable being 30 and 32 msec respectively. In other words, the average intercalation rates for these two DNAs is the same. In contrast, binding to poly (dA-dT) and poly (dI-dC) is much more rapid and the slowest step on the insertion pathway is lost altogether.

Table 5. Dependence of association kinetics of compound **8** on DNA basepair composition.

DNA	Time constants[a] (ms)			Relative amplitudes[b]			F^c	$1/\tau_{av}{}^d$ (s^{-1})
	τ_1	τ_2	τ_3	A1	A2	A3		
Calf thymus	5	24	107	41	43	16	62	33
(dG-dC)	6	16	58	25	30	45	76	31
(dA-dT)	3	16	-	64	36	-	25	130
(dI-dC)	5	20	-	69	31	-	28	103

[a-d] See footnotes to Table 3.

DISSOCIATION KINETICS

The rates of dissociation of the DNA complexes of the amsacrine carboxamides were measured by the surfactant-sequestration method using sodium dodecyl sulphate. The results of identical studies (final [DNA] = 200 µM, final [ligand] = 10 µM, final [SDS] = 10 mM, ionic strength = 0.1, temperature = 20°, pH = 7.0,) for compounds 1 to 3 have been reported previously [26,28]. Spectroscopic measurements confirmed that SDS fully dissociates the amsacrine carboxamide complexes and the absorption characteristics of the micelle-bound ligands are recorded in Table 1. Kinetic measurements were made in the stopped-flow apparatus at, or near, the wavelength of maximum absorption of the ligand in SDS solution where the optical signal is maximal for the dissociation process. Since no time-resolvable optical changes are detectable when DNA-free ligand solutions are mixed with SDS solutions, we infer that the transient absorbance changes measured in the presence of DNA are attributable solely to dissociation of the DNA-ligand complex. Control experiments were performed to show that dissociation rates of the complexes were independent of SDS concentration in the range 5 to 50 mM at constant ionic strength. The methods of data collection and numerical analysis were as described above, the resolved time constants and their associated amplitudes being given in Table 6. It is evident that the sum of the amplitudes of compounds 4 to 6 (in common with 9-aminoacridine and amsacrine) fall short of the equilibrium absorbance change characterising complete dissociation by 20 to 30%. Clearly, for these compounds, there are additional processes in the disociation pathway that occur more rapidly than the dead time of the stopped-flow instrument. However, for the rest of the amsacrine carboxamides the amplitudes of the resolved exponential components sum to at least 88% of the equilibrium absorbance change, indicating, practically speaking, that all of the dissociation processes have been observed.

Table 6. Kinetics of dissociation of calf thymus DNA complexes of the compounds of Table 1.

Compound	Time constants[a] (ms)				Relative amplitudes[b]				F[c]	$1/\tau_{av}$[d] (s^{-1})
	τ_1	τ_2	τ_3	τ_4	A1	A2	A3	A4		
1	3	15	-	-	88	12	-	-	50	430
2	6	28	86	428	15	33	34	18	89	8.6
3	2	6	-	-	75	25	-	-	48	600
4	11	29	-	-	45	55	-	-	79	59
5	81	208	-	-	45	55	-	-	80	8.3
6	58	157	-	-	35	65	-	-	72	11
7	14	103	328	1830	4	35	49	12	88	2.4
8	65	490	2100	6100	7	15	37	41	97	0.30
[e]8	63	434	2160	6170	6	20	43	31	95	0.34
9	79	278	1000	2900	4	14	37	45	100	0.58
10	26	230	570	-	7	55	38	-	92	2.9
11	28	190	960	2900	5	10	45	40	100	0.62

[a] Time constants describing the dissociation spectrum of calf thymus-ligand complexes in 0.10 SHE buffer at 20°. Final [DNA] = 200 µM (basepairs), final [ligand] = 10 µM and final [SDS] = 10 mM. [b,c] See footnotes to Table 3. [d] Here the reciprocal mean harmonic time constant has been corrected for the proportion of the reaction occurring within the dead time of the stopped-flow spectrophotometer. [e] Binding ratio = 0.20.

Table 6 shows that 9-aminoacridine (**1**), amsacrine (**3**) and compounds **4** to **6** have similar dissociation characteristics, insomuch that the kinetic profile of each can be resolved into two components. The 9-anilino substituent of amsacrine weakens the kinetic stability of the complex slightly. Adding the 4-methyl carboxamide (**4**) substantially slows the dissociation rate and brings the large majority of the absorbance change into the stopped-flow time range, and increasing the size of the carboxamide sidechain to include a hydroxyethyl and hydroxypropyl group (compounds **5** and **6**) further slows the dissociation process. Adding another hydroxyl group to **6** to give the propane diol derivative **7** causes a profound change in the kinetic spectrum, there now being four resolvable components with the longest having a time constant of nearly 2 seconds. Clearly, the bulkier sidechain has dramatically increased the kinetic stability of the complex. Replacing the propanediol with the charged dimethylaminoethyl group (**8**) maintains the kinetic complexity but slows the dissociation process even further and gives the slowest dissociating complex among those studied. Modifying the substituents on the terminal nitrogen of **8** to give **9** increases time constants by a factor of about 2, leaving the kinetic spectrum otherwise unperturbed. By contrast, adding a methylene group to the ethyl linker of compound **8**, thereby increasing the distance between the acridine ring and the positively charged nitrogen by one carbon unit (compound **10**), results in much faster dissociation rates, and most interestingly, in the complete loss of the slowest transient. Finally, replacing the sidechain with a glycine amide derivative of the sidechain of **9** to give **11** has but minimal effect on the kinetic spectrum compared to that of compound **9**.

Table 7. Dependence of dissociation kinetics of compound **8** on DNA basepair composition.

DNA	Time constants[a] (ms)				Relative amplitudes[b]				F[c]	$1/\tau_{av}$[d] (s^{-1})
	τ_1	τ_2	τ_3	τ_4	A1	A2	A3	A4		
Calf thy.	65	490	2100	6100	7	15	37	41	97	0.30
(dG-dC)	50	1640	3500	-	2	13	87	-	99	0.30
(dA-dT)	7	38	-	-	19	81	-	-	70	44
(dI-dC)	12	73	193	-	5	17	78	-	85	6.1

[a-d] See footnotes to Table 6.

The dissociation kinetics of compound **8** were also studied as a function of binding ratio and basepair composition. Table 6 includes the result of a measurement made to calf thymus DNA at a binding ratio of 0.2 which shows the kinetic behaviour to be indistinguishable from that found at the lower value of 0.05. Dissociation from poly (dG-dC) (Table 7) is both complex and slow, yielding a spectrum with three components heavily weighted in favour of the slowest one. Whilst it might appear that dissociation is somewhat faster from poly (dG-dC) than from calf thymus DNA, the mean dissociation rates (defined as the reciprocal of the harmonic mean time constants, see legend to Table 3 for definition) from the two polynucleotides are, in fact, identical. This contrasts with dissociation from poly (dA-dT) where the individual steps are very much faster, to the extent that only two of them occur in the stopped-flow time domain. Dissociation from poly (dI-dC) is intermediate between these extremes, there being three resolvable components whose time constants are much smaller than seen with calf thymus or poly (dG-dC) DNA.

Discussion

The amsacrine-4-carboxamides were designed so that each sidechain would necessarily lie in a different groove of the DNA helix when the acridine chromophore is intercalated with it's major axis parallel to the major axes of the basepairs, this being the normal stacking arrangement for intercalated acridines [1,2,31]. The results of the binding, spectroscopic, and circular DNA measurements reported here (Tables 1 and 2) indicate that the amsacrine carboxamides do indeed bind tightly to DNA by an intercalative mechanism. Moreover, the NMR data (Table 2) is direct evidence that these agents have similar stacking geometries to 9-aminoacridine and amsacrine [31]. Thus, we may be confident that the amsacrine carboxamides intercalate via a threading mechanism in which a sidechain penetrates through the helical stack. Many questions about the structure of the resulting DNA complexes and this unusual binding mechanism immediately spring to mind. They include; which sidechain is bound in which groove?, what is the nature of the equilibrium interactions between each sidechain and components of the DNA?, what effect does threading have on the magnitude of binding constants?, what are the consequences of this binding mechanism for base sequence selectivity? and what are the mechanistic details of the threading reaction? The combination of equilibrium and kinetic measurements reported here allow us to begin to address some of these issues and an illuminating place to start is with the last question.

A detailed analytical description of the binding kinetics of the amsacrine carboxamides is beyond the scope of the present study, however, largely because of technical difficulties associated with the complexity of the reaction and the limited nature of the kinetic data. Nevertheless, there is ample information available to make clear some important fundamental characteristics of how these compounds bind to DNA. Although intercalation mechanisms are very complex [34-37], it is generally the case that positively charged intercalating agents bind by a pathway in which the first step is the formation of an 'outside-bound complex'. This complex re-arranges by a series of sequential and/or parallel events, the details of which are ligand-dependent, that leads ultimately to the insertion of the chromophore into an equilibrium position [34-37]. In the context of such a binding mechanism simple 9-aminoacridines intercalate very rapidly with apparent association rates of about $1\text{-}5 \times 10^7$ $(Ms)^{-1}$ [37] and, accordingly, we find that binding of 9-aminoacridine and amsacrine is too fast to measure in our stopped-flow apparatus. By contrast, the amsacrine carboxamides, which also intercalate from an externally bound complex (see **Results**), insert into the helical stack very much more slowly. With the exception of compound **4** which has the smallest carboxamide substituent, they appear to thread through the helix via a common mechanism whose parameters are, to a large but not total extent, independent of sidechain structure (Table 3). If we interpret the reciprocal harmonic mean time constant for association as an indication of the average threading rate, then we see that this parameter varies from 16 to 37 s^{-1} for compounds **5** to **11** with **4** having a value of 230 (Table 3). The slowest compound is the propanediol derivative **7** which has the bulkiest sidechain, when viewed in a truly 3-dimensional sense (i.e. it has a ball-like sidechain), positioned closest to the intercalating chromophore.

This pattern of behaviour is consistent with the view that it is the carboxamide sidechain that threads through the helix, not the 9-anilino substituent (a formal, but sterically unlikely possibility), and that the insertion rate is largely determined by intrinsic dynamic properties of DNA associated with the process of opening a gap between basepairs large enough to facilitate passage of the sidechain. Given that the insertion rate is of the order of a few 10's of ms, which is comparable to basepair breathing rates, it seems likely that disruption of basepair stacking by breathing motions is, in fact, the rate-determining step along the threading pathway. This notion is supported by the mean rates of association of compound **8** to synthetic polynucleotides (Table 5), where binding to poly (dG-dC) is indistinguishable from binding to calf thymus DNA but is much slower than binding to poly (dA-dT) and poly (dI-dC). Thus, for compound **8** at least, the speed of intercalation correlates inversely with the thermal stability of the polymer duplex and basepair breathing rates. Interestingly, increasing the binding ratio of 8, in common with other derivatives (Table 4), has the effect of reducing the intercalation rate. This is, at first sight, somewhat surprising and whilst it may simply reflect the loss of potential intercalation sites by physical occlusion due to the rising concentration of externally bound ligand, it may be a hint that the latter is able to modify basepair breathing motions at a distance. The effect is noticeable when the lattice loading is increased from one ligand molecule per two turns of the helix to one per turn which suggests some form of long-range transmittability, perhaps the externally bound intermediate 'stiffens' the DNA in someway.

Whilst the forward rates of the threading steps on the intercalation pathway are relatively insensitive to sidechain structure, the backward rates are very strongly dependent on the character of the 4-carboxamide substituent (Table 6). This, coupled with the association kinetics data and the principal of microscopic reversibility, is strong evidence that the thermodynamic stability of the complex is determined in substantial part by specific inter-molecular interactions between the 4-carboxamide sidechain and components of the DNA. Indeed, for compounds 7 to 11 there is a direct correlation between the affinity for binding to poly (dG-dC) and the harmonic mean dissociation rate for reaction with calf thymus DNA. Amsacrine carboxamides 4, 5 and 6, although dissociating much more slowly than amsacrine itself, are clearly placed in a different group from compounds 7 to 11, and we may attribute this to the fact that their small carboxamide sidechains are less effective at inter-molecular bonding. To help shed light on what might be the nature of the strongly-stabilising sidechain-DNA interactions for 7 to 11, we turn to the results of our previous study [26] of the kinetics of dissociation of 9-aminoacridine-4-carboxamides. In this earlier work we proposed that the 9-aminoacridine analogue of 8 (the most slowly dissociating amsacrine carboxamide), compound 2 (Table 6), intercalates with it's carboxamide sidechain lying in the DNA minor groove, there being a specific bifurcated hydrogen bonding interaction between the O_2 oxygen atom of a cytosine base adjacent to the intercalation site and the NH atoms of the carboxamide and protonated terminal dimethylamino group of the ligand. For geometrical reasons, this binding mode is available only to those derivatives that possess an ethyl linkage between the carboxamide group and the terminal dimethylamino function. The dissociation spectrum of the calf thymus DNA complex of 2 also has four components (Table 6), the slowest of which we identified as characterising the sidechain-base carbonyl interaction [26]. Modifying the carboxamide sidechain of 2 and 8, making it the same as those in compounds 9 and 10, results in identical changes to the kinetic spectra in both the 9-aminoacridine- and amsacrine carboxamide series [26]. Here the most notable feature of diagnostic value is the loss of the slowest fourth transient on lengthening the distance between the terminal nitrogen and the carboxamide group by one carbon unit. This is compelling evidence that the amsacrine-4-carboxamides orient their sidechains in the DNA minor groove, in the same way as the 9-aminoacridine-4-carboxamides are thought to do [26]. We note, in passing, that molecular models suggest that the propanediol hydroxyls of compound 7, which lacks a positively charged sidechain but nevertheless dissociates slowly, can form a similar bifurcated hydrogen bonding interaction with cytosine carbonyl oxygens.

Measurements of the rate of dissociation of compound 8 from synthetic DNAs (Table 7), like the association measurements, show a strong dependence on base composition that correlates with the thermal stability of the helix and basepair breathing rates. Interestingly, the average rates of dissociation of 8 from calf thymus DNA and poly (dG-dC) are indistinguishable, which echoes the association data for these two DNAs. This suggests that in the calf thymus DNA studies at a binding ratio of 0.05 we are monitoring the kinetics for interaction with GC-rich intercalation sites. However, in contrast to the association measurements, raising the binding level of 8 from 0.05 to 0.2 does not affect the dissociation rates showing that the intercalated ligand seems not to alter DNA dynamics over a distance. Finally, with respect to the kinetic data, we note that the general congruence between the characteristics of the kinetics of binding to calf thymus DNA and the synthetic DNAs of lesser sequence complexity, combined with the insensitivity of the kinetic spectra to the extent of lattice loading strongly suggest that the observed transients describe sequential rather than parallel processes along the reaction pathway.

It is a general finding that appending an uncharged 4-carboxamide sidechain to amsacrine enhances DNA affinity and endows the ligand with selectivity for binding to GC-rich sequences (Table 2). By the same token, it also reduces association and dissociation rates (Tables 3 and 6), although there is no obvious quantitative correlation between the (average) kinetic and thermodynamic parameters. Nonetheless, given that compound 8 intercalates into poly (dA-dT) only four times faster than into poly (dG-dC) but dissociates from the latter 150-fold more slowly, it seems likely that the dissociation rate from GC-rich sequences is the principal determinant of affinity and, hence, selectivity for compounds 4 to 7. If so, since dissociation rates appear to be determined by the dynamic stability of the helix for threading agents, it would seem that simply attaching two inert bulky sidechains to a charged intercalating chromophore, in such a manner that one must pass through the helix, is one strategy for the development of ligands with enhanced affinity, GC-selectivity and slow dissociation rates. By this approach, one is taking advantage of the sequence dependence of the dynamic characteristics of DNA, rather than by designing specific intermolecular interactions between DNA and ligand.

However, our data suggest that if the objective is to develop agents that dissociate from DNA very slowly and have only moderate binding affinity, a better approach would be to design ligands in which the threading element is a positively-charged 'hook' that makes specific interactions with the DNA bases. This is made clear by comparing the equilibrium and kinetic data for compounds 2 and 8 which differ only by the possession of the inert 'blocking' anilino sidechain (Tables 2 and 6). These compounds have equivalent binding constants yet the threading ligand, 8, dissociates 30-fold more slowly. The example of nogalamycin, which has these binding and structural characteristics [14-20], implies that the uncoupling of thermodynamic and kinetic parameters can be further enhanced by making the 'hook' rigid. We note that long DNA residence times and moderate affinity, coupled with the positioning of substantial ligand mass in the DNA minor groove, is a common characteristic of antitumour antibiotics that function as template inhibitors of trancription (e.g., actinomycin, echinomycin, luzopeptin, nogalamycin, chromomycin) [1,2,16,17]. We hope that the above principles and insights will prove useful in the design of novel antitumour agents of this class.

In conclusion, it is fair to say that the present study has been very informative with respect to outlining the general features of the mechanism of DNA binding of the threading amsacrine-4-carboxamides. However, the evidence presented that defines which groove the carboxamide sidechain lies in is circumstantial, as all structural information deduced from thermodynamic and kinetic data must be. Indeed, the model we proposed for binding of the 9-aminoacridine-4-carboxamides, and upon which that for the amsacrine carboxamides heavily relies, has recently been questioned [38]. The two surest routes to resolving this dilemma are NMR spectroscopy and X-ray diffraction studies using oligonucleotide complexes. Despite this uncertainty, it is clear that the amsacrine-4-carboxamides, and compounds embodying their general characteristics, will be useful for delivering reactive functional groups to DNA in a groove specific manner (e.g. refs [8,10,39].

Acknowledgements

This work was supported by the Auckland Division of the Cancer Society of New Zealand, the Medical Research Council of New Zealand, the Peter MacCallum Cancer Institute, Melbourne, Australia and the Australian Research Grants Scheme. We also thank Patricia Chetcuti for expert technical assistance.

References

1. Gale, E.F., Cundliffe, E., Reynolds, P.E., Richmond, M.H. and Waring, M.J., (1981) *The Molecular Basis of Antibiotic Action*, 2nd edn., John Wiley, London, pp 258-401.
2. Wakelin, L.P.G. and Waring, M.J. (1990) DNA Intercalating Agents, in C. Hansch, P.G. Sammes and J.B. Taylor (eds.) *Comprehensive Medicinal Chemistry, Vol 2*, Pergamon Press, Oxford, pp 703-724.
3. Edwards, D.I. (1990) DNA Binding and Nicking Agents, in C. Hansch, P.G. Sammes and J.B. Taylor (eds.) *Comprehensive Medicinal Chemistry, Vol 2*, Pergamon Press, Oxford, pp 725-751.
4. Hertzberg, R.P. (1990) Agents Interfering with DNA Enzymes, in C. Hansch, P.G. Sammes and J.B. Taylor (eds.) *Comprehensive Medicinal Chemistry, Vol 2*, Pergamon Press, Oxford, pp 753-791
5. Baguley, B.C. et al., this volume.
6. Garcia, S.T., McQuillan, A. and Panasci, L. (1988) *Biochem. Pharmacol.* **37**, 3189.
7. Sundquist, W.I and Lippard, S.J. (1990) *Coordination Chemistry Reviews* **100**, 293-322.
8. Gourdie, T.A., Valu, K.K., Gravatt, G.L., Boritzki, T.J., Baguley, B.C., Wakelin, L.P.G., Wilson, W.R. Woodgate, P.D. and Denny, W.A. (1990) *J. Med. Chem.* **33**, 1177-1186.
9. Denny, W.A. (1989) *Anti-Cancer Drug Design*, **4**, 241-263.
10. Palmer, B.D., Lee, H.H., Johnson, P., Baguley, B.C., Wickham, G., Wakelin, L.P.G., McFadyen, W.D. and Denny, W.A. (1990) *J. Med. Chem.* in press.
11. Hurley, L.H., Lee, C.S., McGovren, J.P., Warpehoski, M.A., Mitchell, M.A., Kelly, R.C. and Aristoff, P.A. (1988) *Biochemistry*, **27**, 3886.
12. Hurley, L.H. and Needham-Vandervanter, D.R. (1986) *Accounts of Chemical Research*, **19**, 230.
13. Mitchell, M.A., Johnson, P.D., Williams, M.G., Aristoff, P.A. (1989) *J. Amer. Chem. Soc.*, **111**, 6428.
14. Arora, S.K. (1983) *J. Amer. Chem. Soc.*, **105**, 1328-1332.
15. Collier, D.A., Neidle, S. and Brown, J.R. (1984) *Biochem. Pharmacol.*, **33**, 2877-2880.
16. Searle, M.S., Hall, J.G., Denny, W.A. and Wakelin, L.P.G. (1988) *Biochemistry*, **27**, 4340-4349.
17. Liaw, Y.C., Gao, Y.G., Robinson, H., van der Marel, G.A., van Boom, J.H. and Wang, A.H.J. (1989) *Biochemistry*, **28**, 9913-9918.
18. Fox, K.R. and Waring, M.J. (1984) *Biochim. Biophys. Acta*, **802**, 162-168.
19. Fox, K.R., Brasset, C. and Waring, M.J. (1985) *Biochim. Biophys. Acta*, **840**, 383-392.
20. Das, G.C., Dasgupta, S. and Das Gupta N.N. (1974) *Biochim. Biophys. Acta*, **353**, 274-282.
21. Denny, W.A., Cain, B.F., Atwell, G.J., Hansch, C., Panthananickal, A. and Leo, A.J. (1982) *J. Med. Chem.*, **25**, 276-315.
22. Baguley, B.C., Denny, W.A., Atwell, G.J. and Cain, B. F. (1981) *J. Med. Chem.*, **24**, 170-177.
23. Denny, W.A. and Wakelin, L.P.G. (1987) *Anti-Cancer Drug Design*, **2**, 71-77.
24. Keller, W. (1975) *Proc. Natl. Acad. Sci. U.S.A.*, **72**, 4876-4880.
25. Roos, I.A.G., Wakelin, L.P.G., Hakkennes, J. and Coles, J. (1985) *Anal. Biochem.*, **146**, 287-298.
26. Wakelin, L.P.G., Atwell, G.J., Rewcastle, G.W. and Denny, W.A. (1987) *J. Med. Chem.*, **30**, 855-861.
27. Wilson, W.R., Baguley, B.C., Wakelin, L.P.G. and Waring, M.J. (1981) *J. Molec. Pharmacol.*, **20**, 404-414.
28. Denny, W.A. and Wakelin, L.P.G. (1986) *Cancer Res.*, **46**, 1717-1721.
29. Atwell, J.G., Cain, B.F., Baguley, B.C., Finlay, G.J. and Denny, W.A. (1984) *J. Med. Chem.*, **27**, 1481.

29. Atwell, J.G., Cain, B.F., Baguley, B.C., Finlay, G.J. and Denny, W.A. (1984) *J. Med. Chem.*, **27**, 1481.
30. Waring, M.J. (1976) *Eur. J. Cancer*, **12**, 995-1001.
31. Feigon, J., Denny, W.A., Leupin, W., Kearns, D.R. (1984) *J. Med. Chem.*, **27**, 450-465.
32. Denny, W.A., Atwell, G.J. and Baguley, B.C. (1983) *J. Med. Chem.*, **26**, 1625-1630.
33. Record, M.T., Anderson, C.F. and Lohman, T.H. (1978) *Q. Rev. Biophys.*, **11**, 103-178.
34. Li, H.J. and Crothers, D.M. (1969) *J. Mol. Biol.*, **39**, 461-477.
35. Bresloff, J.L. and Crothers, D.M. (1975) *J. Mol. Biol.*, **95**, 103-123.
36. Wakelin, L.P.G. and Waring, M.J. (1980) *J. Mol. Biol.,* **144**, 183-214.
37. Capelle, N., Barbet, J., Dessen, P., Blanquet, S., Roques, B.P. and Le Pecq, J.B. (1979) *Biochemistry*, **18**, 3354
38. Chen, K.X., Gresh, N. and Pullman, B. (1987) *FEBS Lett.*, **224**, 361.
39. Valu, K.K., Gourdie, T.A., Boritzki, T.J., Gravatt, G.L., Baguley, B.C., Wilson, W.R., Wakelin, L.P.G., Woodgate, P.D. and Denny, W.A. (1990) *J. Med. Chem.*, in press.

AMINOACYL-ANTHRAQUINONES: DNA-BINDING AND SEQUENCE SPECIFICITY

MANLIO PALUMBO and BARBARA GATTO
Department of Organic Chemistry, University of Padova
Via Marzolo, 1
35131 Padova
Italy

ABSTRACT. The binding process of mono- and bis- glycyl-, diglycyl- and triglycyl- derivatives of 1,4-diamino-9,10-anthracenedione to DNA has been studied by means of spectroscopic, chiroptical and hydrodynamic techniques. The thermodynamic parameters for the reversible interaction with the nucleic acid were evaluated at various ionic strength and temperature conditions . Natural as well as synthetic DNAs of different composition were used. Glycyl- and bis-glycyl compounds were shown to undergo a relatively fast intramolecular cyclisation processes in water solvent, spontaneous at neutral pH, forming one or two Shiff bases with the carbonyl oxygens of the anthraquinone moiety. This fact did not allow an accurate evaluation of the binding parameters for the "open" compounds. All other glycyl anthraquinones are stable. For monosubstituted derivatives the interaction with DNA is not cooperative, while the reverse is true for bis-substituted ones. The cooperativity factor drops to unity on raising salt concentration over 0.25 M. The dependence of the intrinsic binding constant upon ionic strength suggests that one side-chain only is on the average involved in electrostatic contacts with the polynucleotide backbone even in bis-substituted compounds. Measurements at variuos temperatures allowed an evaluation of the enthalpic contribution to the process. The entropy changes, negative for the monofunctional glycyl derivative, positive for the corresponding bifunctional one indicate remarkable differences in binding thermodynamics. Experiments with closed circular DNA show that all of the above compounds are able to undergo intercalative processes. In addition circular dichroism measurements show a different tendency to compact DNA by the examined drugs . Experiments in the visible region suggest a geometry in which the transition moment of the planar chromophore lies more or less parallel to the base-pair longest dimension in the intercalation pocket. We thus propose that when two aminoacyl side-chains are present in the anthraquinone derivatives, one will be located in the minor groove of the DNA double helix, and the other in the major groove. Finally, bis-substituted (but not mono-substituted) glycyl anthraquinones exhibited increased affinity for G-C rich sequences. This finding is possibly connected to the more effective stabilisation generated by the interaction of alpha protons of glycyl

B. Pullman and J. Jortner (eds.), Molecular Basis of Specificity in Nucleic Acid-Drug Interactions, 207–224.
© 1990 *Kluwer Academic Publishers. Printed in the Netherlands.*

residues with the nitrogen at position 7 of guanine as compared to N-7 of adenine.

Introduction

Anthracycline derivatives represent an important class of antineoplastic agents whose mechanism of action is related to their ability of forming very stable complexes through intercalation into DNA base pairs (Pigram et al., 1972; Quigley et al., 1980; Arcamone, 1981). Further stabilisation occurs as a result of electrostatic interactions between the positively charged amino-sugar moiety and the negative charge density of the double helix (Quigley et al., 1980; Wang et al., 1987, Patel et al., 1981). Even though anthracyclines are until now widely used in cancer chemotherapy, nonetheless they suffer important limitations due to their cardiotoxicity and to the onset of resistance phenomena (Arcamone, 1981; Fox and Fox, 1984). To overcome these major drawbacks a number of partially or totally synthetic derivatives were recently obtained , sharing the characteristics of a polycyclic intercalating moiety and one or two charged side-chain groups. Among them of particular relevance for its significant clinical activity is the 9,10-anthracenedione derivative mitoxantrone, having two hydroxy-ethylaminoethylamino- residues at positions 1 and 4, and two hydroxyl groups at positions 5 and 8. Its chemical and biological properties have been thoroughly reviewed in a recent book edited by J. W. Lown (1988). However, some of the main biological problems still persist and represent a challenge for the development of new drugs endowed with good activity, yet devoid of adverse side-effects.
In this line of research we have synthesized and investigated in the process of interaction with DNA some glycyl derivatives of 1,4-diammino-9,10-anthracenedione. Their chemical structure is reported in Figure 1. They are labelled as AG , followed by the number of glycine residues at position 1 and 4. They appear interesting for a number of reasons. First of all the side-chains are linked to the planar system through an amide bond. This fact generates a number of electronic and stereochemical effects, which are affecting their mode of interaction with receptor molecules, as well as their redox and pharmacokinetic properties (Palu' et al, 1986; Palumbo et al., 1987 ; Collier and Neidle, 1988; Palumbo et al., 1989). In addition, appropriate peptide sequences could represent the basic elements for specific recognition of given regions in DNA. Thus the above compounds constitute the starting step in the search for specificity in nucleic acid-drug interactions.

Materials and Methods

MATERIALS

DNA from Calf Thymus, E. Coli, C. Perfringens and M. Lysodeikticus were purchased from Sigma Chemical Co. St. Louis (MO) and purified as previously reported (Palumbo and Marciani Magno, 1983). The extinction coefficient values at 260 nm were 6,600, 6,500, 6,400 and 6,900 1/(M cm). Poly [dA-dT] and Poly[dG-dC] were also purchased from Sigma and used without further purification. Molar extinction coefficients at 260 and 254 nm respectively: 6800 and 8400 $M^{-1}cm^{-1}$ (Wells et al., 1970). DNA and polynucleotide solutions were prepared by suspending the nucleic acids overnight at 4 °C in a 10 mM Tris, 1 mM EDTA buffer, pH 7.0. After 1 h shaking and filtration through Millipore filters (3 μm) , the ionic strength was adjusted to the desired value by addition of appropriate amounts of NaCl. The Tris-EDTA-NaCl buffer will be abbreviated as ETN. The plasmid pBR 322 was purchased from Boehringer Mannheim (FRG) and purified by ultracentrifugation in a sucrose gradient (Palu' et al.,1986).

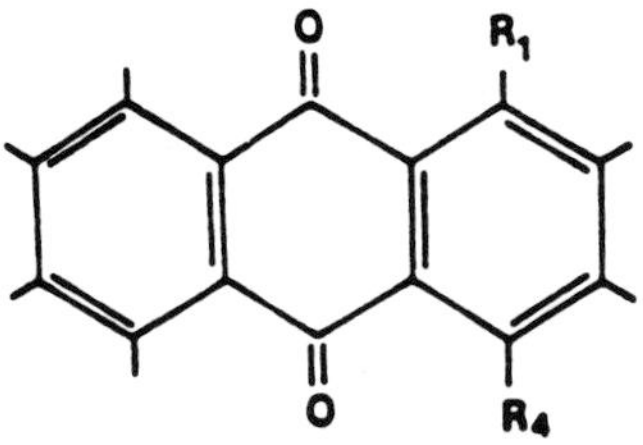

	R_1	R_4
AG10	$-G-H$	$-H$
AG11	$-G-H$	$-G-H$
AG20	$-(G)_2-H$	$-H$
AG22	$-(G)_2-H$	$-(G)_2-H$
AG33	$-(G)_3-H$	$-(G)_3-H$

Figure 1. Chemical structure of examined glycyl derivatives of 1,4-diamino-9,10-anthracenedione. G is the glycyl residue $-CO-CH_2-NH-$.

The glycyl anthraquinones AG10, AG11, AG20, AG22 and AG33 were synthesized in our laboratories, and characterized by NMR, Mass spectrometry, IR , amino acid- and elemental analysis. They were used as the hydrochlorides to ensure high solubility in aqueous media.

METHODS

Spectrophotometric measurements were performed in a Perkin-Elmer Lambda 5 instrument.
Fluorometric titrations were carried out in a Perkin-Elmer MPF 66 apparatus, interfaced to a PE 7500 Data Processor. The excitation vawelength corresponded approximately to the isosbestic point observed in spectrophotometric titrations. In the experiments known amounts (10 to 40 μL) of a concentrated solution of drug were added to a concentrated solution of DNA (0.5 to 2 mM) and to the buffer to reach the same final concentration of anthraquinone derivative (between 1 and

10 μM). Different DNA/drug ratios at constant drug concentration were thus obtained by mixing known volumes of the above two solutions. The amounts of bound (Cb) and free (Cf) drug were obtained according to the following equations :

$$Cb = \frac{I(f) - I}{I(f) - I(b)} \; C \quad ; \quad Cf = C - Cb$$

where I(f) and I(b) are the fluorescence intensities of free and DNA-bound drug, I is the fluorescence response of the mixture of free and bound drug being examined, and C is the total drug concentration.

In the equilibrium dialysis experiments DNA solutions (1.5-3 ml) at various concentrations were dialysed against equal volumes of solvent containing the glycyl anthraquinone derivative. Dialisys tubing (Thomas, U.S.A.) of 12,000 D cutoff were used. Equilibrium was usually reached after overnight shaking of the dialysis apparatus immersed in a thermostatted water bath. Polyelectrolyte concentrations were always low enough to minimize Donnan effects. The concentration of free drug at equilibrium was determined spectrophotometrically in the compartment not containing DNA. The amount of bound drug was evaluated by absorbance readings in the visible region of the solution containing DNA using the corresponding free drug at equilibrium as a reference. In this manner the contribution of the complex alone could be determined. The extinction coefficient of the latter was obtained using solutions of the drug containing large excess (500 to 1,000 times) of DNA.

Binding data were analized according to the neighbor exclusion model (McGhee and Von Hippel, 1974), which allows an evaluation of the intrinsic binding constant (Ki), the number of consecutive base residues covered by a ligand molecule (n) , and the cooperativity parameter (w). A least square fitting program was used on an IBM personal computer.

Electrophoretic dye titration experiments were carried out according to the method described by Espejo and Lebowitz (1976). The plasmid used was pBR322.

Circular dichroism measurements were performed in a Jasco J500 A apparatus equipped with an IBM personal computer for processing the data. Four to eight scans were accumulated for each measurement.

Results

DNA-BINDING PARAMETERS

An evaluation of the affinity of AG derivatives for DNA has been obtained by means of fluorometric titrations and equilibrium dialysis measurements. All examined drugs exhibit intense fluorescence responses in the visible region. These are dramatically modified in the presence of the nucleic acid. Quenching effects are observed, which can be used to determine the amount of free and bound drug.

Cyclisation of AG10 and AG11. The experiments with AG10 and AG11
indicated that the fluorescence response changed as a function of time
and no reliable binding data could be obtained. The same phenomena
occurred in equilibrium dialysis studies. Spectrophotometric experiments
showed that both anthraquinones undergo a structural modification as a
function of time in aqueous media at neutral pH. Figure 2 gives
representative examples thereof.

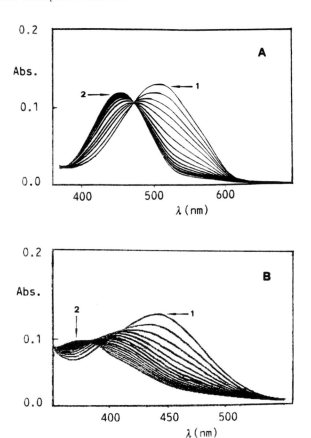

Figure 2. Changes in the absorption spectrum of AG10 (A) and AG11 (B) as
a function of time in ETN 0.1 M, pH 7.0 and 25 °C. 1) Immediately after
dissolution; 2) 24 h after dissolution. Intermediate curves are taken at
1 h intervals.

Kinetic investigations demonstrated that the reaction is monomolecular,
NMR analysis of the reaction product points to the formation of a Schiff
base between the terminal amino group of the glycyl side-chain and the
carbonyl moiety of anthraquinone, to give a seven-membered diazepine-

like ring as shown below:

AG10:

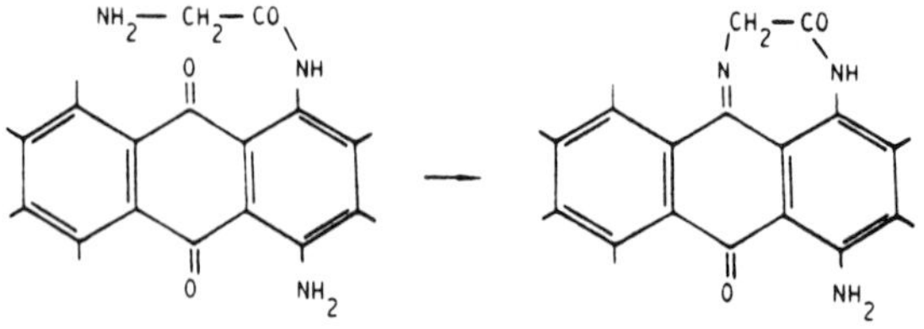

AG11:

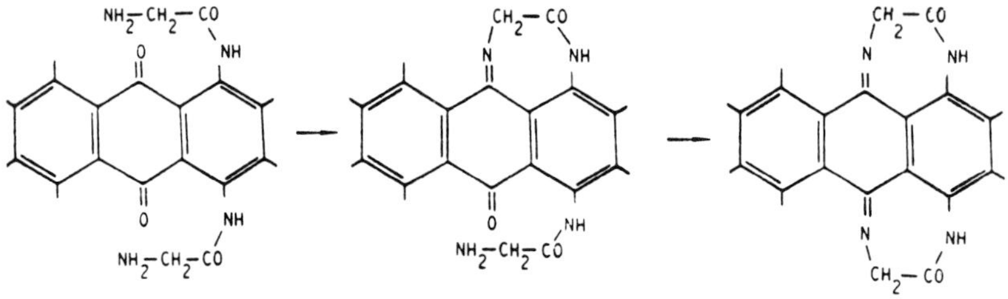

Due to the fact that structural modification of the ligand occurs during titrations with DNA and that the cyclic derivative exhibits substantially lower solubility in ETN than the starting material and tends to precipitate, we did not attempt a precise evaluation of the affinity of AG10 and AG11 for the nucleic acid. In any event K_i at physiological salt concentration and room temperature very probably exceeds 10^5 M^{-1}.

Effects of ionic strength. The binding of all glycyl anthraquinones is substantially affected by ionic strength, as demonstrated by the Scatchard plots reported in Figure 3.
The thermodynamic parameters for the binding of AG20, AG22 and AG33 at various salt concentrations are reported in Table 1.
The affinity of glycyl anthraquinones for DNA is lowered upon increasing salt concentration . The exclusion parameters are only slightly affected. Interestingly, the positive cooperativity at 0.25 M ionic strength vanishes upon raising the latter to 0.50 M and up.
The dependence of the intrinsic binding constant upon ionic strength allows the evaluation of the number of charged side chains able to interact electrostatically with DNA in the complex (Record et al., 1976; Record et al., 1978). A plot of log K_i vs. log (ionic strength) is presented in Figure 4.

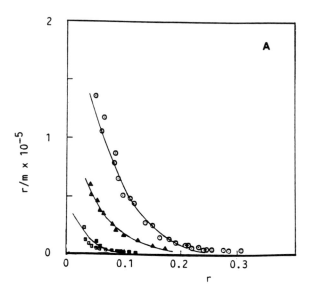

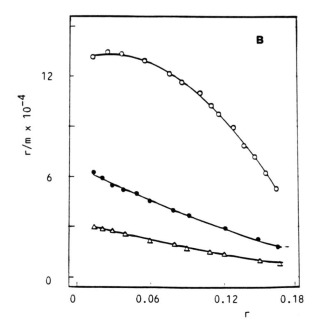

Figure 3. Scatchard plots for the binding of AG20 (A) and AG22 (B) to Calf Thymus DNA at 25 °C, pH 7.0 and various ionic strengths: A) circles 0.1 M, triangles 0.2 M, squares 0.5 M; B) open circles 0.25 M , filled circles 0.5 M, triangles 1 M.

TABLE 1. Thermodynamic parameters for the binding of glycyl anthraquinones to Calf Thymus DNA at 25°C, pH 7.0 and different ionic strength (IS) values.

Drug	IS (M)	Ki ($\times 10^{-5}$)	n	w
	0.10	2.1	3.6	1
AG20	0.20	0.9	3.6	1
	0.50	0.4	5.2	1
	0.25	1.3	4.8	6.7
AG22	0.50	0.7	3.5	1
	1.00	0.3	3.6	1
	0.25	1.3	4.1	6.5
AG33	0.50	0.8	3.1	1
	1.00	0.4	3.1	1

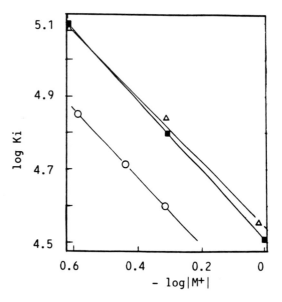

Figure 4. Dependence of log Ki upon log (ionic strength) for the binding of AG20 (circles), AG22 (squares) and AG33 (triangles) to Calf Thymus DNA at 25 °C, pH 7.0.

The slope of the lines presented in Figure 4 is always close to 1. The conclusion follows that one side chain only is effective in generating electrostatic interactions with the polynucleotide even in the case of 1,4 bis-substitution.

Effects of temperature. To evaluate the enthalpic contribution to the binding free energy, Ki was also measured at different temperature values. The results are presented in Table 2.

TABLE 2. Thermodynamic parameters for the binding of glycyl anthraquinones to Calf Thymus DNA at 0.25M ionic strength , pH 7.0 and different temperatures.

Drug	T (K)	Ki $(\times 10^{-5})$	n	w
	283	1.6	3.6	1
AG20	298	0.8	3.6	1
	313	0.5	3.6	1
	283	2.3	3.6	5.8
AG22	298	1.3	4.6	6.7
	313	0.9	4.3	5.0
	283	1.6	3.3	5.2
AG33	298	1.2	4.1	6.5
	313	0.7	3.8	4.5

In all cases the binding reaction is exothermic. From the van't Hoff plot of ln Ki vs. 1/T the following ΔH values were obtained: -32 kJ/mol for AG20 ; -23 kJ/mol for AG22 and -19 kJ/mol for AG33. Considering only minor changes in ΔH in the salt concentration range 0.15 - 0.25 M, the free energy, the non electrostatic component to the free energy , the enthalpy and entropy changes for the binding of glycyl anthraquinones to DNA at physiological salt and pH are summarized in Table 3.

TABLE 3. Changes in free energy (ΔG), free energy corrected for the electrostatic contribution (ΔG^{0}), enthalpy and entropy for the interaction of glycyl anthraquinones to Calf Thymus DNA.

Drug	ΔG (kJ/mol)	ΔG^{0} (kJ/mol)	ΔH (kJ/mol)	ΔS (J/K mol)
AG20	-28.6	-25.4	-32	-11
AG22	-30.3	-26.5	-23	+25
AG33	-30.1	-26.6	-19	+37

Whereas the free energy terms remain almost constant for the three compounds, remarkable differences are observed in the enthalpic and entropic terms when comparing mono- and bis-substituted compounds.

MECHANISM OF BINDING

Due to their planar chromophore and to the large number of similar
compounds thus far examined (Palumbo et al.,1989) glycyl anthraquinones
are likely to intercalate into the double helix of DNA. To confirm this
fact electrophoretic drug titration experiments were performed in the
presence of circular double stranded supercoiled DNA (pBR322). In all
cases the DNA band undergoes the typical changes in superhelical density

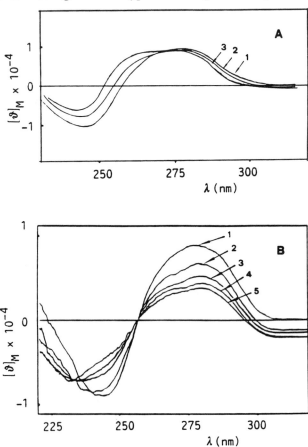

Figure 5. Circular dichroism spectra of Calf Thymus DNA at 0.1 M ionic
strength, 25 °C and pH 7.0 in the presence of increasing amounts of AG20
(A) and AG22 (B). A) DNA concentration 0.199 mM; DNA/drug ratios: 1) no
drug, 2) 9.49, 3) 6.61. B) DNA concentration 0.122 mM; DNA/drug ratios:
1) no drug, 2) 11.0, 3) 5.6, 4) 3.7, 5) 2.8.

which are produced by intercalative agents. These findings are in total agreement with recent reports on the mode of interaction of anthraquinones having amido substituents at positions 1 and 1,4 of the planar ring system (Collier and Neidle, 1988).

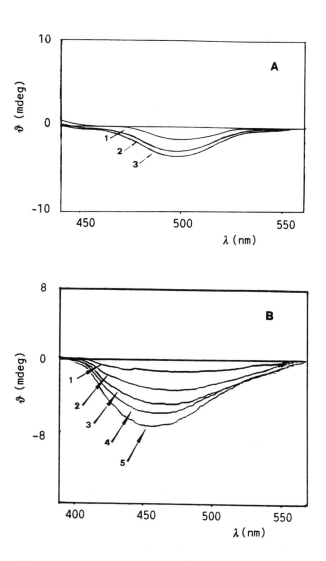

Figure 6. Circular dichroism spectra in the visible for the system AG20-DNA (A) and AG22-DNA (B) at 0.1 M ionic strength, pH 7.0 and 25°C. A) DNA/drug ratios:1) 11.1; 2) 5.5; 3) 4,1. DNA concentration 1.96 mM. B) DNA/drug ratios: 1) 46; 2) 23 ; 3) 15; 4) 13; 5) 9.1. DNA concentration 1.22 mM.

STRUCTURAL EFFECTS AND GEOMETRY OF INTERCALATION

To gain further insight on the structural modifications induced into DNA
by drug binding, circular dichroism studies were performed.
The dichroic spectra of DNA in the presence of increasing amounts of
AG20 are shown in Figure 5A. Only minor changes are observed for the
monosubstituted derivative. In the case of AG22 and AG33 (Figure 5B)
the spectrum is largely modified. In particular the positive dichroism
at 275 nm is reduced to remarkably lower values .
Useful information on the geometry of the intercalation complex was
obtained by examining the circular dichroism response in the visible
region. It refers to the asymmetric induction from the nucleic acid to
the ligand transition. We observed negative dichroism and similar
responses in all cases.The results referring to the system AG22-DNA are
presented in Figure 6.
According to theoretical studies (Schipper et al., 1980; Lyng et al.,
1987) it can be suggested that the transition moment of the
anthraquinone planar moiety lies more or less parallel to the base pair
longest dimension.

SEQUENCE SPECIFICITY

To investigate possible preferences of the examined drugs towards
specific DNA sequences, titration experiments were performed using DNAs
from various sources as well as synthetic polynucleotides. The results
are presented in Table 4 and Figure 7 .

TABLE 4.Thermodynamic parameters for the binding of AG20, of
AG22 and AG33 (ionic strength 0.25 M) to DNAs from various
sources and synthetic polynucleotides at 25 °C and pH 7.0.

Drug	DNA source	GC%	K_i (x 10^{-5})	n	w
AG20	C. Perfringens	28	0.7	3.5	1
	Calf Thymus	40	0.8	3.6	1
	M. Lysodeikticus	72	0.9	3.4	1
AG22	Poly(dA-dT)	0	0.7	5.2	12.8
	C. Perfringens	28	1.0	5.1	7.7
	Calf Thymus	40	1.3	4.8	6.7
	E. Coli	51	1.6	4.6	6.0
	M. Lysodeikticus	72	3.2	4.2	5.9
AG33	C. Perfringens	28	0.7	4.7	8.4
	Calf Thymus	40	1.2	4.1	6.5
	E. coli	51	2.3	3.5	5.3
	M. Lysodeikticus	72	3.9	2.9	3.5

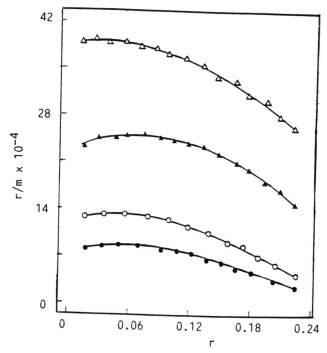

Figure 7. Scatchard plots for the binding of AG33 at 25°C, pH 7.0 and ionic strength 0.25 M. to DNAs from various sources: open triangles, M. Lysodeikticus; full triangles, E. Coli; open circles, Calf Thymus; full circles, C. Perfringens.

A remarkable preference of glycyl anthraquinones for GC rich nucleic acids is immediately evident. The exclusion parameters and cooperativity factors, on the other hand, are not substantially modified using different DNAs.

Discussion

The anthraquinone derivatives examined in this paper are characterized by the presence of glycyl side-chains , the simplest aminoacyl residues, at positions 1 or 1,4 of the tricyclic planar structure. Compounds AG10 and AG11 undergo spontaneous cyclisation in aqueous media to four membered ring systems, which are much less soluble and difficult to investigate. In addition, the lack of charged groups renders the binding to DNA less effective. Although this represents a drawback in an attempt to evaluate the binding parameters to DNA, nonetheless it could be advantageous from other points of view. In fact the cyclised derivative might be able to cross membranes more readily due to its pronounced hydrophobicity. Even more important, it has been shown that the process is reversed under acidic conditions. Considering that usually cancer cells exhibit a pH value lower than normal cells due to accumulation of lactic acid, it is concievable that the original linear structure of the side-chains be restored in the former cells, not in the latter. This could represent an interesting mechanism of selectivity, leading to activation of a drug in neoplastic systems only.

Mono- and bis-oligoglycyl derivatives exhibit very high affinity for DNA, the latter being more effective in binding to the nucleic acid roughly by a factor of 2. An interesting effect is the presence of positive cooperativity at physiological ionic strength for 1,4 bis-substituted compounds, but not for monosubstituted congeners. This fact generates clusters of drug molecules bound along the DNA structure and is indicative of considerable local modifications induced by the drug at the intercalating site.

Only one charged group is involved in complex formation using either AG20 or AG22/AG33 as the ligand. This means that one of the two side-chains of AG22 or AG33 is not interacting electrostatically with the polynucleotide. The explanation to this behaviour could rest on the fact that the protonation constant of the terminal amino group is not the same for both chains. If only one is protonated at neutral pH, the possibility of charged interactions is just one. Protonation of the second chain in the complex with DNA to give a second charged interaction is not likely, as the overall dependence of the apparent binding constant upon ionic strength should evidence this fact too.

The free energy terms exhibit comparable values for all compounds. The enthalpy and entropy contributions however differ remarkably when comparing AG20 to AG22 and AG33. In particular, binding of the former is characterized by a more favourable ΔH value and negative entropy change, while the reverse is true for AG22 and AG33. Thus important differences are revealed by the thermodynamic analysis of the process of interaction to DNA of the various compounds, notwithstanding their close structural similarity.

Chiroptical studies in the nucleic acid absorption region show that minor structural changes of the nucleic acid occur upon addition of AG20, the positive band at 275 nm shifting slightly to the blue. This behaviour is consistent with moderate compaction of the double helix brought about by effective charge neutralisation in the complex. On the other hand AG22 and AG33 cause a dramatic drop in ellipticity at 275 nm, the final values resembling those characteristic of nucleosomes and soluble chromatin (Cowman and Fasman, 1978). In this case cooperativity effects generate condensation areas where the ligand is locally more concentrated and DNA more compact. This arrangement is reminiscent of the beads on a string structure found in chromatin filaments, where alternating nucleosomes and linker DNA are present.

The dichroic spectra in the visible point to geometries of intercalation similar for 1 and 1,4 glycyl anthraquinones. Negative induced dichroism is observed, which is suggestive of a parallel orientation between intercalated chromophore and base-pair longest dimension. Thus, while AG20 (and AG10) can locate their charged side chain either the minor or in the major groove in analogy to other mono amido-anthraquinones (Collier and Neidle, 1988) , AG22 and AG33 (and possibly AG11) to maintain a more or less parallel orientation of the anthraquinone group with reference to base-pair hydrogen bonds must have one chain in the minor and one in the major groove. It has been proposed in a theoretical investigation on intercalating amido-anthraquinones that a 1.4 bis-substituted system cannot present both side chains in the minor groove due to the rigidity and steric hindrance of the amide group (Collier and Neidle, 1988). Two intercalation modes are thus available for this class of compounds: a perpendicular one, having both chains in the major groove which appears to be slightly preferred from the energy calculations, and a parallel one, exhibiting one chain in the major groove and one in the minor. Our experimental data appear to support the latter arrangement for our bis-substituted derivatives. In this connection it has to be pointed out that solvent contributions were not included in the calculations so that the stability of the above binding modes could be reversed in solution. Coming back to the induced circular dichroism spectra and to possible effects of ligand displacement from the center of the helix in the perpendicular mode of intercalation proposed by molecular modelling the planar tricyclic system is slightly protruding towards the major groove. For an AT site such displacement would lead from positive to 0 or to very low negative induced CD, whereas for a GC site it should remain positive (Lyng et al., 1987). In view of the observed preference of glycyl-anthraquinones for the latter type of site, positive CD should be predominant, which is not our case. In addition, the differences in binding enthalpy and entropy found when comparing AG20 to AG22 and AG33, could be interpreted in view of the fact that to generate a spear intercalation process in the latter compounds, one side chain of the bis-substituted system must cross the DNA double helix . This process should require more profound structural deformations of DNA in comparison to 1-substituted anthraquinones, causing a loss of enthalpy, and change the arrangement and number of water molecules bound to the nucleic acid, possibly leading to an increase in entropy.

In any event the binding stereochemistry proposed by us should be considered as a tentative one since more detailed structural information, in particular NMR data, is needed to fully clarify the geometry of the intercalated complex in solution.

A final interesting aspect of glycyl-anthraquinone binding to DNA is the remarkable specificity for GC rich sequences experimentally found in bifunctional but not in monofunctional derivatives. It has been proposed for mitoxantrone that the preference for alternating pyrimidine-purine sequences has to be ascribed to interactions involving side-chain methylene , amino NH and OH groups, and purine nitrogens or phosphate residues (Chen et al., 1986; Pullman, 1989). In particular the stronger attraction exerted by N7 of G, as compared to N7 of A, on the H atoms of the methylene flanking the protonated amino group with N7 of the 5'-purines of the intercalation site, seems to be responsible for the GC preference. The structural characteristics of glycyl side chain groups are compared to those of mitoxantrone in Figure 8.

Clearly numerous similarities can be observed. In particular the H atoms supposed to participate in interactions with DNA are located very closely. Most relevant, the glycine methylene, which is remarkably acidic, corresponds to the methylene group suggested to be responsible for base specificity in mitoxantrone. It should however be enphasized

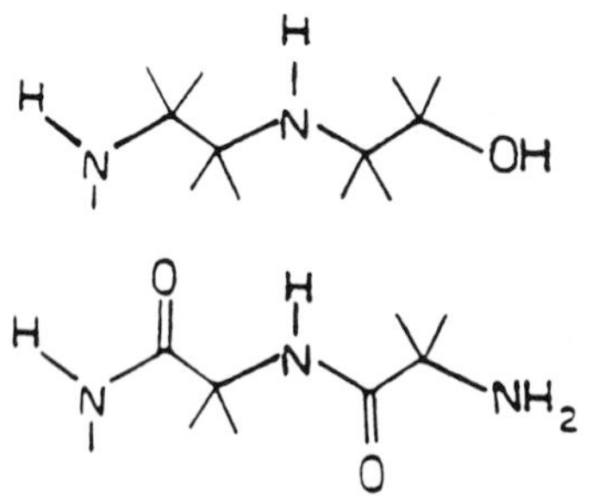

Figure 8. Comparison of a diglycyl side chain with the side chain of mitoxantrone.

that the arrangement proposed by Pullman (1989) for the latter compound includes both side-chains in the major groove. It would be interesting to compare the situation in the spear mode investigated by Collier and Neidle (1988).

Conclusions

The present investigation has demonstrated the ability of glycyl derivatives of 1,4-diamino-anthraquinone AG20, AG22 and AG33 to bind DNA in an effective and base-dependent manner. They constitute the starting compounds to develope peptidyl anthraquinones endowed with recognition elements for given sequences of DNA. In this connection elongation of the side-chain groups with L- or D- amino-acids has been recently performed in our laboratories. As it might be anticipated, preliminary results show distinct differences in the DNA-binding process by the two series of compounds. A thorough investigation of the effects of different chiralities on drug-DNA recognition, the introduction into molecules acting on DNA of short to medium-sized peptides corresponding to specific regions of repressor proteins to allow targeted delivery, as well as a detailed examination of the biological activities exhibited by the new compounds, represent the aim of our future research in this field.

Acknowledgements

The financial support of the Italian Ministry for University and Scientific and Technological Research is gratefully acknowledged.

References

Arcamone, F. (1981) Doxorubicin Anticancer Antibiotics, Academic Press, New York.

Chen, K.X., Gresh, N. and Pullman, B. (1986) 'A theoretical investigation on the sequence selective binding of mitoxantrone to double-stranded tetranucleotides', Nucleic Acids Res. 14, 3799-3812.

Collier, D.A. and Neidle, S. (1988) 'Synthesis, molecular modelling, DNA binding and antitumor properties of some substituted amido-anthraquinones', J. Med. Chem. 31, 847-857.

Cowman, M.K. and Fasman, G. (1978) 'Circular dichroism analysis of mononucleosome DNA conformation ' Proc. Natl. Acad. Sci. USA 75, 4759-4763.

Espejo, R.T. and Lebowitz, J. (1976) ' A simple method for the determination of superhelix density of circular DNAs and for observation of their superhelix density heterogeneity', Anal. Biochem. 7, 95-103.

Fox, B.W. and Fox M. (1984) Antitumor Drug Resistance, Springer-Verlag, New York.

Lown, W.J. (1988) Anthracyclines and Anthracenedione-Based Anticancer Agents, Elsevier Science Publishers, Amsterdam.

Lyng , R., Härd, T. and Norden, B. (1987) 'Induced CD of DNA intercalators: electric dipole allowed transitions.' Biopolymers 26, 1327-1345.

Mc Ghee, J.D. and Von Hippel, P.H. (1974) 'Theoretical aspects of DNA-protein interactions: cooperative and non-cooperative binding of large ligands to one-dimensional homogeneous lattice' J. Mol. Biol. 86, 469-489.

Palu', G., Palumbo, M., Antonello, G.A. and Marciani Magno S. (1986) 'A search for potential antitumor agents: biological effects and DNA binding of a series of anthraquinone derivatives' Mol. Pharmacol. 29, 211-217

Palumbo, M. and Marciani Magno, S. (1983) 'Interaction of deoxy-ribonucleic acid with anthracenedione derivatives' Int. J. Biol. Macromol. 5, 301-307.

Palumbo, M., Palu', G., Gia, O., Ferrazzi, E. and Meloni, G.A. (1987) 'Bis-substituted hydroxy-anthracenediones : DNA binding and biological activity' Anti-Cancer Drug Des. 1, 337-346.

Palumbo, M., Palu', G. and Marciani Magno, S. (1989) 'Chemotherapeutic agents of the anthraquinone structural type: DNA-binding and biological activity' in H. van der Goot, G. Domany, L. Pallos and H. Timmermann (eds), 'Trends in Medicinal Chemistry '88 ' Elsevier Science Publishers, Amsterdam, pp. 757-780

Patel, D.J., Kozlowski, S.A. and Rice, J.A. (1981) 'Hydrogen bonding, overlap geometry and sequence specificity in anthracycline antitumor antibiotic-DNA complexes in solution' Proc. Natl. Acad. Sci. USA 78, 3333-3337

Pigram, W. J., Fuller, W. and Hamilton, L. D. (1972) '
Stereochemistry of intercalation: interaction of daunomycin with
DNA' Nature New Biol. 235, 17-19.

Pullman, B. (1989) 'Molecular mechanisms of specificity in DNA-
antitumor drug interactions' Adv. Drug Res. 18, 1-113.

Quigley, G., Wang, A., Ughetto, G., Van der Marel, G., Van Boom, J.
and Rich, A. (1980) 'Molecular structure of an anticancer drug-
DNA complex: Daunomycin plus d(CpGpTpApCpG)' Proc. Natl. Acad.
Sci. USA 77, 7204-7205.

Record, M.T., Lohman, T.M. and De Haseth, P. (1976) 'Ion effects on
ligand-nucleic acid interactions' J. Mol. Biol. 107, 145-158.

Record, M.T., Anderson, C.F. and Lohman, T.M. (1978) 'Thermodynamic
analysis of ion effects on the binding and conformational
equilibria of proteins and nucleic acids: the roles of ion
association or release, screening and ion effects on water
activity' Q. Rev. Biophys. 11, 103-178.

Schipper,P.E., Norden, B. and Tjerneld, F. (1980) 'Induced circular
dichroism of DNA intercalators' Chem. Phys. Letters 70, 17-21.

Wang, A.H.J., Ughetto, G., Quigley, G.J. and Rich, A. (1987)
'Interactions between an anthracycline antibiotic and DNA:
molecular structure of daunomycin complexed to d(CpGpTpApCpG) at
1.2 Å resolution' Biochemistry 26, 1152-1163.

Wells, R.D., Larsen, J.E., Grant, R.C., Shortle, B.E. and Cantor,
C.R. (1970) 'Physicochemical studies on polydeoxyribonucleotides
containing defined repeating nucleotide sequences' J. Mol. Biol.
54, 465-497.

THE MOLECULAR BASIS OF SPECIFIC RECOGNITION BETWEEN ECHINOMYCIN AND DNA

M. J. WARING
University of Cambridge
Department of Pharmacology
Tennis Court Road
Cambridge CB2 1QJ
England

ABSTRACT. The development of ideas about the process of echinomycin-DNA recognition is reviewed, beginning with solution studies, then nmr investigations, crystallography and footprinting. Evidence drawn from all these areas, as well as structure-activity relations on echinomycin analogues, clearly points to the alanine residues of the octapeptide ring as critical determinants of the observed specificity for CpG sequences in DNA. Hydrogen bonding interactions occur which involve the NH and carbonyl groups of the alanines, together with the 2-amino group and N3 of guanine in the minor groove of the helix. Steric complementarity of fit between the antibiotic and its receptor is important, leading inter alia to the adoption of Hoogsteen base pairing in crystalline complexes between echinomycin or triostin and oligonucleotides. But efforts to detect non-Watson-Crick base pairing in complexes of echinomycin with large DNA molecules in solution have failed.

1. Introduction

It was a quarter of a century ago, in the laboratory of Edward Reich in New York city, that the first observations of echinomycin-DNA interaction were made. They were published, along with similar observations on a number of other DNA-binding substances, in a paper which was to prove seminal in the field of ligand-DNA interactions (Ward et al. (1965)). For the first time it was suggested that the anti-cancer activity of a range of antibiotics might be attributable to their binding to DNA so as to distort its structure and function, a notion which quickly became generally accepted (Waring (1968, 1981); Gale et al. (1981)). It has stood the test of time, though the original assumption that consequential inhibition of transcription and/or replication would be the death of cancer cells is now in doubt (Waring (1981); Denny (1989)).

A couple of years after the work of Ward et al. (1965) a group of Japanese scientists confirmed the DNA-binding property of echinomycin under its synonym quinomycin A (Sato et al. (1967)). Thereafter not much happened until the ability of echinomycin to unwind the DNA helix was tested as part of a comprehensive survey of that property among DNA-binding drugs (Waring (1970)). It was found that echinomycin caused twice the unwinding seen with classical intercalating drugs like ethidium, as well as twice the helix extension (Waring and Wakelin (1974); Wakelin and Waring (1976)). Those findings were the twin pillars upon which the hypothesis was built that echinomycin binds to DNA by a process of bifunctional (or bis-) intercalation, thus giving the antibiotic the unique distinction of being the first bis-intercalator ever identified (Waring and Wakelin (1974)). Synthetic bis-intercalators soon followed (Le Pecq et al. (1975); Wakelin et al. (1978); Wakelin (1986)).

B. Pullman and J. Jortner (eds.), Molecular Basis of Specificity in Nucleic Acid-Drug Interactions, 225–245.
© 1990 *Kluwer Academic Publishers. Printed in the Netherlands.*

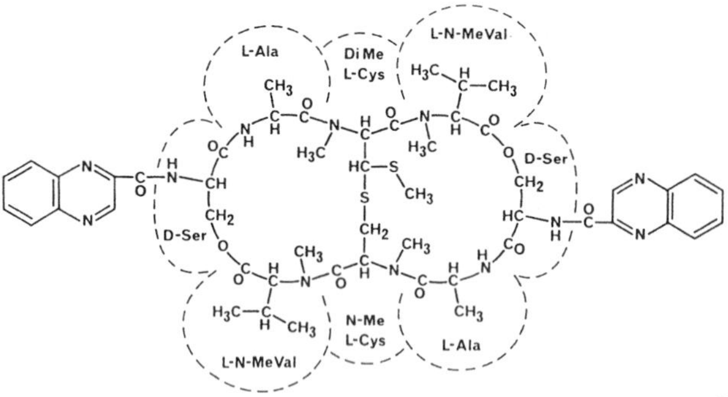

Echinomycin

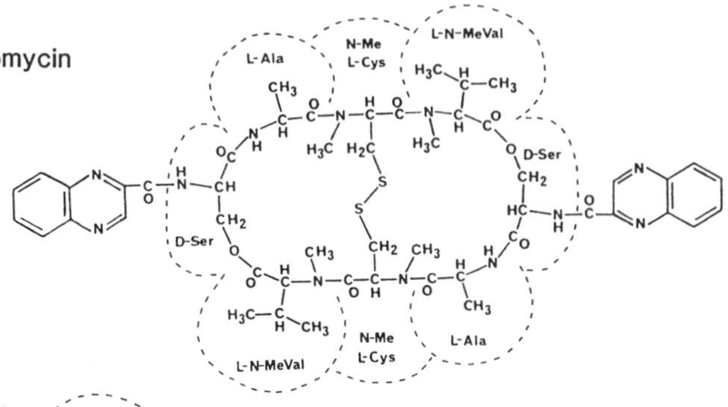

Triostin A

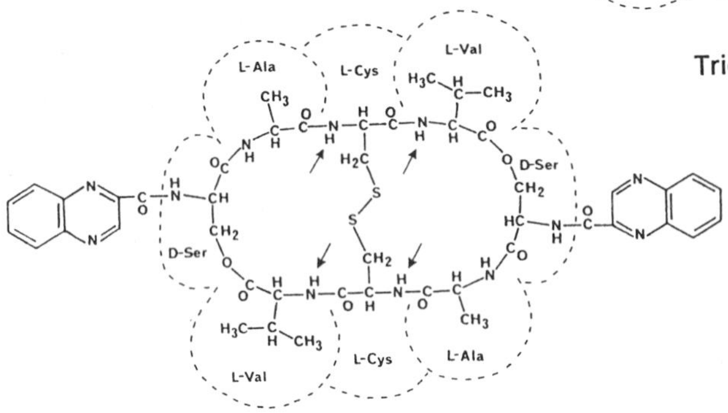

TANDEM

FIGURE 1. Structure of echinomycin and related compounds. Note that echinomycin is one member of a family of antibiotics which are characterised by a cross-linked heterodetic octadepsipeptide ring bearing two quinoxaline-2-carboxamide chromophores (Katagiri et al. (1975); Waring (1979); Olsen (1983)). Echinomycin is identical to quinomycin A. Other quinomycins have conservative amino acid substitutions at the positions of the L-N-methylvaline residues. There are also homologous triostin antibiotics (probably the biosynthetic precursors of quinomycins) in which the cross-link is a regular disulphide rather than a thioacetal. Triostin A is the strict homologue of echinomycin. Olsen (1983) and his colleagues have synthesised a range of quinoxaline depsipeptides including TANDEM (an acronym for des-N-tetramethyl triostin A) which lacks the methyl groups attached to four peptide bonds in the natural antibiotics (arrowed in the illustration).

The broad outlines of the molecular mechanism whereby echinomycin recognises DNA, indeed recognises a particular sequence of nucleotide-pairs within it, were implicit in the bis-intercalation model as originally envisaged in 1974. At a symposium held two years later the following rules were written down to define an acceptable molecular model for echinomycin which would fit the facts:

(1)   the chromophores should lie on the same side of the peptide ring;
(2)   their planes should be approximately parallel;
(3)   the vertical distance between those planes should be an integral multiple (x) of 3.4Å, i.e. the theoretical spacing required to accommodate x base-pairs;
(4)   the space between the chromophores should be essentially free from obstruction by any other substituents attached to the peptide ring (Waring (1977)).

As a result of playing with molecular models based on nmr data (Cheung et al. (1978)) backed up by conformational calculations (Ughetto and Waring (1977)) it was also suggested that the known specificity of echinomycin for binding to GC-rich sequences in DNA might originate from hydrogen bonding interactions between amino acid residues in the peptide ring and acceptor/donor groups on the DNA base-pairs, most likely involving the 2-amino groups of guanine bases exposed in the minor groove of the helix (Lee and Waring (1978); Cheung et al. (1978); Waring (1979)). This prediction demanded no great leap of intellectual insight, for the antibiotic was known to bind well to bacteriophage T2 DNA whose major groove is largely occluded by glucosyl substituents attached to its 5-hydroxymethyl cytosine residues (Waring and Wakelin (1974)), suggesting therefore that the antibiotic must lodge in the minor groove, and the only hydrogen bond donor group in the minor groove of DNA is indeed the 2-amino group of guanine. Moreover, as we now know very well, most small ligands attach to DNA via its minor groove whereas proteins tend to prefer the major groove where a more characteristic sequence-specific pattern of hydrogen bond donors and acceptors is displayed. Of the possible substituents in the peptide ring which might accept a hydrogen bond from the 2-amino group of guanine, the carbonyl group of alanine was considered most likely, though doubt was expressed as to whether both alanine carbonyls could participate in such interactions at the same time (Cheung et al. (1978)). Thus the stage was set for detailed conformational studies of echinomycin and of its interaction with DNA.

2. The Conformation of Echinomycin in Solution

Despite remarks in the early chemical literature suggesting that echinomycin can be recrystallised (Keller-Schierlein et al. (1959)) the structure of the purified antibiotic has not yet been determined by X-ray analysis. Numerous laboratories have attempted over a period of fifteen years to grow crystals suitable for X-ray diffraction, but failed. Perhaps for this reason more than any other it has been subjected to detailed scrutiny by nmr spectroscopy, both ^{1}H and ^{13}C. Unfortunately, the low solubility of echinomycin in water ($\leq 5\mu M$) has precluded all but the most rudimentary spectroscopic studies in purely aqueous environments, though much has been learned about its properties in polar solvents such as dimethyl sulphoxide as well as in frankly non-polar solvent systems such as chloroform where it is amply soluble.

The first attempts to build a molecular model of echinomycin were confounded by an unusual misfortune: the published structural formula (Keller-Schierlein et al. (1959)) was wrong (Dell et al. (1975); Martin et al. (1975); Waring (1979)). Using the corrected formula (Figure 1), and taking note of such conformational information as could be gleaned from nmr

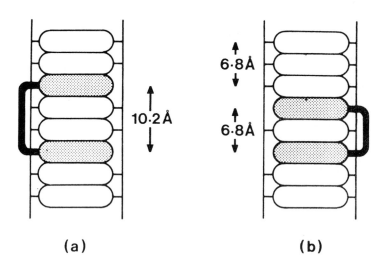

FIGURE 2. Schematic illustration of bis-intercalative binding of a ligand to DNA subject to the constraint of neighbouring site-exclusion (a) or in violation of that principle (b). The intercalated chromophores (shaded) and the DNA base-pairs are shown in edgewise projection as discs 3.4Å thick, with the sugar-phosphate backbones reduced to vertical lines for clarity; the helical twist of the DNA is neglected. The chain of atoms linking the chromophoric moieties is shown as a heavy line; its length effectively determines whether the separation between chromophores is sufficient to allow sandwiching of two base-pairs (a) or one (b). From Denny et al. (1983).

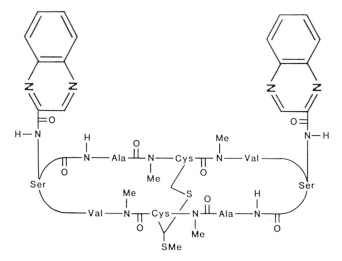

FIGURE 3. Schematic representation of the solution conformation of echinomycin deduced from nmr experiments.

coupling constants, Ughetto and Waring (1977) then embarked upon a semi-empirical computational study to try to determine a minimum-energy conformation for the antibiotic. It transpired that the constraints imposed by the tight packing of atoms within the cross-linked peptide ring were so stringent that the ring was obliged to behave as a more or less rigid disc with the quinoxaline chromophores projecting from one side with their points of attachment separated by about 10Å. There was fairly free rotation about those points of attachment, such that the planes of the chromophores could be placed parallel in an ideal fashion to sandwich two DNA base pairs between them. Thus were satisfied the four rules earlier rehearsed, as well as the requirement of the widely-accepted neighbour exclusion hypothesis which predicts that a successful bis-intercalator must sandwich a minimum of two base pairs (Figure 2) (Le Pecq et al. (1975); Wakelin et al. (1978); Denny et al. (1983)).

There followed a series of more and more sophisticated nmr investigations beginning with that of Cheung et al. (1978). These authors confirmed the rigidity of the peptide ring as well as the trans conformation of all peptide bonds. They found the quinoxalines oriented on the same face of the peptide ring; that face they designated "up", as indicated in Figure 3. From a careful examination of molecular models and reference to the effects of lanthanide shift reagents they succeeded in establishing the orientation of relevant groupings as shown in Figure 3. Both alanine carbonyls and both N-methyl groups of the L-N-methyl valine residues project "up", on the same side as the chromophores. But the carbonyls and N-methyl groups of the L-N-methyl cysteine residues are oriented "down", as also are the carbonyls of the D-serine residues and the S-methyl of the thioacetal cross-bridge. These assignments were confirmed by Williamson and Williams (1981) in a study where they manipulated the nuclear Overhauser effect by the use of a viscous solvent; they also succeeded in establishing the absolute chirality at the asymmetric carbon atom of the thioacetal cross-bridge (S) as well as the conformation of the methylene protons in the cross-bridge. Interestingly, Cheung et al. (1978) noted significant differences in lanthanide-induced deshielding of pairs of structurally equivalent protons in the quasi-symmetrical peptide ring of echinomycin (e.g. the α-hydrogens of the serine residues or the alanine residues). This was assumed to originate from different binding of the lanthanide to the two halves of the antibiotic molecule, which seemed to be conformationally dissimilar.

By contrast, when nmr experiments were undertaken with triostin A (ostensibly a truly symmetrical compound) clear evidence for two different conformations was obtained (Blake et al. (1977); Kawano et al. (1977)). This was attributed to the existence of two distinct symmetrical conformations, slowly interconvertible and separated by an energy barrier around 20-22 kcal/mol. One conformer, observable as the only species in polar solvents such as dimethyl sulphoxide, was designated p and appeared closely similar to echinomycin. The other conformer, denoted n, was found to predominate in non-polar solvents such as chloroform/CCl$_4$. Kalman et al. (1979) concluded that triostin n differed from triostin p by a reversal of the chirality of the disulphide bond in the cross-bridge and also by the formation of intramolecular hydrogen bonds from the alanine NH groups to the quinoxaline carbonyls. Kyogoku et al. (1981) favoured a different explanation, viz. cis-trans isomerisation of the N-methylated peptide bonds. Kyogoku et al. (1978, 1981) and Higuchi et al. (1983) additionally made the interesting observation that triostin p interacts weakly with derivatives of the purine bases such as the nucleosides adenosine and guanosine. They surmised that the interaction was stabilised by mutual stacking of purine and quinoxaline rings as well as hydrogen bonding involving the NH and CO functionalities of the alanine residues. Significantly, the n conformer of triostin was unable to form a complex with the purine derivatives as judged by nmr (Higuchi et al. (1983)), consistent with the idea that an internal hydrogen bond formed by the alanine NH group as postulated by Kalman et al. (1979) would prevent the interaction.

A final thought on the free-solution conformation of echinomycin emerges from phosphorescence and optically detected magnetic resonance studies of echinomycin and its DNA complexes recently undertaken by Alfredson and Maki (1990). They have found evidence for the existence, at low temperature, of two distinct forms of the antibiotic which bind differently to DNA; they speculated that the minor form might contain a weak internal hydrogen bond between the alanine NH group and the quinoxaline ring, perhaps akin to the one detected in crystals of triostin p (Sheldrick et al. (1984)). At all events, the upshot of large amount of detailed work on the comparative properties of quinoxaline antibiotics in solution can be judged from the schematic representation of echinomycin in Figure 3. In the bis-intercalated state we must look to substituents pressed up against the DNA base-pairs in the sandwich to identify the determinants of specificity (barring major conformational changes upon binding to DNA). As regards hydrogen bonding that means the NH and CO groups of the alanine residues.

3. The Nature of the Binding Site

It was clear from early experiments that echinomycin showed some preference for GC-rich binding sites in DNA because its gross binding constant measured with different types of DNA rose with increasing GC content (Wakelin and Waring (1976)). In those days the only sure way to establish a drug's specificity was to determine more or less complete binding curves and analyse them by some sort of Scatchard analysis - a painstaking and tedious process which, at best, would only provide an estimate of the affinity averaged over all the types of binding site present. In the case of echinomycin, efforts to short-cut the classical approach by looking for hints of specific interaction with nucleosides, nucleotides, or the like (as had been done successfully with actinomycin) were to no avail. There seems to be no optical correlate of the interaction between triostin and purine nucleosides detected by Kyogoku et al. (1978, 1981) using nmr methods. That all changed with the advent of footprinting techniques in the early 1980s as briefly outlined below, but credit must be given to Jones et al. (1987) for persevering with optical studies on antibiotic complexes with natural DNA. They made careful circular dichroism first-neighbour analyses of echinomycin complexes with eleven different DNAs from which they were led to conclude that the preferred sites for intercalation contain at least one guanine or cytosine (CpC, GpC, CpG and ApC), with GpC preferred to CpG. The latter conclusion is at variance with the footprinting studies, but the deduced preference for at least one GC base pair is not in dispute.

Footprinting experiments were conducted simultaneously and independently by Low et al. (1984a) and Van Dyke and Dervan (1984) using DNAase I and methidiumpropyl EDTA.Fe(II) cleavage respectively. Happily they were in complete agreement as regards the critical recognition element which constitutes the common denominator of strong binding sites: it is the CpG dinucleotide sequence, though the actual binding site size is at least four base pairs and there is some evidence that AT pairs may be preferred alongside the central CpG. Figure 4 illustrates the pattern of footprinting observed with the tyrT DNA fragment which has served as substrate for many drug binding studies (see, for example, Chaires et al (1987)). Here six or seven regions are protected from DNAase I cleavage by echinomycin, some representing isolated binding sites, others resulting from overlapping closely-spaced multiple sites. Since the publication of these initial results others have confirmed the specificity of echinomycin for sites containing the CpG sequence, employing different DNA substrates and/or novel cleavage reagents (Jeppesen and Nielsen (1988, 1989); Nielsen (1990)).

An unexpected finding which emerged from the footprinting experiments was that certain sequences flanking strong echinomycin binding sites became much more susceptible

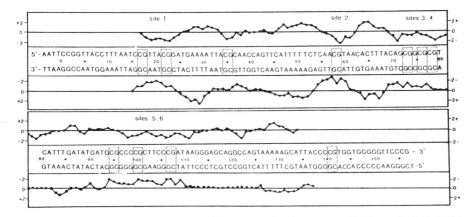

FIGURE 4. Echinomycin-induced differences in the susceptibility of *tyr*T DNA to DNAase I digestion. The upper Watson strand reads 5' to 3' left-to-right, while the lower Crick strand reads 5' to 3' right-to-left. Vertical scales on both sides are in units of $\ln(f_a)-\ln(f_c)$, where f_a is the fractional cleavage at any bond in the presence of 15µM antibiotic and f_c is the fractional cleavage of the same bond in the control, given closely similar extents of overall digestion (approx. 30% of the starting material in both cases). Positive values indicate enhancement, negative values blockage. From Low et al. (1984).

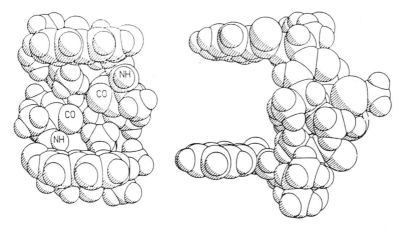

FIGURE 5. Crystal structure of a bis-quinoline analogue of echinomycin (2QN) prepared by directed biosynthesis (Gauvreau and Waring (1984)). In the illustration on the left the molecule is viewed down its quasi-dyad axis with the aromatic chromophores projecting out towards the viewer; this reveals the face of the peptide ring presented towards the DNA base-pairs, with the NH and CO groups of the two L-alanine residues lying in a diagonal array from lower left to upper right. In the illustration on the right the molecule is viewed from the "side", i.e. having been turned through 90° about a vertical axis, so as to show its staple-like arrangement which is ideally suited for bis-intercalation. Small rotations have been applied to the points of attachment of the aromatic rings so as to bring their planes exactly parallel. Unpublished work of G.M. Sheldrick, P.G. Jones, E.F. Paulus and M.J. Waring.

to nuclease cleavage when the antibiotic was present (Low et al. (1984a). Such sequences commonly contained runs of A and T nucleotides and three examples can be seen around positions 27-32, 46-52 and 63-68 in Figure 4. At first it was thought that the phenomenon might be merely an artefact, but when it was observed with other ligands as well as other nucleases (and even chemical probes) the notion was advanced that it reflected the existence of conformational changes propagated into the helical structure as a result of local distortion(s) at the ligand binding site. Changes in the width of the minor groove, the site of nuclease attack, were suggested as the most likely origins of the effect (Low et al. (1984a); Fox and Waring (1984)). Subsequently it was shown that even more far-reaching perturbations of structure could be detected with nucleosome core particles, where echinomycin (in common with other antibiotics) appeared to cause substantial changes in the rotational orientation of DNA on the surface of the core particles (Low et al. (1986a); Portugal and Waring (1987)). The nature and significance of conformational changes produced by binding of drugs to their receptors represent a subject of intense pharmacological interest, not least in the case of echinomycin-DNA interaction as we shall consider below.

Whereas the CpG dinucleotide step seems to represent the element common to DNA sequence recognition by all naturally-occurring quinoxaline antibiotics it is noteworthy that synthetic modification can add to or vary the recognition process. Echinomycin analogues bearing modified chromophores, prepared by directed biosynthesis (Gauvreau and Waring (1984); Williamson et al. (1982)) may bind to DNA more or less tightly than the parent compound in a fashion which speaks for altered sequence recognition (Fox et al. (1980); Cornish et al. (1983)). An analogue bearing 3-amino substituents on the quinoxaline rings seems to demand at least one AT pair flanking the requisite CpG, while analogues containing quinoline rings display enhanced affinity for alternating purine-pyrimidine sequences (Fox et al. (1980); Low et al. (1986b). The bis-quinoline analogue of echinomycin is important by virtue of its crystal structure as described below. However by far the most interesting consequences arise from the structural modifications in TANDEM, prepared by total chemical synthesis (Figure 1). Here the lack of the peptide N-methyl groups causes a considerable diminution in the gross binding constant for natural DNAs and a complete change in specificity such that alternating AT sequences are strongly preferred, perhaps mandatory (Lee and Waring (1978); Fox et al. (1982)). Footprinting and related experiments established that the sequence recognised by TANDEM is more likely TpA than ApT, and that the critical feature of the structure of TANDEM leading to its unique sequence-recognition properties is the lack of N-methylation of the valine residues (Low et al. (1984b, 1986c). The involvement of the alanine NH groups in recognising the AT base-pairs was deduced from the failure to bind to DNA of a TANDEM analogue having L-lactic acid in place of each L-alanine residue (Olsen et al. (1986); Low et al. (1986c); Waring (1987)).

Before we leave the subject of binding sites it is salutary to recall that although footprinting has identified CpG steps as being present in the preferred binding sites for echinomycin these are not the only sites to which the antibiotic can bind. The mere fact that echinomycin (in common with other quinoxaline antibiotics) will bind to polydG.polydC as well as to poly(dA-dT) (Wakelin and Waring (1976)) is sufficient to establish that the CpG step cannot be mandatory and that other sites, presumably of lower affinity, must exist. Evidence of weaker binding sites has been deduced from kinetic studies, prompting the "shuffling" hypothesis which envisages that antibiotic molecules originally become associated with weak binding sites and then migrate to stronger sites over a maturation period subsequent to the initial binding reaction (Fox and Waring (1985)). It seems that recognition is a slow process which often occurs in two stages: first the recognition of a broadly acceptable receptor in the B-form of DNA and then a process of scrutinising base pairs in the minor groove for the ideal (CpG-containing) binding sites.

4. X-Ray Diffraction and Complexes with Oligonucleotides

Recent years have seen the publication of crystal structures for quinoxaline antibiotics (though not echinomycin) and their complexes with oligonucleotides which beautifully substantiate the conclusions as to structure and recognition deduced from solution studies. The structures of four quinoxalines have been determined by direct methods: TANDEM (Viswamitra et al. (1981); Hossain et al. (1982)); triostin A (Sheldrick et al. (1984)); the bis-quinoline analogue of echinomycin, 2QN (G.M. Sheldrick, P.G. Jones, E.F. Paulus and M.J. Waring, unpublished but illustrated on the cover of *Science* 239 (Part II), 12 February 1988) and triostin C (K. Schmidt-Bäse, G.M. Sheldrick and M.J. Waring, unpublished). In each molecule the peptide ring adopts an approximately rectangular shape from which the aromatic chromophores project almost at right angles, their points of attachment separated by 10-12Å and easily rotatable so as to permit their planes to lie parallel. In each structure the sulphur-containing cross-bridge projects away from the peptide ring on the opposite side from the chromophores. These features are in excellent accord with the ideas expressed above. A space-filling representation of the structure of 2QN is illustrated in Figure 5. The functional groups of the peptide bonds are disposed "above" and "below" the peptide ring

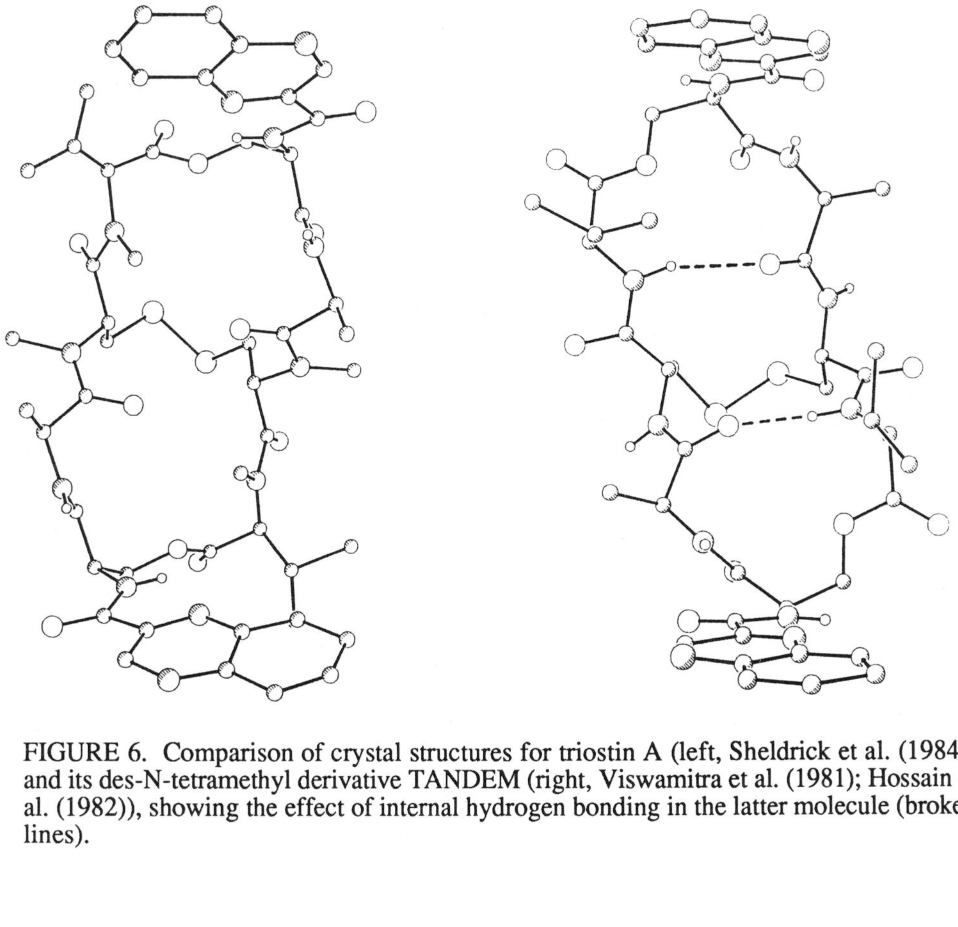

FIGURE 6. Comparison of crystal structures for triostin A (left, Sheldrick et al. (1984) and its des-N-tetramethyl derivative TANDEM (right, Viswamitra et al. (1981); Hossain e al. (1982)), showing the effect of internal hydrogen bonding in the latter molecule (broker lines).

exactly as schematically illustrated in Figure 3, and the same is true for the triostins. There is an important difference in the structure of TANDEM, however, which neatly explains its different DNA-binding characteristics: the peptide ring of TANDEM contains two internal hydrogen bonds from the NH groups of the valine residues to the carbonyl groups of the alanines (Figure 6). These interactions are, of course, not possible in the natural antibiotics because the L-valine residues are N-methylated. The effect of the internal hydrogen bonding, apart from satisfying the acceptor properties of the alanine CO groups, is to bring about minor but significant changes in numerous torsional angles which have the effect of drawing the bulky iso-propyl side-chains of the valine residues forward into a position where they partially occlude the space between the intercalative quinoxaline rings. Thus positioned they must impede interaction between substituents on the DNA base pairs and the alanine carbonyl groups of the depsipeptide ring, most particularly the hydrogen bonding to the 2-NH_2 groups of the guanines which is supposed to confer selectivity for CpG sequences on the natural antibiotics. Thus the gross DNA-binding constant is reduced, and the "residual" selectivity of TANDEM for alternating AT base pairs is presumed to originate from its retained capacity to form hydrogen bonds from the alanine NH groups to acceptors on the base-pairs (Viswamitra et al. (1981)).

A wealth of precise structural information became available when Rich and his collaborators succeeded in co-crystallising echinomycin or triostin A with self-complementary oligodeoxynucleotides containing the CpG recognition element. Their first success came with triostin A and the hexamer d(CGTACG) (Wang et al. (1984)). Next came a complex of echinomycin with the same oligomer (Ughetto et al. (1985)). Then they solved the structure of a complex containing triostin A and an octamer having an additional GC base pair at each end (Quigley et al. (1986); Wang et al. (1986)). All three complexes contained a mini-helix of six or eight base pairs with two symmetry-related antibiotic molecules bis-intercalated around the CpG sequences, as represented in diagrammatic form for the d(GCGTACGC) complex in Figure 7. A high degree of concordance was found among the three structures, with only minor differences arising mainly as a result of the shorter cross-bridge of echinomycin. Since details of this work have been fully discussed elsewhere (Wang (1987) and this symposium) we need only note some of the main features as they relate to the antibiotic-nucleic acid recognition problem. The antibiotic is clearly seen to lie in the minor groove of the mini-helix and it provokes a degree of unwinding, distributed over all the base pairs present, approximately equal to that measured with macromolecular DNA in solution (round about 50° per bound antibiotic molecule; Waring (1979, 1981)). The complex is stabilised by a large number of van der Waals contacts between the peptide ring and the DNA oligomer (19, 21 and 27 for the three complexes respectively) plus stacking interactions involving the quinoxaline rings, though the latter are far from perfect and contribute considerably less to the stabilisation of the entire structure than is commonly supposed for intercalation of drugs into DNA (Ughetto et al. (1985); Wang et al. (1986)). Additional stabilisation comes from hydrogen bonding, and in this regard the L-alanine residues play a commanding role as expected. Their NH groups both form hydrogen bonds to the N3 of guanine moieties in the sandwiched base pairs (Figure 7), though only one of the alanine carbonyl groups of each antibiotic molecule forms a hydrogen bond to the 2-NH_2 of guanine. The distance seems to be too great to permit the other alanine carbonyl to engage in the equivalent interaction with the remaining guanine 2-NH_2, ostensibly because of the lack of two-fold rotational symmetry in the structure of the antibiotic. This is exactly what Cheung et al. (1978) predicted from nmr experiments and model-building with echinomycin. It is, however, a little surprising that similar asymmetry in its structure should prevent triostin A from forming the fourth hydrogen bond, granted that its disulphide cross-bridge is potentially symmetrical and should endow the peptide ring with a little more flexibility anyway.

As is well known, the most striking discovery afforded by the antibiotic-oligonucleotide structures was the rearrangement of all base pairs (other than those sandwiched between the quinoxaline chromophores) from Watson-Crick to Hoogsteen pairing, involving protonation of the cytosine bases at the ends of the stack in the octamer-triostin complex. The fundamental reason for the altered base-pairing seems clear enough: it effectively diminishes the separation between the sugar-phosphate backbones by about 2Å, permitting much better packing between the peptide ring and the nucleic acid, thus generally optimising van der Waals and ring stacking interactions. Bearing in mind the very minor differences in conformation of the antibiotic going from the free state to the oligonucleotide complex we see that the cost of binding in terms of conformational adjustments is a decidedly one-sided affair: practically all the adjustments required to effect steric complementarity are on the part of the nucleic acid "receptor" which must mould itself to fit its rather rigid peptide guest. It is as if the lock had to bend in order to accommodate the key. Does the same thing happen with macromolecular DNA in solution?

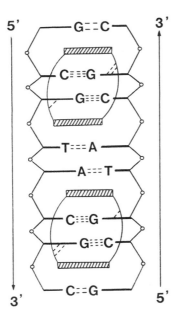

FIGURE 7. Schematic diagram of a triostin A-d(GCGTACGC) complex solved by crystallography. It possesses a dyad axis of molecular symmetry passing through the middle of the structure as drawn. The quinoxaline rings are represented by crosshatched rectangles. The cyclic depsipeptide backbone drawn as an oval is linked by hydrogen bonds from the alanine residues to the guanine bases (broken lines). After Quigley et al. (1986).

5. Probing for conformational changes

The critical feature of Hoogsteen base-pairing, apart from the need to protonate cytosine
where it has to pair with guanine, is the rotation of purine residues about the glycosidic bond
to adopt the syn conformation which allows the purine N7 to accept a hydrogen bond from
the pyrimidine N(3)-H (Figure 8). This entails numerous changes in distances between bases
and sugar protons, some of which can be sensitively measured by nmr using the nuclear
Overhauser effect. Gao and Patel (1988, 1989) have exploited this method in a detailed
survey of the interaction between echinomycin and a variety of self-complementary
oligonucleotides to confirm that Hoogsteen pairing can indeed occur with these mini-helical
structures in solution, but that it is critically sequence- and pH-dependent. Gilbert et al
(1989) went further and showed by one- and two-dimensional nmr that formation of
Hoogsteen base pairs by the central adenine and thymine nucleotides in an echinomycin-
[d(ACGTACGT)]2 complex was also temperature-dependent, such that those adenines

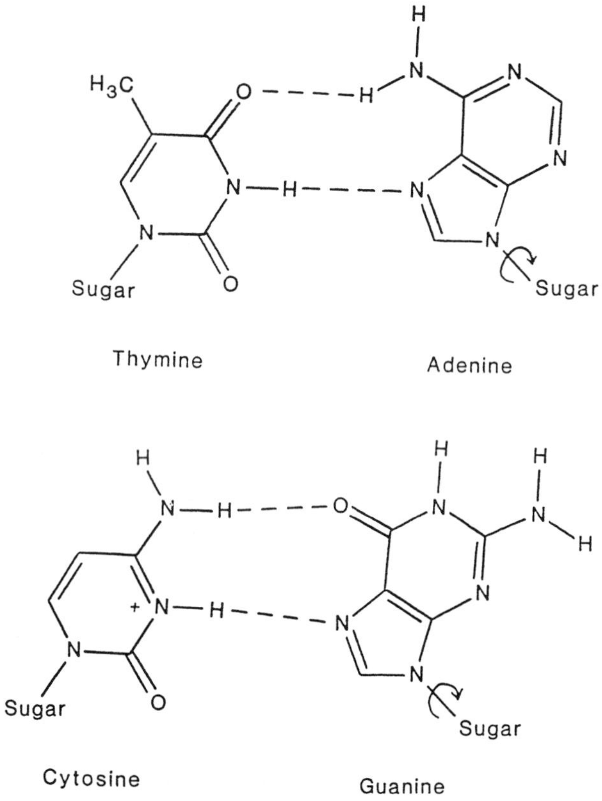

FIGURE 8. Purine-pyrimidine base pairing according to the Hoogsteen scheme. The
placement of the pyrimidines is much the same as in the Watson-Crick scheme, but the
purines are turned over. Note that the C:G Hoogsteen base pairing requires protonation of
the cytosine at N3.

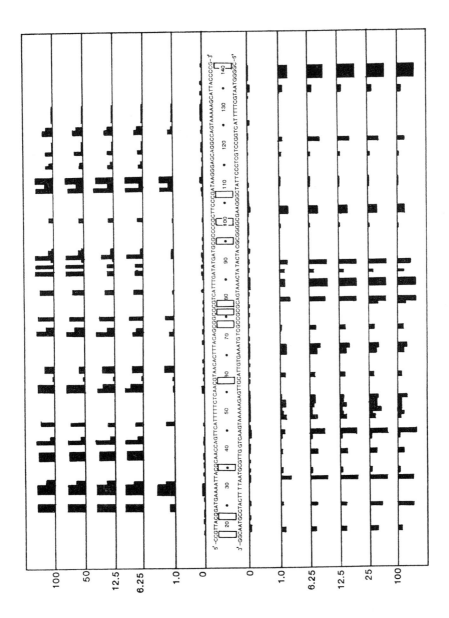

FIGURE 9. Summary map showing sites hyperreactive towards diethylpyrocarbonate (DEPC) on *tyr*T DNA in the presence of different concentrations of echinomycin (μM) as indicated on the ordinate. Reactive sites are indicated by bars whose height is proportional to the enhancement of reaction. Echinomycin binding sites (as deduced from DNAase I footprinting experiments) are located around each of the CpG steps (boxed). From Portugal et al. (1988).

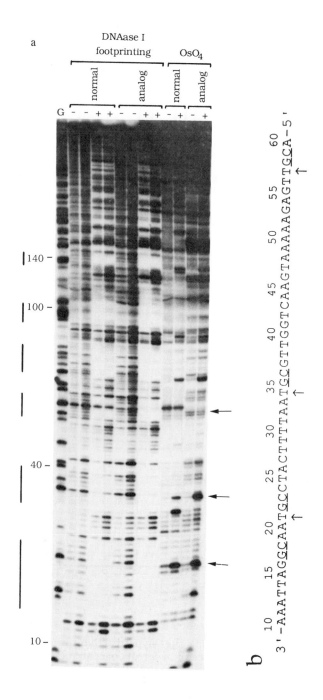

and thymines became unpaired or Watson-Crick paired as the temperature was raised towards the physiological. It appears that, at least in bisintercalator-oligonucleotide complexes, transitions between Hoogsteen and Watson-Crick base pairing may be relatively facile as molecular mechanics calculations predict (Singh et al. (1986)).

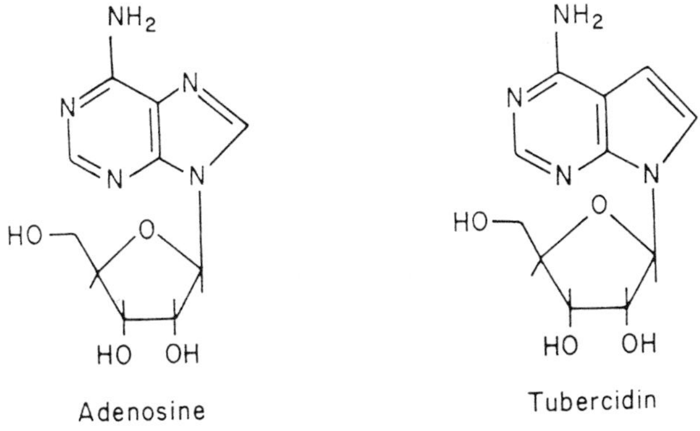

FIGURE 10. Chemical structures of adenosine and tubercidin.

FIGURE 11. (a) Autoradiograph showing the results of DNAase I cleavage of DNA fragments containing deoxyadenosine in both strands (lanes marked "normal") or containing 7-deazadeoxyadenosine in one strand (lanes marked "analog") in the absence (lanes -) or presence (lanes +) of 4μM echinomycin. For each experiment, two samples were run in adjacent lanes, corresponding to 1-min and 5-min digestions with the enzyme. Numbers on the left refer to nucleotide coordinates in the original *tyr*T fragment. Bars highlight footprints or antibiotic-binding sites - i.e. regions of diminished cleavage by DNAase I in the presence of echinomycin. Also shown are the results of OsO4 reactions with these two fragments in the absence (lanes -) or presence (lanes +) of 4 μM echinomycin. The three thymidine residues immediately 3' to echinomycin-binding sites in this strand are marked with arrows. Lane G, dimethyl sulphate/piperidine marker specific for guanine. (b) Sequence of the lower (Crick) strand of the *tyr*T DNA fragment from positions 10-65 for comparison with the autoradiograph in a. Echinomycin-binding sites, centred around CpG steps, are underlined and the same three thymidine nucleotides are marked with arrows. (From McLean et al. (1989)).

However, although these elegant studies with small DNA fragments have thrown fascinating light on the means by which echinomycin recognises its binding site, they leave open the question whether Hoogsteen base pairing is induced by binding of the antibiotic to real DNA. To address that question three groups undertook studies based on the belief that chemical probes should readily be able to distinguish between Watson-Crick and Hoogsteen base-pairs, if they were formed (Mendel and Dervan (1987); Portugal et al. (1988); Jeppesen and Nielsen (1988)). The reagent of choice seemed to be diethylpyrocarbonate (DEPC) which was already known to react sensitively with left-handed Z-DNA and cruciform loops, suggestive of strong reactivity towards any unusual conformation - perhaps particularly towards purines in the syn conformation such as occur in Z-DNA. Indeed, all three groups found that sites hyperreactive towards DEPC appeared on binding of echinomycin to DNA. The great majority of those sites were adenines rather than guanines, among which adenines on the 3' side of the CpG recognition sequence were prominent. The snag was that only a small proportion of the reactive nucleotides were adjacent to the antibiotic binding sites detected by footprinting (Figure 9) and, despite a lot of effort to correlate site-reactivity with affinity for the antibiotic, only meagre concentration-dependence was found (Portugal et al. (1988)). Moreover, the notion that lowering the pH should facilitate the protonation of cytosine required for formation of Hoogsteen GC base pairs, leading to enhanced reaction with DEPC, proved unfounded (Portugal et al. (1988); McLean and Waring (1988)). This prompted a more far-ranging survey of reactivity towards structural probes caused by binding of echinomycin as well as other drugs, unhappily with equally inconclusive results. Jeppesen and Nielsen (1988) showed that hyperreactivity of adenines towards DEPC occurs on binding of acridine bisintercalators (and even mono-intercalators) and agreed with Portugal et al. (1988) and McLean and Waring (1988) that the phenomenon probably reflects increased exposure of the purine N7 in the major groove as a result of intercalation-induced helical unwinding. Whereas the reactivity of DNA towards some probes was totally unaffected by the presence of echinomycin (bromoacetaldehyde, dimethyl sulphate or potassium tetrachloropalladinate, for example) certain thymidines were rendered reactive towards permanganate or osmium tetroxide in a highly sequence-dependent fashion (Jeppesen and Nielsen (1988); McLean and Waring (1988)). Again there was no correlation with the position of known echinomycin binding sites, nor is it obvious why the 5,6 double bond of thymine (with which these oxidising probes react) should be exposed as a consequence of binding echinomycin, unless again it is related to a conformational change such as local helix unwinding or widening of the major groove (McLean and Waring (1988)).

Nevertheless, the echinomycin-induced reactivity towards OsO$_4$ played a crucial part in setting up the last definitive experiment to look for Hoogsteen base pairing in DNA to which echinomycin has bound (McLean et al. (1989)). It was reasoned that deleting the N7 of adenine should preclude the possibility of Hoogsteen pairing to thymine (cf. Figure 8) so that incorporating 7-deaza-2'-deoxyadenosine into DNA would test the twin hypotheses that a transition to Hoogsteen pairing is needed for echinomycin to bind or that OsO$_4$ reactivity signals the existence of Hoogsteen base pairs. Fortunately, 7-deazaadenosine is not an imaginary substance - it occurs naturally as the antibiotic tubercidin (Figure 10) - and its phosphorylated derivatives are known to serve as substrates for incorporation into polynucleotides (Gale et al. (1981); Seela and Kehne (1987)). The results of exposing DNA containing 7-deaza-2'-deoxyadenosine to DNAase I or OsO$_4$ in the presence of echinomycin were refreshingly clear-cut (Figure 11): both the footprinting and the pattern of reactivity towards the probe were indistinguishable from those seen with control DNA (McLean et al. (1989)). It was concluded that Hoogsteen base pairing cannot be the cause of the observed echinomycin-induced hypersensitivity to osmium tetroxide and that it therefore most likely results from local unwinding of the helix. Moreover, preventing the possibility of Hoogsteen base pairing did not preclude echinomycin from binding to the DNA.

6. Conclusions

It has taken a long time from the discovery of echinomycin to the present where we have a fairly clear picture of how the antibiotic recognises its target nucleotide sequence in DNA. Many increasingly sophisticated techniques have been applied to the problem, and no doubt many more will be applied in years to come. Viewed in the context of the whole field of drug-DNA interactions the echinomycin story counts as one of the most detailed, alongside that of actinomycin or the anthracycline antibiotics (Gale et al. (1981); Chaires et al. (1987)). As such it is likely to provide a paradigm against which new techniques and ideas can be tested.

Outstanding questions for future research must centre around whether or not (or to what extent) echinomycin alters the base-pairing in DNA, and how it affects nucleoproteins (Portugal and Waring (1987)). A useful start has been made in these areas, as indicated above. Hopefully the answers to these questions will lead on to a better understanding of how the antibiotic kills cancer cells, but if nothing else they will provide important clues to guide the efforts of chemists and biochemists to improve the selectivity of anti-cancer drugs. The urgency of that goal is as pressing as ever.

ACKNOWLEDGMENTS.

Research in the author's laboratory was supported by grants from the Cancer Research Campaign, the Medical Research Council and the Royal Society. Many collaborators and other colleagues participated in the experimental work and discussions which led to the present state of understanding; their contributions are recorded with deep appreciation. Dean Gentle provided expert and willing technical assistance at all stages. The facilities of Jesus College, Cambridge, were indispensable in preparing the manuscript.

REFERENCES

Alfredson, T.V., Maki, A.H.: 'Phosphorescence and optically detected
    magnetic resonance studies of echinomycin-DNA complexes'. Manuscript submitted
    for publication.
Blake, T.J., Kalman, J.R., Williams, D.H.: 'Two symmetrical conformations of
    the triostin antibiotics in solution'. Tetrahedron Lett. 30, 2621-2624 (1977).
Chaires, J.B., Fox, K.R., Herrera, J.E., Britt, M., Waring, M.J.: 'Site and
    sequence specificity of the daunomycin-DNA interaction'. Biochemistry 26, 8227-
    8236 (1987).
Cheung, H.T., Feeney, J., Roberts, G.C.K., Williams, D.H., Ughetto, G., Waring,
    M.J.: 'The conformation of echinomycin in solution'. J. Amer. Chem. Soc. 100.
    46-54 (1978).
Cornish, A., Fox, K.R., Waring, M.J.: 'Preparation and DNA-binding
    properties of substituted triostin antibiotics'. Antimicrob. Agents Chemother. 23,
    221-231 (1983).
Dell, A., Williams, D.H., Morris, H.R., Smith, G.A., Feeney, J., Roberts, G.C.K.:
    'Structure revision of the antibiotic echinomycin'. J. Am. Chem. Soc. 97, 2497-
    2502 (1975).
Denny, W.A., Baguley, B.C., Cain, B.F., Waring, M.J.: 'Antitumour acridines' in
    Molecular Aspects of Anti-Cancer Drug Action (eds S. Neidle and M.J. Waring),
    Macmillan, London, pp. 1-34 (1983).

Denny, W.A.: 'DNA-intercalating ligands as anti-cancer drugs: prospects for
       future design'. Anti-Cancer Drug Design 4, 241-263 (1989)

Fox, K.R., Gauvreau, D., Goodwin, D.C., Waring, M.J.: 'Binding of quinoline
       analogues of echinomycin to DNA: role of the chromophores'. Biochem. J. 191
       729-742 (1980).

Fox, K.R., Olsen, R.K., Waring, M.J.: 'Equilibrium and kinetic studies on the
       binding of des-N-tetramethyltriostin A to DNA'. Biochim. Biophys. Acta 696, 315
       322 (1982).

Fox, K.R., Waring, M.J.: 'DNA structural variations produced by actinomycin
       and distamycin as revealed by DNAase I footprinting'. Nucleic Acids Res. 12, 9271
       9285 (1984).

Fox, K.R., Waring, M.J.: 'Kinetic evidence that echinomycin migrates between
       potential DNA-binding sites'. Nucleic Acids Res. 13, 595-603 (1985).

Gale, E.F., Cundliffe, E., Reynolds, P.E., Richmond, M.H., Waring, M.J. 'The
       Molecular Basis of Antibiotic Action', 2nd ed., Wiley, London (1981).

Gao, X., Patel, D.J. 'NMR studies of Echinomycin bisintercalation
       complexes with d(A1-C2-G3-T4) and d(T1-C2-G3-A4) duplexes in aqueou
       solution: Sequence-dependent formation of Hoogsteen A1.T4 and Watson-Cricl
       T1.A4 base pairs flanking the bisintercalation site'. Biochemistry 27, 1744-175
       (1988).

Gao, X., Patel, D.J.: 'Antitumour drug-DNA interactions: NMR studies of
       echinomycin and chromomycin complexes'. Quart. Rev. Biophys. 22, 93-13!
       (1989).

Gauvreau, D., Waring, M.J.: 'Directed biosynthesis of novel derivatives of
       echinomycin by Streptomyces echinatus. Part I. Effect of exogenous analogues o
       quinoxaline-2-carboxylic acid on the fermentation'. Can. J. Microbiol. 30, 439-45(
       (1984).

Gauvreau, D., Waring, M.J.: 'Directed biosynthesis of novel derivatives of
       echinomycin. Part II. Purification and structure elucidation'. Can. J. Microbiol. 30
       730-738 (1984).

Gilbert, D.E., van der Marel, G.A., van Boom, J.H., Feigon, J.: 'Unstable
       Hoogsteen base pairs adjacent to echinomycin binding sites within a DNA duplex'
       Proc. Nat. Acad. Sci. U.S.A. 86, 3006-3010 (1989.).

Higuchi, N., Kyogoku, Y., Shin, M., Inouye, K.: 'Origin of slow conformer
       conversion of triostin A and interaction ability with nucleic acid bases'. Int. J
       Peptide Protein Res. 21, 541-545 (1983).

Hossain, M.B., van der Helm, D., Olsen, R.K., Jones, P.G., Sheldrick, G.M.,
       Egert, E., Kennard, O., Waring, M.J., Viswamitra, M.A.: 'Crystal and molecula
       structure of the quinoxaline antibiotic analogue TANDEM (des-N-tetramethyl triostii
       A)'. J. Amer. Chem. Soc. 104, 3401-3408 (1982).

Jeppesen, C., Nielsen, P.E.: 'Detection of intercalation-induced changes in
       DNA structure by reaction with diethyl pyrocarbonate or potassium permanganate
       Evidence against the induction of Hoogsteen base pairing by echinomycin'. FEBS
       Letters 231, 172-176 (1988).

Jeppesen, C., Nielsen, P.E.: 'Photofootprinting of drug-binding sites on DNA
       using diazo- and azido-9-aminoacridine derivatives'. Eur. J. Biochem. 182, 437-444
       (1989).

Jones, M.B., Hollstein, U., Allen, F.S.: 'Site specificity of binding of
       antitumour antibiotics to DNA'. Biopolymers 26, 121-135 (1987)

Kalman, J.R., Blake, T.J., Williams, D.H., Feeney, J., Roberts, G.C.K.: 'The
       conformation of triostin A in solution'. J. Chem. Soc. Perkin I, 1313-1321 (1979).

Katagiri, K., Yoshida, T., Sato, K.: 'Quinoxaline antibiotics' in
    Antibiotics, Mechanism of action of antimicrobial and antitumour agents. Corcoran,
    J.W., Hahn, F.E. (eds.), Vol. III, pp. 234-251. Berlin, Heidelberg, New York:
    Springer (1975).
Kawano, K., Higuchi, N., Kyogoku, Y.: 'Nuclear magnetic resonance studies on
    the conformation of triostin A'. Peptide Chemistry, 93-96 (1977).
Keller-Schierlein, W., Mihailovic, M.L., Prelog, V.: 'Stoffwechsel produkte
    von Actinomyceten'. 'ber die Konstitution von Echinomycin. Helv. Chim. Acta 42,
    305-322 (1959).
Kyogoku, Y., Yu, B.S., Akutsu, H., Watanabe, M., Kawano, K.: 'Stereospecific
    binding of triostin A to nucleic acid purine base derivatives'. Biochem. Biophys.
    Res. Commun. 83, 172-179 (1978).
Kyogoku, Y., Higuchi, N., Watanabe, M., Kawano, K.: 'Conformer equilibria of
    triostin A and its conformer-specific interaction with nucleic acid bases'.
    Biopolymers 20, 1959-1970 (1981).
Lee, J.S., Waring, M.J.: 'Interaction between synthetic analogues of
    quinoxaline antibiotics and nucleic acids. Changes in mechanism and specificity
    related to structural alterations.' Biochem. J. 173. 129-144 (1978)
Le Pecq, J.B., Le Bret, M., Barbet, J., Roques, B.: 'DNA polyintercalating
    drugs: DNA binding of diacridine derivatives'. Proc. Natl. Acad. Sci. USA 72,
    2915-2919 (1975)
Low, C.M.L., Drew, H.R., Waring, M.J.: 'Sequence-specific binding of
    echinomycin to DNA: evidence for conformational changes affecting flanking
    sequences'. Nucleic Acids Res. 12, 4865-4879 (1984). a
Low, C.M.L., Olsen, R.K., Waring, M.J.: 'Sequence preferences in the binding
    to DNA of triostin A and TANDEM as reported by DNAase I footprinting'. FEBS
    Letters 176, 414-420 (1984) b
Low, C.M.L., Fox, K.R., Waring, M.J.: 'DNA sequence-selectivity of three
    biosynthetic analogues of the quinoxaline antibiotics'. Anti-Cancer Drug Design 1,
    149-160 (1986) b.
Low, C.M.L., Fox, K.R., Olsen, R.K., Waring, M.J.: 'DNA sequence recognition
    by under-methylated analogues of triostin A'. Nucleic Acids Res. 14, 2015-2033
    (1986) c.
Low, C.M.L., Drew, H.R., Waring, M.J.: 'Echinomycin and distamycin induce
    rotation of nucleosome core DNA'. Nucleic Acids Res. 14, 6785-6801 (1986) a.
Martin, D.G., Mizsak, S.A., Biles, C., Stewart, J.C., Baczynskyj, L., Meulman,
    P.A.: 'Structure of quinomycin antibiotics'. J. Antibiot. 28, 332-336 (1975).
McLean, M.J., Waring, M.J.: 'Chemical probes reveal no evidence of Hoogsteen
    base-pairing in complexes formed between echinomycin and DNA in solution'. J.
    Molec. Recognition. 1, 138-151 (1988).
McLean, M.J., Seela, F., Waring, M.J.: 'Echinomycin-induced hypersensitivity
    to osmium tetroxide of DNA fragments incapable of forming Hoogsteen base pairs'.
    Proc. Natl. Acad. Sci. USA, 86, 9687-9691 (1989).
Nielsen, P.E.: 'Chemical and photochemical probing of DNA complexes'. J.
    Mol. Recognition 3, 1-25 (1990).
Olsen, R.K.: 'The quinoxaline depsipeptide antibiotics' in Chemistry and
    Biochemistry of Amino Acids, Peptides, and Proteins (ed. B. Weinstein) 7, 1-33
    (1983).
Olsen, R.K., Ramasamy, K. Bhat, K.L., Low, C.M.L., Waring, M.J.: 'Synthesis

and DNA-binding studies of [Lac2,Lac6] TANDEM, an analogue of des-N-tetramethyltriostin A (TANDEM) having L-lactic acid substituted for each L-alanine residue'. J. Amer. Chem. Soc. 108, 6032-6036 (1986).

Portugal, J., Waring, M.J.: 'Analysis of the effects of antibiotics on the structure of nucleosome core particles determined by DNAase I cleavage'. Biochimie 69, 825-840 (1987).

Portugal, J., Fox, K.R., McLean, M.J., Richenberg, J.L., Waring, M.J.: 'Diethyl pyrocarbonate can detect a modified DNA structure induced by the binding of quinoxaline antibiotics'. Nucleic Acids Res. 16, 3655-3670 (1988).

Quigley, G.J., Ughetto, G., van der Marel, G.A., van Boom, J.H., Wang, A.H., Rich, A.: 'Non-Watson-Crick G.C and A.T base pairs in a DNA-antibiotic complex'. Science 232, 1255-1258 (1986).

Sato, K., Shiratori, O., Katagiri, K.: 'The mode of action of quinoxaline antibiotics. Interaction of quinomycin A with DNA.' J. Antibiot. Ser. A 20, 270-276 (1967).

Seela, F., Kehne, A.: 'Palindromic octa- and dodecanucleotides containing 2'-deoxytubercidin: synthesis, hairpin formation, and recognition by the endodeoxyribonuclease Eco R1'. Biochemistry 26, 2232-2238 (1987).

Sheldrick, G.M., Guy, J.J., Kennard, O., Rivera, V., Waring, M.J.: 'Crystal and molecular structure of the DNA-binding antitumour antibiotic triostin A'. J. Chem. Soc. Perkin II, 1601-1605 (1984).

Singh, U.C., Pattabiraman,N., Langridge, R, Kollman, P.A.: 'Molecular mechanics studies of d(CGTACG)2: Complex of triostin A with the middle A.T base pairs in either Hoogsteen or Watson-Crick pairing'. Proc. natn. Acad. Sci. U.S.A. 83, 6402-6406 (1986).

Ughetto, G., Waring, M.J.: 'Conformation of the DNA-binding peptide antibiotic echinomycin based on energy calculations'. Mol. Pharmacol. 13. 579-584 (1977).

Ughetto, G., Wang, A.H., Quigley, G.J., van der Marel, G.A., van Boom, J.H., Rich, A.: 'A comparison of the structure of echinomycin and triostin A complexed to a DNA fragment'. Nucl. Acids Res. 13, 2305-2323 (1985).

Van Dyke, M.W., Dervan P.B.: 'Echinomycin binding sites on DNA'. Science 225, 1122-1127 (1984).

Viswamitra, M.A., Kennard, O., Cruse, W.B.T., Egert, E., Sheldrick, G.M., Jones, P.G., Waring, M.J., Wakelin, L.P.G, Olsen, R.K.: 'The structure of TANDEM, a quinoxaline antibiotic analogue and its implication for bifunctional intercalation into DNA'. Nature 289, 817-819 (1981).

Wakelin, L.P.G., Waring, M.J.: 'The binding of echinomycin to DNA'. Biochem. J. 157, 721-740 (1976)

Wakelin, L.P.G., Romanos, M., Chen, T.K., Glaubiger, D., Canellakis, E.S., Waring, M.J.: 'Structural limitations on the bifunctional intercalation of diacridines into DNA'. Biochemistry 17, 5057-5063 (1978).

Wakelin, L.P.G.: 'Polyfunctional DNA intercalating agents'. Medicinal Research Reviews 6, 275-340 (1986).

Wang, A.H., Ughetto, G., Quigley, G.J., Rich, A.: 'Interactions of quinoxaline antibiotic and DNA: The molecular structure of a triostin A-d(GCGTACGC) complex'. J. Biomolec. Struct. Dynamics 4, 319-342 (1986)

Wang, A.H.: 'Interactions between antitumour drugs and DNA' in Nucleic Acids and Molecular Biology. Vol. I (ed. F. Eckstein, and D.M. Lilley), pp. 53-69. Berlin: Springer (1987).

Wang, A.H., Ughetto, G., Quigley, G.J., Hakoshima, T., van der Marel, G.A., van Boom, J.H., Rich, A.: 'Molecular structure of the DNA-triostin complex'. Science 225, 1115-1121 (1984).

Ward, D., Reich, E., Goldberg, I.H.: 'Base specificity in the interaction of polynucleotides with antibiotic drugs'. Science 149, 1259-1263 (1965).

Waring, M.J.: 'Drugs which affect the structure and function of DNA'. Nature 219, 1320-1325 (1968).

Waring, M.J.: 'Variation of the supercoils in closed circular DNA by binding of antibiotics and drugs: evidence for molecular models involving intercalation'. J. Mol. Biol. 54, 247-279 (1970).

Waring, M.J., Wakelin, L.P.G.: 'Echinomycin: a bifunctional intercalating antibiotic'. Nature 252, 653-657 (1974).

Waring, M.J.: 'Structural and conformational studies on quinoxaline antibiotics in relation to the molecular basis of their interaction with DNA' in Drug action at the Molecular Level. Biological Council Symposium, London 12-13 April 1976 (ed. G.C.K. Roberts) pp. 167-189 (1977). Macmillan, London.

Waring, M.J.: 'Echinomycin, triostin and related antibiotics' in Antibiotics. Vol 5/Part 2, Mechanism of Action of Antieukaryotic and Antiviral Compounds (ed. F.E. Hahn) 173-194 (1979) Springer-Verlag, Heidelberg.

Waring, M.J.: 'DNA Modification and cancer'. Ann. Rev. Biochem. 50,159-192 (1981)

Waring, M.J.: 'Recognition of DNA by quinoxaline antibiotics' in Molecular Mechanisms of Carcinogenic and Antitumour Activity (eds. C. Chagas & B. Pullman) Pontifical Academy of Sciences, Vatican City Scripta Varia 70, 317-337 (1987).

Williamson, M.P., Williams, D.H.: 'Manipulation of the nuclear Overhauser effect by the use of a viscous solvent: the solution conformation of the antibiotic echinomycin'. J. Chem. Soc. Chem. Commun. 165-166 (1981).

Williamson, M.P., Gauvreau, D., Williams, D.H., Waring, M.J.: 'Structure and conformation of fourteen antibiotics of the quinoxaline group determined by 1H NMR'. J. Antibiotics 35, 62-66 (1982).

BIS-PYRROLECARBOXAMIDES LINKED TO INTERCALATING CHROMOPHORE
OXAZOLOPYRIDOCARBAZOLE (OPC): PROPERTIES RELATED TO THE SELECTIVE
BINDING TO DNA AT RICH SEQUENCES

C. AUCLAIR[1], F. SUBRA[1], D. MRANI,[2] G. GOSSELIN[2],
J. L. IMBACH[2] and C. PAOLETTI[1].
[1]Laboratoire de Biochimie-Enzymologie, INSERM U140,
URA 147, Institut Gustave Roussy 94301, Villejuif,
France. [2]Laboratoire de Chimie Bioorganique, CNRS
URA 488, Place Eugène Bataillon, 34095 Montpellier,
Cedex 5, France.

ABSTRACT

We have comparatively investigated some biochemical and
pharmacological properties related to recognition of DNA AT-rich
sequences of two hybrid groove binder-intercalators netropsin-
oxazolopyridocarbazole (Net-OPC) and bis-pyrrolecarboxamide-
oxazolopyridocarbazole (BPC-OPC) in which the amidine moiety of
netropsin has been replaced by a methoxy group (Mrani et al. 1990
submitted). The hybrid molecule Net-OPC binds to poly d(A-T) at two
different sites with Kapp values close to $7 \ 10^6$ and $6 \ 10^8$ M^{-1} (100 mM
NaCl, pH 7.0) whereas it binds to poly d(G-C) to a single binding site
with a lower Kapp of $1.7 \ 10^6$ M^{-1}. In contrast, the molecule devoid of
the amidine moiety BPC-OPC binds to both poly d(A-T) and poly d(G-C)
with quite similar Kapp values of $2.2 \ 10^6$ and $1.4 \ 10^6$ M^{-1} respectively
indicating that the presence of the amidine is required for
recognition of AT-rich sequences. Based on viscometric data, the
binding of Net-OPC is suggested to involve both intercalation and
external binding of the OPC chromophore in AT-rich sequences and
intercalation only in G-C-rich sequences. In similar experimental
conditions, BPC-OPC binds to AT and GC-rich sequences according to an
external mode. Net-OPC and BPC-OPC have been further tested on two
experimental models involving recognition of AT containing sequences
in their process, namely: i) the catalytic activity of the
restriction endonuclease EcoRl which selectively cleave double
stranded DNA at GAATTC sequence and ii) the retroviral cycle of M.MuLV
(Moloney Leukemia Virus) involving the recognition by the integrase
enzymatic system of the proviral DNA sequences TAA-TTA, corresponding
to the integration point and CTTT corresponding to the flanking
region of the integration point (Colicelli & Goff, 1985). Net-OPC acts
as potent and selective inhibitor of both EcoRI activity (EC$_{50}$ 0.1 μM)
in vitro and the M.MuLV retroviral cycle (EC$_{50}$ 5 μM) in 3T3 cells. In
similar experimental conditions, BPC-OPC displays no effect on EcoRI
activity and a markedly lower effect on the retroviral cycle.

B. Pullman and J. Jortner (eds.), Molecular Basis of Specificity in Nucleic Acid-Drug Interactions, 247–260.
© 1990 Kluwer Academic Publishers. Printed in the Netherlands.

1. INTRODUCTION

The artificial regulation of the genomic functions, requires the design of molecules which selectively recognize single stranded or double stranded nucleic acids base sequences. This can be theorically achieved by two family of compounds including i) synthetic antimessenger oligodeoxynucleotides which ensure recognition of RNA complementary sequences (see Izant et al., 1985; Green et al., 1986; Paoletti, 1988 for general reviews) and ii) oligopeptides in the series of polypyrrole (Wartell et al., 1974) and polyimidazolecarboxamides (Kopka et al, 1985; Lown, 1988) which bind respectively to A-T or G-C containing sequences in the minor groove of double stranded DNA (see Pullman, 1989 for general review). From a pharmacological point of view, molecules designed to interfere with the functional activity of the genome should ideally display favourable permeation through cell membranes, high association constant value and favourable accessibility to their presumed target.

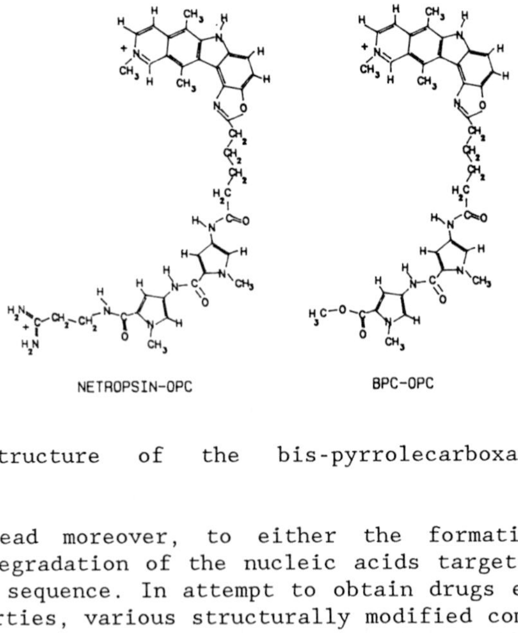

NETROPSIN-OPC   BPC-OPC

Figure 1. Structure of the bis-pyrrolecarboxamide-OPC hybrid molecules.

They should lead moreover, to either the formation of a stable complex or a degradation of the nucleic acids target at the level of the recognized sequence. In attempt to obtain drugs endowed with some of these properties, various structurally modified compounds have been synthesized in the series of oligopeptides. Along this line, polypyrrolecarboxamide containing molecules related to netropsin and distamycin, covalently linked to EDTA have been found to cleave selectively double stranded DNA at definite sequences (Schultz et al., 1982; Taylor et al., 1984) whereas more recently oligopeptides related to netropsin and linked to the intercalating drugs actinomycin D (Dervan, 1986) or acridine have been synthesized and found to recognize AT rich sequences. In this field of investigation, and in view of pharmacological purpose, we have previously synthesized (Mrani et al. unpublished) netropsin-like conjugates in which the guanidine moiety of the netropsin molecule has been replaced by a tetramethylen chain linked to the intercalating chromophore oxazolopyridocarbazole (OPC) (Auclair et al.1984, Gouyette et al. 1985). Among the different molecules synthesized, the compound displaying an oligopeptide chain identical to netropsin (i.e. bis-pyrrolecarboxamide-amidine moiety) (Net-OPC) (Figure 1) was found to exhibit a markedly high selective affinity to AT containing polynucleotides poly d(A-T) and poly dA-dT whereas in contrast, Net-OPC lacking the amidine moiety on the bis-pyrrolecarboxamide chain (BPC-OPC) (Figure 1) displays similar association constants to either AT or GC containing polynucleotides. The present paper describes a comparative study between Net-OPC and BPC-OPC in terms of binding parameters and biological properties related to recognition AT rich sequences.

2. MATERIALS AND METHODS

2.1. SYNTHESIS OF NETROPSIN-OPC CONJUGATES

Netropsine-OPC conjugate (3-{1-Methyl-4-[1-methyl-4-[4-(2-{7, 10, 12-trimethyl-6H-[1,3] oxazolo [5,4-C]pyrido[3,4-g]- carbazole}) butylcarboxamido]pyrrole-2-carboxamido] pyrrole - 2-carboxamido} propionamidine Bis-(acetate)) (figure 1) (Net-OPC) has been prepared as previously described (Mrani et al., submitted). Briefly, Net-OPC was obtained from the conjugation of OPC valerate (2-pentanoic acid-7,10,12-trimethyl(6H-[1,3] oxazolo [5,4-C]pyrido [3,4-g]-carbazole) (OPC-valerate) with the methyl 1-Methyl-4-(1-methyl-amino-pyrrole-2-carboxamido) pyrrole-2-carboxylate. OPC-valerate is prepared from the antitumor agent 2N-methyl-9-hydroxyellipticinium (Celiptium) (NMHE) according to a procedure previously described (Auclair et al. 1984)

2.2. DNAs AND POLYNUCLEOTIDES

DNA from *M. aureus*, *C. perfringens*, poly d(A-T) and poly d(G-C) were obtained from Boehringer, Mannheim, (Germany) and used without further purification.
Preparation of pBR322 DNA: Ecoli HB101 transformed with the monomeric form of the plasmid pBR322 were grown to late log phase in LB media with ampicilline (100 μg/ml) before addition of chloramphenicol (150 μg/ml). After continued growth at 37 °C overnight, cells were harvested by centrifugation and the covalently closed form of the plasmid was purified by centrifugation to equilibrium in cesium chloride-ethidium bromide with 1-butanol satured with water. Concentrations of DNA were evaluated from UV absorbance at 260 nm.

2.3. FLUORESCENCE EXPERIMENTS

Fluorescence spectra were recorded on a SFM 23/B spectrofluorometer (Kontron, Zürich, Switzerland) equipped with a thermostated cell holder. In all cases, fluorescence experiments were performed in quartz cells (1 cm path length) thermostated at 25°C. Scattering of

excitation   radiations   were   removed   using   appropriated   emission
filter.

2.4. VISCOMETRIC EXPERIMENTS

Viscometric   measurements   were   performed   at   25°C   in   a   semi   mi-
crodilution capillary viscometer   linked to a IBM XT computer. The
capacity of tested compounds to increase the length of sonicated calf
thymus DNA was   measured   using operating conditions as described by
Saucier et al. (1971).

2.5. RESTRICTION ENDONUCLEASES ASSAY

Activity of the restriction endonucleases   EcoRI and NruI has been
measured at 37°C using as substrate covalently closed pBR322 DNA which
contain a single site of breakage for both enzymes. Enzymatic activity
in the absence and in the presence of drugs has been estimated by
measuring the pseudo first order rate constants corresponding to the
first (k_1) and to the second (k_2) DNA break using the following
operating conditions: For ECoRI measurement, the experiments have been
performed in a medium composed of 50 mM tris-HCl (pH 7.5), 100 mM
NaCl, 10   mM $MgCl_2$ and 1 mM DTE. For NruI measurement, the medium was
composed of 10 mM tris-HCl, 100 mM NaCl, 5 mM $MgCl_2$ and 1mM Beta-
mercaptoethanol. Drugs to be tested and DNA (3 μg) were added to the
standard incubation mixture (final vol 30 μL) 20 min before the enzyme
addition. In all experiments concentration of enzyme was 2U/ug of DNA.
Samples   (3 μL)   were   removed   from   the   reaction   mixture   at   timed
intervals and immediately mixed with 17 μL of stop solution [20 /(V/V)
of a solution of 0.25 M EDTA pH 8.0, 0.2 /(W/V) sodium dodecyl
sulfate, 0.01 / bromophenol blue and 25/(w/V) sucrose.   The cleavage
products   were   separated   on   1%   agarose   gel   in   a   flat   bed
electrophorese apparatus (15 x 20   cm - gel thickness 4-5   mm) at a
field strength of 6-8 V/cm. The electrophoresis buffer used was 40 mM
Tris-HCl, 2 mM EDTA, (pH 7.4). The gels were then stained under gentle
shaking for 30 min in ethidium bromide solution (500 mcg), and
photographed for 1 min under 354 nm UV light on polaroid 665
positive/negative film. For quantitative evaluation, the negatives
were scanned with a gel scanner. The relative amount of supercoiled,
open circular and linear DNA in each sample was normalized to the
total DNA concentration.

2.6. M.MuLV RETROVIRAL CYCLE ASSAY

This test is a new application (C. Roy et al. unpublished) of the SVX
shuttle   vector   constructed   by   Cepko   et   al.   (1984)   and   will   be
described in detail elsewhere. Briefly 3T3 cells were infected by a
defective M.MuLV recombinant deleted for *gag pol env* genes (Mann et
al. 1983) and containing the *psi* gene required for the packaging
process of the viral RNA and the *neo* gene whose expression results in
the appearance of resistance to the G418 antibiotic (geniticine). The
integration of the viral DNA in the genome of the host 3T3 cells

results therefore in the appearance of cells displaying the resistance phenotype to geniticine and visualized as clones growing on plates in the presence of the antibiotic. The amount of resistant clones is directly related to the extent of proviral DNA integrated. Drugs to be tested were added together with the viral suspension used to infect the 3T3 cells.

2.7. NON LINEAR REGRESSION FITTING

Curves were fitted to data points using programs involving the method of Marquardt. Goodness of fit was quantitated by the least square procedure using as parameter the distance of each point from the curve. Softwares currently used were STATGRAPHICS (Statistical Graphic Corp.), GraphPAD (ISI Software) as well as programs developed in the laboratory.

3. RESULTS

3.1. BINDING PARAMETERS OF Net-OPC AND BPC-OPC TO DNAs AND POLYNUCLEOTIDES

3.1.1. Association constants of Net-OPC and BPC-OPC to poly d(A-T) and poly d(G-C).
Binding of OPC to double stranded polynucleotides results in a strong increase in the fluorescence intensity of the chromophore associated to a shift of the excitation spectrum towards longer wavelengths (Banoun et al. 1985; Auclair et al., 1987). This spectral changes is related to the binding to hydrophobic regions inside helical DNA (Le Pecq & Paoletti, 1967). The fluorescence increase of the OPC chromophore upon binding to DNA or polynucleotides of Net-OPC and BPC-OPC allows the estimation of the amount of drug bound (Db) at the equilibrium using the following relation:

$$Db = dIF/k(V-1) \qquad (1)$$

where, dIF is the difference between the fluorescence of drug in the presence and in the absence of DNA, V is the fluorescence increment resulting from the binding to DNA and k the factor which rely the concentration of free drug to fluorescence intensity of the solution. Binding curves so obtained can be plotted according to Scatchard leading to the estimation of the association constant values (Kapp) and the maximum amount of drug bound per nucleotide at saturating concentration (rmax) which in turn allows the estimation of the apparent size (n) of the binding site. Results so obtained are summarized in table I. The binding of Net-OPC to poly d(A-T) occurs at two different sites with respective association constant values of $5.0 \ 10^7 \ M^{-1}$ and $1.9 \ 10^6 \ M^{-1}$ (cacodylate buffer pH 7.0, containing 500 mM NaCl). In contrast, the binding BPC-OPC to poly d(A-T) involves a single type of binding site with a markedly lower Kapp value. The binding to poly d(G-C) of both Net-OPC and BPC-OPC involves a single type of binding site and is characterized by quite similar Kapp values. It should be noticed that, because of the markedly high affinity of Net-OPC for poly d(A-T), high salt concentra- tions (up to 300 mM) are required to establish the binding curves.

TABLE I. DNA binding parameters of the hybrid molecules Net-OPC and BPC-OPC. a:Kapp were estimated from the binding curves obtained using the fluorescence changes of the OPC chromophore upond binding to the polynucleotides (eq.1). In standard operating conditions each assay medium was composed of cacodylate buffer (pH 7.0) containing 5.5 μM polynucleotide and various NaCl concentrations (unlabeled values: 100 mM NaCl; ** : 500 mM NaCl). b: Slope of the straight line described by the Record equation (eq. 2) which rely the Kapp to the ionic strength and reflecting the number of positive charges involved in the binding process. c: apparent binding site size expressed in base pairs obtained from the binding curves. *: indicates these values were extrapolated using the Record equation and the corresponding slope values. S1 and S2 indicates the two binding sites of Net-OPC to poly d(A-T).

		Poly d(A-T)			Poly d(G-C)		
		K_{app}^{a} $(10^6 \ M^{-1})$	Slope[b]	n[c]	K_{app}^{a} $(10^6 \ M^{-1})$	Slope[b]	n[c]
Net-OPC	S1	600·0* 50·0**	-1.62	10	1.72	-0.64	2.0
	S2	7.1* 1.9**	-0.92	6			
BPC-OPC		2.2	-0.68	8	1.4	-1.0	3.7

In order to compare the association constant values of Net-OPC for poly d(A-T) and poly d(G-C) in identical experi- mental conditions, it has been necessary to determine Kapp values at different salt concentrations. According to Record et al. (1976) the dependence of Kapp on ionic strength is described by the following equation:

$$d \ Log \ Kapp/d \ Log \ [Na^+] = -Zf \qquad (2)$$

where Z is the charge involved in the binding, and f the fraction of counterions associated with each DNA phosphate (for double stranded DNA f = 0.88). The plot Log Kapp versus Log [Na$^+$] corresponding to eq. 2 yields straight lines whose slopes (-0.88 x Z) correspond therefore to the number of charges involved in the binding process. As indicated in table I, it appears clearly that in poly d(A-T), the binding of Net-OPC to the site of higher affinity involves the two positive charges present on the molecule whereas the binding to the site of lower affinity involves only one charge. In poly d(G-C), the binding to the single site involves one positive charge. As expected, the binding of BPC-OPC to either poly d(A-T) or poly d(G-C) involves a single positive charge. Using the slope value of the different straight lines described by eq. 2, Kapp of Net-OPC for poly d(A-T) corresponding to lower salt concentrations can be extrapolated (one star labelled values in table I). For 100 mM NaCl, the extrapolated value of Kapp of Net-OPC for the first site in poly d(A-T) appears to be markedly high compared to the Kapp for poly d(G-C) (ratio=349) This is consistent with the selective binding of the bis-pyrrolecarboxamide moiety to A-T containing sequences. The extrapolated Kapp for the second binding site of lower affinity appears to be in the same order of magnitude (fourfold higher) that the Kapp for poly d(G-C).

3.1.2. Viscometric data

Informations on the nature of the binding to DNA (intercalation versus external binding) can be provided by the viscometric determination of the length increase of sonicated DNA. In this technic, the theorical treatment of the viscometric data has shown that if log $\eta/\eta o$ is plotted vs. log(1+2r), where η and ηo are the intrinsic viscosity of sonicated DNA in the presence and in the absence of the tested drug and r the number of molecules bound per nucleotides the slope value of the straight line so obtained is expected to be near 2.2 for monointercalating agents (Saucier et. al, 1971).

Table II. Slope of the straight line given by the equation Log η/η_0 = k log(1+2r) which account for the increase in the viscosity of sonicated DNA as function of the amount of drug bound (r). In standard operating conditions, the experiments were performed in acetate buffer pH 5.0 containing 100 mM NaCl.

	DNA from	
	C. perfringens (66% AT)	M. aureus (72% GC)
Net-OPC	1.01	1.42
BPC-OPC	0.03	0.07

The present experiments have been performed using natural DNAs which lead to more suitable sonicated preparations than those obtained with the synthetic nucleotides. The slope value obtained (1.01) upon Net-OPC binding to DNA isolated from C. perfringens and containing 66% A-T base pairs (Table II) suggests that OPC is partly intercalated in all binding sites or that typical intercalation occurs in about 50 % of the binding sites only. In similar experimental conditions, BPC-OPC binding did not result in any length increase of DNA indicating that the chromophore is not intercalated between DNA base pairs. Similar behaviour is observed for both drugs in the presence of DNA from *M. aureus* containing 72% G-C bases pairs (Table II). However, the slope value of 1.47 obtained upon binding of Net-OPC is more consistent with a typical intercalation process. Accordingly, the apparent binding site size of two base pairs (Table I) of Net-OPC to poly d(G-C) as determined from the scatchard plot is as well consistent with intercalative binding.

3.2. INHIBITION OF RESTRICTION ENDONUCLEASES

From a biological point of view, selective recognition of a given sequence by drugs should result in the selective alteration of biological functions involving similar sequence recognition. The extent of the selective perturbation following drug interaction can be taken as the reflect of the selective recognition of the DNA sequence. Suitable model in this way is provided by testing the effect of drugs on the catalytic activity of restriction endonucleases which selectively produce strand breakage at discrete sites on DNA. We have tested the inhibitory effect of Net-OPC and BPC-OPC on EcoRI and NruI enzymes which selectively produce double strand break at 5'GAATTC and 5'TCGCGA sequence respectively. In the presence of supercoiled pBR322 DNA (form I) as substrate, the catalytic action of these enzymes results in a double strand break occurring within a single cleavage site and leading to the formation of linear DNA (form III). The reaction occurs in two consecutive steps involving the production of circular DNA (form II) as transient intermediate as follows:

$$\text{Form I} \xrightarrow{k_1} \text{Form II} \xrightarrow{k_2} \text{Form III} \quad (3)$$

were k_1 and k_2 are the rate constants of the 1st and the 2nd strand break respectively. Forms I, II and III can be efficiently separated by gel electrophoresis and the variation of the relative concentration of the three forms during the time course of the enzymatic reaction can be accurately estimated. In appropriated operating conditions, the first and the second nicking reaction occur according to a pseudo first order kinetics (Halford & Goodall, 1988). Under these circumstances, the steady state profile of the variation of the DNA form II concentration as function of time is described by the following exponential association equation:

$$[FII] = A\ 1\text{-}exp^{(-k1t)} + C\ 1\text{-}exp^{(-k2t)} + E \quad (4)$$

where E is the amount of form II initially present before the action of the enzyme. Values of A, k_1, k_2 and C have been estimated from the goodness-of-fit assesed using actual distances provided by a Marquardt procedure. This treatment leads to the estimation of the rate constant k_2 of the second strand break which appears to be the limiting factor of the overall enzymatic reaction. This is in agreement with the occurrence of a two steps consecutive reaction with an obligatory intermediate. In the operating conditions, k_1 and k_2

yielded average values of $0.50 +/- 0.05$ mim^{-1} and $0.34 +/- 0.01$ mim^{-1} for EcoRI and 0.21 min^{-1} and 0.20 min^{-1} in case of NruI.

TABLE III. Effect of Net-OPC and BPC-OPC on the catalytic activity of the restriction endonucleases EcoRI and NruI. Enzymatic activities were measured as described in materials and methods. K_1 and K_2 are the pseudo first order rate constants of the first and the second DNA strand break produced by the enzymes and obtained from the non-linear regression fitting of eq. 3. Values indicated are the concentration of drug expressed in μM which reduced by 50% (ID_{50}) the corresponding rate constant.

	Eco RI		NruI	
	K_1	K_2	K_1	K_2
Net-OPC	0.01	0.02	> 50	> 50
BPC-OPC	> 100	60	> 100	> 100

We have estimated the inhibition of the catalytic activity of the enzymes from the decrease of the rate constants k_1 and k_2 observed in the presence of increasing concentrations of drug. Efficiency of the inhibition is provided by the EC_{50} value which correspond to the concentration of drug which decrease by 50% the rate constant. Values summarized in table III shows that Net-OPC displays a pronounced inhibitory effect on EcoRI enzymatic activity affecting both k_1 and k_2. The selective inhibition of Net-OPC on EcoRI is obvious from the lake of effect of the drug on NruI activity using concentrations up to 50 μM. In similar experimental conditions, BPC-OPC produced no significant effect neither on EcoRI nor NruI activity.

3.3 INHIBITION OF M.MuLV CYCLE

The test consists in the quantification of 3T3 cells displaying resistance to the antibiotic geniticin. The aquired resistance phenotype results from the integration in the host 3T3 cells DNA of the defective proviral DNA containing the neo gene (see materials and methods). The number of resistant clones detected is directly related to the proviral DNA integration efficiency. A decrease in the viral DNA integration score (and in turn in the number of resistant clones) may result from the inhibition of one of the step of the retroviral cycle including the infection of 3T3 cells by the exogenous virus, the reverse transcription of viral RNA to DNA and the integration process itself which is the step involving AT containing sequences recognition. The results summarized in the table IV show that the addition of increasing concentrations of Net-OPC in a medium containing 3T3 cells and infectious defective M.MuLV results in a decrease (ID_{50} = 5 μM) followed by a complete disappearance of geniticin resistant clones.

Table IV. Effect of Net-OPC and BPC-OPC on the number of 3T3 clones resistant to geniticin G418 following the infection of cells by the defective M.MuLV carrying the neo gene (see materials and methods). EC_{50} estimated from the non-linear regression of the sigmoid inhibition curve was 5.0 μM for Net-OPC and 33 μM for BPC-OPC. Values obtained with BPC-OPC were corrected for the toxic effect induced by the drug on 3T3 cells. Each value is the mean of at least three different experiments.

Concentration (10^{-6} M)	Number of clones resistant to G418 in % of the control	
	Net-OPC	BPC-OPC
0	100	100
1	97	112
2.5	89	108
5	62	104
7.5	32	95
10	14	95
20	0	70
25	0	60
50	0	35

It was verified that, at the concentrations used, the drug did not induce any toxic effect neither on 3T3 cells nor on infected 3T3 cells and did not alterate the expression of the neo gene in infected cells. In similar experimental conditions, the addition of BPC-OPC results in markedly lower inhibitory effect (ID_{50} = 33 μM).

4. DISCUSSION

In order to develop for pharmacolgical purpose sequence-specific DNA-binding molecules, we have synthesized various hybrid groove binder-intercalators composed of a bis-pyrrolecarboxamide moiety related to netropsin covalently linked to the intercalating chromophore oxazolopyrido-carbazole (OPC). The rationale supporting the use of OPC in the design of such a molecules is the following: i) OPC displays a high affinity for double stranded DNA which should result in a marked stabilisation of the hybrid molecules-DNA complexes. ii) as hydrophobic cation, OPC should markedly facilitate the diffusion of the hybrid molecules through cell membranes. III) The fluorescence properties of OPC should facilitate as well physico-chemical, biochemical and pharmacological studies. Among the molecules synthesized, Net-OPC and BPC-OPC display DNA binding properties which can be beneficially used in attempt to establish the consequences of the preferential binding of the drugs to AT rich sequences in terms of biochemical and pharmacological properties. As the parent compound netropsin, the hybrid molecule Net-OPC prefe-rentially binds to A-T rich double stranded polynucleotide. Scatchard plot of the binding curves indicates that in poly d(A-T), Net-OPC binds to at least two different binding sites with Kapp values close to $6.0 \ 10^8$ and $7.0 \ 10^6$ M^{-1} in the presence of 100 mM NaCl, whereas in poly d(G-C), the molecule binds to a single type of binding site with a kapp close to $1.7 \ 10^6 \ M^{-1}$. The ratio between the Kapp to the high affinity site in poly d(A-T) and the Kapp to poly d(G-C) is about 350 an reflects a high binding preference to AT containing sequences. In contrast, the BPC-OPC derivative lacking the amidine moiety on the bis-pyrrolecarboxamide chain binds to either poly d(A-T) and poly d(G-C) with similar Kapp values. It should be noticed moreover that the binding of BPC-OPC essentially occurs according to a non-intercalative mode which was obvious from the inhability of the drug to increase the length of the double helix in both AT and GC-rich natural DNAs. In view of tailoring molecules which selectively modify the functional activity of the genome through sequence recognition, it is of interest to study the consequences of the preferential binding to AT sequences compared to the absence of preferential binding as estimated using parameters described above on the respective effect of the molecules in biological models whose process involves at least one step requiring recognition of AT containing sequence(s). Among the various available models, we have decided to use as probes i) the catalytic activity of restriction endonucleases such as EcoRI which selectively cleave the double stranded DNA at GAATTC and NruI as negative control which cleave DNA at TCGCGA sequence and ii) a part of the M.MuLV retroviral cycle involving at the integration step the recognition of either CATT-AATG (integration point) and CTTT (required flanking region) in the process in which the circularized viral DNA acts as the integrable form or CATT and CTTT in the process in which the linear viral DNA is integrated. The results obtained using these models clearly show that the Net-OPC molecule displaying a preferential binding to AT-rich sequences strongly inhibits both the catalytic activity of EcoRI and the retroviral cycle whereas in contrast the related molecule BPC-OPC which is devoid of binding preference to AT-rich sequences displays no detectable effects. The striking observation is the absence of inhibitory effect on the activity of the enzyme NruI which strongly suggest that the inhibition of EcoRI by Net-OPC is in direct relation with the sequence recognition. Similar assumption can be formulated regarding the results obtained using the retroviral cycle. However, in the present case, a non-selective action on either the viral infection of the 3T3 cells or the reverse transcriptase activity cannot be ruled out. Moreover, a lower but significant inhibition of the appearance of resistant clones is observed following the addition of PBC-OPC. Nevertheless, it can be concluded that the design of molecules which preferentially bind to AT-rich sequences on double stranded DNA with a high affinity leads to molecules able to produce a selective alteration of biological processes involving the recognition of AT containing sequences.

5. REFERENCES

Auclair, C., & Paoletti, C. (1981) Bioactivation of the anti-tumor drug 9-hydroxyellipticine and derivatives by a peroxidase hydrogen peroxide system J. Med. Chem. 24, 289-295.

Auclair, C., Voisin, E., Banoun, H., Paoletti, C., Bernadou, J. & Meunier, B. (1984) Potential anti-tumor agents: synthesis and biological properties of aliphatic amino-acid-9-hydroxyellipticinium derivatives. J. Med. Chem. 27, 1161-1166.

Auclair, C., Schwaller, M. A., René, B., Banoun,H., Saucier, J. M., & Larsen, A. K. (1988) Relationships between physicochemical and biological properties in the series of oxazolopyridocarbazole derivatives (OPCd); comparison with related anti-tumor agents. Anti-Cancer Drug Design 3, 133-144.

Banoun, H., Le Bret, M. and Auclair, C. (1985) Accessibility to bacterial nucleic acids of the intercalating drugs aliphatic amino acids ellipticinium derivatives in *Escherichia coli* and *Salmonella typhimurium*. Biochemistry 24, 701-707.

Cepko, C.L. Roberts, B.E. and Mulligan, R.C. (1984) Construction and application of a highly transmissible murine retrovirus shuttle vector. Cell 37, 1053-1062.

Colicelli, J. and Goff, S.P. (1985) Mutants and pseudorevertants of Moloney murine leukemia virus with alterations at the integration site Cell 42, 573-580.

Dervan, P.B. (1986) Design of sequence-specific DNA-binding molecules. Science 232, 464-471.

Dryer, G.B. and Dervan, P.B. (1985) Sequence-specific cleavage of single-stranded DNA: Oligonucleotide-EDTA-Fe(II). Proc. Natl. Acad. Sci. USA 82, 968-972.

Goppelt, M., Langowski, J., Pingoud, A., Haupt, W., Urbank, C., Mayer, H. and Maass, G. (1981) The effect of several nucleic acid binding drugs on the cleavage of d(GGAATTCC) and pBR322 by the Eco RI restriction endonuclease. Nucleic Acids Res. 9, 6115-6127.

Gouyette, A., Auclair, C., & Paoletti, C. (1985) A revised structure of the anti-tumor drug ellipticinium-amino(acid) adducts. Biochem. Biophys. Res. Commun. 131, 614-619

Green, P.J., Pines, O. and Inouye, M. (1986) The role of antisens RNA in gene regulation. Annual Review of Biochemistry 55, 569.

Halford, S.E. and Goodall, A.J.(1988) Modes of DNA Cleavage by the EcoRV restriction endonuclease. Biochemistry 27, 1771-1777.

Izant, J.G. and Weintraub, H. (1984) Inhibition of thymidine kinase gene expression by antisense RNA: a molecular approach to genetic analysis. Cell 36, 1007.

Izant, J.G. and Weintraub, H. (1985) Constitutive and conditional suppression of exogenous genes by antisense RNA. Science 229, 345.

Kopka, M.L., Yoon, C., Goodsell, D., Pjura, P. and Dickerson, R.E. (1985) The molecular origin of DNA-drug specificity in netropsin and distamycin. Proc. Natl. Acad. Sci. USA, 82, 1376-1380.

Kopka, M. L., Yoon, C., Goodsell, D., Pjura, P., & Dickerson, R.E. (1985) Binding of an antitumor drug to DNA: netropsin and CGCCAATT[br]CGCG. J. Mol. Biol. 183, 553.

Le Pecq, J. B., & Paoletti, C. (1967) A fluorescent complex between ethidium bromide and nucleic acids. J. Mol. Biol. 27, 87-106.

Lown, J.W., Krowicki, K., Baht, U.G., Skorobogaty, A., Ward, B. and Dabrowiak, J.C. (1986) Molecular recognition between oligopeptides and nucleic acids: novel imidazole containing oligopeptides related to netropsin that exhibit altered DNA sequence specificity. Biochemistry 25, 7408.

Lown, J.W. (1988) Lexitropsins: rational design of DNA sequence reading agents as novel anti-cancer agents and potential cellular probes. Anti-Cancer Drug Design 3, 25-40.

Mann, R., Mulligan, C.M. and Baltimore, D. (1983) Construction of a retrovirus packaging mutan(t and its use to produce helper-free defective retrovirus. Cell 33, 153-159.

Marky, L.A. and Breslauer, K.J. (1987) Origin of netropsin binding affinity and specificity: correlations of thermodynamic and structural data. Proc. Natl. Acad. Sci. USA 84, 4359.

Mrani, D., Gosselin, G., Auclair, C., Auclair, C., Balzarini, J., De Clercq, E., Paoletti, C. and Imbach, J.L. Synthesis, DNA binding and biological activity of oxazolo-pyridocarbazole-netropsin hybrid molecules. Submitted.

Paoletti, C. (1988) Anti-sense oligonucleotides as potential antitumour agents: prospective views and preliminary results. Anti-Cancer Drug Design 2, 325-331.

Pullman, B. (1989) Molecular mechanism of specificity in DNA-antitumour drug interactions. In Advances in drug research. Academic Press Limited, Vol. 18 pp. 1-113.

Record, M. T., Lohman, T. M., & De Haseth, P. L. (1976) Ion effects on ligand-nucleic acid interactions J. Mol. Biol. 107, 145-158.

Saucier, J. M., Festy, B., & Le Pecq, J. B. (1971) The change of the torsion of the DNA helix caused by intercalation. Biochimie 53, 973-977.

Schultz,P.G.,Taylor,J.S. and Dervan,P.B. (1982) Design and synthesis of a sequence-specific cleaving molecule. (Distamycin-EDTA)iron II. J. Am. Chem. Soc.104, 6861-6863.

Sequin, U. (1974) Nucleosides and nucleotides. Part 7. Four Dithymidine monophosphates with different anomeric configurations their synthesis and behaviour towards phosphodiesterases. Helvetica Chimica Acta 57, 68-73.

Taylor, J.S., Schultz, P.G. and Dervan, P.B. (1984) Sequence specific cleavage of DNA by distamycin-EDTA.FeII and EDTA-distamycin FeII Tetrahedron, 40: 457-465.

Toulmé, J.J., Krish, H.M., Loreau, N., Thuong, N.T. and Hélene, C. (1986) Specific inhibition of mRNA translation by complementary oligonucleotides covalently linked to intercalating agents. Proc. Natl. Acad. Sci. USA 83, 1227.

Wartell, R.M., Larson, J.E., and Wells, R.D. (1974) Netropsin: a specific probe for AT regions of duplex deoxyribonucleic acid J. Biol. Chem. 249, 6719.

PARALLEL-STRANDED NUCLEIC ACIDS AND THEIR INTERACTION WITH INTERCALATING
AND GROOVE BINDING DRUGS

JOHAN H. VAN DE SANDE, BERND W. KALISCH AND MARKUS W. GERMANN
Department of Medical Biochemistry
The University of Calgary
Calgary, Alberta
Canada T2N 4N1

ABSTRACT. DNA structures with parallel strand alignment can be formed
from oligonucleotides containing exclusively AT base pairs.
Intramolecular duplexes, containing polarity reversal 5'-5'
phosphodiester linkages, and intermolecular duplex DNA containing either
homo-oligomeric or alternating A-T sequences can exist in vitro as
parallel-stranded (PS) DNA. The formation of stable PS duplex structures
is demonstrated by the following criteria: (i) electrophoretic
mobilities under native conditions are similar to those of antiparallel-
stranded (APS) duplexes. (ii) spectroscopic properties are characteristic
for base paired structures but are different from those of the
corresponding APS duplex. (iii) PS duplexes undergo thermally induced
helix to coil transitions, however, the melting temperature is
considerably lower than that of APS DNA. (iv) NMR measurements support
a normal phosphodiester backbone conformation for PS DNA as well as base
pairing and base stacking between T and A residues similar to those in
APS DNA. Ethidium bromide binds more strongly to parallel stranded DNA
with a higher fluorescence quantum yield than that for the binding to APS
DNA. In contrast, the minor groove specific ligand Hoechst 33258
exhibits a markedly reduced affinity for PS DNA compared to the binding
of this drug to control APS duplex DNA. The reverse Watson-Crick base
pairing between the T and A residues of parallel stranded DNA results in
the presence of the 5-CH$_3$ group in the minor groove of the DNA which could
sterically inhibit the binding of ligands such as Hoechst 33258.

Introduction

Common properties of the three major polymorphic families of duplex DNA,
A-, B-, and Z-DNA, are the antiparallel orientation of complementary
strands and Watson-Crick base pairing. Alternatively, PS duplex
structures can be formed if bases are appropriately chemically modified,
protonated or by the presence of unusual sugar-phosphate backbones
(reviewed in 1). In triplex polynucleotide structures formed between
polypurine and polypyrimidine sequences, the parallel orientation of one
polypyrimidine strand and the polypurine strand of the former duplex is
a consequence of Hoogsteen base pairing (2). Recently, we have

261

B. Pullman and J. Jortner (eds.), Molecular Basis of Specificity in Nucleic Acid-Drug Interactions, 261–274.
© 1990 Kluwer Academic Publishers. Printed in the Netherlands.

demonstrated the existence of an alternative helical structure containing two unmodified complementary oligodeoxynucleotide strands, e.g. - parallel-stranded DNA (1). Force field calculations predicted the stability of such parallel-stranded right-handed structures with reverse Watson-Crick type base pairing (3).

We have shown that DNA hairpin molecules containing polarity reversal phosphodiester bonds in the loops, have the strands aligned in a parallel orientation (1). Intramolecular PS hairpin structures are formed from AT sequences in either a homo oligomeric or an alternating dinucleotide orientation (4,5). The PS hairpins form duplexes of thermal stability comparable to those of the antiparallel stranded (APS) hairpins of identical sequence. Parallel stranded DNA is, however, not limited to modified hairpin structures but can also be formed from conventional complementary linear oligonucleotides having the appropriate sequence design (6-10). The identical primary sequence of PS DNA and APS control DNA permits a comparison of the relative stabilities and properties of the two different duplex structures. Differences in helical structures is reflected by the differential binding of intercalating agents and minor groove binding ligands to PS and APS duplex DNA structures of identical sequence. The reduced affinity of minor groove binding ligands to parallel stranded DNA is most likely a consequence of the reverse Watson-Crick base pairing in these structures.

Materials and Methods

MATERIALS

Oligodeoxynucleotides were synthesized by using automated phosphoramidite chemistry on a DNA synthesizer (Applied Biosystems, Model 380A). 5'-Phosphoramidites were synthesized as described previously (1). The tritylated synthesis products were purified by reverse-phase HPLC on a PRP-1 column (11), detritylated, and desalted on a Sephadex G-25 column. Molar extinction coefficients of the pure products were estimated from the extinction coefficients of the mononucleotides at 90 °C in 5 M NaClO$_4$ (12).

Hoechst 33258 was gift from Dr. H. Loewe, Hoechst Laboratories, Frankfurt, West Germany. Ethidium bromide was purchased from Boehringer-Mannheim. Extinction coefficients of 4.2 x 10^4 M^{-1} cm^{-1} at 338 nm (Hoechst 33258) and 5.45 x 10^3 M^{-1} cm^{-1} at 480 nm (Ethidium bromide) were used (13,14).

SPECTROSCOPY

Ultraviolet (UV) absorption and thermal denaturation profiles were recorded with a Varian 2280 spectrophotometer equipped with temperature regulated cuvette holders. Fluorescence measurements were obtained on a Perkin Elmer 560-10S fluorescence spectrophotometer, equipped with a temperature regulated cuvette holder. The excitation and emission wavelengths were as follows: Hoechst 33258, λ_{ex} 355 nm, λ_{em} 480 nm; Ethidium bromide, λ_{ex} 525 nm, λ_{em} 600 nm.

THERMAL DENATURATION ANALYSIS

Denaturation was followed at 260 nm, using a temperature gradient of 0.5 °C/min. Absorbance readings were collected at 0.5°C intervals, corrected for volume expansion, and were fitted to a concerted 2-state model by using a Simplex procedure (4). The melting point is defined as the temperature at which 50% of the duplex is denatured. Enthalpies were assumed to be independent of the temperature within the range considered. Entropies are calculated from $\Delta S° = \Delta H°_{vH}/T_m - R \ln(C_T/4)$, where R is the gas constant, T_m is the melting temperature (Kelvin), and C_T is the total strand concentration of the constituent oligonucleotides (M).

GEL ELECTROPHORESIS

Analysis of deoxyoligonucleotides was carried out on 15% polyacrylamide gels run under renaturing conditions in TBM buffer (90 mM Tris-borate, pH 8.3) containing 10 mM $MgCl_2$. Bands were visualized by either UV shadowing or by staining with ethidium bromide or Hoechst 33258.

Results and Discussion

STRUCTURES

The sequences of the oligonucleotides studied are shown in Figure 1. The hairpin oligonucleotides A, B and C can form either intramolecular hairpin structures or intermolecular dimeric structures.

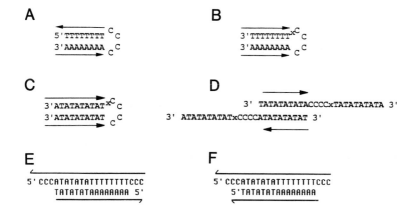

Figure 1. (A) Intramolecular homopolymeric antiparallel hairpin structure (APS). (B) Intramolecular homopolymeric parallel hairpin structure (PS). (C) Intramolecular alternating dinucleotide parallel hairpin structure. (D) Possible intermolecular antiparallel (dimeric) structure for the alternating dinucleotide parallel hairpin structure C. (E) Linear antiparallel-stranded duplex structure. (F) Linear parallel stranded duplex structure.

The monomer-dimer equilibrium of DNA hairpin structures is shifted to the monomer hairpin form at low DNA concentrations and high temperature. The presence of hairpin loops at the center of oligonucleotides A, B and C decreases the tendency to form the dimer and, therefore, hairpin structures predominate even at high strand concentration (15). Potential antiparallel dimeric and concatameric structures can be identified by their concentration-dependent thermodynamic and spectroscopic properties. Concentration independence of physical properties is consistent with the formation of parallel-stranded hairpins for structures B and C. The linear duplex structures E and F have a common template strand of 21-nucleotides containing 3C residues at both the 5' and 3' ends. Complementary parallel-stranded and antiparallel-stranded oligomers of 15 nucleotides have the identical nucleotide sequence if read in the opposite polarity. The structures are designed to minimize potential formation of alternative structures, especially for the parallel-stranded duplex F. The linear duplex DNA structures contain both homooligomeric dA.dT and alternating d(A-T) segments. The presence of the alternating d(A-T) segment provides a differential in stability between the desired parallel-strand alignment of the complementary strands and any other partially base paired intra- or intermolecular structures.

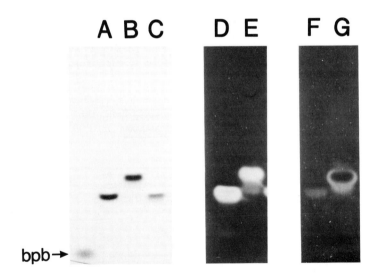

Figure 2. Polyacrylamide gel electrophoretic analysis of PS and APS hairpins. TBM gel was run at 4°C. (A) $3'dT_8xC_4A_8 3'$. (B) $5'dT_8C_4A_8 3'$. (C) d(GTAC)$_2$. (D, E) Lanes A and B stained with ethidium bromide (0.5 μg/ml^{-1}) at 4°C. (F, G) Lanes A and B stained with H-33258 (20.0 μg/ml^{-1}) at 4°C. Lanes A, B, C visualized by ultraviolet shadowing. Lanes D, E, F, and G visualized by fluorescence. bpb is bromophenol blue marker.

CHARACTERIZATION AND PROPERTIES OF PARALLEL-STRANDED STRUCTURES

The putative parallel-stranded structures were analyzed in comparison
with the antiparallel duplexes of identical sequence. The PS and APS
oligomers have similar gel electrophoretic mobilities under native
conditions and migrate exclusively as single bands, as shown in Figure
2 for oligonucleotides A and B.
      In general, the PS duplex has a slightly faster mobility than the
APS control sequence for both the hairpin as well as the linear duplex
structures. Strong staining of both APS and PS structures was observed
with ethidium bromide while only the APS structure stained strongly with
Hoechst 33258. Gel electrophoretic analysis also support the parallel-
stranded nature of the hairpin oligonucleotides with 8 and 10 AT base
pairs in the stem at strand concentration smaller than 30 mM.
Alternating AT hairpin structures with 12 or 14 AT base pairs in the stem
show bands corresponding to the formation of dimeric and trimeric
antiparallel species (structure D).   The length and concentration
dependence of the PS-APS equilibrium for alternating AT sequences is
presented elsewhere (5).
      Diagnostic ultraviolet spectra are observed for parallel stranded
DNA duplexes containing A-T base pairs (1,4,5).   The ultraviolet
absorption spectra for hairpins A and B are presented in Figure 3.
      The higher absorbance for the PS duplex is accompanied by a blue
shift of 4 nm from that of the APS complement.   The same spectral
difference is also observed for the linear duplexes E and F.   The
hyperchromicity profile (Figure 3B) indicates a high hyperchromicity
between 240 and 260 nm and a sharp decrease at higher wavelength for the
APS structure.   In contrast, the PS hairpin structure exhibits a low
hyperchromicity below 260 nm and reaches the maximum hyperchromicity
(35%) at 270 nm.   Such differences between PS and APS structures have
been observed for all the duplexes examined.

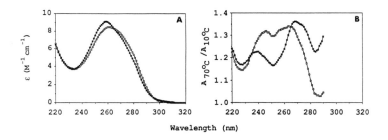

Figure 3:   Ultraviolet absorption spectra of $3'dT_8xC_4A_83'$ (PS) and
$5'dT_8C_4A_83'$ (APS).   (A) Absorption spectra of native PS (■) and APS (□)
hairpin structures in 100 mM NaCl, 0.1 mM EDTA, 10 mM phosphate pH 7.0
at 15° C.   Extinction coefficients are given per residue.   (B)
Wavelength dependent hyperchromicity of PS (■) and APS (□) DNA hairpins.
Absorbances at 70° C were divided by the absorbance of the same sample
at 15° C.

Hypochromicity results froma closed, stacked alignment of the transition dipoles of the bases, therefore the stacking arrangement in the PS and APS structures must be significantly different. Circular dichroism spectra, including vacuum CD, for both the antiparallel-stranded and parallel-stranded helices are similar. Both show the characteristic bands expected for right-handed B-type helices (1,4,5). All of the parallel-stranded duplexes undergo thermally induced transitions from base stacked double helices to denatured coil forms. The thermodynamic parameters are summarized in Table 1.

TABLE 1. Thermodynamic Parameters for Helix-Coil Transitions

Duplex	Tm(°C)	$\Delta H°$ (kJ mol^{-1})	$\Delta S°$ (kJ mol^{-1}K^{-1})
5'-dT$_8$C$_4$A$_8$-3'	45.6[a]	190	0.60
3'-dT$_8$xC$_4$A$_8$-3'	35.2[a]	154	0.50
5'-d(AT)$_5$C$_4$(AT)$_5$-3'	57.2[b]	244	0.74
3'-d(AT)$_5$xC$_4$(AT)$_5$-3'	39.3[b]	137	0.44
3'-d(AT)$_4$xC$_4$(AT)$_4$-3'	36.3[b]	106	0.34
5'-dC$_3$(AT)$_4$T$_7$C$_3$-3' 3' -d(TA)$_4$A$_7$ -5'	41.0[c]	487	1.42
5'-dC$_3$(AT)$_4$T$_7$C$_3$-3' 5' -d(TA)$_4$A$_7$ -3'	19.1[c]	309	0.94

Melting temperatures (Tm), $\Delta H°$ (van 't Hoff enthalpy) and $\Delta S°$ (melting entropy) were determined in 0.1mM EDTA, 10 mM sodium phosphate, pH7.0 containing [a]0.1 M NaCl, [b]0.8 M NaCl, [c]0.4 M NaCl. x denotes 5'-5' internucleotide bond.

As can be seen, the homopolymeric PS structure is closest in thermal stability to its APS control. The presence of alternating A-T sequences destabilizes the PS helix as can be seen from the larger differential in melting temperature between the parallel-stranded sequence and its antiparallel complement. However, the difference in thermal stability between the linear duplexes E and F is smaller in the presence of MgCl$_2$ (42.7°C and 24.5°C respectively). This indicates that a preferential stabilization of a PS duplex by divalent ions occurs and that the effect of cations and ligands on parallel-stranded duplex stability needs to be determined. Thermally induced helix-coil transitions for all these structures are monophasic and reversible and can be fitted with high accuracy to a 2-state model. Melting enthalpies are consistently lower (20-40%) for the PS structures than for the APS complements.

Structural details of parallel-stranded DNA have also been obtained by proton and ^{31}P NMR analysis (4). Spectra obtained for structures A and B differ considerably and, in particular, show a smaller degree of chemical shift dispersion in the aromatic resonances and thymine methyl groups of the PS hairpin than those of the APS structure. This suggests that the PS hairpin forms a more symmetrical structure than what is

observed for the APS control APS DNA. [31]P chemical shifts indicate that
the backbone conformation of the stem of the PS hairpin is not
dramatically different from that of the control. We have also analyzed
the conformation of the PS hairpin structure B by comparison of observed
NOESY cross-peak patterns to those observed for A, B or Z-DNA (4,16).
The bases in the stem are in an anti position and the sugar puckering is
in the 2'-endo range. Our observation of NOE effects between the T imino
protons and AH-2 protons demonstrate that base pairing between the A and
T residues occurs in either Watson-Crick or reverse Watson-Crick, but not
Hoogsteen, orientation. Hoogsteen base pairing was also excluded because
of the identical pattern of chemical methylation by dimethyl sulfate for
the PS and APS hairpin molecules (1). Since Watson-Crick type base
pairing is not compatible with the formation of parallel-stranded duplex
DNA, we conclude that the base pairing in PS DNA occurs in the reverse
Watson-Crick orientation as originally predicted by Pattabiraman (3).
The difference in base pairing orientation is depicted in Figure 4.

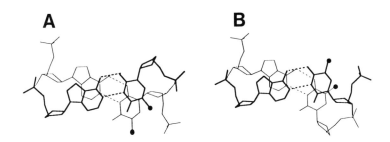

Figure 4. Base pairing and base stacking in parallel- and antiparallel-
stranded DNA. The coordinates of the APS (B) structure were those of B-
DNA (17) and the PS (A) coordinates were obtained by implementing a 180°
rotation of the thymidine monophosphate around the N_3-C_6 axis.

DRUG-DNA INTERACTIONS

Parallel-stranded DNA was previously characterized by an increase in the
fluorescence of EBr with the fluorescence of bound ethidium bromide
compared to the antiparallel control (1). The homo-oligomeric AT hairpin
structures A and B are used to determine whether the increase in the
ethidium bromide fluorescence is due to: (i) increased affinity for
parallel-stranded DNA; (ii) increase in the quantum yield of ethidium
bromide bound to the parallel-stranded structure; or, (iii) the parallel-
stranded structure does not follow the nearest neighbour exclusion
principle observed for ethidium bromide binding to antiparallel stranded
DNA (14). To determine the rationale for the increase in fluorescence
enhancement of ethidium bromide binding to PS DNA, a titration experiment
was carried out as shown in Figure 5.

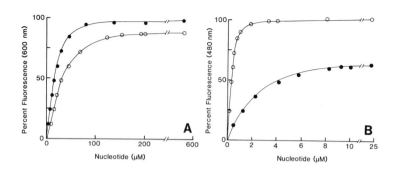

Figure 5. Fluorescence titration of ethidium bromide and Hoechst 33258 with 5'dT$_8$C$_4$A$_8$3' (APS) and 3'dT$_8$xC$_4$A$_8$3' (PS) DNA hairpins. (A) Titration of 0.25 μM ethidium bromide in 100 mM NaCl, 0.1 mM EDTA, 10 mM Tris, pH 7.0 at 10 °C with PS (■) and APS (□) DNA hairpins. (B) Titration of 20 nM Hoechst 33258 in 400 mM NaCl, 0.1 mM EDTA, 10 mM Tris, pH 7.0 at 10° C with PS (■) and APS (□) DNA hairpins. The measured fluorescence was normalized to 1 in both A and B. DNA concentrations are given per nucleotide residue.

     The ethidium bromide titrations show midpoints at 1.6 x 10^{-5} for the PS DNA and 2.6 x 10^{-5} M for the APS DNA hairpin. At higher DNA concentrations, both binding curves reach a plateau with the APS hairpin exhibiting only slightly lower fluorescence (10%) than that for the PS DNA. At the titration end point the DNA concentration is at least ten-fold greater than the ethidium bromide concentration, and therefore, each hairpin would bind less than 1 dye molecule. The emission spectrum of ethidium bromide bound to the APS and PS hairpin respectively was recorded at the end point of the titration as shown in Figure 6.
     The only difference in the emission spectrum of ethidium bromide bound to the two DNA hairpins is the lower fluorescence intensity obtained for the APS structure. This is indicative of a lower EBr quantum yield and demonstrates that variations at the intercalation sites are experienced by the ethidium bromide molecule. Hyperchromicity plots have indicated that the stacking pattern in the parallel-stranded DNA is different from that of the antiparallel-stranded duplex which would account for a difference in hydrophobic environment experienced by the ethidium bromide molecule upon binding to the PS DNA as compared to that of the APS structure. The increase in ethidium bromide fluorescence can also be observed in the gel staining in lanes D and E of Figure 1. The PS duplex shows a considerably greater fluorescence than the APS control upon staining with ethidium bromide. The difference in fluorescence emission, however, is too small to account for the nearly 2-fold fluorescence enhancement obtained at lower DNA concentrations for the PS structures. Therefore, we conclude that the intercalator ethidium bromide binds stronger to parallel stranded DNA than to the APS complement.

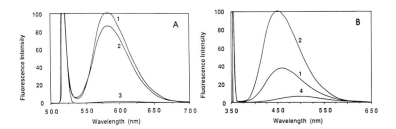

Figure 6. Fluorescence emission spectra of ethidium bromide and Hoechst 33258 bound to 5'dT$_8$C$_4$A$_8$3' (APS) and 3'dT$_8$xC$_4$A$_8$3' (PS) DNA hairpins. (A) Uncorrected fluorescence emission of ethidium bromide at 10°C in complexes with PS (1) and APS (2) hairpins and free in solution (3). (B) Uncorrected fluorescence emission of Hoechst 33258 at 10°C in complexes with PS (1) and APS (2) DNA hairpins and free in solution (4). DNA samples were from the end point of the titration shown in Figure 5. The excitation was at 520 nm for ethidium bromide and at 355 nm for Hoechst 33258.

The minor groove binding A-T specific bis(benzimidazole) drug Hoechst 33258 shows a considerably weaker staining of the PS hairpin compared to that of the APS hairpin (Figure 1, Lanes F and G). Titration of Hoechst 33258 with the PS and APS hairpins also indicate a drastically reduced fluorescence enhancement for the parallel-stranded DNA hairpin (Figure 5B). Two modes of binding to DNA can be differentiated for Hoechst 33258. The low affinity binding (emission at 490 nm) is insensitive to base composition and is greatly reduced by increasing the ionic strength (13). The high affinity binding mode (emission at 460 nm) is predominant for A-T base pair regions in DNA and is essentially independent of the salt concentration (18). Both the PS and APS hairpins display the spectral properties and the salt independent binding (0.4-2.0 M NaCl) associated with the high affinity binding mode expected for DNA containing exclusively AT base pairs (Figure 6B). X-ray defraction studies, as well as footprinting experiments, indicate that H-33258 requires at least four consecutive A-T base pairs for the high affinity binding (19,20). The drug makes hydrogen bond contacts in the minor groove with N3 of adenine and O2 of thymine. Hairpin molecules containing 8 A-T base pairs in the stem can, therefore, provide only single H-33258 binding sites and the binding can be analyzed using a simple bimolecular model. Using Scatchard plots and curve fitting, binding constants of 8 x 10^6 and 5.5 x 10^8 M^{-1} for the PS and APS hairpins respectively. The binding constant calculated for the binding of Hoechst 33258 to the APS hairpin is greater than the range reported previously (10^6-10^7 M^{-1}) determined at higher drug concentrations and for polynucleotides with differing AT contents (21,22). However, the polynucleotide poly d(A-T) and the APS hairpin bind the drug with similar affinity and, therefore, the larger binding constant obtained for the APS hairpin is not due to structural peculiarities of this oligonucleotide. The high affinity binding observed for the APS structure could suggest that the H-33258 could potentially stabilize dimeric or multimeric forms of the PS hairpin. The binding curves obtained for the PS structure are not compatible with such a model, which would be strongly dependent on the strand concentration and would require the presence of a large amount of dimeric or multimeric structures even at low concentrations. Also, the equilibrium constant calculated at a 10-fold lower drug concentration does not differ from that at the higher drug concentration. Again, the greatly reduced affinity of the PS hairpin for the H-33258 ligand suggests that the helical parameters of the PS structure are significantly different from those of B-DNA (1,4).

The binding of ligands to DNA is often accompanied by a pronounced stabilization of the DNA to thermal denaturation (23). The effect of ethidium bromide and Hoechst 33258 on the thermal stability of the APS and PS DNA hairpins was determined as shown in Figure 7.

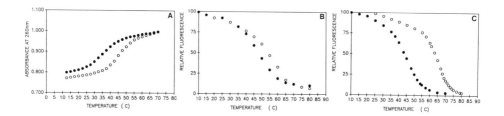

Figure 7. Thermal denaturation of 5'dT$_8$C$_4$A$_8$ 3' (APS), (O) and 3'dT$_8$ x C$_4$ A$_8$ 3' (PS), (•). (A) Absorbance at 260 nm in 0.1 M NaCl. (B) Relative fluorescence of ethidium bromide DNA complexes in 0.4 M NaCl. (C) Relative fluorescence of Hoechst 33258-DNA complexes in 0.1 M NaCl.

A monophasic, fully reversible, thermally induced transition from helix to coil was observed for both the PS and APS duplex. The thermal denaturation was concentration independent. The thermal denaturation of the ligand DNA complexes was measured by loss of fluorescence at increasing temperatures. Again, a monophasic transition was observed for the denaturation of the ethidium bromide-DNA complex for both the PS and APS DNA hairpin. The differential between the thermal stability of the two DNA structures was considerably less than that observed in the absence of the drug. The thermal denaturation of the Hoechst 33258-DNA complexes, again, indicated cooperative transitions, however, with a greater melting point differential between the parallel and antiparallel stranded DNA hairpins. These data are summarized in Table 2.

TABLE 2 - Drug Binding Effect on Melting Temperature (Tm)

	5'-dT$_8$C$_4$A$_8$-3'		3'-dT$_8$xC$_4$A$_8$-3'
Tm-no added drug [a]	45.6°C		35.2°C
ΔTm		10.4°C	
Tm-ethidium bromide [b]	54.5°C		49.0°C
ΔTm (EBr)		5.5°C	
Tm-Hoechst 33258 [c]	65.5°C		45.0°C
ΔTm(H-33258)		20.5°C	

[a]Melting temperatures (Tm) were determined in 0.1 mM EDTA, 10 mM sodium phosphate, pH 7.0 containing a 0.1 M NaCl; [b]Ethidium bromide (1.25μM) fluorescence emission was determined at 600 nm in 0.4 M NaCl; [c]Hoechst 33258 (0.5μM) fluorescence emission was measured at 480 nm in 0.1 M NaCl. DNA concentration was 2.5μM.

The increase of the melting temperature of the PS and APS hairpin in the presence of ethidium bromide is considerably greater than expected for the increase in salt concentration from .1 M to .4 M. In the absence of any ligand, the difference in melting temperatures between the PS and APS structures is 10.4° C. In the presence of ethidium bromide, this difference decreases to 5.5° C indicating a preferential stabilization of the parallel-stranded structure. This is in good agreement with the higher binding affinity of ethidium bromide for the PS structure as projected from the titration experiments. In the presence of Hoechst 33258, a large increase is seen in the thermal stability of the APS hairpin, while the PS hairpin shows only half the increase. Thus, the difference in thermal stability between the two hairpins has increased to 20.5° C, again supporting the lower affinity of Hoechst 33258 for the parallel-stranded DNA structures.

The difference in affinity of Hoechst 33258 for parallel-stranded DNA could be a consequence of the different base pairing, reverse Watson-Crick in the PS DNA structures. The linear PS duplex F also exhibits the dramatically reduced fluorescence with Hoechst 33258 in comparison to that of the APS linear duplex E. The presence of reverse Watson-Crick base pairing in PS DNA structures could suggest a possible explanation for the lower affinity of H-33258 for such helical structures (Figure 8).

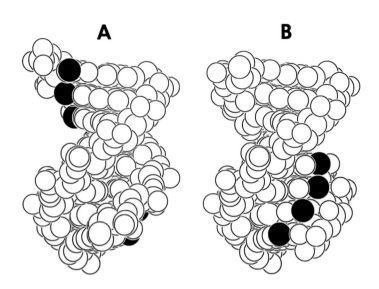

Figure 8.   Antiparallel and parallel structure for $dA_8 \cdot dT_8$.   Space filling drawings of APS (A) and PS (B) were generated on a MacIntosh Plus using the B-DNA coordinates and rotation described in the legend of Figure 4.   Filled in circles represent the 5-CH_3 groups of thymine residues.

      The minor groove in the antiparallel stranded DNA is accessible for the ligand Hoechst 33258 to interact with the appropriate contact sites on A and T residues.   In contrast, little distinction can be seen between the minor and major grooves of the model for parallel stranded DNA. Reverse Watson-Crick base pairing necessitates the projection of the 5-CH_3 groups of the thymine residues into the formal minor groove.   Most likely, this would provide for considerable steric difficulty in the binding of Hoechst 33258 to the contact sites required for hydrogen bonding.   We are currently determining the validity of this model by the use of parallel stranded oligonucleotides containing modified bases to have less steric crowding in the binding site for Hoechst 33258.   The detailed mechanism of drug DNA interaction with parallel stranded helical structures awaits the solution structure for such helices as is currently being determined by NMR analysis.

Acknowledgements

We thank Mrs. D. Hunt for expert assistance in the preparation of this manuscript.   This work was supported by the Medical Research Council of Canada and the Alberta Heritage Foundation for Medical Research.

References

1.   van de Sande, J.H., Ramsing, N.B., Germann, M.W., Elhorst, W., Kalisch, B.W., Kitzing, E.V., Pon, R.T., Clegg, R.C. and Jovin, T.M. (1988) 'Parallel Stranded DNA', Science 241, 551-557.

2.   Lee, J.S., Johnson, D.A., and Morgan, A.R. (1979) 'Complexes formed by (pyrimidine)$_n$-(purine)$_n$ DNAs on lowering the pH are three stranded', Nucleic Acids Res. 6, 3073-3091.

3.   Pattabiraman, N. (1986), 'Can the Double Helix Be Parallel?', Biopolymers 25, 1603-1606.

4.   Germann, M.W., Vogel, H.J., Pon, R.T., and van de Sande, J.H. (1989), 'Characterization of a Parallel-Stranded Hairpin', Biochemistry 28, 6220-6228.

5.   Germann, M.W., Kalisch, B.W., Pon, R.T. and van de Sande, J.H. (1990) 'Length Dependent Formation of Parallel Stranded DNA in Alternating AT Segments', manuscript submitted.

6.   Germann, M.W., Kalisch, B.W. and van de Sande, J.H. (1988) 'Relative Stability of Parallel- and Antiparallel-Stranded Duplex DNA', Biochemistry 27, 8302-8306.

7.   Ramsing, N.B. and Jovin, T.M. (1988) 'Parallel Stranded Duplex DNA', Nucleic Acids Res. 16, 6659-6676.

8.   Ramsing, N.B., Rippe, K. and Jovin, T.M. (1989) 'Helix-Coil Transition of Parallel-Stranded DNA. Thermodynamics of Hairpin and Linear Duplex Oligonucleotides', Biochemistry 28, 9528-9535.

9.   Rippe, K., Ramsing, N.B. and Jovin, T.M. (1989) 'Spectroscopic Properties and Helix Stabilities of 25-nt Parallel-Stranded Linear Duplexes', Biochemistry 28, 9536-9541.

10. Rippe, K. and Jovin, T.M. (1989) 'Substrate Properties of 25-nt Parallel-Stranded Linear DNA Duplexes' Biochemistry 28, 9542-9549.

11. Germann, M.W., Pon, R.T. and van de Sande, J.H. (1987) 'A General Method for the Purification of Synthetic Oligodeoxyribonucleotides containing Strong Secondary Structure by Reversed-Phase High-Performance Liquid Chromatography on PRP-1 Resin', Anal. Biochem. 165, 399-405.

12. Germann, M.W. and van de Sande, J.H. unpublished results.

13. Latt, S.A. and Wohlebb, J.C. (1975) 'Optical Studies of the Interaction of 33258 Hoechst with DNA, Chromatin and Metaphase Chromosomes', Chromosoma (Berlin) 52, 297-316.

14. Waring, M.J. (1965) 'Complex Formation between Ethidium Bromide and Nucleic Acids', J. Mol. Biol. 13, 269-282.

15. Wolk, S.K., Hardin, C.C., Germann, M.W., van de Sande, J.H. and Tinoco, I., Jr. (1988) 'Comparison of the B- and Z-Form Hairpin Loop Structures Formed by d(C6)$_5$T$_4$ (C6)$_5$', Biochemistry 27, 6960-6967.

16. Cohen, J.S. (1987) '2D-NOESY of DNA: As Easy as A,B,Z', Trends Biochem. Sci. 12, 132-135.

17. Arnott, S. and Hukins, D.W.L. (1972) 'Optimized Parameters For A-DNA and B-DNA', Biochem. Biophys. Res. Commun. 47, 1504-1509.

18. Stokke, T. and Steen, H.B. (1985) 'Multiple Binding Modes for Hoechst 33258 to DNA', J. Histochem. Cytochem. 33, 333-338.

19. Jorgenson, K.F., Varshney, U. and van de Sande, J.H. (1988) 'Interaction of Hoechst 33258 with Repeating Synthetic DNA Polymers and Natural DNA', J. Biomol. Struct. Dyn. 5, 1005-1023.
20. Pjura, P.E., Grzeskowiak, K. and Pickerson, R.E. (1987) 'Binding of Hoechst 33258 to the Minor Groove of B-DNA', J. Mol. Biol. 197, 257-271.
21. Bontemps, J., Houssier, C., and Fredericq, E. (1975) 'Physico-chemical study of the complexes of 33258 Hoechst with DNA and Nucleohistone' Nucleic Acids Res. 2, 971-984.
22. Mikhailov, M.V., Zasedatelev, A.S., Krylov, A.S., and Gurskii, G.V. (1981) 'Mechanism of the "Recognition" of AT Pairs in DNA by Molecules of the Dye Hoechst 33258' Mol. Biol. (Engl. Transl.) 15, 541-554.
23. Kalisch, B.W and van de Sande, J.H. (1979) 'The Effect of Antibiotics on the T_4 polynucleotide ligase catalyzed template dependent polymerization of oligodeoxythymidylates', Nucl. Acids Res. 6, 1881-1894.

DESIGN OF BIFUNCTIONAL NUCLEIC ACID LIGANDS

T. MONTENAY-GARESTIER[1], J.S. SUN[1], J. CHOMILIER[1],
J.L. MERGNY[1], M. TAKASUGI[1], U. ASSELINE[2], N.T.
THUONG[2], M. ROUGEE[1] and C. HELENE[1]

(1) *Laboratoire de Biophysique, INSERM U.201, CNRS UA.481,
Muséum National d'Histoire Naturelle, 43 rue Cuvier,
75005 PARIS*

(2) *Centre de Biophysique Moléculaire,
CNRS, 45071 ORLEANS CEDEX 02*

ABSTRACT. Homopyrimidine oligodeoxynucleotides have been covalently linked to intercalating agents. These bifunctional nucleic acid ligands bind to the major groove of DNA at homopurine-homopyrimidine sequences when they form triple helices. Spectroscopic studies and molecular modelling shows that the intercalating agent inserts its aromatic ring at the triplex-duplex junction. A strong stabilization of triple helical structures is observed. Energy transfer can occur between two derivatized oligonucleotides bound to neighboring sequences on both single-stranded and double-stranded DNA. Bifunctional oligonucleotide-intercalator conjugates provide new tools for a selective control of gene expression and for the detection of close proximity between two sequences on nucleic acids.

Introduction

Regulation of gene expression in living organisms is usually achieved by specific nucleic acid binding proteins [1]. More recently, it has been demonstrated that nucleic acids could also play a regulatory role [2]. A new strategy using synthetic oligonucleotides has been developed to control gene expression in an artificial way [3]. When targeted to messenger RNAs, oligonucleotides inhibit translation (the so-called "antisense strategy") [3]. Oligonucleotides can also be targeted to DNA, thereby inhibiting transcription (the so-called "anti-gene strategy") [5].

An optimal length of the oligonucleotide has to be choosen in order to ensure a high specificity of biological effects. If the oligonucleotide is too short it will find several identical complementary sequences, if it is too

B. Pullman and J. Jortner (eds.), Molecular Basis of Specificity in Nucleic Acid-Drug Interactions, 275–290.

long it might hybridize to mismatched sequences [6]. The optimal length might not correspond to a strong enough complex with the target sequence under physiological conditions. Therefore we have tried to increase the stability of the oligonucleotide-target complex by attaching a DNA-intercalating agent at one end of the oligomer [7].

These oligonucleotide-intercalator conjugates behave as bifunctional nucleic acid ligands : the oligonucleotide recognizes its complementary sequence and the intercalator provides an additional binding energy that leads to complex stabilization. This paper summarizes some of the physico-chemical data obtained with single-stranded and double-stranded targets.

1.   Oligonucleotide-intercalator conjugates : complex stability

The idea of attaching an intercalating agent to an oligonucleotide rests upon the assumption that the free energy of binding of the conjugate will be the sum of the free energies of the two separate components [8]

(1) $\Delta G_{ONBI} = \Delta G_{ON} + \Delta G_I - T\Delta S$

where ΔG_{ONBI} is the free energy for complex formation by the OligoNucleotide-Bridge-Intercalator, ΔG_{ON} and ΔG_I are the free energies of binding of the oligonucleotide and the intercalator, respectively. The entropy term $T\Delta S$ takes into account the limited space in which the intercalating agent can move once it is attached to the oligomer.

The equilibrium association constant $[\exp(-\Delta G/RT)]$ is given by equation (2) :

(2) $K_{ONBI} = \alpha \, K_{ON} \times K_I$

where $\alpha = \exp(\Delta S/R)$.

The oligonucleotide-intercalator conjugate can bind to single-stranded or double-stranded nucleic acids. Some of the characteristics of these complexes are described below.

1.1. SINGLE-STRANDED TARGETS

Thermodynamic data for the binding of ONBIs to single-stranded nucleic acids have been previously reported [7]. As shown in Figure 1 attachment of an acridine derivative to short oligonucleotides stabilizes the complexes formed with the complementary sequence. The gain in stability depends on the base sequence at the end of the oligonucleotide. This is due to the fact that the acridine can engage intramolecular interaction with terminal bases of the oligomer.

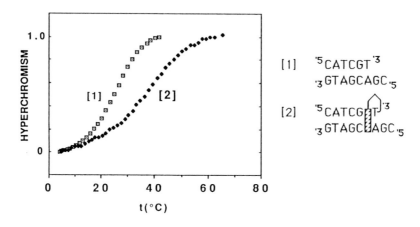

Figure 1. Comparison of melting curves for binding of 5'CATCGT3' with and without acridine to 3'GTAGCAGT5'. 2-methoxy, 6-chloro, 9-aminoacridine was attached to the 3' phosphate of the hexamer via a pentamethylene linker. Absorption measurements were carried out in a pH 7 buffer containing 10 mM sodium cacodylate and 0.1 M sodium chloride at a 10^{-5} M concentration.

1.2. DOUBLE-STRANDED TARGETS

Homopyrimidine oligonucleotides can bind to homopurine-homopyrimidine sequences of double-helical DNA. Thymine and protonated cytosine form two hydrogen bonds with A.T and G.C base pairs, respectively (Figure 2). These hydrogen bonds involve the purine bases of Watson-Crick base pairs. They are located in the major groove of the double helix. In such triple helices the third strand has a parallel orientation with respect to the homopurine sequence.

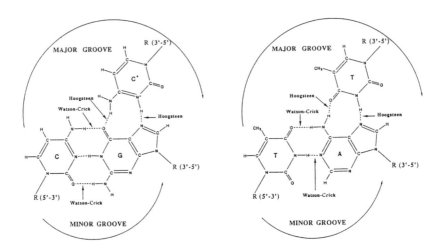

Figure 2. Hoogsteen hydrogen bonding interactions between thymine and a Watson-Crick A.T base pair and between protonated cytosine and a Watson-Crick G.C base pair.

As shown in Figure 3 and Table 1, the stability of a triple helix is enhanced when the oligonucleotide is attached to an intercalating agent. A previous report has described the property of the 5'-substituted 11-mer under different experimental conditions [9]. The gain in stability is higher when an acridine is linked to the 5'-side of the oligonucleotide. Changing the length of the linker on the 3'-side influences complex stability. These differences can be explained in structural terms taking into account the distortion that occurs at the 5'- and 3'-junctions between triplex and duplex structures, as described below.

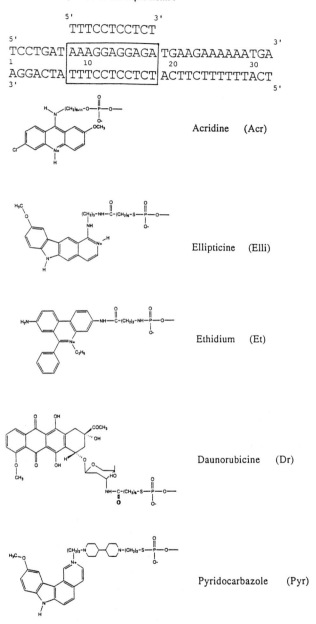

Sequence of the triple helix :

Acridine   (Acr)

Ellipticine   (Elli)

Ethidium   (Et)

Daunorubicine   (Dr)

Pyridocarbazole   (Pyr)

Figure 3. Different substitutents attached to the 11-mer oligonucleotide. The linkers between the aromatic ring and the terminal phosphate group to the 5'- or 3'-OH group of the oligonucleotide are indicated.

TABLE 1. Temperatures of half-dissociation of triple helices
formed between a 32-mer double helix and different substituted
11-mers (see Figure 3). Measurements were carried out in a
10 mM sodium cacodylate buffer (pH 6.0) in the presence of
0.1 M NaCl and 1 mM spermine. All oligonucleotide
concentrations (32-mer and 11-mer) were 1 μM.

	Tm (°C)
11mer	19.3
Acr-m_5-5'11mer	37.4
11mer3'-m_5-Acr	25.5
11mer3'-m_{11}-Acr	30.0
Et-5'-11-mer	29.0
11-mer-3'-Et	17.0
11-mer-3'-Elli	21.0
11-mer-3'-Pyc	22.0
11-mer3'-Dr	24.0

1.3. MOLECULAR MODELLING OF TRIPLEX-DUPLEX JUNCTIONS

By means of molecular modelling using the AMBER 3 software, the
structure of the 11-mer bound to a 32-bp double helix and the triplex-
duplex junction have been studied. For reasons of decent computer time,
the complex formed by the oligonucleotide with its target was split in two
parts. One is denoted 5' junction and is composed of six triplets and six base
pairs. It corresponds to the 5' end of the pyrimidine strand bound by
Hoogsteen hydrogen bonds to the target. The 3' junction is also composed
of six base pairs and six triplets, at the 3' end of the Hoogsteen-bonded
strand. These sequences are reported on Figure 4.

The coordinates of the atoms used to construct both junctions were
taken from Arnott X-ray diffraction data obtained on triple helix fibers [10]
for all atoms in the triplets, and from the canonical B-DNA data [11] for
atoms in the base pairs. Molecules were generated with the JUMNA
program [12], which is using helicoidal parameters [13], especially suited for
the generation of nucleic acids.

To build the junctions bound to the acridine derivative, the rise
parameter was doubled at the intercalation site, and the twist was

subsequently reduced in order to accommodate the drug. The acridine derivative was generated with the Insight interactive software for drug design and was manually linked to the Hoogsteen strand of the triple helix. Two linkers of different lengths were generated. One is five methylene group long (referred to as m_5 from hereon) and a second one is an eleven methylene group bridge (denoted m_{11}), corresponding to the synthesized conjugates (Table 1). Charges were taken from the literature [14]. For purpose of comparison, the computation was also carried out with the free drug, i.e. without the linker, located in the same site.

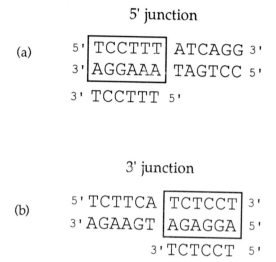

Figure 4. Sequences at the 5' (a) and 3' (b) junctions investigated by molecular graphics.

Three intercalation sites were investigated: one places the acridine derivative in between the last two triplets of the triple helix, another one is at the very junction between the triple and the double helix, and the last one is in between the two first base pairs of the double helix. These positions are respectively abbreviated as TH, TH/DH and DH.

Minimization of these structures was performed with AMBER [15]. In tables 2 and 3 are given the computed energies for intercalation of the acridine derivative (2-methoxy, 6-chloro, 9-aminoacridine) either free or covalently attached to the 11-mer oligonucleotide. In these two tables the first term (Eisc) represents the energy which has to be provided to create the intercalation site between two base pairs (DH), two triplets (TH) or one base pair and one triplet (DH/TH). The second term represents the stacking interaction energy between the acridine and the neighboring base pairs and/or triplets; Ei is the total energy for intercalation of acridine at the indicated sites and is the sum of the two previously defined terms. Calculations were performed both for the 5'- and the 3'-side of the bound 11-mer oligonucleotide.

TABLE 2. Computed energies at the 5' junction for various Acr intercalation sites. Energies are in kcal/mol. Es is the stacking energy, Eisc the site creation energy and Ei the intercalation energy.

Intercalation site	Linker	Eisc	Es	Ei
DH	-	14.5	-43.0	-28.5
DH/TH	-	18.6	-48.6	-30.0
TH	-	31.2	-51.9	-20.7
DH	m_5	24.1	-28.6	-4.5
DH/TH	m_5	25.2	-40.2	-15.0
TH	m_5	37.7	-41.7	-4.0
DH	m_{11}	22.2	-30.3	-8.1
DH/TH	m_{11}	24.0	-37.3	-13.3
TH	m_{11}	38.3	-50.4	-12.1

TABLE 3. Computed energies at the 3' junction for various Acr intercalation sites. Energies are in kcal/mol. Es is the stacking energy, Eisc the site creation energy and Ei the intercalation energy.

Intercalation site	Linker	Eisc	Es	Ei
DH	-	16.5	-42.5	-26.0
DH/TH	-	20.5	-44.7	-24.2
TH	-	38.3	-49.1	-10.8
DH	m_5	21.1	-17.2	38.3
DH/TH	m_5	25	-26.4	-1.4
TH	m_5		No convergence	
DH	m_{11}	18.7	-37.3	-18.6
DH/TH	m_{11}	26.6	-46.8	-20.2
TH	m_{11}	59.3	-56.3	9.0

1.3.1 *Free acridine.* When acridine is a free intercalator, the stacking energy increases in absolute value, i.e. the stability increases, when going from the DH site to DH/TH and to the TH site, at both junctions. One can argue that this is consistent with an increased stacking of the acridine ring, when going from a position between two base pairs to a position between two base triplets. This increase of stacking energy alone would favor the location of the drug inside the triple helix. However, from the point of view of site creation energy, the opposite behavior is encountered. Forcing a triple helix to increase the rise between two triplets to accommodate an intercalated drug, is more expensive, in terms of energy, than forcing the same rise to increase within a double helix. This variation of Es and Eisc with respect to the site is independent of the junction studied, 3' or 5'. The total intercalation energy is the sum of these two energy terms, and finally the least favorable site, for free acridine, is located inside the triple helix, for both junctions.

The intercalation energy is of the same order of magnitude for the free drug in the DH or DH/TH sites, for both junctions. Nevertheless, the 5' junction favors the DH/TH site compared to the DH site, while an opposite behavior is obtained for the 3' junction. This can be explained by the fact that the stacking energy at the DH/TH site on the 5' side is more favorable than for the other end of the oligomer. One might argue that the stacking at the 5' end is more important than at the 3' end. Actually, at the 5' junction, the direction of the twist for a right handed double helix, leads the first base pair of the double helix to stack very favorably with the last triplet of the triple-stranded part. This induces an increase of the stacking energy of the drug with the nucleic acid. On the contrary, at the 3' junction, the first base pair is twisted in such a way that it stacks very moderately with the three bases of the triple helix, thus the stacking energy decreases.

1.3.2. *Acridine Linked at the 5' junction.* When the acridine derivative is linked by a covalent linker, m_5 or m_{11}, the site creation energy varies, as a function of the intercalation site, in the same way as stated above for free acridine, and this holds for both junctions.

At the 5' junction, the stacking energy, for the m_5 and the m_{11} linkers, varies as with the free acridine, i.e. the most favorable stacking is inside the triple helix, although this energy is higher than in the case of free acridine at the same intercalation site, accounting for the linker interactions. At the TH site, the stacking energy for the m_{11} linker is almost the same as for free acridine, but for the m_5 linker it is very different. This indicates that the m_{11} linker is suitable to allow for intercalation at that position whereas the m_5 linker is probably too short. At the DH/TH site, m_5 is slightly more favorable than m_{11}, which indicates that five methylene groups are sufficient at this site. The difference of energies between m_5 and m_{11} linkers is larger for a TH site than for a DH site. Thus, acridine intercalation is less unfavored at the DH site than at the TH site for a m_5 linker as compared to a m_{11} one. Globally, the intercalation energy values show that both m_5 and m_{11} linkers are possible at the DH/TH site, but for the other two sites the m_{11} linker is more favorable.

1.3.3. *Acridine Linked at the 3' junction.* For the 3' junction, in contrast to the 5' junction, the stacking energy does not vary as it does with free acridine as a function of the intercalation site position. Obviously, the only case that can be retained for the m_5 linker is at the DH/TH site, since DH leads to a positive intercalation energy, and no convergence occurs for the TH site. This result evidences that at the 3' junction the m_5 linker is too short to be placed anywhere else than at the junction. For the m_{11} linker, both DH/TH and DH are feasible, but TH must be rejected.

From these studies, it is clear that a systematic study of the linker must be done to optimize it in order to obtain the highest possible stabilization of the triple helix.

2. Oligonucleotide-chromophore conjugates : energy transfer

The chromophores which were linked to oligonucleotides in order to provide an additional interaction energy are usually nucleic acid intercalators, as previously mentioned, but non- intercalating ligands of nucleic acids can also be linked to oligomers. Even if attachment does not lead to any stabilization of the complex, these oligonucleotide-chromophore conjugates can be used to probe the close proximity of two target sequences on the same single- or double-stranded nucleic acid fragment through excitation energy transfer [16]. Here are presented two different examples of such applications with oligonucleotides linked to chromophores and forming either double or triple helices with their targets .

2.1. SINGLE-STRANDED TARGETS

Chromophores involved in this study were a coumarin [17] and an ethidium derivative. Coumarin acts as an excitation energy donor and ethidium as an energy acceptor. An undecamer [d5'(TTTCCTCCTCT)3'] linked to the coumarin derivative kindly provided by Pr. B. Valeur, was synthesized by Dr U. Asseline in Orléans and two different undecamers of thymines linked to ethidium at the 5'- or 3'-end were synthesized in

Novossibirsk by Dr. Bulychev and Lebedev. Sequences are presented on Figure 5.

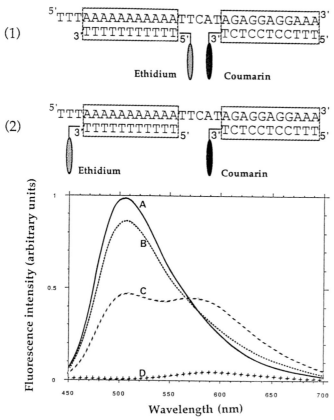

Figure 5. Top : Sequence of the 30 nt single-stranded target and of the two derivatized oligonucleotides. Ethidium was attached to the 5' end of (dT)11 (1) or to the 3'-end of (dT)11 (2). Bottom : Fluorescence spectra. A : The coumarin-substituted 11-mer bound to the 30 nt single-stranded target. B : (dT)11 with ethidium attached to the 3'-end and coumarin-11-mer bound simultaneously to the 30 nt target. C : (dT)11 with ethidium attached to the 5'-end and coumarin-11-mer bound simultaneously to the 30 nt target. D : (dT)11 with ethidium attached to the 5'-end bound alone to the 30 nt target (in the absence of coumarin-11-mer).

Detailed studies (absorption, static and dynamic fluorescence measurements) have been carried out with these conjugated oligomers including their separate interaction with the 29-mer target. When ethidium-undecamers are bound to single-stranded targets, the spectroscopic properties of the ethidium chromophore are not strongly altered leading to the conclusion that it is not intercalated in the double helix. Dissociation of the coumarin-oligomer from its target followed by absorption spectroscopy reveals that coumarin linked to the undecamer is destabilizing the association as compared to the free undecamer- target complex. This allows us to draw the conclusion that coumarin is not intercalated in the double helix formed when the coumarin-undecamer is associated to its complementary sequence as schematized on Figure 5.

Two different experiments were carried out using the oligomer-coumarin conjugate as an excitation energy donor and two different acceptor oligomers with the same oligonucleotide sequence and ethidium either attached to the 5' or 3' phosphate and used as an excitation energy acceptor (Figure 5.1 and 5.2).

Fluorescence emission spectra of the 29-mer separetely associated with each of the oligomer-chromophore conjugates and with both simultaneously associated were recorded. Figure 5 presents fluorescence emission spectra of the oligomer-donor conjugate alone and in the presence of the two different ethidium- oligomers at an excitation wavelength (480 nm) where the ethidium chromophore itself is not markedly excited demonstrating an efficient excitation energy transfer from coumarin to ethidium when ethidium is linked at the 5'-end, i.e., when ethidium is closer to coumarin. Analysis of coumarine conjugate fluorescence intensity as a function of ethidium conjugate concentration shows that spectroscopic effects reach a plateau at a 1:1 ratio of the two oligonucleotides. A detailed analysis of the transfer efficiency as a function of temperature showed that it decreases from 42 % at 0° C when both oligomers are bound to their target, to 0 at 45°C when the ethidium-oligomer is dissociated from its target. Simultaneously a shortening of the coumarin-oligomer fluorescence decay was observed in presence of the ethidium-oligomer.

2.2. DOUBLE-STRANDED TARGETS

It was interesting to check whether excitation energy transfer could occur when oligopyrimidines linked to convenient chromophores are associated to their homopurine- homopyrimidine targets where they form triple helices.

A 11-mer homopyrimidine oligonucleotide was covalenty linked to an acridine derivative via its 5'-phosphate (Acr-11-mer) and a 13-mer to an ethidium derivative via its 3' phosphate (13-mer-Et). Each of them formed a triple helix with a 31 bp DNA fragment containing two homopurine-homopyrimidine sequences, 11bp and 13bp in lengths, separated by 3 base pairs (Figure 6). When both oligonucleotides were bound to the 31 bp DNA fragment, energy transfer was observed from acridine to ethidium, as revealed by a quenching of acridine fluorescence and a sensitized ethidium emission. Transfer was temperature-dependent and occurred only when both oligonucleotides were simultaneously bound to the DNA matrix. On Figure 6 are shown the melting profiles for the dissociation of the triple helices formed by the two substituted oligonucleotides. The transfer efficiency decreases when either one of the two oligonucleotides is dissociated.

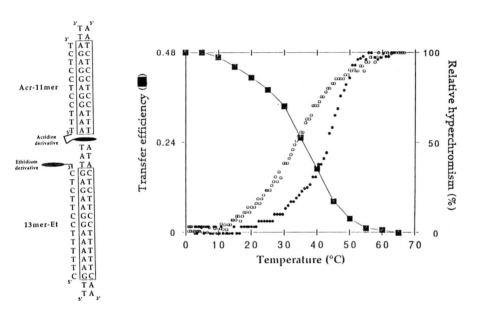

Figure 6. Left : Sequence of the 31 bp DNA target with two homopurine-homopyrimidine sequences and of the two derivatized oligonucleotides. Right : Transfer efficiency (■) from Acr-11-mer to 13-mer-Et as a function of temperature (left scale). The melting curves for dissociation of the triple helices formed by Acr-11-mer (●) and 13-mer-Et (○) are also shown.

Conclusion

Complexes involving oligonucleotides can be strongly stabilized by attaching an intercalating agent to one or the other end. When the target is a double-stranded DNA sequence, intercalation takes place at the triplex-duplex junction. These bifunctional nucleic acid ligands are highly sequence-specific and offer new possibilities to control gene expression. Energy transfer experiments have demonstrated that derivatized oligonucleotides are useful tools to demonstrate that two different sequences are close to each other either on a single-stranded or a double-stranded nucleic acid.

Acknowledgements

This work was supported by INSERM, CNRS, the Muséum National d'Histoire Naturelle and Rhône-Poulenc Santé. The Association pour le Développement des Sciences Biophysiques is also acknowledged for its support.

References

[1]   Hélène, C. and Lancelot, G. (1982) 'Interactions between functional groups in protein-nucleic acid associations', Prog. Biophys. Molec. Biol., 39, 1-68.

[2]   Green, P.J., Pines, O. and Inouye, M. (1986) 'The role of antisense RNA in gene regulation', Ann. Rev. Biochem., 55, 569-597.

[3]   Cohen, J.S. Ed. (1989) 'Oligodeoxynucleotides. Antisense inhibitors of gene expression' in Topics in Molecular and Structural Biology, volume 12, MacMillan Press.

[4]   Hélène, C. and Toulmé, J.J. (1990) 'Control of gene expression by antisense, sense and anti-gene nucleic acids' Biochim. Biophys. Acta (in press).

[5]   Cooney, M., Czernuszewicz, G., Postel, E.H., Flint, S.J. and Hogan, M.E. (1988) 'Site specific oligonucleotide binding represses transcription of human c-myc gene *in vitro*', Science, 241, 456-459.

[6] Hélène, C. and Toulmé, J.J. (1989) 'Control of gene expression by oligodeoxynucleotides covalently linked to intercalating agents and nucleic acid-cleaving reagents' In "Oligodeoxynucleotides. Antisense inhibitors of gene expression" in Topics in Molecular and Structural Biology, volume 12, Cohen, J.S. Ed., MacMillan Press.

[7] Asseline, U., Thuong, N.T. and Hélène, C. (1983) 'Nouvelles substances à forte affinité spécifique pour des séquences d'acides nucléiques : oligodésoxynucléotides liés de façon covalente à un agent intercalant', C. R. Acad. Sc. Paris, 297 (série III), 369-372.

Asseline, U., Delarue, M., Lancelot, G., Toulmé, F., Thuong, N.T., Montenay-Garestier, T. and Hélène, C. (1984) 'Nucleic acid-binding molecules with high affinity and base sequence specificity : intercalating agents covalently linked to oligodeoxynucleotides', Proc. Natl. Acad. Sci. USA, 81, 3297-3301.

Asseline, U., Toulmé, F., Thuong, N.T., Delarue, M., Montenay-Garestier, T. and Hélène, C. (1984) 'Oligodeoxynucleotides covalently linked to intercalating dyes as base sequence-specific ligands. Influence of dye attachment site', EMBO J., 3, 795-800.

[8] Thuong, N.T., Asseline, U. and Montenay-Garestier, T (1989) 'Oligodeoxynucleotides covalently linked to intercalating and reactive substances : synthesis, characterization and physicochemical studies' In "Oligodeoxynucleotides. Antisense inhibitors of gene expression" in Topics in Molecular and Structural Biology, volume 12, Cohen, J.S. Ed., MacMillan Press.

[9] Sun, J.S., François, J.C., Montenay-Garestier, T., Saison-Behmoaras, T., Roig, V., Chassignol, M., Thuong, N.T. and Hélène, C. (1989) 'Sequence-specific intercalating agents. Intercalation at specific sequences on duplex DNA via major-groove recognition by oligonucleotide intercalator conjugates', Proc. Natl. Acad. Sci. USA, 86, 9198-9202.

[10] Arnott, S., Bond, P. J., Selsing, E. and Smith, P. J. C. (1976) 'Models of triple stranded polynucleotides with optimised stereochemistry', Nucleic Acids Res., 3, 2459-2470.

[11] Arnott, S., Chandrasekaran, R., Birdsall, D. L., Leslie, A. G. W. and Ratliff, R. L. (1980) Nature, 283, 743-746.

[12] Lavery, R., Sklenar, H., Zakrzewska, K. and Pullman, B. (1986) 'The flexibility of the nucleic acids: (II) The calculation of internal energy and applications to mononucleotide repeat DNA', J. Biomol. Struct. and Dyn., 3, 989-1014.

[13] Lavery, R. and Sklenar, H. (1989) 'Defining the structure of irregular nucleic acids: conventions and principles', J. Biomol. Struct. and Dyn., 6, 655-667.

[14] Cieplak, P., Rao, S. N., Hélène, C., Montenay-Garestier, T. and Kollman, P. (1987) 'Conformations of duplex structures formed by oligodeoxynucleotides covalently linked to the intercalator 2-methoxy-6-chloro-9-aminoacridine', J. Biomol. Struct. and Dyn., 5, 361-382.

[15] Weiner, S. J., Kollman, P. A., Nguyen, D. T. and Case, D. A. (1986) 'An all atom force field for simulations of proteins and nucleic acids', J. Comput. Chem., 7, 230-252.

[16] Morrisson, L.E. (1988) in Gene Probe Technology II San Diego.

Morrisson, L.E., Halder, T.C. and Stols, L.M. (1989) 'Solution-phase DNA assay using fluorescent probes', Anal. Biochem, 183, 231-244.

Cardullo, R.A., Agrawal, S., Flores, C., Zamecnik, P.C. and Wolf, D.E. (1988) 'Detection of nucleic acid hybridization by nonradiative fluorescence resonance energy transfer', Proc. Natl. Acad. Sci. USA, 85, 8790-8794.

[17] Le Bris, M.T., Mugnier, J., Bourson, J. and Valeur, B. (1984) 'Spectral properties of a new fluorescent dye emitting in the red : a benzoxazinone derivative', Chem. Phys. Letters, 106, 124-127.

Dupuy, F., Rullière, C., Le Bris, M.T. and Valeur, B. (1984) 'A new class of laser dyes : benzoxazinone derivatives', Optics communications, 51, 36-40.

SEQUENCE-SPECIFIC RECOGNITION AND CLEAVAGE OF DUPLEX DNA BY DERIVATIZED OLIGONUCLEOTIDES

C. HELENE[1], J.C. FRANCOIS[1], C. GIOVANNANGELI[1],
T. SAISON-BEHMOARAS[1], U. ASSELINE[2] and
N.T. THUONG[2]

[1] *Laboratoire de Biophysique, INSERM U.201, CNRS UA.481, Muséum National d'Histoire Naturelle, 43 rue Cuvier, 75005 PARIS*

[2] *Centre de Biophysique Moléculaire, CNRS, 45071 ORLEANS CEDEX 02*

ABSTRACT. A homopyrimidine 11-mer oligodeoxynucleotide selectively recognizes a 11-bp homopurine.homopyrimidine sequence in duplex DNA of simian virus 40 (SV40). Binding occurs in the major groove via Hoogsten base pairing of thymine and protonated cytosine to Watson-Crick A.T and G.C base pairs, respectively. The 11-mer oligonucleotide was covalently linked to 5-amino-1,10-phenanthroline via different linkers. Efficient cleavage of SV40 circular DNA was observed in the presence of Cu^{2+} and a reducing agent when a pentamethylene carboxamide linker was used to tether phenanthroline to a 5'-thiophosphate group of the 11-mer oligonucleotide. The distribution of the cleavage sites on the two strands was asymmetric. They were shifted towards the 3'-side indicating that cleavage occurred from the minor groove. Recognition of the major groove by the oligonucleotide and intercalation of phenanthroline at the triplex-duplex junction account for the observed sequence-specific cleavage reaction from within the minor groove. An ellipticine derivative was covalently attached to the 3'-end of the 11-mer oligonucleotide. Upon irradiation at wavelengths longer than 300 nm the DNA target was cleaved at the positions expected if the oligonucleotide binds in a parallel orientation to the homopurine sequence. In addition it was shown that triple helix formation by the unsubstituted 11-mer oligonucleotide targeted a photo-induced cleavage by free ellipticine derivatives at the triplex-duplex junctions. Molecular modeling and energy minimization studies revealed that the distortion at the triplex-duplex junctions could account for these results.

B. Pullman and J. Jortner (eds.), Molecular Basis of Specificity in Nucleic Acid-Drug Interactions, 291–299.

Introduction

Specific control of gene expression can be achieved by oligodeoxynucleotides and their derivatives (1). The more popular targets have been messenger RNAs whose translation can be blocked as a result of oligonucleotide binding. The efficiency of the biological effect depends upon the activity of an endogenous enzyme, RNase H, which cleaves the mRNA in the mRNA-oligodeoxynucleotide hybrid (2). When the oligonucleotide target is located upstream of the initiation codon, translation inhibition might also involve an RNase H-independent contribution due to the inhibition of ribosome assembly. We previously showed that covalent attachment of an intercalating agent to one or both ends of an oligonucleotide strongly increased its inhibitory effect (2,3). *In vitro* this increase was ascribed to enhanced binding of the oligonucleotide to its target sequence as a result of an additional binding energy provided by the intercalating agent (4,5). *In vivo* two additional effects contributed to the enhanced biological activity : i) substitution of the 3'-end by an intercalating agent protected the oligonucleotide against degradation by 3'-exonucleases (6,7) and ii) the intercalating agent favored uptake of the oligonucleotide by living cells in culture (7,8). Such oligonucleotide-intercalator conjugates were shown to selectively kill trypanosomes (7) and to inhibit the cytopathic effect of influenza virus in MDCK cells (8).

Double-stranded DNA can also be recognized by oligonucleotides via Hoogsteen base pairing of homopyrimidine oligonucleotides to homopurine.homopyrimidine sequences in DNA. Thymine and protonated cytosine form two hydrogen bonds with Watson-Crick A.T and G.C base pairs, respectively (figure 1). Attachment of a photoactive group to the oligonucleotide allowed us to photo-induce cross-linking of the oligonucleotide to each strand of duplex DNA and to cleave the DNA strands at the cross-linked sites by piperidine treatment (9,10). DNA cleaving reagents such as Fe-EDTA (11,12) and Cu-phenanthroline (13-15) were shown to induce site-specific double-strand cleavage upon addition of a reducing agent. Sequence-specific intercalation occurs when an intercalating agent is covalently linked to a homopyrmidine oligonucleotide (16). When phenanthroline is chosen as the intercalating agent the cleavage reaction induced upon addition of a reducing agent to the copper chelate occurs in the minor groove even though oligonucleotide binding takes place in the major groove (15). Clevavage can be photo-induced by an ellipticine derivative bound to the triple helix-forming oligonucleotide (17).

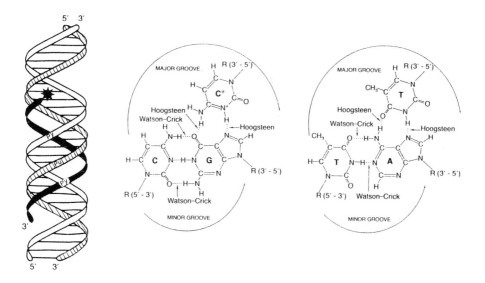

Figure 1. Oligonucleotide binding to duplex DNA. <u>Left</u> : the third strand homopyrimidine oligonucleotide (black) binds to the major groove of duplex DNA at a homopurine.homopyrimidine sequence. The star represents a reactive group that can induce irreversible reactions on both strands of DNA. <u>Right</u> : Hoogsteen hydrogen bonding of T and protonated C to A.T and G.C Watson-Crick base pairs, respectively.

Recognition and cleavage of double helical DNA at specific sites by Cu-phenanthroline tethered to an oligonucleotide

An 11 bp sequence with all purines on the same strand was chosen as a target for an oligonucleotide-phenanthroline conjugate (see figure 3 for the sequence). The oligonucleotide sequence was $d^{5'}(TTTCCTCCTCT)^{3'}$. It was synthesized to bind in a parallel orientation to the homopurine sequence. Phenanthroline was attached to the 5'-end of the 11-mer homopyrimidine oligonucleotide. Different linkers were used to tether the 5-amino group of 5-amino, 1,10-phenanthroline to a 5'-thiophosphate group of the oligonucleotide. In the presence of Cu^{2+} and a reducing agent (β-mercaptopropionic acid) cleavage was observed on the two strands of a 32 bp duplex DNA fragment containing the 11 bp target sequence (15). The most efficient cleavage was obtained with a pentamethylene carboxamide linker. The cleavage sites on the two strands had an asymmetric distribution. They were shifted towards the 3'-side indicating that the cleavage reaction had taken place from the minor groove. The species responsible for cleavage are believed to be OH˙ radicals. These diffusible radicals react within a restricted domain as indicated by the limited number of bonds which undergo cleavage.

The asymmetric distribution of the cleavage sites observed with Cu-phenanthroline tethered to the 11-mer oligonucleotide indicated that cleavage occurred from the minor groove even though phenanthroline was brought into the major groove by oligonucleotide binding. This result can be explained if phenanthroline intercalates at the triplex-duplex junction as previously described for an acridine derivative tethered to the 11-mer oligonucleotide (16). The linker was attached to the C-5 position of the phenanthroline ring. Therefore intercalation brings the nitrogen atoms (N-1 and N-10) in the minor groove where copper chelation occurs followed by oxidative attack of the deoxyribose and strand cleavage (figure 2).

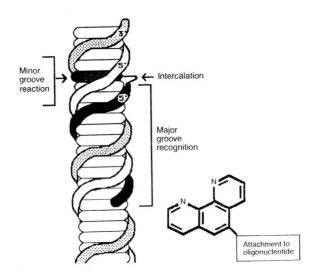

Figure 2. Oligonucleotide-intercalator conjugates recognize the major groove of duplex DNA at specific sequences (see figure 1). If phenanthroline is used as an intercalator, cleavage of the two strands of DNA is observed (ref. 15). Cleavage can also be photo-induced by an ellipticine derivative (ref. 17).

The target sequence of the 11-mer oligopyrimidine was contained within the DNA of simian virus SV40. There is a single copy of the 11 bp sequence among the 5243 bp of the viral DNA. The phenanthroline-substituted oligonucleotide was shown to cleave SV40 DNA at a single site both on linear and circular DNA (13,15). Double-strand cleavage was observed by agarose gel electrophoresis. The cleaved phosphodiester bonds were then identified by polyacrylamide gel electrophoresis of denatured SV40 fragments obtained after cleavage by the phenanthroline-oligonucleotide conjugate. They were identical to those observed on the 32-mer duplex DNA as described above. The efficiency of double-strand cleavage reached about 70 % at 20°C in a pH 7.4 buffer containing 10 mM Na phosphate, 0.1 M NaCl and 1 mM spermine. Cleavage was measured as a function of temperature. The effect was reduced by 50 % at 30°C. The oligonucleotide-phenanthroline conjugate formed a much more stable triple helix than the unsubstituted 11-mer as a result of both phenanthroline intercalation and copper chelation which locks the complex in place from within the minor groove. At low temperatures (< 10°C) a secondary cleavage site was observed which corresponded to binding of the oligonucleotide to a mismatched sequence (15).

Photo-induced cleavage of duplex DNA by an oligonucleotide-ellipticine conjugate

An ellipticine derivative (1-amino-5,11-dimethyl-9-methoxy ellipticine) was covalently attached via its 1-amino group to the 3' phosphate of the 11-mer oligonucleotide described in the preceeding paragraph. This oligonucleotide was bound to a 32 bp DNA fragment containing the 11 bp homopurine.homopyrimidine target sequence where a triple helix is formed. Upon UV irradiation (λ > 310 nm) cleavage of the two strands of duplex DNA was observed where expected if the 11-mer oligonucleotide is bound in a parallel orientation with respect to the homopurine strand (17). Photo-induced cross-linking of the oligonucleotide-ellipticine conjugate to the target sequence was also observed upon irradiation. Neither cleavage nor cross-linking was observed when the triplex was dissociated at high temperature.

Site-specific binding of an ellipticine derivative at a triplex-duplex junction

The free ellipticine derivative used in the studies described above binds to duplex DNA, most probably by intercalation. Irradiation of the ellipticine-DNA complex leads to photo-induced cleavage on both strands. The yield of the reaction is slightly higher at G.C base pairs which might reflect either a G.C preference for ellipticine binding or an increased photochemical reactivity at these sites. When the 11-mer oligonucleotide was bound to the 32 bp duplex DNA fragment to form a triple helix, the cleavage reaction photo-induced by ellipticine was strongly enhanced at the triplex-duplex junctions. When the ellipticine concentration was decreased down to 0.1 μM the strong cleavage on the 5'-side of the triple-strand-forming oligonucleotide was still observed while that on the 3'-side disappeared (figure 3). This result indicated that a strong binding site ($K > 10^7$ M^{-1}) was created at the triplex-duplex junction on the 5'-side of the third strand while a moderately strong site ($K \sim 10^6$ M^{-1}) was created at the 3'-junction. These experimental results are in good agreement with molecular modelling studies (Montenay-Garestier et al., this volume).

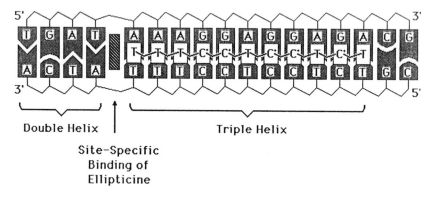

Figure 3. Sequence-specific binding of an ellipticine derivative at the triplex-duplex junction on the 5'-side of the third strand 11-mer homopyrimidine oligonucleotide. Upon irradiation cleavage is observed at the T and A nucleotides that flank the intercalation site on both strands of duplex DNA (ref. 17).

Conclusion

Homopyrimidine oligonucleotides bind to the major groove of DNA at homopurine.homopyrimidine sequences forming local triple helices. An aromatic ring (acridine, phenanthroline ellipticine) attached to the oligonucleotide via an appropriate linker may intercalate at the junction between the double helix and the triple helix. The complex is stabilized by the additional binding energy provided by this interaction. In the case of phenanthroline, intercalation provides the required structure for double-strand cleavage by the copper complex in the minor groove even though the reaction is targeted to a specific sequence by recognition of the major groove. The negatively charged oligonucleotide prevents intercalation of acridine, phenanthroline or ellipticine at DNA sequences where the oligonucleotide does not find its binding sequence. In other words, non-specific binding of the intercalating agent to any DNA sequence is lost. Intercalation takes place only at sequences which are recognized by the oligonucleotide. Therefore oligonucleotide-intercalator conjugates are highly sequence-specific DNA intercalating agents that could be used to control gene expression at the DNA level. We (18) and others (19,20) have shown that local triple helix formation inhibits restriction enzyme cleavage and binding of transcription factors (19). Preliminary data also indicate that DNA replication might be inhibited. Recognition of homopurine.homopyrimidine sequences of DNA by homopyrimidine oligonucleotides might not be the only recognition code for triple helix formation. There is evidence that such sequences can also be recognized by homopurine oligonucleotides which are able to block transcription (21). Further applications of artificial sequence-specific double-strand nucleases can be envisaged, for example, in gene mapping experiments on long DNA fragments or for site-directed mutagenesis. The cleavage reaction can be chemically induced (e.g., Cu-phenanthroline "activated" by a reducing agent) or photochemically induced (e.g., ellipticine). These reactions can take place in the low-melting agarose used to prepare long DNA macromolecules such as chromosomes. Pulse-field electrophoresis can then be used to separate the cleaved fragments. Site-directed mutagenesis can also be contemplated after induction of sequence-specific modifications in DNA *in vitro* or *in vivo*, using nuclease-resistant oligonucleotides such as oligo-[α]-deoxynucleotides (9,10).

Acknowledgements

This work was supported by INSERM, CNRS, the Muséum National d'Histoire Naturelle and Rhône-Poulenc Santé. The Ligue Nationale Française contre le Cancer and the Fondation pour la Recherche Médicale are also acknowledged for their support.

References

(1) Toulmé, J.J. and Hélène, C. (1988) *Gene* <u>72</u>, 51-58.

(2) Cazenave, C., Loreau, N., Thuong, N.T., Toulmé J.J. and Hélène, C. (1987)*Nucleic Acids Res.* <u>15</u>, 4717-4736.

(3) Toulmé, J.J., Krisch, H.M., Loreau, N., Thuong, N.T. and Hélène, C. (1986) *Proc. Natl. Acad. Sci. USA* <u>83</u>, 1227-1231.

(4) Asseline, U., Thuong, N.T. and Hélène, C. (1983) *C.R.Acad .Sci. Paris* <u>297</u>(III), 369-372.

(5) Asseline, U., Delarue, M., Lancelot, G., Toulmé, F., Thuong, N.T., Montenay-Garestier, T. and Hélène, C. (1984) *Proc. Natl. Acad. Sci.* <u>81</u>, 3297-3301.

(6) Cazenave, C., Chevrier, M., Thuong, N.T and Hélène, C. (1987) *Nucleic Acids Res.* <u>15</u>, 10507-10521 .

(7) Verspieren, P., Cornelissen, A.W.C.A., Thuong, N.T., Hélène, C. and Toulmé J.J. (1987) *Gene* <u>61</u>, 307-315.

(8) Zérial, A., Thuong, N.T. and Hélène, C. (1987) *Nucleic Acids Res.* <u>15</u>, 9909-9919.

(9) Le Doan, T., Perrouault, L., Praseuth, D., Habhoub, N., Decout, J.L., Thuong, N.T., Lhomme, J. and Hélène, C. (1987) *Nucleic Acids Res.* <u>15</u>, 7749-7760.

(10) Praseuth, D., Perrouault, L., Le Doan, T., Chassignol, M., Thuong, N.T. and Hélène, C. (1988) *Proc. Natl. Acad. Sci. USA* <u>85</u>, 1349-1353.

(11) Moser, H.E. and Dervan, P.B. (1987) *Science* <u>238</u>, 645-650.

(12) Strobel, S.A., Moser, H.E. and Dervan, P.B. (1988) *J. Am. Chem. Soc.* <u>110</u>, 7927-7929.

(13) François, J.C., Saison-Behmoaras, T., Chassignol, M., Thuong, N.T. and Hélène, C. (1988) *C. R. Acad. Sci. Paris* <u>307</u>(III), 849-854.

(14) François, J.C., Saison-Behmoaras, T., Chassignol, M., Thuong, N.T. and Hélène, C. (1989) *J. Biol. Chem.* 264, 5891-5898.

(15) François, J.C., Saison-Behmoaras, T., Barbier, C., Chassignol, M., Thuong, N.T. and Hélène, C. (1989) *Proc. Natl. Acad. Sci.* 86, 9702-9706.

(16) Sun, J.S., François, J.C., Montenay-Garestier, T., Saison-Behmoaras, T., Roig, V., Chassignol, M., Thuong, N.T. and Hélène, C. (1989) *Proc. Natl. Acad. Sci.* 86, 9198-9202.

(17) Perrouault, L., Asseline, U., Rivalle, C., Thuong, N.T., Bisagni, E., Giovannangeli, C., Le Doan, T. and Hélène, C. (1990) *Nature* 344, 358-360.

(18) François, J.C., Saison-Behmoaras, T., Thuong, N.T. and Hélène, C. (1989) *Biochemistry* 28, 5891-5898.

(19) Maher III, J.L., Wold, B. and Dervan P.B. (1989) *Science* 245, 725-730.

(20) Hanney, J.C., Shimizu, M. and Wells, R.D. (1990) *Nucleic Acids Res.* 18, 157-161.

(21) Cooney, M., Czernuszewicz, G., Postel, E.H., Flint, S.J. and Hogan, M.E. (1988) *Science* 241, 456-459.

BIS(PLATINUM) COMPLEXES. CHEMISTRY, ANTITUMOR ACTIVITY AND DNA-BINDING.

J.D. Hoeschele[a], A.J. Kraker[a], Y. Qu[b], B. Van Houten[c] and N. Farrell[b*]

a: Warner-Lambert-Parke Davis Pharmaceutical Research Division, 2800 Plymouth Rd., Ann Arbor, MI 48105 and Departments of (b) Chemistry and (c) Pathology, University of Vermont and The Vermont Regional Cancer Center, Burlington, VT 05405, USA.

Abstract

Bis(platinum) complexes of general formula $[\{PtX_nL_{3-n}\}_2(\text{diamine})]^{(2-n)+}$ (X = for example Cl, L = for example NH_3, n = 1,2, and where diamine is most generally represented as $H_2N\text{-}R\text{-}NH_2$, with R a linear or substituted aliphatic $-(CH_2)_m-$, m usually 2 - 6) represent a unique class of complexes with novel antitumor activity and DNA-binding. In the above formula variation of n (1 or 2) give complexes with either mono- or bidentate coordination spheres respectively. The complexes display good antitumor activity toward both murine and human tumor cell lines resistant to the clinically used cisplatin. The principal feature of the DNA binding of bis(platinum) complexes is that they form unique interstrand cross-links through binding of each Pt atom to opposite strands of DNA. In contrast to cisplatin, alternating purine/pyrimidine sequences, especially $(CG)_n$ or $(GC)_n$, are favored DNA-binding sites of the bis(platinum) complexes. The interstrand cross-linking results in a conformational alteration of the bis(platinum)-modified DNA as judged by Circular Dichroism spectra. This distortion is unique to DNA modified by the bis(platinum) complex containing monodentate coordination spheres. The relevance of this new DNA binding mode for platinum complexes to the biological activity is discussed.

B. Pullman and J. Jortner (eds.), Molecular Basis of Specificity in Nucleic Acid-Drug Interactions, 301–321.
© 1990 Kluwer Academic Publishers. Printed in the Netherlands.

Introduction

The clinical utility of the simple inorganic complex cis-[PtCl$_2$(NH$_3$)$_2$)] (Cisplatin, cis-DDP) and its carboxylate derivative [Pt(CBDCA)(NH$_3$)$_2$)] (Carboplatin, CBDCA = 1,1-cyclobutane-dicarboxylate) in cancer treatment is by now well established.[1] Two disadvantages of the platinum complexes are, however, that they have limited activity against many common human cancers and that they are susceptible to the phenomenon of acquired drug resistance. The molecular mechanism of action of cis-DDP is accepted to be by interaction with DNA, with binding in the major groove.[2,3] The principal adduct of cis-DDP on DNA is the intrastrand link between two adjacent guanine (GG) or adenine/guanine (AG) bases. These lesions constitute a block to both replication and transcription. Extensive NMR and X-ray studies on cis-DDP-adducted di-, tri- and oligonucleotides have shown that the intrastrand crosslink causes a kink or bend in the helix (Summarised in Refs. 1-3). This conformational distortion is thus the origin of the biological effects manifested.

The empirical structure-activity relationships originally delineated for platinum complexes stressed the necessity for the cis-[PtX$_2$(amine)$_2$] structure, where X is a leaving group such as chloride and amine represents ammonia or a primary monodentate or bidentate amine. The trans isomer of cis-DDP, *trans*-[PtCl$_2$(NH$_3$)$_2$)] (trans-DDP), was considered inactive. At the molecular level, a possible explanation for this difference between isomers is that *cis* compounds form platinum-DNA adducts which inhibit DNA replication to a greater extent than those formed from *trans*-DDP.[4] Alternatively, DNA adducts formed by *trans* compounds may be repaired more rapidly.[5] In this respect, a recent noteworthy result is that *trans* complexes of planar ligands such as pyridine are more cytotoxic than their *cis* isomers and show equivalent cytotoxicity to cis-DDP itself.[6]

In analogue development, much emphasis has been placed upon complexes containing a 1,2-diaminocyclohexane (dach) ligand as such complexes are active in vitro and in vivo against cell lines rendered resistant to cisplatin.[7] The adducts formed by dach complexes, and probably all simple cis-diamine(Pt) analogues, however, are similar to those of cis-DDP[8,9] and it is unclear whether these compounds will have activity complementary or superior to cisplatin in the clinic.

Bis(platinum) Complexes

The understanding of the mechanism of action of cis-DDP raises questions as to whether greater conformational distortion by structurally different Pt complexes or greater DNA affinity can be reflected in increased antitumor activity. Enhanced DNA-binding affinity and conformational changes may also be associated with more difficult repair of the drug-DNA lesion. Indeed, enhanced repair activity contributes to cis-DDP resistance in murine leukemia L1210 cells.[10] Further, complexes acting by different mechanisms might also display a broader spectrum of clinical activity.

To address some of these points we originally prepared bis(platinum) complexes containing two cis-Pt(amine)$_2$ units linked by a variable length diamine chain, [{cis-

PtCl$_2$(NH$_3$)}$_2$(H$_2$N(CH$_2$)$_n$NH$_2$)].[11] Our ongoing studies on the mechanism of action and possible clinical utility of these complexes has resulted in study of further structurally different bis(platinum) complexes containing monodentate coordination spheres, [trans-{PtCl(NH$_3$)$_2$}$_2$H$_2$N(CH$_2$)$_n$NH$_2$].[12] In general, the complexes display a spectrum of antitumor activity somewhat different to that of cis-DDP and a significant feature is their activity toward cis-DDP-resistant cells.[13,14,15] The modes of DNA-binding of bis(platinum) complexes are also more diverse than their monomeric analogues. In particular, DNA-binding studies have shown that bis(platinum) complexes form unique interstrand cross-links by binding of one Pt atom to each strand of DNA.[16] This chapter reviews our studies to date on the chemistry, antitumor activity and DNA-binding of these interesting complexes, and their interrelationship.

Chemistry of Bis(platinum) Complexes

Monomeric platinum-amine complexes represent a series [PtX$_{4-n}$(am)$_n$]$^{(n-2)+}$ (X = halide, am = amine, n = 0-4). The total number of possibilities for all combinations of chloro-amine bis(platinum) complexes are summarised in Table 1. As with monomeric complexes, the most interesting from the point of view of biological activity are those compounds containing monodentate and/or bidentate coordination spheres (i.e. 1 or 2 Cl bound to Pt). However, the synthesis and study of many of the other possible bis(platinum) complexes is not simply pedantic- many are useful intermediates and serve as models for elucidation of specific chemical and biochemical aspects. For a convenient abbreviation we have adopted a system where the numbers refer to the number of chlorides (or anionic leaving groups) on each platinum atom. Where there are two chlorides on the same Pt, the lettering specifies their mutual geometries (cis or trans). For those possibilities where there is only one chloride in a coordination sphere, the lettering refers to the geometry with respect to the nitrogen of the bridging diamine. Once these two parameters are specified, the geometry of the overall complex is automatically fixed. Figure 1 shows the structures of the three 2,2 isomers (**2,2/c,c**, **2,2/t,t** and **2,2/c,t** respectively).

We have now reported on the synthesis and characterisation of complexes containing both platinum atoms in cis-[PtCl$_2$(amine)$_2$], trans-[PtCl$_2$(amine)$_2$] and Pt-tetra-amine coordination spheres[17], the mixed **cis/trans**-[PtCl$_2$(amine)$_2$] complex[13] as well as the complex containing two monodentate coordination spheres.[12] The synthesis of the mixed **2,2/c,t** complex is noteworthy for its use of a dangling amine as a precursor for the bridging diamine- in this way highly specific asymmetric bis(platinum) complexes may be designed:

Table 1.   Possible Isomers for Chloro-amine Bis(platinum) Complexes.

COORDINATION SPHERES	POSSIBLE ISOMERS			#
$[PtCl_3(am)]/[PtCl_3(am)]$	--			1
$[PtCl_3(am)]/[PtCl_2(am)_2]$	3,2/c	3,2/t		2
$[PtCl_3(am)]/[PtCl_2(am)_2]$	3,1/c	3,1/t		2
$[PtCl_3(am)]/[PtCl(am)_4]$	--			1
$[PtCl_2(am)_2]/[PtCl_2(am)_2]$	2,2/c,c	2,2/c,t	2,2/t,t	3
$[PtCl(am)_3]/[PtCl_2(am)_2]$	1,2/c,c	1,2/t,c		4
	1,2,c/t	1,2/t,t		
$[PtCl(am)_3]/[PtCl(am)_3]$	1,1/c,c	1,1/c,t	1,1/t,t	3
$[Pt(am)_4]/[PtCl_2(am)_2]$	0,2/c	0,2/t		
$[Pt(am)_4]/[PtCl(am)_3]$	0,1/c	0,1/t		2
$[Pt(am)_4]/[Pt(am)_4]$	--			1

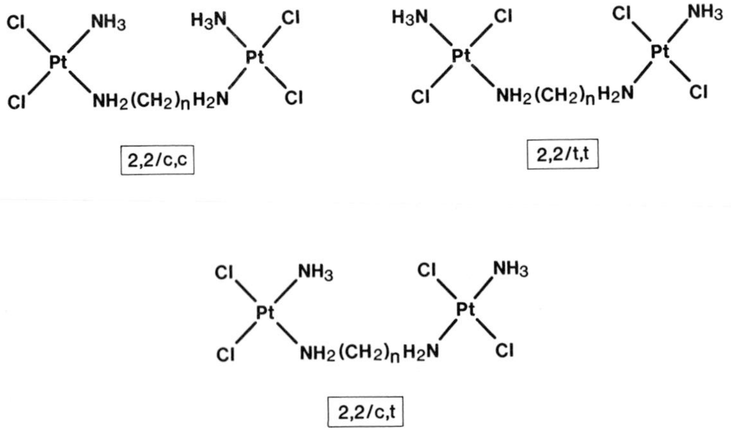

Figure 1. Structures of the linkage isomers of bis(platinum) complexes where both platinum atoms contain bidentate coordination spheres. From Ref. 13.

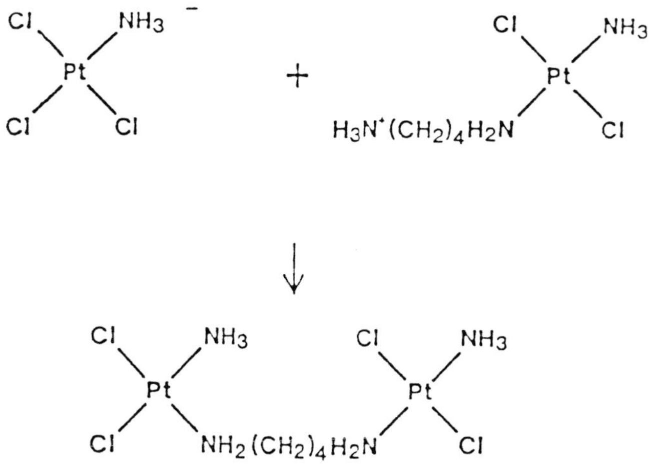

Cl / NH₃ ⁻ Pt Cl / Cl   +   Cl / NH₃ Pt H₃N⁺(CH₂)₄H₂N / Cl

↓

Cl / NH₃ Pt Cl / NH₂(CH₂)₄H₂N   Cl / NH₃ Pt / Cl

2,2/c,t

The synthetic schemes for the complexes to be discussed in this review are outlined in Figure 2 and the complexes offer a rich area of platinum chemistry. The products are dependent on the nature of the starting platinum complex and the length of the diamine. For example, ethylenediamine and propanediamine-bridged 2,2/c,c complexes cannot be prepared directly, as double salts are formed (Route C).[18] On the other hand the 1,1/t,t complexes are readily prepared with the short chain diamines (Route D). The 2,2/c,c complexes are best prepared by Route B, an adaptation of the Dhara method for cis-DDP.[19]

Treatment of the iodo complexes with Ag⁺ gives the corresponding aqua species in solution, which upon treatment with Cl⁻ precipitates the chloro complexes. The nature of the aqua species has been studied and is highly pH-dependent with both inter- and intramolecular hydroxo bridge formation.[20] These initial studies indicate that intramolecular hydroxo bridge formation between the two platinum atoms of the same bis(platinum) molecule is not a major feature of the hydrolysis. The 2,2/c,c chloride complexes are in general not very soluble in water but derivatization using dicarboxylate leaving groups such as malonate etc. is relatively straightforward.

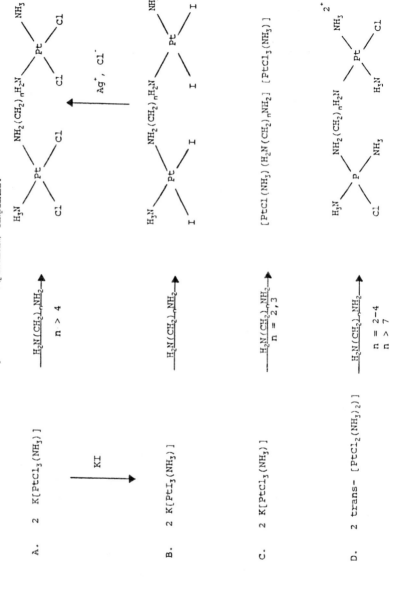

Figure 2. Synthetic Schemes for Preparation of Bis(platinum) Complexes.

Antitumor Activity of Bis(platinum) Complexes.

The principal complexes studied so far are the 2,2 series and the 1,1/t,t derivative. The selected antitumor data in Tables 2-4 confirm and expand the early results. Studies on both murine and human tumors have been undertaken. The activity may be summarised:

1) The complexes have high antitumor activity *in vitro* against murine cis-DDP-resistant lines, Table 2.

2) The complexes are also active *in vitro* against L1210 cell lines resistant to $[PtCl_2(dach)]$, Table 2. Thus, the bis(platinum) complexes represent a unique class non-cross-resistant to both cis-DDP and Pt-1,2-diaminocyclohexane complexes.

3) The activity, and especially the resistance factor, is somewhat dependent on chain length. The resistance factor is defined as the ID_{50}(resistant) divided by the ID_{50}(sensitive) value.

4) Alteration in the diamine backbone also affects biological activity, Table 3. Again it is of interest to note that alteration of the diamine backbone may affect the resistance factor.

5) In *in vivo* testing, the most potent compound of the class $[\{cis\text{-}PtCl_2(NH_3)\}_2((H_2N(CH_2)_nNH_2)]$ is the n = 5 derivative. The complexes show activity against a range of leukemias and solid tumors (L1210, P388, B16, M5076, C26). Representative %T/C values (4mg/kg, inj. ipip, d3,7,11) are 218 (L1210S) and 204 (L1210/DDP5).[14] Typical values for cis-DDP under the same conditions (but at 5mg/kg) are 225 (L1210S) and 106 (L1210/DDP5).

6) Bis(platinum) complexes are also active in human tumor lines both sensitive and resistant to cis-DDP, Table 4.

7) The structural feature of a monodentate bis(platinum) complex is sufficient, *of and by itself*, to produce activity in murine tumors resistant to cis-DDP, Table 2. Indeed, the 1,1/t,t complex is at least as cytotoxic toward resistant than to sensitive cells, and in some cases, depending on source of cell line, actually more so (See Ref. 12). This is of particular interest because monodentate complexes such as $[PtCl(NH_3)_3]^+$ or $[PtCl(dien)]^+$ are definitively not cytotoxic. Representative values for *in vivo* tests (5 mg/kg, inj. ipip, d3,7,11) gave a %T/C of 130 (L1210S) and 146 (L1210/DDP5).[14] Thus, although less potent than the 2,2/c,c series, this complex is actually more potent in the resistant line. The combined results on this compound also confirm our previous observations that the presence of at least one cis-Pt(amine)$_2$ group is essential to obtain antitumor activity equivalent to cis-DDP in cis-DDP <u>sensitive</u> cell lines.

8) Bis(platinum) complexes containing both Pt coordination spheres in the *trans* configuration do not have dramatically increased cytotoxicity over the simple monomer.[13]

Table 2: Growth Delay Effects of Bis(platinum) and Related Complexes in Murine Leukemia Cell Lines Sensitive and Resistant to Cisplatin[a]

ID_{50} (μM)[b]

Cell Line	2,2/c,c[c] (n=4)	2,2/c,c[c] (n=5)	2,2/c,c[c] (n=6)	1,1/t,t (n=4)	cisDDP	$PtCl_2$(DACH)[d]
L1210S	0.5	0.38	0.45	3.0	0.92	0.21
L1210/DDP5	0.9(1.7)	2.1(5.5)	2.2(4.9)	4.05(1.35)	17.0(14)	0.38(1.8)
L1210/DACH	0.8(1.4)	1.1(1.6)	1.2(2.7)		2.1(1.7)	9.5(45)
P388S	0.15	0.30	0.17	5.25	0.70	0.21
P388 PtR4	1.3(9)	1.9(6.3)	1.6(9.4)	5.86(1.1)	23.3(33)	2.2(10)

a: L1210S: Sensitive to cisplatin. L1210/DDP5: Resistant to cisplatin. L1210/DACH: Resistant to $PtCl_2$(DACH) P388S: Sensitive to cisplatin. P388/PtR4: Resistant to cisplatin.

b: Numbers in parentheses, fold resistance defined as ID_{50}(resistant line) − ID_{50}(sensitive line). 72 hour growth delay determined with cells in continuous contact with drug. All values calculated as per Ref. 14.

c: 2,2/c,c: $[\{cis\text{-}PtCl_2(NH_3)\}_2 H_2N(CH_2)_nNH_2]$; 1,1/t,t: $[\{trans\text{-}PtCl(NH_3)_2\}_2 H_2N(CH_2)_nNH_2]Cl_2$

d: DACH = R,R-1,2-diaminocyclohexane.

Table 3: Effect of Diamine Linker in [{cis-PtCl$_2$(NH$_3$)}(H$_2$N-R-NH$_2$)] on Growth Delay in Murine Leukemias Sensitive and Resistant to Cisplatin[a]

R[c]	ID$_{50}$ (μM)[b]			
	L1210S	L1210/DDP5	P388S	P388PtR$_4$
(CH$_2$)$_5$	0.69	1.5(2.2)	0.19	1.2(6.4)
(CH$_2$)$_2$-CF$_2$-(CH$_2$)$_2$	0.42	9.53(22.7)	0.71	>6
(CH$_2$)$_2$-C(CH$_3$)$_2$-(CH$_2$)$_2$	0.24	1.25(5.2)	0.22	2.3(10.7)
CH$_2$-C(CH$_3$)$_2$-(CH$_2$)$_3$	0.26	1.58(6.1)	0.24	4.0(16.7)
(CH$_2$)$_2$-CH(CH$_3$)-(CH$_2$)$_2$	0.29	1.11(3.8)	0.25	3.67(14.6)
CH$_2$-CH(CH$_3$)-(CH$_2$)$_3$	0.25	1.85(7.4)	>0.45	7.0
(CH$_2$)$_2$-CH(OH)-(CH$_2$)$_2$	0.031	0.10(3.4)	0.094	0.44(4.6)
(CH$_2$)$_2$-O-(CH$_2$)$_2$	0.49	2.39(4.8)	0.29	1.94(6.7)
(CH$_2$)$_2$-N$^+$(CH$_3$)$_2$-(CH$_2$)$_2$[d]	3.18	>6	5.76	>6
(CH$_2$)$_2$-SO$_2$-(CH$_2$)$_2$	0.84	10.9(12.9)	-	-

a: As per Table 1. b: As per Table 1. c: All linkers have a total atom connectivity of 5 between the NH$_2$ terminal groups.

Table 4: GROWTH DELAY EFFECTS OF BIS(PLATINUM) COMPLEXES [(cis-PtCl$_2$(NH$_3$))$_2$H$_2$N(CH$_2$)$_n$NH$_2$] IN HUMAN TUMOR LINES IN VITRO

ID_{50} (μM)

Tumor Line[b]	2,2/c,c[a]			cis-DDP
	(n = 4)	(n = 5)	(n = 6)	
HCT-8	.25	0.31	0.17	1.3
H23	0.07	0.11	0.06	0.31
H125	0.46	0.42	0.28	0.70
H520	0.20	0.36	0.16	0.47

a: 2,2/c,c = [(cis-PtCl$_2$(HN$_3$))$_2$H$_2$N(CH$_2$)$_n$NH$_2$].

b: Cells were in contact with drug for the entire period of the assay. HCT-8 is a human colon tumor; H23 is an adenocarcinoma of the lung; H125 is also a lung adenocarcinoma; H520 is a squamous cell carcinoma.

DNA-Binding of Bis(platinum) Complexes.

Similar to cis-DDP, the 2,2/c,c complexes selectively inhibit DNA synthesis and it is reasonable to conclude that bis(platinum) complexes will exert their cytotoxic effects at the DNA level. Our major goal is to determine what are the differences betweeen the DNA binding of bis(platinum) complexes and cis-DDP, and especially how these relate to the differences in biological activity. If the activity of bis(platinum) complexes in cis-DDP-resistant cells has an explanation in DNA binding, then a rational approach to overcoming cis-DDP resistance may be achieved from the molecular level. Design of platinum complexes with an altered spectrum of clinical anticancer activity in comparison to the presently used agents would then be a distinct possibility. The distinct differences in the <u>pattern</u> of antitumor activity between the two types of bis(platinum) complexes studied to date strengthen this approach.

The dominant difference in DNA-binding between bis(platinum) complexes and their monomeric analogues is the formation of interstrand cross-links by binding of one Pt atom of each molecule to opposite strands of DNA. Current studies in our laboratories include quantitation of DNA adducts in cells and determination of the efficiency of bis(platinum) cytotoxicity per adduct in sensitive and resistant cells. We may further obtain much insight from study of isolated DNA and the conformational distortion and sequence specificity of bis(platinum)-modified DNA. This section reviews our results to date. The specific complexes to be discussed are shown in Figure 3, along with their respective monomeric analogues. The presence of monodentate or bidentate coordination spheres in bis(platinum) complexes may result in altered patterns of interstrand cross-linking. These differences could be reflected in the efficiency of cross-link formation, its sequence specificity or the conformational distortion induced A preliminary study showed the presence of interstrand cross-links in genomic DNA from cultured cells, Figure 4.

Interstrand Cross-linking

The cross-linking was assayed using the (Dde I - EcoR I)$_{65}$ fragment of pBR322. Complexes I-IV were incubated with this fragment in the presence of CT DNA (4 ug/mL) at varying concentrations for 1h. Upon electrophoresis under denaturing conditions, non-cross-linked 5'labelled fragments migrate as 65 base single strands whereas cross-linked fragments migrate as a higher molecular weight species.[12,16] Table 5 summarises the cross-linking efficiency and their quantitation per adduct. Cross-linking occurs at concentrations as low as 0.5 uM for the 1,1/t,t complex I. At intermediate concentrations up to 50 uM distinct bands are also seen in the gel corresponding to structurally different species. The frequency of cross-links for I increases with concentration in the range 1 - 10 uM whereas the frequency for II rises to a maximum with a subsequent plateau in the range 5 - 100 uM. In contrast, the r_b for II increases with greater Pt concentration whereas the r_b for I rises to a maximum and then levels off. As expected, no interstrand cross-linking could be observed for either monomeric complex at concentrations up to 100 uM.

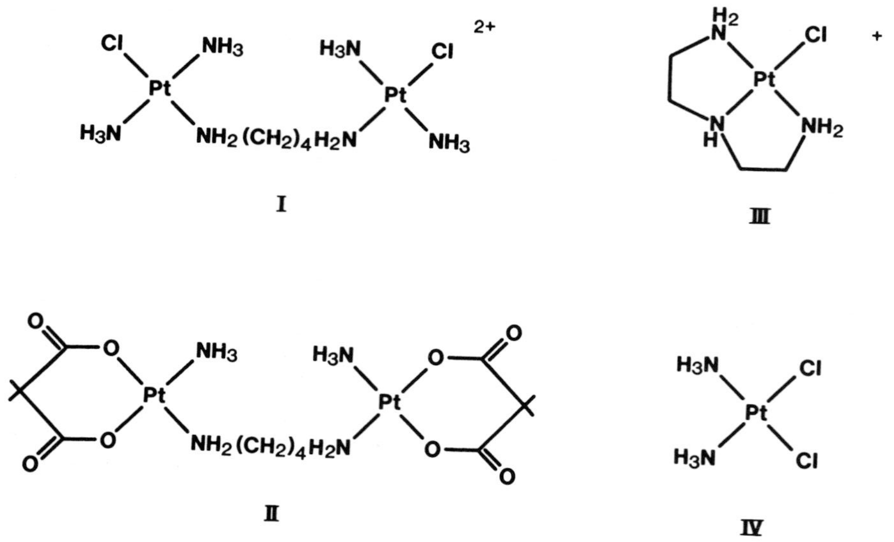

Figure 3. The structures of bis(platinum) complexes with monodentate and bidentate coordination spheres and their monomeric analogues. I, [trans-(PtCl(NH₃)₂)₂H₂N(CH₂)ₙNH₂]Cl₂, 1,1/t,t; II, [{Pt(mal)(NH₃)}₂H₂N(CH₂)₄NH₂], 2,2/c,c; III [PtCl(dien)]Cl, Pt(dien); IV, cis-[PtCl₂(NH₃)₂], cis-DDP. From Ref. 12.

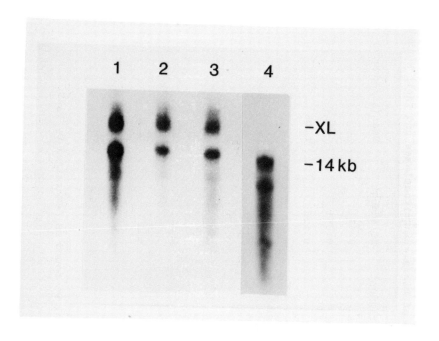

Figure 4. **Formation of interstrand cross-links by bis(platinum) complexes in cultured L1210 cells.**

Mouse L1210 cells were treated with 150, 100, 50 and 0 uM [trans-{PtCl(NH₃)₂}₂H₂N(CH₂)ₙNH₂] (1,1/t,t, Lanes 1,2,3 and 4 respectively) at 37°C. Total cellular DNA was extracted and digested with Eco R1. the DNA was then denatured by heating to 90°C for 5 min. in 30mM NaOH and 1mM EDTA and loaded on to a 0.5% alkaline agarose gel. The DNA was electrophoresed at 90 mAmps and 35 volts for 18 h. in 30mM NaOH and 1mM EDTA, and transferred to a nylon membrane by the method of Southern.[21] The blot was probed with ³²P-labeled mitochondrial DNA, yielding a 14kB fragment. Cross-linked DNA (XL) may be seen as a band which mirates more slowly than the 14kb fragment. It is apparent that the 1,1/t,t complex produced significant amounts of cross-linked DNA. At the concentrations of complex used, the r_b values for Pt bound to DNA range from 2.5×10^{-4} to 1.5×10^{-3}.

Table 5: Frequency of Interstrand Crosslink Formation of Bis(platinum)Complexes and Their Monomeric Analogues.[a]

Complex (Type)[b]	Conc. (μM)[c]	r_b[c]	XL/Fragment	XL/Adduct	Mean(XL/Adduct)
I, 1,1/t,t	1	0.01	0.018	0.014	
	5	0.028	0.184	0.051	
	10	0.037	0.60	0.125	0.06
II, 2,2/c,c	5	0.01	0.025	0.019	
	10	0.0155	0.055	0.027	
	50	0.071	0.18	0.020	
	100	0.135	0.28	0.016	0.02
III, [PtCl(dien)]Cl	100	NC	---	---	---
IV, cisDDP	100	NC	---	---	---

a: Adapted from Refs. 12 and 16.

b: Refers to complexes of Figure 3.

c: Concentration is for complex incubated. r_b is Total Pt bound as determined by FAAS. XL = interstrand cross-links.

Conformational Alteration of DNA By Bis(platinum) Complexes. Circular Dichroism Spectra of Bis(platinum)-modified Calf Thymus DNA.

The origin of the conformational alterations of platinum-adducted double-stranded oligonucleotides lie in the displacement from planarity of the bases, loss of base-pair hydrogen bonding and switches in the sugar conformation (N, 3'-endo to S, C2'-endo) and base configuration (from the anti to the syn form).[22,23] In the case of interstrand cross-links the combination of these chemical effects are likely to produce a conformational distortion distinctly different from the intrastrand cross-link of cis-DDP. Analysis of the adducts induced by cis-DDP and related complexes show that the principal sites of attack of cis-DDP on DNA are adjacent guanine/guanine or adenine/guanine bases forming intrastrand cross-links. Bis(platinum) complexes may target larger or different sequences and react at sites inaccessible to cis-DDP.

Circular Dichroism gives information on the overall conformational changes induced by platination. The major changes for cis-DDP or the monodentate [PtCl(dien)]Cl at low r_b, and for a variety of DNAs, are a slight increase in ellipticity of the positive band but no major alteration of the spectrum occurs.[24,25] The CD spectra of DNA upon reaction with Complexes I and II are shown in Figure 5. The CD spectrum of DNA modified by the 2,2/c,c complex II is very similar to that of cis-DDP-modified DNA. The spectrum of the 1,1/t,t complex I is quite different. There is a decrease in the intensity of the negative band centered at 246 nm with increasing r_b, Figure 5b, although the positive band centered at 270 nm is essentially unaltered. In the case of I the interstrand cross-linking, which in its simplest form is two monodentate binding sites on opposite strands, produces a significantly altered structure in comparison to the purely monodentate binding of III.

Sequence Specificity of Bis(platinum) Complexes

The 3'-> 5' exonuclease activity of T4-DNA polymerase has been used by us to probe the sequence specificity of bis(platinum) complexes. The 3'-> 5' exonuclease activity of T4 polymerase has been shown to stop one or two bases prior to the platinated nucleotide.[26] The T4-exonuclease product (with a 3'-OH terminal group) will run 1.5-2 bases slower than the Maxam-Gilbert fragment (with a 3'-P terminal group) of the same length, Figure 6. There are several interesting features and four distinct types of stop site may be distinguished:

(i) stop sites common to all complexes of Figure 3,

(ii) sites common to both complexes with monodentate coordination spheres (the 1,1/t,t and Pt(dien),

(iii) sites common to both bis(platinum) complexes and finally,

(iv) sites attacked only by the 1,1/t,t complex are observed. These are summarised briefly below:

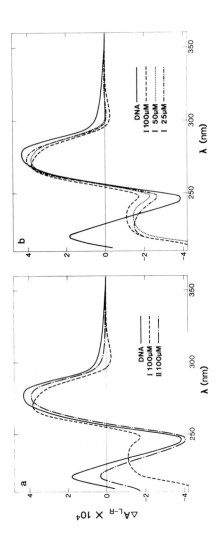

Figure 5. Circular Dichroism spectra of Calf Thymus DNA modified by (bis(platinum) complexes.**I**, 1,1/t,t; **II**, 2,2/c,c, as in Figure 3. From Ref. 12.

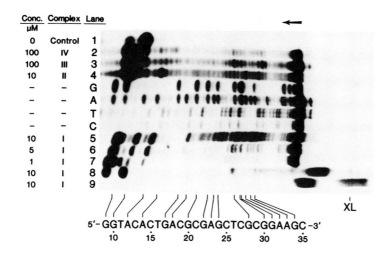

Figure 6. Sequence specificity of the bis(platinum) complexes. From Ref. 12. The 49 bp duplex was synthesized as two separate oligonucleotides on a Dupont automatic DNA synthesizer (Generator DNA Synthesizer) and purified by eletrophoresis on 12% polyacrylamide gels under denaturing conditions (8M Urea). The fully duplex DNA was annealed in standard fashion and purified on a 12% polyacrylamide native gel. The 49 bp duplex which had been 5'-labeled with ^{32}P was treated with the complexes in Figure 3. Following digestion by the exonuclease activity of T4-polymerase the DNA was treated with KCN to remove all bound platinum (Lanes 1-7). The resulting samples were analzyed by electrophoresis on a 12% polyacrylamide sequencing gel. The dried gel was used for autoradiography. Lane 8 contains DNA which had been treated with 10 uM 1,1/t,t and treated with exonuclease, but not KCN. Lane 9 contains DNA which was only treated with 10 uM 1,1/t,t. I = 1,1/t,t, II = 2,2/c,c, III = Pt(dien), IV = cis-DDP (From Fig. 3). A,G,T,C refers to the A+G, G, T+C and C Maxam and Gilbert sequencing reactions. The arrow indicates the direction of electrophoresis. XL = DNA interstrand cross-links.

All Pt   1,1/Bis(Pt)   All Bis(Pt)

5'-G A C T A C T T G G T A C A C T G A C G C C G A G C T C G C G G A A G C T C A T T C C A G T G C G C -3'

3'-C T G A T G A A C C A T G T G A C T G C G C T C G A G C G C C T T C G A G T A A G G A C A C G C G -5'

1.....5......10......15.....20......25......30.....35......40......45.....

An interesting feature is the class of stop sites unique to the cross-linking agents. These clearly occur within alternating purine/pyrimidine CGCG sequences, although further refinement is required to distinguish (CG) from (GC) binding sites. An example of this can be seen for the stop site at $G_{28}C_{29}$ corresponding to an inter-strand cross-link at $G_{28}G'_{29}$. Unique binding sites of the 1,1/t,t complex occurs in the C_{19} (or G'_{19}) to G_{22} region, a likely candidate for interstrand cross-links involving G'19 and G'21. A further unique site for the 1,1/t,t complex is a possible $A'_{16}G_{17}$ cross-link.

Discussion

In the most general sense, bis(platinum) complexes represent the first examples of a large class of structurally novel antitumor agents - bimetallic interstrand cross-linking agents. The antitumor activity of bis(platinum) complexes is highlighted by their cytotoxicity toward cis-DDP-resistant cells, in both murine and human tumor models. The activity is dependent on the nature and chain length of the diamine linker. Our studies to date further indicate that the pattern of antitumor activity, especially in comparison to cisplatin, is dependent on the specific structure of the complex.

In DNA binding studies, the dominant features are the interstrand cross-linking, with concomitant conformational changes and sequence specificity. Cis-DDP and its monomeric analogues display a distinct preference for GG and AG sequences on the same strand. The binding is nonrandom and is dependent to some extent on neighboring base sequence. Interstrand cross-linking of bis(platinum) complexes results in enhanced affinity for alternating purine/pyrimidine sequences such as $(-GC-)_n$ or $(-CG-)_n$. Bis(platinum) complexes containing monodentate coordination spheres, and capable of "pure" interstrand cross-links, induce unique conformational changes in DNA, as judged by circular dichroism spectra. The change in CD spectrum consists of a loss of the negative band and results in a non-conservative spectrum. The CD spectrum of DNA modified by those complexes containing bidentate coordination spheres is, on the other hand, similar to that of the cis-DDP adducts.

The interstrand cross-linking of bis(platinum) complexes is reminiscent of alkylating agents such as the nitrogen mustards, CC1065, and mitomycin C. The potential for rational development of bis(platinum) complexes with an altered spectrum of antitumor activity to cis-DDP may be appreciated if we consider that the biological activity of the 2,2/c,c complexes is best considered as the sum of both cis-DDP-like intrastrand cross-links and bis(platinum) interstrand cross-links. On the other hand, these latter lesions must be principally responsible for the activity of 1,1 complexes which are structurally incapable of behaving like cis-DDP. Interstrand cross-links formed by 1,1 complexes results in "saturation" of the monodentate platinum coordination spheres (Cl displaced by guanine). The formation of a cross-link by the bidentate platinum atoms in 2,2 complexes may proceed in one of two limiting ways: an interstrand cross-link is initially formed followed by subsequent closure of the intrastrand cross-links on each strand or alternatively an intrastrand cross-link could be formed first, which is followed by the interstrand cross-link formation upon reaction of the second platinum atom. These competing processes, and our ability to alter the resultant combination of lesions formed, may affect both sequence specificity and ultimately antitumor activity.

Acknowledgements. The work at UVM is supported by an American Cancer Research grant CH-463. We thank W. Elliott for supplying **in vivo** results. We especially thank Dr. H. Showalter for suggesting this topic to the Symposium organisers.

REFERENCES
1. Farrell, N. "Transition Metal Complexes as Drugs and Chemotherapeutic Agents" in "Catalysis by Metal Complexes" James, B.R.; Ugo, R. Eds. Kluwer Academic Publishers (Dordrecht) 1989, v. 11 pp. 46-66.

2. Sherman, S.E. and Lippard, S.J., Chem. Rev., 1987, **87**, 1153.

3. Reedijk, J., Fichtinger-Schepman, A.M.J., van Oosterom, A.T., and van de Putte,, P., Structure & Bonding (Berlin), 1987, **67**, 53.

4. Johnson, N.P., Lapetoule, P., Razaka, H.; Villani, G., in Biochemical Mechanisms of Platinum Antitumor Drugs (eds. D.C.H. McBrien and T. Slater) IRL Press, Oxford 1986, p.1.

5. Ciccarelli, R.B., Solomon, M.J., Varshavsky, A., Lippard, S.J.: Biochemistry, 1985, **24**, 7533.

6. Farrell, N., Ha, T.T.B., Souchard, J.-P., Wimmer, F.L., Cros, S., Johnson, N.P.: J. Med. Chem., 1989, **32**, 2240.

7. Burchenal, J.H., Kalaher, K., Dew, K., Lokys, L., and Gale, G.: Biochimie, 1978, **60**, 961.

8. Jennerwein, M.M., Eastman, A. and Khokhar, A.: Chem. Biol. Inter., 1989, **70**, 39.

9. Page, J.D., Husain, I., Sancar, A., and Chaney, S.G.: Biochemistry, 1990, **29**, 1016-1024.

10. Eastman, A. and Schulte, N.: Biochemistry, 1988, **27**, 4730.

11. Farrell, N.P., de Almeida, S.G. and Skov, K.A.: J. Am. Chem. Soc., 1988, **110**, 5018.

12. Farrell, N., Qu, Y., Feng, L. and Van Houten, B.: Biochemistry, Submitted.

13. Farrell, N., Qu, Y. and Hacker, M.P.: J. Med. Chem. In Press.

14. Kraker, A.J., Hoeschele, J.D., Moore, C.W., Farrell, N. and Elliott, W.L.: Proc. AACR, 1990, **31**, 1991.

15. Hoeschele, J.D., Kraker, A.J., Sercel, A.D., Showalter, H.D.H., Elliott, W.L. and Farrell, N.P.: Proc. AACR, 1990, **31**, 1960.

16. Roberts, J.D., Van Houten, B., Qu, Y. and Farrell, N.P.: Nuc. Acids Res., 1989, **17**, 9719.

17. Farrell, N. and Qu, Y.: Inorg. Chem., 1989, **28**, 3416.

18. Farrell, N. de Almeida, S.G. and Qu, Y.: Inorg. Chim. Acta In Press.

19. Dhara, S.C.: Indian J. Chem., 1970, **8**, 193.

20. Qu, Y. and Farrell, N.: J. Inorg. Biochem. Submitted.

21. Southern, E.: J. Mol. Biol., 1975, **98**, 503.

22. den Hartog, J.H.J., Altona, C., Chottard, J.-C., Girault, J.-P., Lallemand, Y., de Leeuw, F.A.A.M., Marcelis, A.T.M., Reedijk, J.: Nuc. Acids. Res., 1982, **10**, 4715.

23. Sherman, S.E., Gibson, D., Wang, A.H.-J. and Lippard, S.J.: Science, 1985, **230**, 412.

24. Macquet, J.P., Butour, J.L. and Johnson, N.P.: in "Platinum, Gold, and Other Metal Chemotherapeutic Agents" ACS Symposium Series, Lippard, S.J., Ed., ACS, Washington, 1983, **209**, 75.

25. Marrot, L. & Leng, M., Biochemistry, 1989, **28**, 1454.

26. Royer-Pokora, B., Gordon, L.K., and Haseltine, W.A.: Nuc. Acids Res., 1981, **9**, 4595.

INTERACTION OF CALICHEAMICIN WITH DNA

NADA ZEIN, WEI-DONG DING, GEORGE A. ELLESTAD
Infectious Disease Research Section
American Cyanamid Company, Medical Research Division
Lederle Laboratories, Pearl River, New York 10965

ABSTRACT. The interaction of the potent antitumor agent calicheamicin with DNA has been investigated by analyzing the cleavage of plasmids, restriction fragments, and synthetic dodecamers. Proton NMR analysis of the aromatized product from the reaction of calicheamicin with unlabeled and deuterium-labeled DNA has provided additional insight into the molecular events involved in strand cleavage by the diyne-ene moiety of this novel natural product.

Introduction

For the past several years we have been studying the remarkable DNA cleavage properties of calicheamicin γ^1 (1), an extremely potent antitumor antibiotic isolated from fermentations of the soil microorganism *Micromonospora echinospora* ssp. *calichensis*.[1,2] This natural product has been shown to be some 1000 times more active than adriamycin against murine tumors, causes mutagenesis in *Escherichia coli*, and chromosome aberrations in human diploid lung fibroblasts. Calicheamicin, along with esperamicin[3] (2), neocarzinostatin[4] (3), and dynemicin[5] (4), belongs to a unique class of natural products that contain unusual diyne-ene systems. Although the exact biochemical basis for the cytotoxic action of these drugs remains speculative, cleavage of cellular DNA appears to be essential for their activity. All of these compounds have been found to cleave duplex DNA by a diradical intermediate which, when bound in the minor groove, abstracts hydrogens from a deoxyribose on both strands to initiate oxidative strand cleavage. Goldberg and colleagues have studied the cleavage properties of neocarzinostatin for a number of years now and these results have provided the basis for a great deal of our understanding of how these compounds interact with DNA.[6]

Cleavage Chemistry

Calicheamicin γ_1 exhibits a marked preference for cleaving supercoiled, duplex plasmid DNA at very low concentrations (7nM) to give single-strand and a high percentage ($\geq 50\%$) of double-strand breaks consistent with the proposed bifunctional nature of the cleaving agent. Indeed, it is likely that this double-strand cleavage is responsible for the extreme cytotoxicity exhibited by these compounds.[7] The chemistry believed to be responsible for cleavage is outlined in Scheme 1.

B. Pullman and J. Jortner (eds.), Molecular Basis of Specificity in Nucleic Acid-Drug Interactions, 323–330.
© 1990 *Kluwer Academic Publishers. Printed in the Netherlands.*

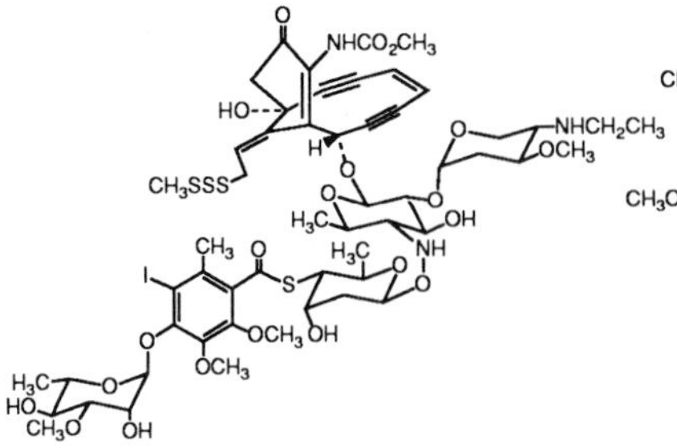

1 calicheamicin

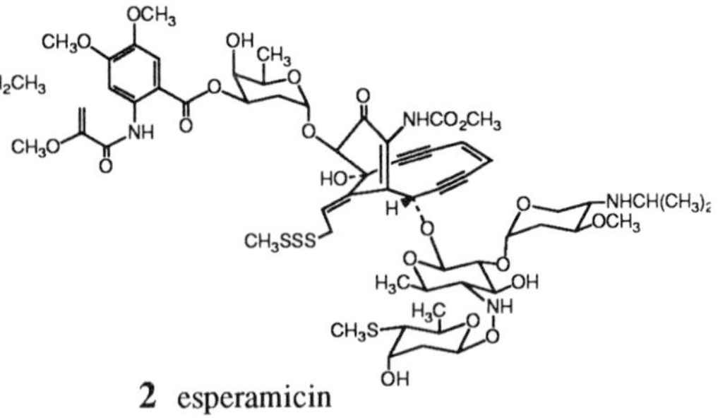

2 esperamicin

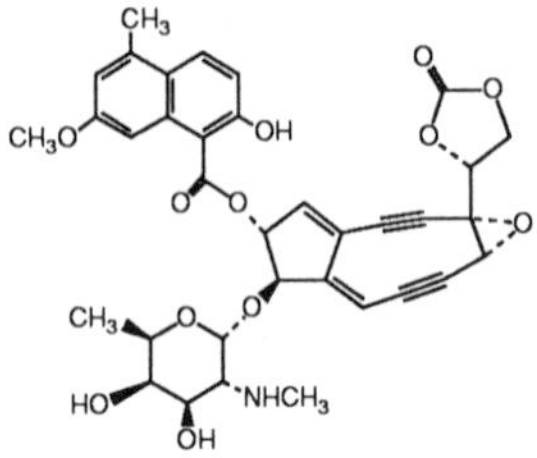

3 neocarzinostatin

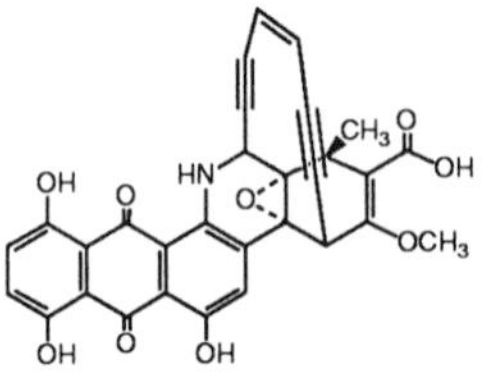

4 dynemicin

Cleavage is catalysed by thiols such as β-mercaptoethanol, dithiothreitol, and glutathione which cause reductive scission of the allylic trisulfide to give a thiolate (**5**) which undergoes a Michael addition to the α,β,-unsaturated ketone leading to the dihydrothiophene (**6**). This brings the two acetylenes closer together thus facilitating Bergman aromatization via the transient but extremely reactive 1,4-diyl intermediate (**7**).[8]

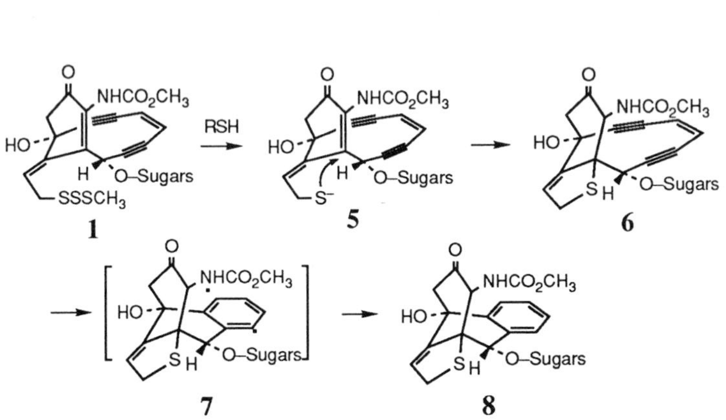

Scheme 1. Mechanism of diradical formation in calicheamicin.

By excluding oxygen from the reaction mixture, cleavage was dramatically inhibited.[2] But strand scission was not affected when carried out in the presence of excess superoxide dismutase or catalase. Thus, reduced oxygen species are apparently not involved. These results, in conjunction with the specificity of the cuts, argue against a diffusible radical species and support, instead, a non-diffusible carbon-centered radical similar to observations with neocarzinostatin. A possible cleavage mechanism is presented in Scheme 2.

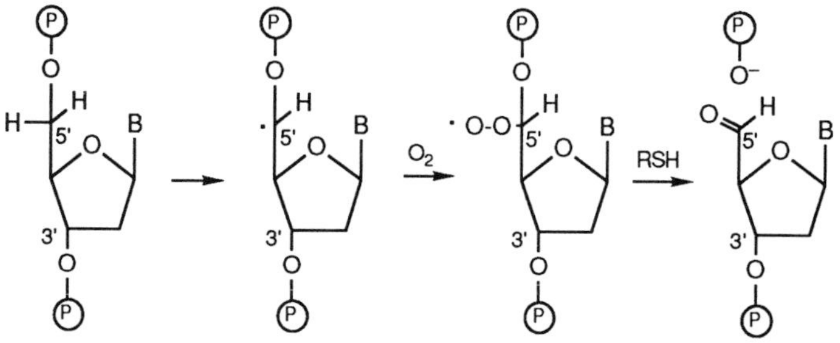

Scheme 2. Probable mechanism of DNA cleavage at the 5'-carbon of the preferred deoxyribose.

Cleavage Site Specificity

One of the striking features of the cleavage properties of calicheamicin is the pronounced sequence dependence on reaction with DNA. This is especially remarkable for a molecule of only 1367 daltons. Cutting experiments with a number of 5'- and 3'-end-labeled restriction fragments showed very specific cleavage at homopyrimidine/homopurine sequences such as TCCT/AGGA and CTCT/GAGA where the 5' cytidine and thymidine of the pyrimidine strands are the principal targets. This results in fragments ending in 3'-phosphates and most likely oligomers ending in 5' aldehydes similar to neocarzinostatin as determined by analysis of the electrophoretic mobilities of the fragments on polyacrylamide sequencing gels.

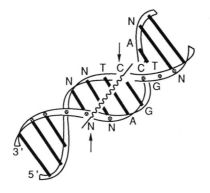

Figure 1. Depiction of minor groove DNA cleavage by calicheamicin. Arrows denote the positions of hydrogen abstraction by diradical intermediate.

The cleavage pattern is asymmetric with an offset two nucleotides to the 3' side of the complementary AGGA tract indicating the minor groove to be the binding site (Figure 1).[2] This was further confirmed by pre-incubation of the DNA with netropsin, a known minor groove binder at AT rich regions, which altered the cutting specificity of calicheamicin. Also, T4 phage DNA in which the major groove is blocked with carbohydrate, was cut as efficiently as normal DNA.

The identity of the hydrogen abstracted on the deoxyribose of the 5' cytidine of the TCCT site was inferred to be one of the geminal hydrogens at the 5' carbon (probably the pro S based on models). This was based on the electrophoretic mobilities of fragments obtained from 3'-end-labeled segments of DNA. The principal radioactive band migrated as if this DNA oligomer was two nucleotides longer than the expected one matching the 5'-phosphate-ended fragment from the Maxam-Gilbert reaction. Alkaline treatment of the reaction mixture then yielded a radioactive fragment which comigrated with the anticipated chemically produced marker. This is consistent with the formation of a 5'-aldehyde terminus which, in the presence of base, undergoes a β-elimination, to give the 5'-phosphate-ended oligomer.[2a] Goldberg and coworkers first observed this chemistry with neocarzinostatin.[6]

NMR Evidence for Nonexchangeable Hydrogen Abstraction

Unequivocal chemical evidence that calicheamicin does indeed abstract nonexchangeable hydrogen atoms from the DNA backbone came from NMR experiments with calf thymus DNA.[9] In one experiment, calicheamicin was reacted with deuterium-exchanged, sonicated calf thymus DNA in the presence of deuterated methyl thioglycolate ($DSCH_2CO_2CH_3$) as the reducing thiol, $10\%C_2D_5OD$ for solubilizing the antibiotic, and Tris / DCl / D_2O at pD 7.8. The proton NMR spectrum of the aromatized product (8) (aromatic portion shown in Figure 2a) shows no deuterium incorporation from the medium. This indicates that hydrogen incorporation at C1 and C4 occurred and that DNA is the sole source of both protons.

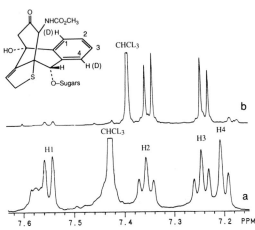

Figure 2. ^{1}H NMR of aromatic region of calicheamicin ε (8). a) Reaction carried out in the presence of deuterated medium and deuterium-exchanged calf-thymus DNA. b) Reaction in the presence of deuterated medium without DNA.

When the same reaction was carried out as above but without the DNA, the NMR of the product showed over 98% deuterium incorporation at C1 and C4 (Figure **2b**). Furthermore, these results strongly suggest that both hydrogens must be abstracted from both DNA strands since monodeuterated products were not observed. It is noteworthy that the high yield of **7** in the presence of DNA (65%) compared to 20% without DNA, suggests that the DNA might act as a template for this reaction. Thus the two strands of the DNA may move the two acetylenes closer together in the minor groove to facilitate aromatization.

Experiments with Synthetic Dodecamers

Analysis of cutting experiments on sequencing gels with 5' and 3' end-labeled synthetic dodecamers containing the T<u>C</u>CT cleaving site showed the same specificity as observed with large restriction fragments.[2b] This suggested that synthetic dodecamers specifically labeled with deuterium might be useful for studying the sites of hydrogen abstraction by calicheamicin using the above mentioned NMR methodology. In the event, reaction of calicheamicin with dodecamers **I** and **II** specifically labeled with deuterium at the C5' of the 5'-cytidine in the T<u>C</u>CT sequence resulted in a remarkably specific atom-transfer to C4 of calicheamicin ε (**7**).[10] This confirms the C5' of the target cytidine deoxyribose as a site of hydrogen abstraction. Obviously, this technique can be used to probe the site of hydrogen abstraction on the complementary AGGA strand.

5' GGGGT<u>C</u>CTGGGG   GGGT<u>C</u>CTAAATT
   CCCCAGGACCCC 5'   CCCAGGATTTAA

I   **II**

In addition, these experiments provide important insight into the geometry of the calicheamicin/DNA complex.[2b,11] In the case of these two dodecamers, the calicheamicin apparently associates in such a manner that the carbohydrate tail binds toward the 3'/5' side of the TCCT/AGGA cutting site. A re-interpretation may be necessary of the cleavage experiments in which the TCCT unit in dodecamer substrates was cleaved only when the sequence was incorporated at the 3'-end as opposed to the 5'-end.[2b] Whether or not the binding orientation is sequence dependent will depend on additional labeling experiments with other sequences.

Hydrophobic and Electrostatic Binding Interaction with DNA

The question as to why calicheamicin associates with the DNA is intriguing. Calicheamicin is extremely hydrophobic with essentially no solubility in water while the DNA is very hydrophilic. However, the wealth of recent molecular recognition studies in aqueous media between small hydrophobic molecules with receptors such as cyclodextrins and various synthetic cyclophanes containing hydrophobic cavities have revealed hydrophobic effects to be important binding factors.[12] Based on these results, we propose that an important binding parameter in the present case is the hydrophobic association between the water insoluble calicheamicin and the low-dielectric environment of the interior of the minor groove of the DNA which can be considered to be the hydrophobic cavity of the host.[13] A precedent for such hydrophobic interaction between DNA and a small ligand has recently been reported where a hydrophobic interaction between acetone and the hydrophobic interior of the DNA appears to dominate the photochemical reaction at high ionic strength.[14] The exact role of the amino sugar in the calicheamicin/DNA interaction is unknown at this time but cleavage experiments with a derivative lacking the ethylaminosugar (9), show identical site-specific cutting to that of calicheamicin γ_1 but with considerably less efficacy.[2b] It appears that the ethylaminosugar contributes, along with the glycosylated hydroxylamino sugar, to general electrostatic binding with the DNA.

9

References

1. Lee, M. D., Dunne, T. S., Siegel, M. M., Chang, C. C., Morton, G. O., and Borders, D. B. (1987), 'Calicheamicins, a novel family of antibiotics. 1. Chemistry and partial structure of calicheamicin γ_1' *J. Am. Chem. Soc.* **109**, 3464-3466; Lee, M . D., Dunne, T .S., Chang, C. C., Ellestad, G. A., Siegel, M. M., Morton, G. O., McGahren, W. J., and Borders, D. B. (1987), 'Calicheamicins, a novel family of antibiotics. 2. Chemistry and structure of calicheamicin γ_1' *J. Am. Chem. Soc.*, **109**, 3466-3468.

2. a) Zein, N., Sinha, A. M., McGahren, W. J., and Ellestad, G. A. (1988), 'Calicheamicin γ_1,: An antitumor antibiotic that cleaves double-stranded DNA site specifically' *Science (Washington, D. C.)* **240**, 1198-1201; b) Zein, N., Poncin, M., Nilakantan, R., and Ellestad, G. A. (1989) 'Calicheamicin γ_1 and DNA: Molecular recognition process responsible for site-specificity' *Science (Washington, D. C.)* **244**, 697-699.

3. Golik, J., Clardy, J., Dubay, G., Groenewold, G., Kawaguchi, H., Konishi, M., Krishnan, B., Ohkuma, H., Saitoh, K., and Doyle, T. W. (1987) 'Esperamicins, a novel class of potent antitumor antibiotics. 2. Structure of esperamicin X' *J. Am. Chem. Soc.*, **109**, 3461-3462; Golik, J., Dubay, G., Groenewold, G., Kawaguchi, H., Konishi, M., Krishnan, B., Ohkuma, H., Saitoh, K., and Doyle, T. W. (1987) 'Esperamicins, a novel class of potent antitumor antibiotics.3. 'Structures of esperamicins A$_1$, A$_2$, and A$_{1b}$' *J. Am. Chem. Soc.*

109, 3462-3464; Long, B. H., Golik, J., Forenza, S., Ward, B., Rehfuss, R., Dabrowiak, J. C., Catino, J. J., Musial, S. T., Brookshire, K. W., and Doyle, T. W. (1989) 'Esperamicins, a class of potent antitumor antibiotics: mechanism of action' *Proc. Natl. Acad. Sci. USA* **86**, 2-6; Sugiura, Y., Uesawa, Y., Takahashi, Y., Kuwahara, J., Golik, J., and Doyle, T. W. (1989) 'Nucleotide-specific cleavage and minor-groove interaction of DNA with esperamicin antitumor antibiotics' *Proc. Natl. Acad. Sci. USA* **86**, 7672-7676; Fry, D. W., Shillis, J. L., and Leopold, W. R. (1986) 'Biological and biochemical activities of the novel antitumor antibiotic PD 114,759 and related derivatives' *Investigational New Drugs* **4**, 3-10.

4. Edo, K., Mizugaki, M., Koide, Y., Seto, H., Furihata, K., Otake, N., and Ishida, N. (1985) 'The structure of neocarzinostatin chromophore possessing a novel bicyclo [7, 3, 0] dodecadiyne system' *Tetrahedron Lett.* **26**, 331-334.

5. Konishi, M., Ohkuma, H., Matsumoto, K., Tsuno, T., Kamei, H., Miyaki, T., Oki, T., Kawaguchi, H., VanDuyne, G. D., and Clardy, J. (1989), 'Dynemicin A. a novel antibiotic with the anthraquinone and 1,5-diyne-3-ene subunit' *J. Antibiotics.* **42**, 1449-1452; Semmelhack, M. F., Gallagher, J., and Cohen, D. (1990), 'Bioreductive alkylation as a trigger for toxic effects of dynemicin' *Tetrahedron Lett.* **31**, 1521-1522; Snyder, J. P., and Tipsword, G. E., (1990), 'A proposal for blending classical and biradical mechanisms in antitumor antibiotics: Dynemicin A.' *J. Am. Chem. Soc.* (in press).

6. Kappen, L. S. and Goldberg, I.H. (1989) 'Identification of 2-deoxyribonolactone at the site of neocarzinostatin-induced cytosine release in the sequence d(AGC)', *Biochemistry* **28**, 1027-1032. For a general review of the earlier literature see Goldberg, I. H. (1987) 'Free radical mechanisms in neocarzinostatin-induced DNA damage' *Free Radical Biology & Medicine* **3**, 41-54. For the chemistry of neocarzinostatin diradical formation see: Myers, A. G., Proteau, P. J., and Handel, T. M. (1988), 'Stereochemical assignment of neocarzinostatin chromophore. Structures of neocarzinostatin chromophore-methyl thioglycolate adducts', *J. Am. Chem. Soc.* 110, 7212-7214; Myers, A. G., and Proteau, P. J. (1989) 'Evidence for spontaneous, low-temperature biradical formation from a highly reactive neocarzinostatin chromophore-thiol conjugate' *J. Am. Chem. Soc.* **111**, 1146-1147.

7. Povirk, L. F., Houlgrave, C. W., and Han, Y. (1988) 'Neocarzinostatin-induced DNA base release accompanied by staggered oxidative cleavage of the complementary strand', *J. Biol. Chem.* **263**, 19263-19266. Zhao, B., Konno, S., Wu, J. M., and Oronsky, A. L. (1990) 'Modulation of nicotinamide adenine dinucleotide and poly(adenosine diphosphoribose) metabolism by calicheamicin γ_1 in human HL-60 cells', (in press).

8. Bergman, R. G. (1973) 'Reactive 1,4-dehydroaromatics' *Acc Chem. Res.* **6**, 25-31; Nicolaou, K. C., Ogawa, Y., Zucarello, G., and Kataoka, H. (1988) 'DNA cleavage by a synthetic mimic of the calicheamicin-esperamicin class of antibiotics' *J. Am. Chem. Soc.* **110**, 7247-7248; Nicolaou, K. C., Zuccarello, G. Ogawa, Y., Schweiger, E. J., and Kumazawa, T. (1988) 'Cyclic conjugated enediynes related to calicheamicins and esperamicins: calculations, synthesis, and properties' *J. Am. Chem. Soc.* **110,** 7247-7248; Magnus, P. and Carter, P. A. (1988) 'A model for the proposed mechanism of action of the potent antitumor antibiotic esperamicin A₁' *J. Am. Chem. Soc.* **110**, 1626-1628; Snyder, J. P. (1989) 'The cyclization of calicheamicin-esperamicin analogues: A predictive biradicaloid

transition state' *J. Am. Chem. Soc.* **111**, 7630-7632; Mantlo, N. B., and Danishefsky, S. J. (1989) 'A core system that simulates the cycloaromatization and DNA cleavage properties of calicheamicin-esperamicin: A correlation experiment' *J. Org. Chem.* **54**, 2781-2783; Haseltine, J. N., Danishefsky, S. J. and Schulte, G. (1989) 'Experimental modeling of the priming mechanism of the calicheamicin/esperamicin antibiotics: Actuation by the addition of intramolecular nucleophiles to the bridgehead double bond' *J. Am. Chem. Soc.* **111**, 7638-7640; Cabal, M. P., Coleman, R. S., and Danishefsky, S. J. (1990) 'Total synthesis of calicheamicinone: A solution to the problem of the elusive urethane' *J. Am. Chem. Soc.* **112**, 3253-3255.

9. Zein, N., McGahren, W. J., Morton, G. O., Ashcroft, J. and Ellestad, G. A. (1989) 'Exclusive abstraction of nonexchangeable hydrogens from DNA by calicheamicin γ_1' *J. Am. Chem. Chem. Soc.* **111**, 6888-6890.

10. Townsend, C, A, DeVoss, J. J., Zein, N., Ding, W., Morton, G. O., Ellestad, G. A., Tabor, A. B,.and Schreiber, S. L. (1990) 'Site specific atom transfer from DNA to a bound ligand defines the geometry of a DNA-calicheamicin γ_1 complex. (1990) *Science (Washington, D.C.)* (submitted for publication).

11. Hawley, R. C., Kiessling, L. L., and Schreiber, S. L. (1989) 'Model of the interactions of calicheamicin γ_1 with a DNA fragment from pBR 322' *Proc. Natl. Acad. Sci.* USA **86**, 1105-1109.

12. Shepodd, T. J., Petti, M. A. and Dougherty, D. A. (1988) 'Molecular recognition in aqueous media: Donor-acceptor and ion-dipole interactions produce tight binding for highly soluble guests' *J. Am. Chem. Soc.* **110**, 1983-1985 and references therein.

13. Conrad, J., Troll, M., and Zimm, B. H. (1988) 'Ions around DNA: Monte Carlo estimates of distribution with improved electrostatic potentials' *Biopolymers* **27**, 1711-1732.

14. Rokita, S. E., Prusiewicz, S., and Romero-Fredes, L. (1990) 'The effect of ionic strength on the photosensitized oxidation of d(CG)$_6$' *J. Am. Chem. Soc.* **112,** 3616-3621..

THE EFFECTS OF LIGAND STRUCTURE ON BINDING MODE AND SPECIFICITY IN THE INTERACTION OF UNFUSED AROMATIC CATIONS WITH DNA

W. DAVID WILSON, FARIAL A. TANIOUS, HENRYK BUCZAK,
M.K. VENKATRAMANAN, B.P DAS AND DAVID W. BOYKIN
Department of Chemistry
Georgia State University
Atlanta, Georgia 30303   USA

ABSTRACT.   We have synthesized four diphenylfuran derivatives with terminal amidine, imidazoline, amine and piperazylamide groups.   The molecules have the same unfused central aromatic ring system and the same charge.   Spectroscopic, hydrodynamic and kinetics results indicate that these four closely related derivatives interact with DNA by three completely different binding modes.   The amidine and the imidazoline derivatives binds to AT sequences in a minor groove complex but intercalate at GC sequences in DNA.   The amine derivative binds to both AT and GC sequences by a traditional intercalation model.   The piperazylamide binds to both AT and GC sequences in a threading intercalation model.   These four molecules, thus, provide fundamental information on how ligand molecular structure and substituents interact with local DNA sequences to define binding modes.

1.   Unfused   Aromatic   Cation-DNA   Interactions

1.1   BACKGROUND

Aromatic amidines interact with DNA and display clinically useful activity against several diseases, particularly those of parasitic origin (Das and Boykin, 1977; Steck et al., 1981, 1982; Walzer et al., 1988).   The aromatic diamidines such as DAPI (Figure 1), berenil and stilbamidine have a very strong minor groove binding mode in consecutive sequences of three or more AT base pairs (Zimmer and Wahnert, 1986; Neidle et al., 1987:   Wilson, 1990).   Such complexes appear similar in structure and interaction energetics to the DNA complex of the aliphatic amidine, netropsin, which has been studied in molecular detail (Kopka et al., 1985; Coll et al., 1989).   Lown and co-workers (1988)   have   been   successful   in   engineering   varieties   of   sequence recognition properties into lexitropsin analogs of netropsin.

We have for sometime been interested in the design and synthesis of both intercalators and groove binding molecules with enhanced recognition properties   for   nucleic   acid   sequences   and   structures.   The   very   strong binding and relatively simple structure of the aromatic amidines makes this type of system especially attractive for preparation of analogs.   The highly detailed   structural   information   available   on   the   similar   oligonucleotide complexes of the dicationic amidines, netropsin and DAPI (Kopka et al., 1985;

B. Pullman and J. Jortner (eds.), Molecular Basis of Specificity in Nucleic Acid-Drug Interactions, 331–353.
© 1990 Kluwer Academic Publishers. Printed in the Netherlands.

Coll et al., 1989; Larsen et al., 1989), provides a paradigm for use in interpreting interaction variations with DNA which occur with structural analogs of these model systems.   It is particularly important to investigate analogs of DAPI in detail since we have recently found that, in addition to its well characterized minor groove complex in AT regions of DNA, DAPI binds strongly in GC regions of DNA by intercalation (Wilson et al., 1989a, 1990a,b). The switch in binding mode for DAPI is dependent on DNA sequence, but we have found that a range of other unfused aromatic compounds with terminal basic functions bind to both AT and GC sites of DNA by intercalation (Wilson et al., 1989b; Strekowski et al. 1987).   Clearly both DNA sequence and ligand structure are affecting the binding mode and sequence selectivity of unfused aromatic cations.

1.2   COMPOUND DESIGN

In order to develop a better understanding of the influence of ligand structure on the interaction of unfused aromatic dications with DNA, the diphenylfuran derivatives in Figure 1 were designed and synthesized.   These molecules have specific modifications in their structure and chemical groups which allow the systematic probing of their interactions with DNA.   The diphenylfuran amidine and imidazoline derivatives (Figure 1) are clearly similar to DAPI, but they have a hydrogen-bonding accepting group in the center of the molecule rather than the indole N-H donating group in DAPI. The importance and binding restrictions of the amidine group are probed with the imidazoline, amine and piperazylamide substitutions.   Steric restriction on the amidine function in its DNA complexes are probed with the chemically similar imidazoline.   The amine is actually smaller in size than the amidine but is significantly different in structure and charge delocalization. The piperazylamide moves the charges farther apart, changes the overall steric bulk of the system, and has a terminal group more similar to that of Hoechst 33258, another type of AT specific minor groove binding agent, than to DAPI.

For such relatively minor changes in structure, the molecules display an almost unbelievable variety of effects in their interactions with DNA. They vary significantly both in base pair specificity and in binding mode in their DNA complexes.   The apparently simple change, for example, of the amidine to the amine group with the diphenylfurans (Figure 1) converts the molecule from a very specific minor groove binding agent in AT sequences to an intercalator.   Such dramatic differences in binding mode and specificity for compounds so similar in structure are unprecedented.

2.   Materials   and   Methods

2.1   MATERIALS.

Synthesis of the amidine and imidazoline have been described   (Das and Boykin, 1977), the synthesis of the other two diphenylfurans will be reported elsewhere.   Purity of all compounds was verified by NMR and elemental analysis.   DNA polymer samples were sonicated, purified and dialysed against the desired buffer as previously described (Wilson et al., 1990a).   Closed

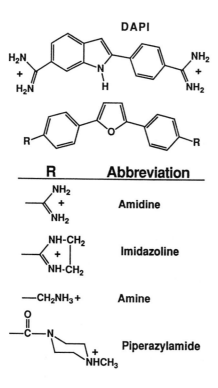

Figure 1: Structures of diphenylfuran derivatives.

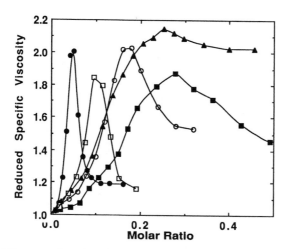

Figure 2: Viscometric titrations of closed circular supercoiled DNA with ethidium (●), amidine (▲), imidazoline (○), amine (□), and piperazylamide (■). The reduced specific viscosity ratio is plotted as a function of molar ratio of compound to DNA base pairs. Experiments were conducted in MES buffer at 30°C.

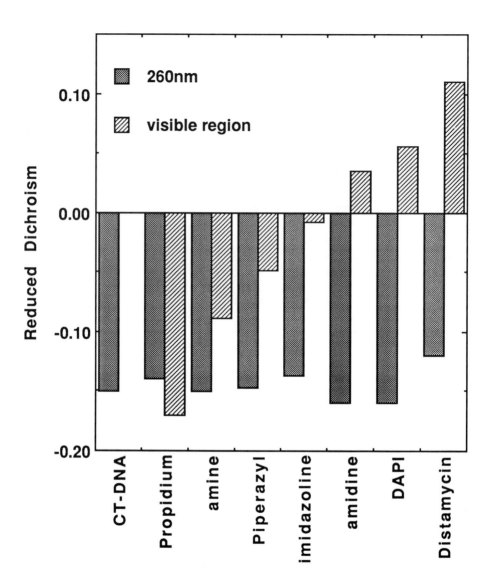

Figure 3: A histogram of the reduced dichroism for CT-DNA and CT-DNA complexes with propidium, amine, piperazylamide, imidazoline, amidine, DAPI, and distamycin. Flow dichroism results are given at the DNA maximum wavelength 260 nm (solid bar) and at the bound ligand maximum wavelength (diagonal line).

circular superhelical DNA was prepared as previously described (Jones et al., 1980). All studies were conducted in 0.01M MES (2-(N-morpholino) ethane-sulfonic acid), 10^{-3}M EDTA, pH 6.2 adjusted to the desired salt concentration by the addition of NaCl.

2.2   METHODS

Procedures for viscosity, linear dichroism, spectroscopy, equilibrium dialysis, stopped-flow kinetics and molecular modeling have recently been described (Wilson et al., 1990a). Modeling studies were conducted with the MACROMODEL program by Professor Clark Still of Columbia University using the AMBER force field developed by Kollman and co-workers (Weiner et al., 1986).

3.   Results

3.1.   SUPERHELICAL DNA VISCOSITY.

Titration of closed circular supercoiled DNA with ethidium and the furan derivatives are compared in Figure 2A. Ethidium and the amine give the unwinding, viscosity maximum at the open circular state and rewinding with viscosity decrease expected of a classical intercalator (Waring, 1981; Wilson, 1990a). The other derivatives also produce intercalation type titration curves with supercoiled DNA, however, the curves are generally more broad and are shifted to higher ratios than with ethidium and the amine. The variation in the position of the viscosity maximum can occur due to differences in unwinding angle, to multiple binding modes (eg. groove binding modes in AT sequences) and to differences in binding constant (Waring, 1981). Clearly, however, the unwinding of supercoiled DNA by these compounds suggests that they all intercalate with DNA in at least some sequences.

3.2.   LINEAR DICHROISM.

Linear dichroism studies on flow oriented calf thymus DNA alone and complexed with the furan derivatives were conducted at 260 nm where the absorption due to DNA dominates, and at the maximum wavelengths above 300 nm for the DNA complexes (Figure 3). The results are compared to results for the minor groove binding compounds DAPI and distamycin and with the intercalator propidium at a bound ratio of 0.05 (compound/base pair). Calf thymus DNA alone and complexed with all compounds exhibits a significant negative reduced dichroism ($^{red}D = A_{||} - A_{\perp} / A_o$; where $A_{||} - A_{\perp}$ is the difference in absorbence for light polarized parallel and perpendicular to the direction of flow and A_o is the absorbence of the stationary sample) at 260 nm. The furan amidine, DAPI, and distamycin have positive ^{red}D values above 300 nm where only the induced dichroism of the bound molecule is being monitored. On the other hand, the amine and piperazylamide, and propidium have negative ^{red}D values above 300 nm. The imidazoline has ^{red}D values close to zero. At higher ratios of compound to DNA, the ^{red}D values of the amidine and imidazoline compounds become more negative.

top

middle

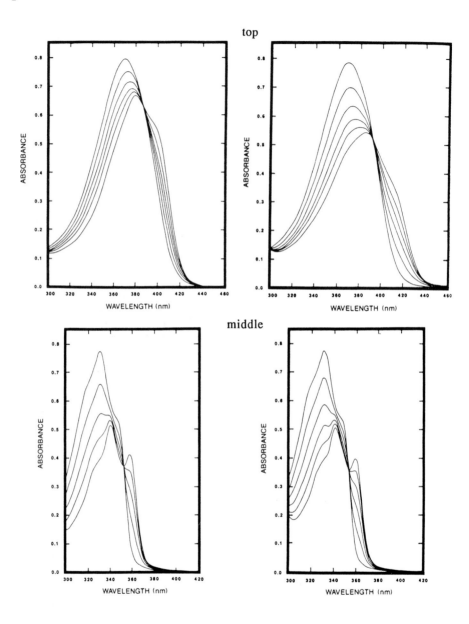

bottom

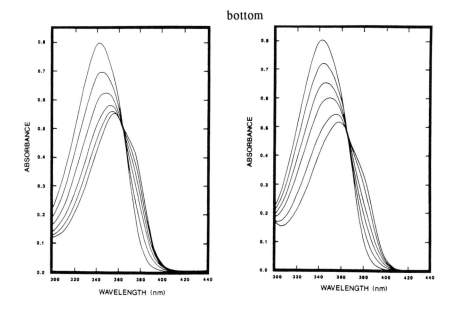

Figure 4:   Spectrophotometric titrations of diphenylfuran compounds with polyd(A-T)$_2$ (left side) and polyd(G-C)$_2$ (right side).   The top spectra are for the imidazoline, the middle are for the amine and the bottom are for the piperazylamide.   Titrations with the amidine were very similar to those with the imidazoline.   All titrations were conducted in 1 cm cell in MES buffer at 25°C.

In the top panels the concentrations are as follows: 1.87×10^{-5}M   imidazoline, polyd(A-T)$_2$ concentrations   in base pairs of zero, 1.2×10^{-5}, 2.4×10^{-5}, 3.5×10^{-5}, 4.7×10^{-5}, and 7.0×10^{-5} and polyd(G-C)$_2$ concentrations in base pairs of zero, 1.0×10^{-5}, 2.0×10^{-5}, 3.0×10^{-5}, 3.9×10^{-5}, and 5.9×10^{-5} respectively from the top to the bottom curves at 370 nm.

In the middle panels the concentrations are 2.74×10^{-5}M amine,   polyd(A-T)$_2$ concentrations of zero, 1.2×10^{-5}, 2.4×10^{-5}, 3.5×10^{-5}, and 5.8×10^{-5} and polyd(G-C)$_2$ concentrations of zero, 1.0×10^{-5}, 2.0×10^{-5}, 3.0×10^{-5}, 3.9×10^{-5}, and 5.9×10^{-5} respectively from the top to the bottom curves at 331 nm.

In the bottom panels the concentrations 2.27×10^{-5}M piperazylamide,   polyd(A-T)$_2$ concentrations of zero, 1.2×10^{-5}, 2.4×10^{-5}, 3.5×10^{-5}, 4.7×10^{-5}, and 7.0×10^{-5} and polyd(G-C)$_2$ concentrations of 1.0×10^{-5}, 2.0×10^{-5}, 3.0×10^{-5}, 3.9×10^{-5}, 4.9×10^{-5} and 7.9×10^{-5} respectively from the top to the bottom curves at 342 nm.

3.3   SPECTRAL CHANGES: DNA POLYMERS

The viscosity and linear dichroism results strongly suggest that the diphenylfurans can bind to heterogeneous DNA by both intercalation and groove-binding modes but they do not provide specific information on how the DNA sequence influences the binding mode. To obtain information on the sequence-specific interactions of these four compounds with DNA, we have first focused on their binding with the two well-characterized alternating sequence polymers, polyd(A-T)$_2$ and polyd(G-C)$_2$.

3.3.1 *UV-Visible Absorption.* Absorption spectral results for titration of the furan derivatives with the alternating sequence polymers polyd(A-T)$_2$ and polyd(G-C)$_2$ are shown in the left and right panels, respectively, of Figure 4. The four free compounds have significantly different absorption spectra. The amidine and imidazoline derivatives have similar extended conjugated systems and they have similar spectra. The shape of the absorption spectrum for the piperazylamide is similar to that for the amidine and the imidazoline but is shifted to shorter wavelength (Figure 4). The amine derivative has only the diphenylfuran conjugated system and its spectrum has a quite different shape.
Addition of both polyd(A-T)$_2$ and polyd(G-C)$_2$ causes a decrease in extinction coefficients at the compound maximum wavelength (hypochromic effects) and shifts of the spectra to longer wavelengths for all compounds (Figure 4). At the concentrations of our experiments, all compounds have isosbestic points in their DNA titrations. Comparison of the bound spectra in Figure 4 reveals an interesting point, the amidine and imidazoline derivatives have very different extinction coefficients when bound to polyd(A-T)$_2$ and to polyd(G-C)$_2$ while the extinction coefficients of the amine and piperazylamide are similar when bound to the two polymers. At the wavelength maximum, the amidine and imidazoline have ~20% hypochromicity when bound to polyd(A-T)$_2$ and ~45% when bound to polyd(G-C)$_2$. The amine and piperazylamide compounds have 47$\pm$5% hypochromicity when bound to both polymers. Thus, the hypochromicity on binding of the amidine and imidazoline to polyd(A-T)$_2$ is anomolously low when compared to all of the other complexes. These results indicate that in the GC complex the diphenylfuran aromatic system interacts strongly with base pairs. The amine and the piperazylamide show similar interactions in their complexes with AT base pairs. The small hypochromicity change for the amidine and imidazoline on complex formation at AT base pairs is characteristic of groove-binding interactions.

3.3.2 *Fluorescence.* Fluorescence titrations of the diphenylfuran amidine with polyd(A-T)$_2$ and polyd(G-C)$_2$ are shown in (Figure 5). The free amidine has a strong fluorescence maximum at 465 nm. Addition of polyd(A-T)$_2$ causes a slight decrease in the uncorrected fluorescence and a significant shift of the peak to lower wavelengths (Figure 5). Addition of polyd(G-C)$_2$, on the other hand, causes a very slight shift of the peak maximum to longer wavelengths and causes a large decrease in fluorescence intensity (Figure 5).

Fluorescence titration results with the imidazoline are very similar to those for the amidine.

The diphenylfuran amine has a very different fluorescence spectrum (Figure 5) and addition of both polyd(A-T)$_2$ and polyd(G-C)$_2$ cause large decreases in fluorescence intensity with little change in the peak wavelengths. The decreases in fluorescence intensity at a given ratio of polymer to amine are larger for addition of polyd(A-T)$_2$ than for polyd(G-C)$_2$.

The piperazylamide compound has a fluorescence spectrum very similar to that for the amidine (Figure 5). Addition of polyd(A-T)$_2$ to this compound causes slight decreases in fluorescence intensity and shifts of the fluorescence maximum to a shorter wavelength. Addition of polyd(G-C)$_2$ causes pronounced decreases in fluorescence intensity with little shift in the wavelength of maximum fluorescence. Thus, addition of polyd(G-C)$_2$ to all four diphenylfurans causes similar effects, a large decrease in fluorescence intensity with little shift in the peak wavelengths. Addition of polyd(A-T)$_2$ to the amine results in similar spectral changes. Addition of polyd(A-T)$_2$ to the other three diphenylfurans causes quite different spectral changes, slight decreases in fluorescence intensity and shifts of the fluorescence maximum to shorter wavelengths.

3.3.3 *Circular Dichroism (CD)*.

None of the diphenylfuran derivatives has an intrinsic CD spectrum. Addition of polyd(G-C)$_2$ to the amidine results in a weak positive CD signal at 370 nm (Figure 6). With polyd(A-T)$_2$ the spectral intensity is greater, the spectral shape changes significantly with ratio and two different limiting spectra are obtained. At molar ratios above 0.2, the spectrum has a band centred at 390 nm with a shoulder at 370 nm. At ratios below 0.2 the spectrum is shifted to lower wavelength and the CD band is centered at 370 nm with a shoulder at 390nm. The imidazoline gives spectral changes similar to those obtained for the amidine. For the amine, addition of either polymer did not give any significant induced CD spectrum.

With the piperazylamide compound, addition of polyd(G-C)$_2$ results in a weak induced CD signal at 350nm (Figure 6). Addition of polyd(A-T)$_2$ to this compound results in a similar but larger CD signal which is centered at 355 nm. Thus, two different types of behaviour are obtained with these four compounds. Weak to no induced CD signals are seen on binding of all compounds to polyd(G-C)$_2$ and for the amine and piperazylamide on binding to polyd(A-T)$_2$. On the other hand, strong, ratio dependent CD spectra are observed when the amidine and imidazoline bind to polyd(A-T)$_2$.

3.3.4 *NMR*.

As in our previous studies with DAPI and other unfused aromatic cations (cf. Wilson et al., 1988; 1989a,b; 1990a,b), NMR results provide key information for determination of the influence of DNA sequence on binding mode. The shifts of the aromatic protons of the four diphenylfurans on binding to polyd(A-T)$_2$ and polyd(G-C)$_2$ are summarized in Table I. Three distinct types of NMR behaviour are obtained for the four compounds indicating that they have three distinct types of binding behaviour. The aromatic protons of the amidine and imidazoline exhibit small, mixed upfield and downfield shifts on binding to polyd(A-T)$_2$ and significant upfield shifts

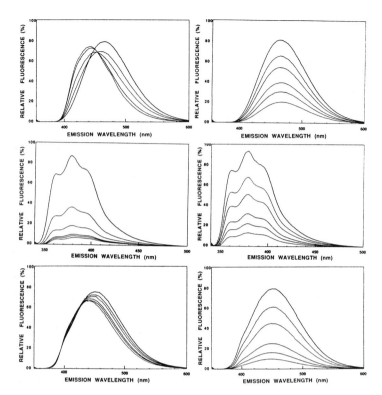

Figure 5: Fluorescence emission spectral titrations of diphenylfuran derivatives $(6 \times 10^{-6} M)$ with polyd(A-T)$_2$ (left side) and polyd(G-C)$_2$ (right side). The top panels are for the amidine, the middle are for the amine and the bottom are for the piperazylamide. The imidazoline titration was similar to that for the amidine. All titrations were conducted at 25°C in MES buffer with different salt concentrations, higher salt concentrations being used for the compounds which bind more strongly (Table II).

The free amidine has a strong fluorescence maximum at 465 nm (top spectrum). Molar ratios of the amidine to polyd(A-T)$_2$ in base pairs are 0.92, 0.46, 0.23, and 0.11 (MES buffer with 0.3M NaCl) respectively from the second to the bottom curves at 465 nm. Molar ratios of the amidine to polyd(G-C)$_2$ in base pairs are 0.92, 0.46, 0.23, 0.18 and 0.11 (MES buffer with 0.1M NaCl).

The free amine has a strong fluorescence maximum at 378 nm. Molar ratios of the amine to both polymers in base pairs are 0.92, 0.46, 0.23, 0.18 and 0.11 respectively from the second to the bottom curves at 378 nm (MES buffer with 0.05M NaCl).

The free piperazylamide (bottom panels; $6 \times 10^{-6} M$) has a strong fluorescence maximum at 450 nm. Molar ratios of the piperazylamide to polyd(A-T)$_2$ in base pairs are 0.92, 0.46, 0.23, and 0.11 respectively from the second to the bottom curves at 450 nm (MES buffer with 0.05M NaCL). Molar ratios of the piperazylamide to polyd(G-C)$_2$ in base pairs are 0.92, 0.46, 0.23, 0.18 and 0.11 respectively from the second to the bottom curves at 450 nm (MES buffer with no added NaCl).

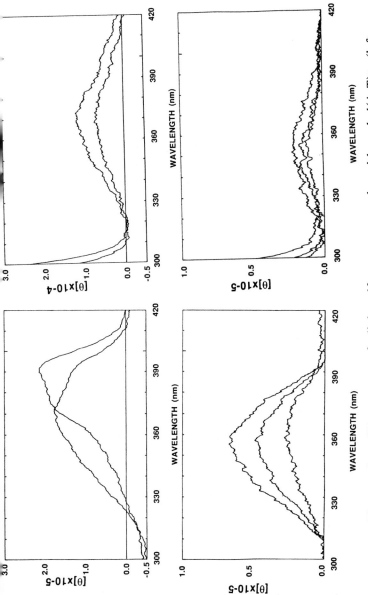

Figure 6: CD spectra of diphenylfuran compounds with polyd(A-T)2 (left side) and polyd(G-C)2 (right side). The top panels are for the amidine and the bottom panels for the piperazylamide.
Two ratios of amidine/polymer in base pairs are plotted: 0.34 which has CD maximum at 390 nm and 0.04 which has CD maximum at 370 nm for polyd(A-T)2; and 0.45 (upper curve) and 0.20 (lower curve) for polyd(G-C)2.
Three ratios of piperazylamide/polymer are plotted: 0.45, 0.23, and 0.11 respectively from the bottom to the top curves at 355 nm for both polyd(A-T)2 and polyd(G-C)2.

Table I. Diphenylfuran aromatic proton chemical shifts for the compounds free and bound to poly d(A-T)$_2$ and poly d(G-C)$_2$

R	H	No DNA *M.R.= 0 δ	Poly d(A-T) M.R. = 0.3 δ	Δδ	Poly d(G-C) M.R. = 0.3 δ	Δδ
Amidine	2',6'	8.05	7.90	-0.15	7.27	-0.78
	3',5'	7.89	7.79	-0.10	7.40	-0.49
	3,4	7.22	7.32	+0.10	6.34	-0.88
Imidazoline	2',6'	7.99	7.75	-0.24	7.22	-0.77
	3',5'	7.85	7.69	-0.16	7.35	-0.50
	3,4	7.20	7.35	+0.15	6.42	-0.78
Amine	2',6'	7.92	7.00	-0.92	6.94	-0.98
	3',5'	7.56	7.16	-0.40	7.03	-0.53
	3,4	7.05	6.07	-0.98	6.11	-0.94
Piperazyl- amide	2',6'	7.97	7.32	-0.65	7.58	-0.39
	3',5'	7.59	7.25	-0.34	7.36	-0.23
	3,4	7.12	6.54	-0.58	6.66	-0.46

*M.R.: molar ratio of compound to DNA base pairs

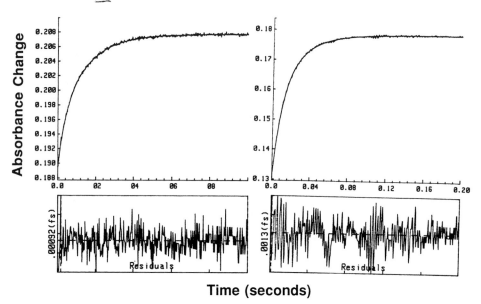

Time (seconds)

Figure 7: Stopped flow kinetic traces for the SDS-driven dissociation of the amidine from polyd(A-T)$_2$ (left side) and from polyd(G-C)$_2$ (right side). The experiments were conducted at 20°C in MES buffer with [Na$^+$] = 0.035 at a ratio of 1:20 compound to polymer base pair. The concentration of the compound after mixing was 1.25×10^{-5} M. The smooth lines in the panels are the two exponential fit values to the experimental data. Residual plots for both data sets are shown under the experimental plots. Note that the time scales in the two experiments differ by a factor of 50.

on binding to polyd(G-C)$_2$, as previously observed with DAPI   (Wilson et al., 1990a,b) with these two polymers.   The aromatic protons of the amine, however, exhibit large and similar upfield shifts on binding to both polyd(A-T)$_2$ and polyd(G-C)$_2$.   The piperazylamide aromatic protons also have upfield shifts on binding to both polymers but the magnitude of the shifts is significantly less, ~ two-thirds to one-half, that observed with the amine.   The three types of behaviour on binding of these compounds to the two polymers can easily be seen in Table I: (i) small shifts of the amidine and imidazoline with polyd(A-T)$_2$;   (ii)   large upfield shifts of the amidine and imidazoline with polyd(G-C)$_2$ and of the amine with both polymers;   (iii) smaller upfield shifts of the piperazylamide with both polyd(A-T)$_2$ and polyd(G-C)$_2$.

3.4   STOPPED-FLOW DISSOCIATION KINETICS: DNA POLYMERS

The absorption spectral changes observed for the diphenylfurans on binding to polyd(A-T)$_2$ and polyd(G-C)$_2$ can also be used to follow the kinetics of binding.   As with the NMR results, three distinct types of behaviour are observed with the dissociation kinetics results for the four diphenylfurans of Figure 1.   The combination of NMR and dissociation kinetics results provides the critical information needed to define the multiple, different interaction modes of the four different diphenylfurans.

Typical absorption versus time stopped-flow kinetics traces for the SDS-driven dissociation of the amidine from polyd(A-T)$_2$ and polyd(G-C)$_2$ are compared in Figure 7.   Two exponential fits to the data are also shown along with plots of residuals.   One exponential fits gave unsatisfactory residuals and significantly lower RMS deviations than the two exponential fits.   Going to three exponential fits did not significantly improve the residuals or RMS deviations.   With polyd(A-T)$_2$ the amplitude for the fast reaction accounts for 20-30% of the total under all conditions used while with polyd(G-C)$_2$ the fast phase amplitude was 50-60% of the total.   The dissociation rate increases by over a factor of 30 in going from polyd(A-T)$_2$ to polyd(G-C)$_2$.

To obtain more details on the mechanism of complex formation, the dissociation kinetics of the complexes of all of the diphenylfurans with polyd(A-T)$_2$ and polyd(G-C)$_2$ were analyzed in MES buffers of different salt concentrations   (Lohman, 1985;   Wilson et al., 1985).   For the purposes of comparison, a dissociation lifetime (τ) and an apparent rate constant (k_{app} = $1/\tau$) were calculated from the observed rate constants and amplitudes as suggested by Denny et al. (1985):

$$\tau = 1/ (A_1k_1+A_2k_2) \qquad\qquad (1)$$

where A and k values refer to the computer derived amplitudes and rate constants for the two exponential fits to the dissociation results.   The amplitudes and rate constants in the two exponential fits are highly correlated and are quite sensitive to slight changes in the noise and baselines. This is not true of the lifetimes or apparent rate constants calculated from eq. (1) and we feel that these values are most useful for comparing results for several molecules of a series such as the diphenylfurans. The $\log k_{app}$ values are plotted versus -log [Na$^+$] in Figure 8 and values for the dicationic

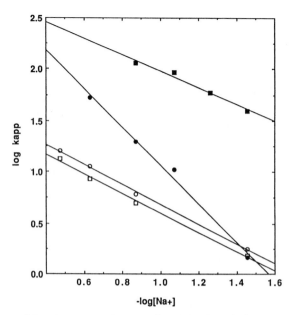

Figure 8: Plots of log k_{app} vs -log [Na$^+$] for dissociation of the amidine from polyd(A-T)$_2$, (●); polyd(G-C)$_2$, (■); and of propidium from polyd(A-T)$_2$, (○); polyd(G-C)$_2$, (□). Experiments were conducted in MES buffer at different ionic strengths in the manner described in Figure 7. The propidium results are from (Wilson et al. 1985b, 1986).

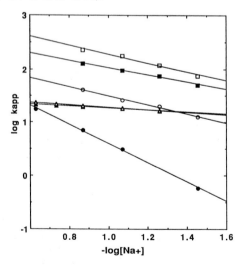

Figure 9: Plots of log k_{app} vs -log [Na$^+$] for dissociation of the imidazoline from polyd(A-T)$_2$, (●); polyd(G-C)$_2$, (○); the amine from polyd(A-T)$_2$, (■); polyd(G-C)$_2$, (□); and the piperazylamide from polyd(A-T)$_2$, (▲); polyd(G-C)$_2$, (Δ⁻) . Experiments were conducted in MES buffer at different ionic strengths in the manner described in Figure 7.

phenanthridinium intercalator, propidium, are included for reference. For the SDS driven dissociation of the amidine from polyd(G-C)$_2$ the slope is 0.8 $\pm$ 0.1 and with polyd(A-T)$_2$ a slope of 1.8 $\pm$ 0.1 is obtained. The slope for propidium with both polymers is ~0.8 (Figure 8).

Similar experiments were conducted with the imidazoline and the dissociation kinetics curves also required a two exponential fit. Relative to the amidine, the imidazoline dissociates approximately 3 times more slowly from polyd(A-T)$_2$ and from polyd(G-C)$_2$. With polyd(A-T)$_2$ the amplitude for the fast reaction accounts for 25-35% of the total under all conditions used while with polyd(G-C)$_2$ the fast phase amplitude was 65-75% of the total. The dissociation rate increases by over a factor of 30 in going from polyd(A-T)$_2$ to polyd(G-C)$_2$. Dissociation results for the imidazoline complex with polyd(A-T)$_2$ and polyd(G-C)$_2$ were also monitored at several salt concentrations (Figure 9) and with polyd(G-C)$_2$ the slope is 0.8 $\pm$ 0.1 and with polyd(A-T)$_2$ a slope of 1.9$\pm$ 0.1 is obtained.

SDS driven dissociation experiments with the amine also required a two exponential fit. With polyd(A-T)$_2$ the amplitude for the fast reaction accounts for 35-65% of the total under all conditions used while with polyd(G-C)$_2$ the fast phase amplitude was 40-70% of the total. The dissociation rate increases by only a factor of 2 in going from polyd(A-T)$_2$ to polyd(G-C)$_2$. Dissociation reactions of the amine complex with polyd(A-T)$_2$ and polyd(G-C)$_2$ were also monitored at several salt concentrations (Figure 9) and with polyd(G-C)$_2$ the slope is 0.7 $\pm$ 0.1 and with polyd(A-T)$_2$ a slope of 0.8 $\pm$ 0.1 is obtained.

The piperazylamide gives dissociation kinetics behaviour which is distinctly different than observed with the other diphenylfurans. First, the dissociation profiles for the piperazylamide compound with both polyd(A-T)$_2$ and polyd(G-C)$_2$ are fitted quite well by single exponential curves. Second, with the other diphenylfurans the rates of dissociation are faster with polyd(G-C)$_2$ than with polyd(A-T)$_2$, but with the piperazylamide the rates are the same within experimental error for the two polymer complexes. Third, the slopes in logk$_{app}$ versus log[Na$^+$] plots for the piperazylamide complex are 0.25 $\pm$0.05 with both polyd(A-T)$_2$ and polyd(G-C)$_2$. These slopes are only ~ one-half those observed with the amine. Thus, the four diphenylfurans give three distinct types of behaviour in their logk$_{app}$ versus log[Na$^+$] plots and all three types are illustrated in Figure 9. Large slopes, 1.7-1.8, are observed in the plots for dissociation of the amidine and imidazoline complexes with polyd(A-T)$_2$. Slopes of 0.7-0.8 are observed for the amidine and imidazoline complexes with polyd(G-C)$_2$ and with the amine complexes with both polymers. With the piperazylamide smaller slopes, ~0.25, are observed with both polymers.

3.5   RELATIVE BINDING STRENGTHS: DNA POLYMERS

Spectral changes on complex formation and equilibrium dialysis were also used to measure the binding constants (Keq) for the four diphenylfurans with polyd(A-T)$_2$ and polyd(G-C)$_2$. As previously observed with DAPI (Wilson et al., 1990a), the amidine and imidazoline display curve shapes indicative of cooperative binding in their Scatchard plots with polyd(A-T)$_2$. No such behaviour is seen with polyd(G-C)$_2$ or with either polymer with the amine or

Table II: Binding Constants at AT/GC Sites

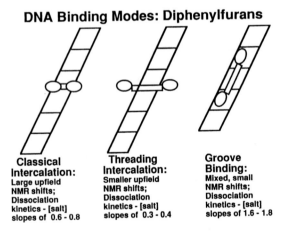

R	~K^*_{AT}	~K^*_{GC}
NH$_2$ / +NH$_2$	1.0	1.0
NH-CH$_2$ / +NH-CH$_2$	4.0	5.0
—CH$_2$NH$_3$+	0.01	0.8
—C(=O)—N / +NHCH$_3$	0.0003	0.01

K^* is normalized to the amidine

$K_{AT}/K_{GC} \sim 400$ for the amidine

DNA Binding Modes: Diphenylfurans

Classical Intercalation:
Large upfield NMR shifts; Dissociation kinetics - [salt] slopes of 0.6 - 0.8

Threading Intercalation:
Smaller upfield NMR shifts; Dissociation kinetics - [salt] slopes of 0.3 - 0.4

Groove Binding:
Mixed, small NMR shifts; Dissociation kinetics - [salt] slopes of 1.6 - 1.8

Figure 10: Cartoon for the three possible interaction modes for the diphenylfuran derivatives of Figure 1. The diphenylfuran ring system is shown as a rectangle with terminal basic groups (ovals). The DNA helix is shown as an unwound ladder. In the classical model for intercalation both substituents are in the major groove and in the threading model one substituent is in each groove. In the groove binding model the molecule lies along the floor of the minor groove and the diphenylfuran fits between the walls of the groove. Qualitative predictions for induced NMR chemical shift changes on binding and for log k_{app}-log [Na$^+$] slopes for dissociation reactions are given for each model.

with the piperazylamide compounds. For consistency of analysis, the amidine and imidazoline compounds with polyd(A-T)$_2$ will be compared to the other results at bound ratios past the peak in the cooperative binding curve (eg. past the cooperative transition region). The binding constants for all compounds with both polymers decrease as the salt concentration is increased and the slopes in logKeq versus log[Na$^+$] plots are ~2 as expected for the interaction of dications with DNA.

The relative magnitudes of the binding constants are compared in Table II. As can be seen from the Table, the binding constants for these quite similar compounds vary over several orders of magnitude. Binding of the amidine and imidazoline to polyd(A-T)$_2$ is strongest, a Keq of 6-7x10^7 M^{-1}, for the imidazoline and ~one-half that much for the amidine in 0.1M [Na$^+$]. Binding of the amidine and imidazoline to the GC polymers is 300-400 times weaker than for the AT polymer but is in the range observed for common dicationic intercalators such as the acridine, quinacrine (Wilson et al., 1989a). Binding of the amidine is slightly weaker than with DAPI with both polymers while binding of the imidazoline is slightly stronger. The piperazylamide binds most weakly to both polyd(A-T)$_2$ and polyd(G-C)$_2$ (Table II). The amine binds approximately a factor of 100 more weakly to polyd(A-T)$_2$ than the amidine but binds very similarly to the amidine with polyd(G-C)$_2$. All compounds bind more strongly to polyd(A-T)$_2$ than to polyd(G-C)$_2$.

4.   Discussion

Interpretation of experimental results for the diphenylfurans in terms of binding models for complex formation with DNA samples of different base pair composition requires predicted results for the different models. We have evaluated the results for the four diphenylfurans in terms of the three general interaction models illustrated in Figure 10: classical intercalation; a threading model for intercalation and minor-groove binding (Wilson, 1990). Classical intercalators such as proflavine and ethidium typically bind with their long axis approximately parallel to the long axes of the base pairs at the binding site. This orientation results in large upfield shifts of the aromatic protons of the bound molecule (Wilson and Li, 1990). Such intercalators also typically dissociate rapidly from their DNA complexes and, for dications, have logk$_{app}$ versus log[Na$^+$] slopes of 0.6-0.8 (Wilson, 1985; Wilson et al., 1990a). Minor groove binding dications such as netropsin and DAPI interact strongly in AT rich regions of DNA and dissociate more slowly than simple intercalators (Wilson et al., 1990a). These molecules are much less affected by the base pair aromatic ring currents, and chemical shift changes in their aromatic protons on complex formation are significantly smaller than with intercalators (small, mixed upfield and downfield shifts are frequently seen in groove binding) (Wilson and Li, 1990). Slopes in logk$_{app}$ versus log[Na$^+$] plots for dications which bind in the minor groove are 1.6-1.8 (Wilson, 1985; 1989a; Wilson et al., 1990a).

The threading intercalation model is characteristic of aromatic molecules such as nogalamycin and naphthalene diimides (Yen et al., 1982) which have bulky and/or charged groups which must pass through the intercalation site from one groove to the other for final complex formation (Wilson, 1990). Shifts of the aromatic protons of these type molecules are typically smaller than for classical intercalators and $\log k_{app}$ versus $\log[Na^+]$ slopes for threading dications, such as the naphthalene diimides, are ~0.3 (Tanious et al., 1990). These slopes are only about one-half that for classical dicationic intercalators and the difference is presumably due to the fact that ionic interactions are released during dissociation in a concerted fashion with traditional intercalators but are released in a sequential fashion with the threading molecules. The combination of NMR analysis of complex formation with DNA polymers and of analysis of the effects of salt concentration on dissociation kinetics provides a high resolution method for screening a series of analogs for variation in binding mode as a function of DNA sequence. These are the only definitive methods for determination of variation in binding mode at different DNA sequences.

The four diphenylfurans of Figure 1 have the same aromatic nucleus and molecular mechanics calculations indicate that, in the low energy conformation, the three unfused aromatic rings are essentially planar in all four compounds. The aromatic systems do have torsional freedom about the phenyl-furan bonds, and in NMR spectra in water the 2',6' and 3',5' protons are equivalent down to 0 °C. These NMR results indicate that fast rotation about the phenyl-furan bonds occurs in the unbound molecule. In addition to having the same ring system, the four derivatives are all dications at normal pH values. $LogKeq-log[Na^+]$ plots indicate that all four compounds form two ion pairs in their DNA complexes. This result indicates that similar electrostatic interactions and ion release are obtained for all compounds on complex formation with DNA. The dramatic differences in the interactions of the diphenylfurans with $polyd(A-T)_2$ and $polyd(G-C)_2$ must then be due to differences in the type, shape, and positioning of their terminal basic groups.

Molecular mechanics studies of DAPI and the amidine derivative (Figure 1) indicate that the amidine plane is twisted with respect to the attached phenyl plane by ~20°. The amidine is clearly similar to DAPI (Figure 1) but with the indole N-H hydrogen-bond donating group of DAPI replaced by the furan oxygen hydrogen-bond accepting group. As with DAPI (Wilson et al., 1990a), the amidine binds very strongly and with positive cooperativity to $polyd(A-T)_2$. Both the amidine and DAPI have small chemical shift changes in their aromatic protons on binding to the AT polymer and have $\log k_{app}$ versus $\log[Na^+]$ of 1.7-1.8 on dissociation from the $polyd(A-T)_2$ complex. As discussed above, these latter two observations are key characteristics of groove-binding interactions. The imidazoline compound is larger, slightly more planar, and has more restrictions on its hydrogen bonding groups than the amidine. It actually binds slightly more strongly and dissociates more slowly than the amidine with $polyd(A-T)_2$. The imidazoline and the amidine have very similar $\log k_{app}$ versus $\log[Na^+]$ slopes and aromatic proton chemical shift changes on binding to $polyd(A-T)_2$. Clearly both of these derivatives bind very similarly to $polyd(A-T)_2$ in the well established AT specific minor groove complex seen with netropsin and DAPI (Kopka et al., 1985; Larsen et al., 1989). The changes from indole (DAPI) to phenyl-furan or from amidine to imidazoline make little difference in the binding strength or mode of these compounds with $polyd(A-T)_2$. These observations indicate that the replacement of the indole N-H of DAPI and that the extra size and reduced hydrogen bonding capacity of the imidazoline group relative to the amidine are not detrimental to complex formation at AT minor groove sites.

The unwinding of supercoiled DNA by all of the diphenylfurans shows that they can bind to at least some DNA sequences by intercalation (Waring, 1981). The groove-binding mode for the amidine and imidazoline at AT sequences suggests that the intercalation binding must occur in GC or mixed sequences. The amidine and imidazoline have significant upfield shifts for their aromatic protons on complex formation with polyd(G-C)$_2$ (Table I) and $\log k_{app}$ versus $\log[Na^+]$ slopes of 0.6-0.8 for dissociation of their complexes with polyd(G-C)$_2$. Such results have been previously observed with the DAPI complex with polyd(G-C)$_2$ (Wilson et al., 1989a, 1990a) and they fit the criteria discussed above for the traditional or classical intercalation model (Figure 10). The imidazoline binds slightly more strongly to the GC polymer than DAPI which binds slightly more strongly than the amidine (Table II). A similar order of binding was observed with polyd(A-T)$_2$ although binding to the AT polymer is 300-400 times stronger than to the GC polymer. It should be emphasized at this point that DAPI and the diphenylfuran amidine and imidazoline derivatives bind to GC sequences in DNA in the range normally seen for strong binding dicationic intercalators such as the acridine, quinacrine (Wilson et al., 1989a, 1990a). The minor-groove binding mode for these type compounds at AT sequences is simply particularly strong and dissociation rates are much slower from the AT than from the GC complexes under the same conditions. This strong binding and slow dissociation probably accounts for the favourable footprinting patterns in AT regions observed at low ratios of these small molecules bound to DNA (Portugal and Waring, 1988; Wilson et al., 1990a).

As indicated, molecular mechanics calculations on DAPI, the amidine and the imidazoline indicate that the low energy conformations are relatively planar. With the amine, however, the predicted stable conformation is with the C-N bond of the methylene-amino group ~perpendicular to the phenyl plane. Steric clash between the methylene and phenyl protons leads to this out-of-plane twist. There is also no long range conjugation of the amino group with the diphenylfuran aromatic system, as there is with the amidine and imidazoline groups, to help maintain a planar conformation. Even with this difference in conformation, however, the amine and amidine bind very similarly to polyd(G-C)$_2$. The amidine binds slightly more strongly and dissociates more slowly than the amine but they have similar $\log k_{app}$ versus $\log[Na^+]$ slopes and NMR chemical shift changes on complex formation.

Given the very similar binding of the amidine and amine at GC sites, the dramatic differences between their complexes at AT sites provide fundamental information for understanding how very similar ligands select different binding modes at particular DNA sequences. The amine binds to AT sites only slightly more strongly than to GC sites and the complex with AT has NMR and kinetics characteristics of intercalation, but not of a groove binding mode as observed with the amidine. The complete change in binding mode at AT sites with relatively small substituent variations emphasizes two important points: (i) the amidine or imidazoline groups are key structures in the selection of groove-binding interactions of DAPI and diphenylfurans; and (ii) the stereoelectronic restrictions on groove-binding interactions are much more stringent than in intercalation complexes. This no doubt accounts for the fact that most intercalators show weaker binding, faster dissociation and much less specificity in their DNA complexes than groove-binding molecules such as DAPI and netropsin.

Molecular mechanics calculations with the piperazylamide indicate that the amide plane is twisted to an angle of ~50° with respect to the attached phenyl plane. This twist value is in good agreement with experimental studies on similar compounds (Baumstark et al., 1987). The molecular twist, which is in contrast to the more planar amidine and imidazoline compounds, is caused by steric clash of protons on the piperazylamide group with those on the phenyl ring in a planar conformation. The piperazylamide has a conjugated substituent attached to the phenyl ring, as with the amidine and imidazoline, and its spectral characteristics are more similar to those two compounds than to the amine. The piperazyl substituent is the same as the cationic group on the AT specific groove-binding molecule, Hoechst 33258. Our initial ideas were that the piperazylamide derivative might be a minor groove-binding agent with some enhanced GC specificity due to the furan and piperazyl groups. The results with both poly d(A-T)$_2$ and poly d(G-C)$_2$, however, indicate that this molecule chooses neither the minor-groove binding mode nor traditional intercalation but instead binds to both polymers by the threading intercalation mode.

The piperazylamide has small upfield aromatic proton chemical shift changes (Table 1) and low log kapp -log[Na$^+$] slopes for both its complex with poly d(A-T)$_2$ and with poly d(G-C)$_2$ (Figure 9). The piperazylamide thus, binds to both the AT and GC polymers by intercalation, as with the amine, but it binds to both significantly more weakly than the amine. Even though it has the lowest binding constant of all four compounds (Table II) it has anomolously slow dissociation kinetics (Figure 9). It dissociates significantly more slowly from both poly d(A-T)$_2$ and poly d(G-C)$_2$ then the amine. This combination of results suggests that dissociation of the piperazylamide from DNA is inhibited by some step not involved with dissociation of the amine. This is exactly what is expected for the threading model of intercalation. In this model the DNA must open to allow one piperazylamide group to slide between base pairs before the compound can bind or dissociate (Figure 10). The requirement for such a large amplitude breathing motion of the double helix significantly reduces the association and dissociation rate contents. The threading intercalation binding model is supported by the NMR chemical shift results and log kapp -log[Na$^+$] slopes.

The four diphenylfurans of Figure 1, thus, display three completely different types of binding interactions in their DNA complexes. The amidine and imidazoline display the strongest binding in AT specific minor groove complexes which are similar to the DAPI complex in AT regions of DNA. The amidine is similar in structure to DAPI but, in addition to the indole to phenyl-furan change, has greater molecular curvature then DAPI. We believe that the similar binding strength of DAPI and the amidine in AT sequences, even with loss of the indole N-H hydrogen-bonding group in the furan, is due to a more optimum fit of the diphenylfuran curvature with the DNA groove than can occur with DAPI (cf. Goodsell and Dickerson, 1986). Thus, conversion of the indole N-H to a furan might lower the binding in AT sequences but the more optimum curvature of the amidine would allow a better fit of that molecule into the DNA minor groove and, perhaps, better hydrogen-bonding of the amidine groups than can occur with DAPI. The compensating effects result in very similar binding of DAPI and the diphenylfuran amidine in AT sequences.

The diphenylfuran amidine and amine have similar hydrogen-bonding potential with AT base pairs but, while the amidine binds strongly in the minor groove in AT regions, the amine binds much more weakly in an intercalation complex. Due to intramolecular steric clash, however, the amine groups are rotated out of the diphenylfuran plane and are, thus, not in the optimum position for hydrogen-bonding to DNA. This clearly seriously lowers the interaction free energy for the minor groove mode and intercalation becomes the favoured binding mode. The amine, thus, appears to intercalate in all DNA sequences with very low base pair specificity. The strong, highly specific, minor-groove binding mode in AT regions, thus, requires molecules with hydrogen-bond donating groups located at the correct molecular positions and with the molecular twist set to a fine tolerance (Goodsell and Dickerson, 1986). The molecule may have the correct curvature of its unfused ring system, but if the hydrogen-bond donating groups are not correctly positioned, the minor groove binding free energy drops considerably.

As with the amine, the hydrogen-binding N-H groups of the diphenylfuran piperazylamide are moved away from their optimum positions for the AT specific minor groove complex. This suggests that the piperazylamide should bind at both AT and GC sequences by intercalation, and that is exactly what is observed. With the piperazylamide, however, the intercalation complex is entirely different from the amine intercalation complex. Molecular modeling studies indicate that DAPI (Wilson et al., 1990a), the amidine and the amine bind in intercalation complexes (only at GC sequences with DAPI and the amidine) with both of their charged groups oriented into the major groove. This is the type of intercalation geometry favoured by many unfused aromatic intercalators (Wilson et al., 1988; Wilson, 1990). The piperazylamide, however, binds in both the AT and GC sequences of DNA by the threading intercalation model (Figure 10). The classical and threading intercalation models are available to all intercalators but most intercalators examined to date bind with their long axis approximately parallel to the long axis of the base pairs at the intercalation site in a classical mode. The increased length of the piperazylamide system apparently makes this molecule too long to fit optimally into a classical intercalation geometry and it selects a threading intercalation binding mode.

Based on the results with these four unfused aromatic molecules, we clearly see that binding mode selection is extremely sensitive to both ligand structure and substituents as well as DNA sequence. We are now in a position to rank all of the binding modes in order of selection by a particular ligand. For example, with the diphenylfuran amidine the order is: AT-minor groove > GC-classical intercalation ~ AT-classical intercalation > GC-threading intercalation ~ AT-threading intercalation > GC-minor groove ~ simple electrostatic interactions. The order can change significantly with small changes in ligand substituents. The diphenylfuran amine, for example, binds more strongly by intercalation than by groove-binding at AT sites. This information will be extremely valuable in designing new molecules to select specific sequences and structures in their DNA complexes.

ACKNOWLEDGMENT. This work was supported by NIH Grant NIAID AI-27196.

REFERENCES

Baumstark, A. L., Balakrishnan, P., Dotrong, M., McCloskey, C. J., Oakley, M. G. and Boykin, D. W. (1987) *J. Am. Chem. Soc. 109*, 1059.

Coll, M., Aymami, J., van der Marel, G. A., van Boom, J. H., Rich, A., and Wang, A. H.-J. (1989) *Biochemistry 28*, 310.

Das, B. P. and Boykin, D. W. (1977) *J. Med. Chem. 20*, 531.

Denny, W. A., Atwell, G. J., Baguley, B. C., and Wakelin, L. P. G. (1985) *J. Med. Chem. 28*, 1568.

Goodsell, D. and Dickerson, R .E. (1986) *J. Med. Chem. 29*, 727.

Jones, R. L., Lanier, A. C., Keel, R. A., and Wilson, W. D. (1980) *Nuc. Acids Res. 8*, 1613.

Kopka, M. L., Yoon, C., Goodsell, D., Pjura, P. and Dickerson, R. E. (1985b) *Proc. Natl. Acad. Sci. U.S.A. 82*, 1376.

Lohman, T. M. (1985) *CRC Crit. Rev. Biochem. 19*, 191.

Lown, J. W. (1988) *Anti Cancer Drug Des. 3*, 25.

Neidle, S., Pearl, L. H., Skelly, J. V. (1987) *Biochem J. 243*, 1.

Portugal, J., and Waring, M. J. (1988) *Biochem. Biophys. Acta 949*, 158.

Steck, E. A., Kinnamon, K. E., Davidson, Jr., D. E., Duxbury, R. E., Johnson, A. J., and Masters, R. E. (1982) *Experimental Parasitology 53*, 133.

Steck, E. A., Kinnamon, K. E., Rane, D. S., and Hanson, W. L., (1981) *Experimental Parasitology 52*, 404.

Strekowski, L., Strekowska, A., Watson, R. A., Tanious, F. A., Nguyen, L. T., and Wilson, W. D. (1987) *J. Med. Chem. 30*, 1415.

Tanious, F. A., Yen, S.-F. and Wilson, W. D. (1990) *Biochemistry*, submitted.

Walzer, P. D., Kim, C. K., Foy, J., Linke, M. J. and Cushion, M. T., (1988) *Antimicrobial Agents and Chemotherapy 32*, 896.

Waring, M. J. in Gale, E. F., Cundiffe, E., Reynolds, P. E., Richmond, M. H., and Waring, M. J., (eds.) (1981) *The Molecular Basis of Antibiotic Action, 2nd Edn.*, pp. 258-401, Wiley, London.

Weiner, S. J., Kollman, P. A., Nguyen, D. T., and Case, D. A. (1986) *J. Comput. Chem.* 7, 230.

Wilson W. D., Krishnamoorthy, C. R., Wang, Y. H., and Smith, J. C. (1985a) *Biopolymers* 24, 1941.

Wilson, W. D., Wang, Y. H., Krishnamoorthy, C. R., and Smith, J. C. (1985b) *Biochemistry* 24, 3991.

Wilson, W. D., Wang, Y. H., Krishnamoorthy, C. R., and Smith, J. C. (1986) *Chem.-Biol. Interact.* 58, 41-57.

Wilson, W. D. (1990) in *Nucleic Acids in Chemistry and Biology* (Blackburn, M. and Gait, M., eds.) Chapter 8, IRL Press Ltd., Oxford.

Wilson, W. D., and Li, Y. (1990) in *Advanced in DNA Sequence Specific Agents* (Hurley, L., Eds.) JAI Press (in press).

Wilson, W. D., Strekowski, L., Tanious, F., Watson, R., Mokrosz, J. L., Strekowska, A., Webster, G., and Neidle, S. (1988) *J. Am. Chem. Soc.* 110, 8292.

Wilson, W. D., Tanious, F. A., Barton, H. J., Strekowski, L., Boykin, D. W. and Jones, R. L. (1989a) *J. Am. Chem. Soc.* 111, 5008.

Wilson, W. D., Tanious, F. A., Watson, R., Barton, H. J., Strekowska, A., Harden, D. B., and Strekowski, L. (1989b) *Biochemistry* 28, 1984.

Wilson, W. D., Tanious, F. A., Barton, H. J., Jones, R. L., Fox, F., Wydra, R. L. and Strekowski, L. (1990a) *Biochemistry*, 29, 8452.

Wilson, W. D., Tanious, F. A., Barton, H. J., Wydra, R. l., Jones, R. L., Boykin, D. W. and Strekowski, L. (1990b) *Anti Cancer Drug Design*, 5, 31.

Yen, S.-F., Gabbay, E. J., and Wilson, W. D. (1982) *Biochemistry* 21, 2070.

Zimmer, Ch. and Wahnert, U. (1986) *Prog. Biophys. Mol. Biol.* 47, 31.

MODULATION OF PROTEIN-DNA INTERACTIONS BY INTERCALATING AND NONINTERCALATING AGENTS

Bruce C. Baguley, Karen M. Holdaway and Graeme J. Finlay

Cancer Research Laboratory
University of Auckland Medical School
Auckland
New Zealand

ABSTRACT. The cytotoxicity of a series of DNA-binding drugs was measured in a number of both mouse and human cell lines and some multidrug resistant sublines. The intercalating drugs used were derivatives of 9-anilinoacridine, including amsacrine. The nonintercalating drugs were phenylbisbenz-imidazole derivatives, including the Hoechst dyes H33258 and H33342. Comparison of the ability of the intercalators to inhibit the growth of cultured L1210 cells showed that the presence of certain substituents on the anilino moiety, such as a methoxy group or a benzenesulphonamide group in the correct place, greatly enhances the degree of growth inhibition for a given amount of DNA binding. Divergence between activity and DNA binding was also found with two multidrug resistant Jurkat sublines but in this case the anilino substituents for optimal activity were quite different. The Hoechst dyes at low concentrations did not inhibit cell growth, but some did strongly potentiate the cytotoxicity of the amsacrine analogue CI-921. In this case, potentiation was strongly dependent on the substituents on the terminal phenyl ring. The results can be explained if it is assumed that both the anilino substituents of the intercalators and the phenyl substituents of the Hoechst dyes interact with an enzyme receptor while the remainder of the molecule binds to DNA. The most likely enzyme receptors are isozymes of the topoisomerase II isozymes.

1. Introduction

The cellular processes necessary for life depend ultimately on the control of gene expression and of genome replication. This control is accomplished predominately by interaction between DNA sequence-specific DNA-binding proteins. The addition of DNA-binding drugs to cells might be therefore be expected to lead to modulation, and particularly inhibition, of the interaction between DNA and DNA-binding proteins, as a result of drug-induced changes to DNA conformation or of shielding of protein recognition sites. However, in a number of cases, drug binding may augment protein-DNA interaction. The simplest mechanism for this augmentation involves the binding of a drug to a macromolecular pocket comprising both DNA and protein, thus forming and stabilizing a ternary complex. Some examples of such proposed complexes from the literature are shown in Table 1.

The physical study of protein-drug-DNA complexes is difficult because of the complexity of a system incorporating three variables, and to date the evidence for such complexes is circumstantial, often based on structure-activity relationships for a series of drugs. In contrast, drug-protein interactions and drug-DNA interactions have been studied very intensively in the past few years because of improvements in technology, and are comparatively well understood. The recognition by a DNA-binding drug of a sequence-selective DNA-binding protein enhances not only the binding constant but also the sequence selectivity of the drug. Protein-DNA

355

B. Pullman and J. Jortner (eds.), Molecular Basis of Specificity in Nucleic Acid-Drug Interactions, 355–367.

binding may also be enhanced, and if the enzyme is a homodimer the presence of the symmetrical drug binding sites will amplify this effect.

TABLE 1. Examples of possible DNA-drug-protein ternary complexes

Ligand	Enzyme-DNA complex	Reference
Quinolone antibacterials	Gyrase	Shen et al., 1989
Amsacrine	Topo II	Nelson et al., 1984
Actinomycin D	RNA polymerase	White and Phillips, 1989
Ditercalinium	UvrABC endonuclease	Lambert et al., 1989

In this communication, we would like to consider the evidence that the enzyme topoisomerase II (topo II) is a possible target for ternary complex formation, using both intercalative and non-intercalative drugs as examples. Topo II is a chromatin-associated enzyme which is capable, by creating a temporary, protein-masked, double-stranded break in one molecule, of passing one double-stranded DNA molecule through another. The enzyme is responsible for maintaining DNA in a topologically appropriate form in the cell, and is active during the transcription, replication and recombination of DNA (Liu, 1989). Recently, evidence for two genes for topo II, one coding a 170-kDa isozyme (topo IIα) and one a 180-kDa isozyme (topo IIβ), has been published. The proportions of these forms varies in some multidrug-resistant cell lines and also in sensitive lines during the cell division cycle (Chung et al., 1989). These observations indicate that there are two distinguishable topo II targets for drug design (Finlay et al., 1990a).

Two series of compounds have been used to investigate possible examples of ternary complex formation. The first comprises derivatives of 9-anilinoacridine (see Figure 1 for structures) first synthesised by Cain and coworkers in the early 1970's (Cain et al., 1971) and includes the clinical antileukaemia agent amsacrine (Cain and Atwell, 1974). This drug not only has clinical utility in combination with other agents (Arlin et al., 1983) but is the first drug to have been shown to act as a poison of the enzyme topo II (Nelson et al., 1984). The association of amsacrine with the enzyme stabilizes it in a form known as the cleavable complex (Nelson et al., 1984) where the enzyme homodimers are each covalently joined to the DNA 5'-ends via a phosphotyrosyl group (Liu et al., 1983). Since there are preferred points of enzyme cleavage on the DNA (Pommier et al., 1983) the association of amsacrine with topo II essentially converts it into a sequence-selective DNA cutting agent. A number of analogues of amsacrine have also been found to stimulate formation of the topo II cleavable complex, and the degree of stimulation is related to their cytotoxic activity (Covey et al., 1988). One of these analogues, CI-921 is more potent than amsacrine and has been subjected to clinical trial (Hardy et al., 1988).

The second series comprises the phenylbisbenzimidazole derivatives (Figure 1), also known as Hoechst dyes (Loewe and Urbanietz, 1974) are well known for their DNA binding and cell staining properties. Pibenzimol (H33258), which has undergone Phase I clinical trial (Kraut et al., 1988) has been crystallised as a drug-oligodeoxynucleotide complex, where it binds in the minor groove and is selective for adenine-thymine rich DNA (Pjura et al., 1987). Pibenzimol inhibits the condensation of adenine-thymine rich chromosome regions of growing cells (Marcus et al., 1979) and has found many uses in cell biology. It has been recently reported to inhibit the action of both topoisomerase I (McHugh et al., 1989) and topo II (Woynarowski et al., 1989) in cell-free systems. Recently, a series of Hoechst compounds, variously substituted on the phenyl ring, has been shown to inhibit the cytotoxic effects of several topo II poisons (Finlay et al., 1990b). More interestingly, one of these compounds has been shown to potentiate the cytotoxic effect of some topo II poisons on cultured Lewis lung carcinoma cells at low concentrations (Finlay et al., 1990b).

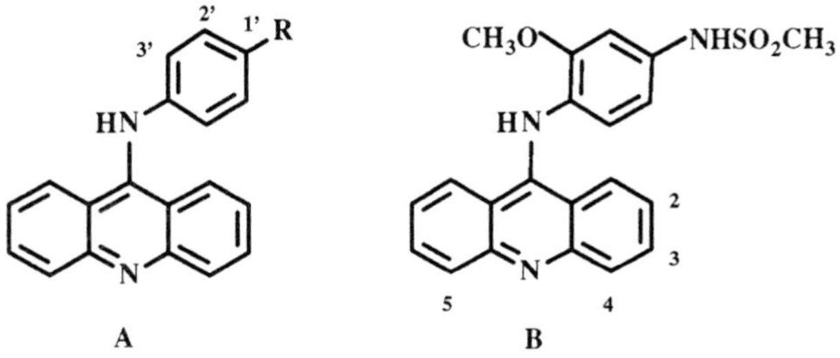

Figure 1.   Chemical structures of 9-anilinoacridine (A) and amsacrine (B). CI-921 is the 4-methyl,   5-(N-methyl)carboxamide   derivative   of   amsacrine.   SN   8551   is   the   1'-benzenesulphonamide derivative of 9-anilinoacridine)

Although the evidence is circumstantial, the results discussed here provide suggestive evidence for interactions between the substituents on the anilino group of 9-anilinoacridines, and on the benzene group of the phenylbisbenzimidazole dyes, and topo II or other protein. These findings may allow the design and synthesis of further active antitumour agents.

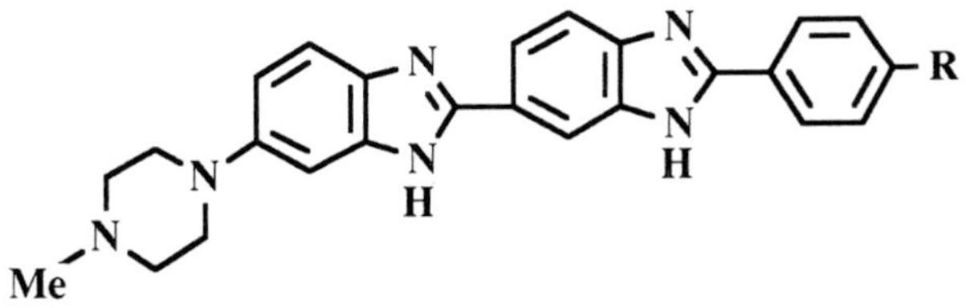

Figure 2. Chemical structures of the Hoechst dye series used in the present study.

2.   Materials and Methods

2.1.   CHEMICALS

Bisbenzimidazoles were kindly provided as chloride salts by Dr. H. Loewe of Hoechst A.G., Frankfurt, Germany. Amsacrine and its 4-methyl, 5-methylcarboxamide analogue CI-921 were provided as isethionate salts by the Parke-Davis Division of the Warner-Lambert Company, Ann Arbor, USA. All other 9-anilinoacridine derivatives were synthesised in this laboratory by Drs. W.A. Denny, G.J. Atwell, G.J. Rewcastle and the late Dr. B.F. Cain. MTT (4,5-dimethylthiazole-2-yl)-2,5-diphenyltetrazolium bromide was from the Sigma Chemical Co.

2.2.   CELL LINES

Jurkat leukaemia cells were originally obtained from Professor J.D. Watson, Department of Molecular Medicine, University of Auckland School of Medicine. The Jurkat JL clone of this line, and the JL/DOX and JL/AMSA lines (selected for resistance to doxorubicin and amsacrine, respectively) were derived by Dr. K. Snow (Finlay et al., 1990a). A tissue culture-adapted subline of Lewis lung carcinoma (LLTC) cells was obtained from Dr R.C. Jackson, Warner-Lambert Company, Ann Arbor, USA. L1210 mouse leukaemia cells were obtained from Arthur D Little Inc, Boston, USA. P388 cells were obtained from the Developmental Therapeutics Program of the National Cancer Institute, USA.

2.3. CULTURE OF CELL LINES

Human cells were cultured as previously described in a humidified atmosphere of 5% CO_2 in air at 37° (Finlay et al., 1990a). Growth inhibition assays defined the IC_{50} value as the concentration of drug which inhibits cell growth to 50% of that in control (drug-free) cultures. IC_{50} values were determined using duplicate cultures on at least three separate occasions for each compound. Cultures in αMEM (Gibco, Grand Island, NY) supplemented with foetal calf serum (10%), penicillin (100 units/ml) and streptomycin (100 μg/ml) in 96-well microculture plates (Linbro) were established (3750 cells/well) in 135 μl volumes and incubated for 4 h prior to adding cytotoxic compounds diluted in growth medium. Cultures were incubated for 4 days and growth was determined by colorimetric analysis using MTT. Absorbance was determined using a microplate reader (Dynatech MR 600) using a sample wavelength of 570 nm and a reference of 630 nm.

LLTC cells were cultured as above but growth was quantitated by staining cultures with methylene blue. Mouse leukaemia cells were cultured in 24-well tissue culture dishes, using RPMI-1640 medium instead of αMEM (Baguley and Finlay, 1988). Growth medium was supplemented as above but with the addition of 2-mercaptoethanol (50 μM). Exponentially growing cells (3 x 10^4 cells per well in 1 ml) were cultured for three days in the presence of drugs before being counted with an electronic cell counter.

2.4. CLONOGENIC ASSAYS

Clonogenic assays using LLTC cells were performed as described (Finlay and Baguley, 1989). Briefly, cells were cultured in 100 mm dishes at 1.5 x 10^6 cells/15 ml/dish for 18 h, after which they were trypsinized (0.07% w/v trypsin in 134 mM KCl, 15 mM trisodium citrate), and collected by centrifugation. They were resuspended to 10^5 cells/ml in growth medium in polystyrene tubes (5 ml/tube) for exposure to Hoechst dyes and CI-921. Preincubation with dyes was performed for 1 h in a 37°C water bath, after which topo II-directed drugs were added for a further 1 h. Cells were then collected by centrifugation, washed twice, resuspended in growth medium (containing 5% v/v FBS) and different dilutions plated in 60 mm dishes (5 ml/dish). After 9 days colonies were fixed and stained using methylene blue in 50% ethanol and those containing at least 50 cells counted. The potentiation factor was determined by dividing the survival fraction for CI-921 alone (6.2% in this experiment) by the surviving fraction for CI-921 in the presence of the Hoechst dye.

2.5. DNA BINDING ASSAYS

DNA binding constants for DNA intercalating drugs were measured by displacement of the DNA-binding fluorochrome ethidium in 0.01 ionic strength buffer, pH 7.0, according to previously published methods (Baguley et al., 1981; Baguley, 1982). A correction for drug-induced quenching of the fluorescence of DNA-bound ethidium was necessary for 9-anilinoacridine derivatives, and this was performed in a separate assay. The method has been validated by comparison of results obtained in the displacement assay with those obtained by spectrophotometric determination of the binding constant for a series of acridine derivatives. A second assay employed the minor groove binder SN 6999 (5 μM), a non-fluorescent minor groove DNA binder (Baguley, 1982) as the displaceable species. Binding of the Hoechst dyes to DNA was measured by the fluorescence enhancement.

3.   Results

3.1.   RELATIONSHIP BETWEEN DNA BINDING AND CYTOTOXICITY OF 9-ANILINOACRIDINE ANALOGUES AGAINST CULTURED TUMOUR CELLS

Growth inhibitory concentrations (IC_{50} values) and DNA association constants have now been obtained for a large series of 9-anilinoacridine derivatives. The data, plotted in Figure 3, show that alteration of substituents of 9-anilinoacridine result in a five $\log_{10}$ range in IC_{50} values and a two $\log_{10}$ range in DNA binding affinities. Although there is no significant correlation between these two parameters, inspection of the points below the diagonal line in Figure 3 demonstrates three common patterns which appear in the compounds with high biological activity for a given degree of DNA binding.

   (a)   The presence of 3'-methoxy group (together with a 1'-alkylsulphonamide group.
   (b)   The presence of a 3'-dimethylamino group.
   (c)   The presence of a 1'-benzenesulphonamide substituent.

It is noteworthy that compounds with electron withdrawing 1'-substituents are outliers above the diagonal line. This suggests that the electron density on the anilino ring is a further important criterion for activity.

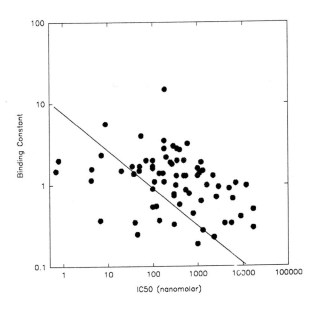

Figure 3.   Relationship between the DNA association constant (units M^{-1} $x10^6$), as estimated by the ethidium displacement method, and growth inhibitory activity against cultured L1210 leukaemia cells, as measured by IC_{50} values.

A smaller series of compounds was selected for study using a panel of cell lines including L1210 and P388 murine leukaemia lines, the Lewis lung LLTC murine carcinoma line, the Jurkat JL human leukaemia line and two Jurkat multidrug-resistant sublines, JL/DOX and

JL/AMSA. A correlation matrix for logarithmic IC_{50} values for a number of tumour cell lines is shown in Table 2. The squared correlation coefficients provide an indication of the fraction of variance in the data which is common between the lines. A very high degree of correlation is observed between the three parent leukaemia lines whereas the two multidrug resistant lines were very different and the LLTC line was intermediate. The cell line showing the greatest differences from L1210 cells was JL/DOX. A plot of IC_{50} values against DNA binding was made for this cell line and is shown in Figure 4. The most effective compound in terms of activity for a given degree of DNA binding was the 3'-methoxy derivative of the 1'-methylcarbamate.

TABLE 2. Squared correlation matrix for logarithmic IC_{50} values for six cultured cell lines

	P388	JL	LLTC	JL/DOX	JL/AMSA
L1210	0.96	0.94	0.69	0.10	0.22
P388		0.94	0.62	0.01	0.24
JL			0.67	0.09	0.25
LL				0.11	0.35
JL/DOX					0.70

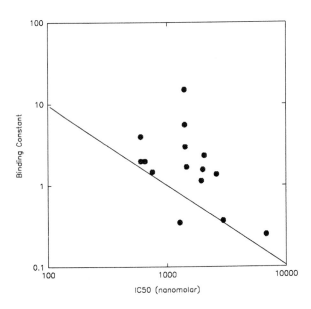

Figure 4. Relationship between the DNA association constant (units $M^{-1} \times 10^6$), as estimated by the ethidium displacement method, and growth inhibitory activity against cultured JL/DOX human leukaemia cells, as measured by IC_{50} values.

3.2. RELATIONSHIP BETWEEN DNA BINDING AND POTENTIATION OF TOPO II-DIRECTED CYTOTOXICITY FOR HOECHST DYES

In order to study the relationship between the nature of the phenyl substituent on the Hoechst dye and its ability to potentiate the effect of a topo II poison, a series of compounds were tested in the presence of the amsacrine analogue and potent topo II poison CI-921. The results are shown in Table 3. The dye concentrations used in the potentiation studies had no cytotoxic effect themselves. Potentiation was not observed with H33258 (R = OH), but was observed with both the alkoxy- and phenoxy-substituted compounds. Both the maximum degree of potentiation and the range of concentrations leading to potentiation increased with the size of the substituent. It is apparent from Table 3 that potentiation is critically dependent on the size of the substituent R (Figure 2) and has no relationship to DNA binding.

Direct measurement of DNA binding appeared to be feasible since the compounds are highly fluorescent and fluorescence is enhanced by DNA binding. However, attempts to measure binding directly showed only that the binding constant was very high. Compounds were compared by their ability to displace intercalated ethidium (Baguley et al., 1981). The results (Table 3) indicate that the substituents had only a minor effects. DNA binding was also compared by measuring the ability of these compounds to displace another AT-specific but non-fluorescent minor groove binder, SN 6999 (Baguley, 1982; Leupin et al., 1986), again showing that the compounds had similar binding affinities (results not shown). One factor which could have a bearing in the potentiation experiments is the rate of uptake of dyes by the LLTC cells. Since the drugs are highly fluorescent, a study was made of the dependence of rate of uptake on substitution pattern using flow cytometry (data not shown). Uptake rate increased with the lipophilicity of the substituent. It was very rapid for the alkoxy and phenoxy compounds but 10- to 30-fold slower for the hydroxy compound H33258.

TABLE 3. Relationship between DNA binding coefficient, as measured by displacement of ethidium, and the degree of potentiation of cytotoxicity of the amsacrine analogue CI-921

Compound	Substituent (R)	Optimal Conc. μM	Potentiation factor	Relative DNA Binding
H33258	OH	(1-20)	1.0x	1
H33342	OC_2H_5	0.20	5.6x	1.05
H33378	OC_3H_7	0.15	6.5x	1.18
H33293	OC_4H_9	0.30	8.4x	0.74
H33777	O-Phenyl	0.45	9.9x	0.49

It was of interest to determine whether the potentiation of cytotoxicity observed with the Hoechst dyes could be shown by means of an IC_{50} assay. The results shown in Table 4 indicate that the addition of H33777 to cultures, under the same conditions (continuous exposure) that were used for the data in Figure 3, reduced IC_{50} values for two Jurkat lines significantly. Furthermore, potentiation of the multidrug-resistant JL/DOX subline occurred to approximately the same extent as it did for the parent line.

TABLE 4. IC$_{50}$ values for SN 8551 (acridinylaminobenzenesulphon-anilide), amsacrine and CI-921 for two cell lines in the presence and absence of the Hoechst dye H33777 (100 nM).

Cell Line	Drug	IC$_{50}$ Value	
		- H33377	+ H33377
LLTC	SN 8551	7.1	3.9
	Amsacrine	10.3	5.6
	CI-921	6.7	3.6
Jurkat JL	SN 8551	34	24
	Amsacrine	62	31
	CI-921	19	12
Jurkat JL/DOX	SN 8551	3100	1300
	Amsacrine	3300	1800
	CI-921	2100	600

4. Discussion

4.1. RELATIONSHIP BETWEEN DNA BINDING AND GROWTH INHIBITION FOR 9-ANILINOACRIDINE DERIVATIVES

The results in Figures 3 and 4 show two quite different relationships between DNA binding and biological activity for derivatives of 9-anilinoacridine in two different cell lines. The data are difficult to interpret since two factors which contribute to drug variations in IC$_{50}$ values may vary from drug to drug. Firstly, the rate of cellular uptake of drug has not been measured. However, the uptake of one of these derivatives, amsacrine, has been found to be extremely rapid (Robbie et al., 1988) and the L1210 cells used do not show signs of transport-related multidrug resistance (Baguley and Finlay, 1988). Furthermore, the experiments are conducted with long times for drug exposure, allowing time for equilibration. A second problem is that the stability of the drugs in culture may vary. Drug breakdown in culture will give rise to higher IC$_{50}$ values. It is known that the stabilization of amsacrine and its derivatives by the addition of ascorbate to the medium reduces the observed IC$_{50}$ values by a factor of two- to three-fold (Finlay and Baguley, 1988) and IC$_{50}$ values used for the 3'-methoxy and 3'-dimethylamino substituted compounds used in this analysis have been carried out using ascorbate-containing medium. However, compounds containing 3'-methylamino substituents are not stabilized by ascorbate and are therefore less active according to this assay. Despite these reservations, a clear relationship is found in Figure 3 between the presence of certain substituents and activity. It is notable that only anilino substituents give rise to the outliers which fall below the diagonal line in Figure 3. The benzenesulphonamide group, identified in these experiments as one of the substituents which potentiate activity, has also been identified from in vivo studies where it was accommodated in structure-activity calculations as an indicator variable (Denny et al., 1982).

A possible explanation of the results can be obtained by considering the nature of the interaction between 9-anilinoacridine derivatives and their target. Amsacrine is known to bind to DNA by intercalation with a DNA unwinding angle of 20.5° (Waring, 1976). The crystal structure of amsacrine shows the plane of the anilino side chain to be almost perpendicular (80°) to that of the acridine, giving it the correct geometry to bind in one of the grooves of the partially unwound DNA double helix. It has been argued both on the basis of analogy with the structures of known drug-oligonucleotide complexes (Wilson et al., 1981; Sakore et al., 1979) and from theoretical calculations (Chen et al., 1988) that the minor groove is the preferred binding site. Substituents on 1'- and 2'-positions of the anilino group therefore project from the minor groove in a position to interact with a second macromolecule (Baguley and Nash, 1981; Wilson et al., 1981). As argued in the introduction, this second macromolecule is most likely to be topo IIα, which is bound to the DNA in the cleavable complex form It is noteworthy that alterations in the anilino moiety of amsacrine change in vitro activity by a factor of nearly 100,000 without substantial change in DNA binding (Figure 3). On the other hand, alteration of the acridine substitution pattern varies in vitro potency approximately in parallel with DNA binding affinity (Baguley and Cain, 1982). These results would be explained if it is assumed the changes in the anilino portion of amsacrine alter its interaction with topo II while changes in the acridine moiety change the interaction with DNA.

Possible interactions between the anilino moiety and topo II include the following:

(a)   The interaction between an electron-rich area on the anilino ring and an electron-poor region (e.g. a histidine residue) on the enzyme. The evidence for this interaction is based on the ability of active antitumour drugs in this series to interact with electron-poor DNA-bound ethidium, apparently forming of a charge transfer complex (Baguley, 1990).

(b)   The formation of a bond between a hydrogen bond donor on the enzyme and an acceptor on the 3'-methoxy or 3'-dimethylamino group of the 9-anilinoacridine. The 3'-methylamino group, which is a hydrogen bond donor, does not appear to augment activity for a given degree of DNA binding, even when short-term drug exposures are used to counter the instability of these compounds (results not shown).

(c)   The hydrophobic interaction of a phenyl substituent of the 1'-benzenesulphonamide with an 'oily pocket' on the enzyme.

4.2.   RELATIONSHIP BETWEEN DNA BINDING AND GROWTH INHIBITION OF MULTIDRUG RESISTANT CELLS

The correlation matrix in Table 2 indicates that, for a series of 15 derivatives, very similar results are obtained for mouse L1210 and P388 cells, and for human Jurkat leukaemia cells. On the other hand, little correlation is found between IC_{50} values for the sensitive and resistant lines. If IC_{50} values are plotted against DNA binding constants, it is possible, as in the case of L1210 cells, to look for substituents which increase biological activity for a given degree of DNA binding. The situation with the Jurkat JL/DOX line is very different to that with L1210. The presence of a 3'-methoxy group decreases, rather than increases activity of the series of compounds with a 1'-methanesulphonamide. It is noteworthy that the presence of the 1'-benzenesulphonamide group does not augment selectivity in the JL/DOX line, but in another Jurkat multidrug-resistant lines (JL/AMSA) the presence of this group does increase activity (Finlay et al., 1990a). The best substituent for optimal activity appears to be the 1'-methylcarbamate. Addition of the 3'-methoxy group does not increase activity, but because it decreases DNA binding it gives rise to a compound with the greatest activity for a given amount of DNA binding (Figure 4).

One explanation for these results is that the JL/DOX line is expressing an altered topo II enzyme, perhaps topo IIβ, which because of its altered structure has a different requirement for interaction with the drug in a ternary complex, i.e. there is no contact for the combination of the 3'-methoxy group with the 1'-methanesulphonamide, and no 'oily pocket'. It is noteworthy that the compound which shows the best selective activity in this assay is also the compound which has the highest antiviral activity against Herpes simplex virus (Goldwater et al., 1985). This raises the question of whether topo IIß is important for Herpesvirus replication.

4.3. RELATIONSHIP OF POTENTIATION OF CYTOTOXICITY TO DNA BINDING PROPERTIES OF HOECHST DYES

The series of Hoechst dyes illustrated in Figure 2 bind strongly to double stranded DNA. It is likely that all bind by the same mode, i.e. in the minor groove as proposed from the results of Pjura et al. (1987). It is unlikely that the variations observed in Table 3 for the series of dyes reflects variation in uptake since with the exception of H33258 all drugs are taken up rapidly by cells and should equilibrate during the one hour pre-incubation time. In the case of H33258, even a 20-fold higher drug concentration, which should allow an equivalent cellular uptake of drug, did not cause any potentiation. It is also unlikely that variations in DNA binding constant are responsible for the differences in activity (Table 3). The most reasonable explanation for these results is that the active compounds stabilize the cleavable complex of topo II with DNA. This is supported by the observation that the frequency of DNA breaks induced by CI-921 is substantially increased by the presence of the Hoechst dye in which R = phenoxy (Dr. W.R. Wilson, University of Auckland Medical School; personal communication). It appears that although the Hoechst dyes cannot selectively stabilize the cleavable complex of topo II with DNA by themselves, they can bind to pre-formed cleavable complex and further stabilize it.

It is tempting to propose that this stabilization is effected by the formation of a quaternary complex where the Hoechst dye bisbenzimidazole moiety binds in the DNA minor groove while the alkoxy or phenoxy group binds in an 'oily pocket' of the cleavable complex of topo IIα, CI-921 and DNA. This raises the question of whether this 'oily pocket' is the same as that recognised by the benzenesulphonamide-substituted anilinoacridine, SN 8551. However, experiments performed with Jurkat leukaemia cell lines using the same methodology as for the 9-anilinoacridine studies suggest otherwise. The H33777 dye (Fig. 2; R = phenoxy) decreases the IC_{50} value (i.e. potentiates cytotoxicity) of both CI-921 and SN 8551 against the resistant Jurkat L/DOX line as well as it does against the parent Jurkat line (Table 4).

4.4. CONCLUSIONS

Three biological systems have been described where DNA binding drugs positively affect a biological phenomenon involving a complex of DNA and protein. The assumption that the observed structure-activity relationships reflect differences in drug-protein interactions can only be validated by some type of direct measurement. However, the adoption of this assumption as a working hypothesis may allow the design and synthesis of increasingly selective topo II inhibitors, and consequently more effective clinical antitumour drugs.

5.   Acknowledgements

The authors are grateful to Dr. H. Loewe for the Hoechst dyes and to Lynden Wallis for preparing the manuscript.

6.   References

Arlin Z. (1983) 'Current status of amsacrine combination chemotherapy programs in acute leukemia', Cancer Treat. Rep. 67, 967-970.

Baguley, B.C. (1982) 'Nonintercalative DNA binding antitumour compounds', Cell Molec. Biochem. 43, 167-181.

Baguley, B.C. (1990) 'The possible role of electron transfer complexes in the action of amsacrine analogues', Biophys. Chem., in press.

Baguley, B.C. and Cain, B.F. (1982) 'Comparison of the *in vivo* and *in vitro* antileukemic activity of monosubstituted derivatives of 4'-(9-acridinylamino)methanesulfon-m-anisidide (m-AMSA)', Molecular Pharmacol. 22, 486-492.

Baguley, B.C. and Finlay, G.J. (1988) 'Derivatives of amsacrine: determinants required for high activity against the Lewis lung carcinoma', J. Natl. Cancer Inst. 80, 195-199.

Baguley, B.C. and Nash, R. (1981) 'Antitumour activity of substituted 9-anilino acridines - comparison of *in vivo* and *in vitro* testing systems', Eur. J. Cancer 17, 671-679.

Baguley, B.C., Denny, W.A., Atwell, G.J., and Cain, B.F. (1981) 'Potential antitumor agents. Part 34. Quantitative relationships between DNA binding and molecular structure for 9-anilino-acridines substituted in the anilino ring', J. Med. Chem. 24, 170-177.

Cain, B.F. and Atwell, G.J. (1974), 'The experimental antitumour properties of three congeners of the acridinyl methanesulphonanilide (AMSA) series', Eur. J. Cancer 10, 539-549.

Cain, B.F., Atwell, G.J. and Seelye, R.N. (1971) 'Potential antitumor agents. Part 11. 9-Anilinoacridines', J. Med. Chem. 14, 311-315.

Chen, K.X., Gresh, N., and Pullman, B. (1988), 'Energetics and stereochemistry of DNA complexation with the antitumor AT-specific intercalators tilorone and m-AMSA', Nucleic Acids Res. 16, 360-373.

Chung, T.D.Y., Drake, F.H., Tan, K.B., Per, S.R., Crooke, S.T., and Mirabelli, C.K. (1989) 'Characterization and immunological identification of cDNA clones encoding two human DNA topoisomerase isozymes', Proc. Natl. Acac. Sci. USA 86, 9431-9435.

Covey, J.M., Kohn, K.W., Kerrigan, D., Tilchen, E.J., and Pommier, Y. (1988) 'Topoisomerase II-mediated damage produced by 4'-(9-acridinylamino)-methanesulfon-m-anisidide and related acridines in L1210 cells and isolated nuclei: relation to cytotoxicity', Cancer Res. 48, 860-865.

Denny, W.A., Atwell, G.J., Cain, B.F., Hansch, C., Leo, A. and Panthananickal, A. (1982) 'Potential antitumor agents. Part 36. Quantitative relationships between antitumour potency, toxicity and structure for the general class of 9-anilinoacridine antitumour agents', J. Med. Chem. 25, 276-315.

Finlay, G.J. and Baguley, B.C. (1989) 'Selectivity of N-[2-(dimethylamino)ethyl]acridine-4-carboxamide towards Lewis lung carcinoma and human tumour cell lines *in vitro*', Eur. J. Cancer Clin. Oncol. 25, 270-277.

Finlay, G.J., Baguley, B.C., Snow, K., and Judd, W. (1990a) 'Multiple patterns of resistance of human leukemia cell sublines to analogs of amsacrine', J. Natl. Cancer Inst., 82, 662-667.

Finlay, G.J. and Baguley, B.C. (1990b) 'Potentiation of cytotoxicity of topoisomerase II-directed anti-cancer drugs by DNA minor groove binders', Eur. J. Cancer Clin. Oncol., submitted.

Goldwater, P.N., Flynn, K.E., Gunn, C.S., and Baguley, B.C. (1985) '9-Anilinoacridines: novel compounds active against Herpes simplex virus', Chem-Biol. Interactions 54, 377-382.

Hardy, J.R., Harvey, V.J., Paxton, J.W., Evans, P.C., Smith, S., Grillo-Lopez, A., Grove, W., and Baguley, B.C. (1988) 'A phase I trial of the amsacrine analog 9-[(2-methoxy-4-methyl-sulfonylamino)-phenylamino]-N,5-dimethyl-4-acridinecarboxamide (CI-921)', Cancer Res. 48, 6593-6596.

Kraut, E., Malspeis, L., Bakerzak, S., and Grever, M. (1988) 'Evaluation of pibenzimol (NSC 322921) in refractory solid malignancies', Proc. Amer. Soc. Clin. Oncol. 7, 62.

Lambert, B., Jones, B.K., Roques, B.P., and Le Pecq, J-B. (1989) 'The non-covalent complex between DNA and the bifunctional intercalator ditercalinium is a substrate for the Uvr ABC endonuclease of Escherichia coli', Proc. Natl. Acad. Sci. USA 86, 6557-61.

Leupin, W., Chazin, W.J., Hyberts, S., Denny, W.A., Stewart, G.M. and Wüthrich, K. (1986) '1D and 2D NMR study of the complex between the decadeoxyribonucleotide d(GCATTAATGC)$_2$ and a minor groove binding drug', Biochemistry 25, 5902-5910.

Liu, L.F. (1989) 'DNA topoisomerase poisons as antitumor drugs', Ann. Rev. Biochem. 58, 351-375.

Liu, L.F., Rowe, T.C., Yang, L., Tewey, K.M., and Chen, G.L. (1983) 'Cleavage of DNA by mammalian DNA topoisomerase II', J. Biol. Chem. 258, 15365-15370.

Loewe, H. and Urbanietz, J. (1974) 'Basisch substituierte 2,6-bis-benzimidazolderivate, eine neue chemotherapeutisch active Körperklasse', Arzneimittel-Forsch. 24, 1927-1933.

Marcus, M., Nattenberg, A., Gotein, R., Nielsén, K., and Gropp, A. (1979) 'Inhibition of condensation of human Y chromosome by the fluorochrome Hoechst 33258 in a mouse-human cell hybrid', Hum. Genet. 46, 193-198.

McHugh, M.M., Woynarowski, J.M., Sigmund, R.D., and Beerman, T.A. (1989) 'Effect of minor groove binding drugs on mammalian topoisomerase I activity', Biochem. Pharmacol. 38, 2323-2328.

Nelson, E.M., Tewey, K.M., and Liu, L.F. (1984), 'Mechanism of antitumor drug action: Poisoning of mammalian topoisomerase II on DNA by 4'-(9-acridinylamino)methanesulfon-m-anisidide', Proc. Natl. Acad. Sci. USA 81, 1361-1364.

Pjura, P.E., Crzeskowiak, K., and Dickerson, R.E. (1987) 'Binding of Hoechst 33258 to the minor groove of DNA', J. Mol. Biol. 197, 257-271.

Pommier, Y., Zwelling, L. A. Mattern, M. R., Erickson LC, Kerrigan, D., Schwartz R., and Kohn, K.W. (1983), 'Effects of dimethyl sulfoxide and thiourea upon intercalator-induced DNA single-stranded breaks in mouse leukemia L1210 cells', Cancer Res. 43, 5718-5724.

Robbie, M.A., Baguley, B.C., Denny, W.A., Gavin, J.B., and Wilson, W.R. (1988) 'Mechanism of resistance of non-cycling mammalian cells to 4'-(9-acridinylamino)methanesulphon-m-anisidide: comparison of uptake, metabolism and DNA breakage in log- and plateau-phase Chinese hamster fibroblast cell cultures', Cancer Res. 48, 310-319.

Sakore, T.D., Reddy, B.S., and Sobell, H.M. (1979), Visualization of drug-nucleic acid interactions at atomic resolution. IV. Structure of an aminoacridine dinucleoside monophosphate crystalline complex, 9-aminoacridine-5-iodocytidylyl-(3'-5') guanosine', J. Mol. Biol. 135, 763-785.

Shen, L.L., Mitscher, L.A., Sharma, P.N., O'Donnell, T.J., Chu, D.W.T., Cooper, C.S., Rosen, T., and Pernet, A.G. (1989), 'Mechanism of inhibition of DNA gyrase by quinolone antibacterials: a coooperative drug-DNA binding model', Biochemistry 28, 3886-3894.

Waring, M.J. (1976) 'DNA binding characteristics of acridinylmethanesulfonanilide drugs: comparison with antitumour properties', Eur. J. Cancer 12, 995-1001.

White, R.J., and Phillips, D.R. (1989), 'Drug-DNA dissociation kinetics. *In vitro* transcription and sodium dodecyl sulphate sequestration', Biochem. Pharmacol. 38, 331-334.

Wilson, W.R., Baguley, B.C., Wakelin, L.P.G., and Waring, M.J. (1981) 'Interaction of the antitumor drug m-AMSA (4'-(9-acridinylamino)methanesulfon-m-anisidide) and related acridines with nucleic acids', Molecular Pharmacol. 20, 404-414.

Woynarowski, J.M., Sigmund, R.D., and Beerman, T.A. (1989) 'DNA minor groove binding agents interfere with topoisomerase II mediated lesions induced by epipodophyllotoxin derivative VM-26 and acridine derivative m-AMSA in nuclei from L1210 cells', Biochemistry 28, 3850-3855.

ANTITUMOR ANTIBIOTICS ENDOWED WITH DNA SEQUENCE SPECIFICITY

FEDERICO ARCAMONE
Menarini Ricerche Sud, Spa
Via Tito Speri 10, Pomezia (Rome)
00040 Italy

ABSTRACT. DNA binding appears to be a necessary albeit not sufficient condition for antitumor activity in the anthracyclines. Apparently, a step different from that of intercalation, as for instance interaction of the complex with an enzyme like topoisomerase II, should be responsible with bioactivity. The latter might be related with the presence of GC rich sequences. This hypothesis stems from both experimental and computational results coupled with structure-activity relationships. Distamycin and related compounds owe their pharmacological effects to an interaction at the level of transcription and this property is related with the high affinity for AT/TA rich sequences. Derivatives containing alkylating functions are endowed with high antitumor activity. Among the different biochemical consequences of the peculiar DNA binding ability of distamycin the interference with the binding of nuclear transacting factors to cis elements of the eukariotic genes is exemplified in recent experimental investigations.

1. INTRODUCTION

It is well known that antiviral and anticancer chemotherapic agents presently in clinical use are characterized by severe side effects. These effects limit their use in a great number of cases that would otherwise benefit from the therapy. Also, major advances are needed in the treatment of clinical important solid tumors, such as lung, breast, colorectal, ovarian cancers (to cite only a few), that fail to respond adequately to any known treatment when in the metastatic stage. A prerequisite for a therapeutic advancement in this field is the discovery of compounds endowed with molecular properties that might improve the selectivity for inhibition of tumor cells in

B. Pullman and J. Jortner (eds.), Molecular Basis of Specificity in Nucleic Acid-Drug Interactions, 369–383.
© 1990 Kluwer Academic Publishers. Printed in the Netherlands.

respect to normal ones. Different important anticancer
agents in clinical use have cell DNA as their major site of
pharmacological action. Now, can selectivity of DNA binding
antibiotics be improved?

I: R=OH; II: R=H   III

IV   V

In this presentation I shall review the developments
that have recently taken place at this regard in the field
of two important classes of antibiotics, namely the
anthracyclines and the distamycins, in view of the
possibility that a modulation of sequence specificity in the
formation of nucleic acid complexes by compounds belonging
to these well known chemical types might lead to better
anticancer and/or antiviral drugs. In order to proceed in

this direction we need first of all to confirm the relevance
of DNA binding to the antitumor activity in the classes of
drugs under scrutiny. Secondly it will be necessary to
ascertain the available evidence concerning the
relationships between sequence specificity and
pharmacological behavior. Once the said points have been
established it will be possible to develop methods aimed at
the evaluation of new derivatives designed on the basis of
an improved selectivity for given DNA sequences.

It is well known that doxorubicin (Adriamycin, I) is an
important cancer chemotherapic agent because of its wide
spectrum of activity. Other anthracycline glycosides in
current clinical usage are daunorubicin (II), epirubicin
(III), idarubicin (IV), pirarubicin (V) (1). New compounds
of potential clinical interest are, among others, cytorhodin
S (rhodorubicin, (VI)) (2), MX-2 and SM5887 (VII and VIII)
(3).

Distamycin (IX) is an antiviral antibiotic with strong affinity for DNA. It has been the object of chemical, biochemical and pharmacological investigations (4,5).

IX

2.THE ANTHRACYCLINES

The antitumor anthracyclines are currently considered to act pharmacologically through a still undefined mechanism that however involves binding of the drugs to cell nuclear DNA. Other hypotheses involving either the formation of toxic oxygen radical species or effects at the membrane levels are invalidated by structure-activity considerations as well as by the absence of pharmacological evidence (6). Clinically useful anthracyclines such as doxorubicin (I) and 4'-deoxy-4'-iododoxorubicin, when added to the culture medium of K 542 cells (a human leukemia cell line) reduced to 50 percent the doubling number of the cells in exponential phase at the same values of intranuclear DNA bound drug concentrations, independently of cellular phenotype, whether sensitive or resistant, and also independently of the C-4' substitution. The said concentrations where determined according to a non-destructive quantitative microspectrofluorometric technique (7). It appears therefore that intranuclear DNA bound drug accumulation is responsible for the cytotoxicity within this class of drugs. In fact the extent of DNA breaks, due to the proved interference of anthracyclines with the DNA-topoisomerase II reaction, together with the persistence of the same, are correlated with the cytotoxic effects of doxorubicin and some strictly related analogs (8).

These conclusions are in agreement with the relatioship between affinity for double helical B-DNA and the level of optimal therapeutic doses in tumor bearing mice of different anthracyclines. Those compounds that have reached the clinical stage after a careful pharmacological and toxicological selection belong to the group showing the highest affinity (9). However, the value of the DNA binding constant is not always related with cytotoxicity. According to a recent publication (2) a high correlation is found between the cytotoxic activity of different anthracycline glycosides and lipophilicity and binding affinity to DNA if both parameters are taken into account simultaneously. The case of 9-deoxydoxorubicin (X), a compound with high apparent binding constant, is a notable exception. The pharmacological data indicate that this compound is distinctly less active than the parent (10). It appears therefore that the presence of the 9-OH determines a peculiar type of interaction of importance for the exhibition of bioactivity. The orientation of the 9-hydroxyl in 9-deacetyldaunorubicun towards the other side of the tetracyclic ring system had been found to reduce both the affinity for DNA and the antitumor activity as compared with the epimer possessing the natural orientation. The importance of a hydrogen bond donating substituent at C-9 is shown also by the lack of antitumor activity of 9-deoxy-9-methylidarubicin and of 9,10-anhydrodaunorubicin (10). Another compound deviating from the rule is 4-demethyl-6-O-methyl doxorubicin, a derivative showing a 20-fold reduction in the binding affinity when compared with doxorubicin but only moderately (two fold) less potent than the parent (11).

The structure and stereochemistry of the glycosidic portion of the anthracycline molecule is a major factor determining the affinity of the interaction. In particular, the higher is the number of the sugar residues, the highest is the stabilization of the complex: this is shown by the high value of the binding constant of aclacinomycin (12) and by the greater affinity for DNA of beta-rhodomycin II as compared with beta-rhodomycin I and iremycin. The higher binding constant of doxorubicin when compared with daunorubicin is possibly related with the formation of a hydrogen bond with a phosphate anion of DNA (13). Although the biological potency was strongly dependent on the sugar moiety in a range of modified daunorubicinone and adriamycinone glycosides, the antitumor effect in the P388 murine leukemia test was practically related with the substitution at C-14, being adriamycinone derivatives almost invariably more effective than the corresponding daunomycinone glycosides in the *in vivo* antitumor tests (10,14,15).

An investigation concerning the comparison of the circular dichroism (CD) spectra of different anthracyclines in free and DNA bound form has shown that structural modifications of the sugar moiety alter the binding affinity without influencing aglycone (daunomycinone or adriamycinone) conformation in the binding site. On the other hand a different binding site geometry was deduced for those anthracyclines lacking the 4-methoxy group as in

idarubicin or carminomycin (interestingly both are more
"potent" cytotoxic agents when compared with doxorubicin).
Two compounds endowed with a lower bioactivity, namely 13-
dihydrodoxorubicin and 9-deoxy-doxorubicin, have CD spectra
(and affinity for DNA) similar to those of the parent drug.
This suggests that the said chemical modifications interfere
with the mechanism responsible for bioactivity in a step
either preceding or subsequent to that of intercalation
(16). Correlation of cytotoxicity data with both association
and dissociation rates of the anthracyclines was not found
significant but doxorubicin displayed the slowest
dissociation rate of the intercalation complex (17).
Clearly, a different class of anthracyclines are those
compounds that may form covalent adducts to DNA such as the
3'-morpholino derivatives (18).

Sequence specificity of anthracycline-DNA binding. Following
the demonstration by Chaires et al (19) on the basis of
spectroscopic and footprinting experiments of the binding
preference of daunorubicin for a triplet consisting of two
adjacent GC base pairs flanked at the 5' end by a AT base
pair, results obtained in solution using different
experimental methodologies agree in the conclusion of a GC
specificity in doxorubicin intercalation complex, the GpC
being the most preferred site (20-22). A DNA conformation
specificity has also been demonstrated and is causative for
the discrimination, i.e. higher affinity for DNAs with
alternating as compared to those with non alternating
purine-pyrimidine sequences (23). Different relative binding
affinity revealed at different temperatures by NMR
measurements on the complexes of 2-fluoro-4-demethoxy-
daunorubicin with d(G-C)$_5$ and d(A-T)$_5$ is indicative of the
influence of experimental conditions destabilizing the
double helix on complex formation with different DNA
sequences (24).

X   XI

We have compared the equilibrium and kinetic aspects of
the interaction of four anthracyclines, namely doxorubicin,
daunorubicin, 9-deoxydoxorubicin (X) and 3'-deamino-3'-
hydroxy-4'-epidoxorubicin (XI) with the d(CGTACG) duplex,
identical to that used in the X-ray diffraction study of
Wang et al. (25), and with the d(CGCGCG) duplex. Profit was
made of the different fluorescence yields of anthracyclines
intercalated at CpG site (complete quenching) in respect to
ApT sites (large residual fluorescence). The analysis
indicated preferential intercalation of the first two
compounds at CpG site in d(CGTACG), as compared with the
same sites in d(CGCGCG), whereas the 9-deoxy derivative
showed a lower affinity in both cases. On the other hand the
reduction in the affinity of binding of the deamino analog
was higher in the former case than it was in the latter,
thus leading to a shift in favor of sequences richer in GC
pairs when compared with the parent drugs (26). These
results are in full agreement with the theoretical
investigations. The computations showed that, inter alia,
the main determinants of the site preferences are the 9-OH
group and the aminosugar moiety. The former is involved in
hydrogen bonding interactions with the 2-amino hydrogens and
with N-3 of the guanine at the intercalation site. The
stabilization energy involved is in the order of 5-6
kcal/mole. The second is responsible for a repulsive
interaction with a guanine adjacent to the intercalation
site, the energy differences involved being of the order of
20 kcal/mole. On the other hand, energy differences for the
interaction of the chromophore with the different
intercalation sites studied do not exceed, in general, 1
kcal/mole. The 9-OH interactions are well documented by the
X-ray analysis of the complexes of daunorubicin with
d(CpGpTpApCpG)$_2$ (25), and with (CpGpApTpCpG)$_2$ (27). Other
interactions evidenced in these complexes are those
involving also water molecules, namely those due to the 3'-
NH$_2$ itself, to the C-13 carbonyl and to the oxygen atoms at
C-4 and C-5. In a recent study (28), the binding affinities
for poly d(A-T) and poly d(G-C) of doxorubicin, 9-
deoxydoxorubicin, 4-demethyl-6-O-methyl-doxorubicin, 3'-
deamino-3'-hydroxy-4'-epidoxorubicin and 6-deoxy-6-amino-4-
demethoxydaunorubicin were determined spectrophotometrically and compared with the stabilization energies
computed for the intermolecular interactions of the same
compounds with oligonucleotides d(CGCGCG)$_2$, d(TATATA)$_2$, and
d(CGTACG)$_2$. The higher affinity of the 6-amino compound for
AT sequences is interpreted in terms of a more favorable
chromophore-backbone interaction as compared with
doxorubicin. On the other hand, the GC preference of the 3'-
deamino analog is related with the interaction of the two
sugar hydroxyls with the C of the intercalation site and the

G immediately upstream of it, this interaction being more stabilizing the complex than that of the corresponding T and A in d(TATATA). The increased preference of the 9-deoxy compound for AT sequences is expected on the basis of the already mentioned interactions of the 9-hydroxyl, whereas the dramatic loss of DNA affinity of the 6-O-methyl derivative is related with a different geometry of intercalation. The latter, however, is not incompatible, as we have seen, with the exhibition of biological activity (11). Finally we should mention footprinting experiments (29) showing that daunorubicin and the heavily glycosylated ditrisarubicin display different sequence preference. In fact the best binding sites for ditrisarubicin often contain the dinucleotide GpT (ApC).

The relevance of sequence specificity as a determinant of biological effects of the anthracyclines has not yet been studied. However, it is known that the topoisomerase-mediated cleavage of DNA, an established molecular site of anthracycline action, occurs at specific sites on the DNA and that anthracyclines cause a characteristic pattern of cleavage sites of the macromolecule (30). Also, intercalators like the anthracyclines may act as allosteric effectors converting Z-DNA, a DNA conformation with a presumed role in gene expression, to an intercalated right handed B-form under conditions that would otherwise favor the Z conformation (31). Examination of the limited data available would suggest that a shift to higher affinity for the ATA (TAT) from the CGC (GCG) sequence would correspond with a reduction of antitumor efficacy in the P 388 leukemia mouse test. Compounds such as 9-deoxydoxorubicin and the disaccharide analogs showing preference for the TAT sequence afford low protection to tumor bearing mice, whereas the deamino analog (XI) endowed with an enhanced GCG over TAT preference gives much better results (10,32).

3. THE DISTAMYCINS

It is a generally accepted notion that distamycin (IX) and related antibiotics (the best known of which in netropsin) owe their outstanding and diverse biological effects to their ability of binding double stranded DNA. This binding takes place with high preference for AT rich sequences (equilibrium constants of the binding of netropsin to polyd(A-T) is in the nanomolar range) (5). In cell coltures, the antiviral activity, either in terms of inhibition of viral proliferation or of viral DNA synthesis, of the drug is demonstrated exposing the cells to a medium containing 10^{-6} M (for vaccinia virus infected cultures) or 10^{-5} M (for Herpes viruses or Moloney Sarcoma Virus)

(4,5,33) after that virus adsorption has taken place. However, no direct information is available as regards the rate of drug uptake in the cells nor its intracellular concentration, and therefore no direct proof is available that distamycin or the related compounds interact at such an internal target as, for instance, the cell nucleus. According to Orlowski et al (34) cell electro-permeabilization increased 200-fold the cytotoxicity of netropsin as deduced from the EC_{50} values on cultured chinese hamster lung fibroblasts. This result might indicate that cell penetration is a major limiting factor of drug activity and that the target of netropsin is an intracellular one.

Indirect evidence that bioactivity of distamycin was related with a mechanism involving an interaction at the level of transcription was deduced from the property that the drug was effective in inhibiting the expression of inducible enzymes in *E. coli* but not of constitutive enzymes, an observation that also implies a selectivity of action whose molecular mechanism is not yet understood (35). In the same direction points the relationship of affinity for calf thymus DNA of distamycins with different number of residues of 1-methylpyrrole-2-carboxylic acid and antiviral activity of the same compounds (32). In fact the initial investigations on the structure activity relationships of distamycin derivatives and analogs showed the favorable effect of increasing the number of the pyrrole residues for the exhibition of antiviral activity. On the other hand different side chains bearing an amidino function were compatible with high activity, but the N-formyl group was superior to other acyl groups. Distamycin analogs in which a p-formylaminobenzoyl or a formylamino substituted heterocyclic residue was put in the place of the formyl group were also studied. Compounds in which the added ring was the benzene ring or the thiophene ring exhibited remarkable inhibition of the proliferation of herpes virus and of Moloney sarcoma virus. The high bioactivity was accompanied by a strong affinity for poly d(A-T), similarly to the parent antibiotic (36). More recently distamycin derivatives bearing bis-(2-chloroethyl) substituents in the place of the formyl group (nitrogen mustards) have been synthetized and studied. These derivatives combine the AT sequence specificity with alkylating properties and appear endowed with outstanding antitumor activity. The most effective one in preclinical tests was the compound corresponding to structure XIII that however was not able to give covalent adducts with DNA in an in vitro situation (37). The hypothesis can be made that, because of the requirement of the nitrogen mustard function for high biological potency, a molecule other than DNA but in some

way associated with its AT rich regions might be the
pharmacologically important reactant. On the other hand it
has been shown that distamycin itself does indeed interfere
(presumptively in a reversible manner) with proteins
associated with DNA such as topoisomerase II (38,39), RNA
polymerase (40,41), DNA polymerases (42), and.trans-acting
factors (43,44). A shift to an irreversible interaction
would explain the high cytotoxicity of XII and its
congeners.

XII

In the study by Gambari et al. (44) the effect of
distamycin on the binding of nuclear factors to a sinthetic
oligonucleotide (GTATA/IFN-γ) mimicking the portion -278/-
256 from the start of transcription of the human HLA-DRα
gene. This region contains a sequence (GTATA) that is
required for nuclear protein binding and that is likely to
interact with distamycin. Distamycin inhibits the
interaction between the nuclear factors and the synthetic
oligonucleotide, thus suggesting that the drug might alter
the binding of transacting factors to cis-elements
containing AT/TA sequences, and that this mechanism be of
relevance for the biological activities of distamycin and
its congeners.
     One or more of the above mentioned molecular
interactions of distamycin might be used for (1) the
determination of pharmacological relevance of sequence
selectivity within this class of DNA binding agents and (2)
the development of derivatives directed at selected targets.
Another experimental approach would be the detailed analysis
of the effects on simple viral systems such as the one
represented by the SV40 in mammalian CV1 cells. In this
system distamycin has no effect on the expression of the
early gene leading to the synthesis of large T-antigen at
concentrations that completely suppress viral proliferation,
whereas it is able to selectively inhibit the synthesis of

viral DNA (but not that of cell DNA) that takes place in the infected cultures at a later time (45).

REFERENCES

1. Arcamone, F. and Penco, S. (1988) "Synthesis of New Doxorubicin Analogs" in J. W. Lown (ed.), Anthracycline and Anthracenedione-Based Anticancer Agents, Elsevier, Amsterdam, pp. 1-53.
2. Hoffman, D., Berscheid, H. G., Boettger, D., Hermentin, P., Sedlacek, H. H., and Kraemer, H. P. (1990) "Structure-Activity Relationship of Anthracyclines in Vitro", J. Med. Chem. 33, 166-171.
3. Ohe, Y., Nakagawa, K., Fojiwara, Y., Sasaki, Y., Minato, K., Bungo, M., Niimi, S., Horichi, N., Fukuda, M., and Saijo N. "In Vitro Evaluation of the New Anticancer Agents KT6149, MX-2, SM5887, Menogaril, and Liblomycin Using Cisplatin- or Adriamycin-resistant Human Cancer Cell Lines" Cancer Res. 49, 4098-4102.
4. Arcamone, F. (1989) "Oligo (N-Methylpyrrolecarboxamide) Antibiotics" in Atta-Ur-Rahman (Ed.), Studies in Natural Products Chemistry, Vol. 5, Structure Elucidation (Part B), Elsevier, Amsterdam, pp. 549-588.
5. Zimmer, Ch. and Waehnert, U. (1986) "Non Intercalating DNA Binding Ligands: Specificity of the Intercalation and Their Use as Tools in Biophysical, Biochemical and Biological Investigation of Genetic Material" Prog. Biophys. Mol. Biol. 47, 31-112.
6. Arcamone, F. and Penco, S. (1989) "Relationship of Structure to Anticancer Activity and Toxicity in Anthracyclines" in T. Takeuchi, K. Nitta, N. Tanaka (Eds.) Antitumor Natural Products, Basic and Clinical Research, Gann Monograph on Cancer Research No. 36, Japan Sci. Soc. Press, Tokyo and Taylor & Francis Ltd. London and Bristol, pp. 81-94.
7. Gigli, M., Rasonaivo, T. W. D., Millot, J.-M., Jeannesson, P., Rizzo, V., Jardillier J.-C., Arcamone, F., and Manfait, M. (1989) "Correlation between Growth Inhibition and Intranuclear Doxorubicin and 4'-Deoxy-4'-iododoxorubicin Quantitated in Living K562 Cells by Microspectrofluorometry", Cancer Res. 49, 560-564.
8. Capranico, G., De Isabella, P., Penco, S., Tinelli, S., and Zunino, F. (1989) "Role of DNA Breakage in Cytotoxicity of Doxorubicin, 9-Deoxydoxorubicin, and 4-Demethyl-6-deoxydoxorubicin in Murine Leukemia P388 Cells", Cancer Res. 49, 2022-2027.
9. Valentini, L., Nicolella, V., Vannini, E., Menozzi, M., Penco, S., and Arcamone, F. (1985) "Association of

Anthracycline Derivatives with DNA: a Fluorescence Study", Il Farmaco Ed. Sci. 40, 377-390.

10. Arcamone, F. (1981) "Doxorubicin, Anticancer Antibiotics" Medicinal Chemistry Series Vol. 17, Academic Press, New York.

11. Zunino, F., Barbieri, B., Bellini, O., Casazza, A. M., Geroni, C., Giuliani, F., Ciana, A., Manzini, G., and Quadrifoglio, F. (1986) "Biochemical and Biological Activity of the Anthracycline Analog, 4-Demethyl-6-O-methyldoxorubicin", Invest. New Drugs 4, 17-23.

12. Katenkamp, U., Stutter, E., Petri, I., Gollmick, F. A., and Berg, H. (1983) "Interaction of Anthracyccline Antibiotics with Biopolymers. VIII. Binding Parameters of Aclacinomycin A to DNA", J. Antibiotics (Tokyo) 36, 1222-1227.

13. Fritzsche, H. and Berg, H. (1987) "Analysis of Equilibrium, Kinetic and Structural Data of Anthracycline-DNA Interaction", Gazz. Chim. Ital. 117, 331-352.

14. Bargiotti, A., Casazza, A. M., Cassinelli, G., Di Marco, A., Penco, S., Pratesi, G., Supino, R., Zaccara, A., Zunino, F., and Arcamone, F. (1983) "Synthesis, Biological and Biochemical Properties of New Anthracyclines Modified in the Aminosugatr Moiety", Cancer Chemother. Pharmacol. 10, 84-89.

15. Cassinelli, G., Ballabio, M., Arcamone, F., Casazza, A. M., aand Podestà, A. (1985) "New Anthracycline Glycosides Obtained by the Nitrous Acid Deamination of Daunorubicin, Doxorubicin and Their Configuration Analogues", J. Antibiotics (Tokyo) 38, 856-867.

16. Rizzo, V., Penco, S., Menozzi, M., Geroni, C., Vigevani, A., and Arcamone, F. (1988) "Studies of Anthracycline-DNA Complexes by Circular Dichroism", Anti-Cancer Drug Design 3, 103-115.

17. Rizzo, V., Sacchi, N., and Menozzi, M. (1989) "Kinetic Studies of Anthracycline-DNA Interaction by Fluorescence Stopped Flow Confirm a Complex Association Mechanism", Biochemistry 28, 274-282.

18. Lau, D. H. M., Lewis A. D., and Sikic, B. I. (1989) "Association of DNA Cross-Linking with Potentiation of the Morpholino Derivative of Doxorubicin by Human Liver Microsomes", J. Nat. Cancer Inst. 81, 1034-1038.

19. Chaires, J. B., Fox, K. R., Herrera, J. E., Britt, M., and Waring, M. J. (1987) "Site and Sequence Specificity of Daunomycin-DNA Interaction", Biochemistry 26, 8227-8236.

20. Jones, M. B., Hollstein, U., and Allen, F. S. (1987) "Site Specificity of Binding of Antitumor Antibiotics to DNA", Biopolymers 26, 121-135.

21. Eriksson, M., Norden, B., and Eriksson, S. (1988) "Anthracycline-DNA Interactions Studied with Linear Dichroism and Fluorescence Spectroscopy", Biochemistry 27, 8144-8151.
22. Phillips, D. R., White, R. J., abd Cullinane, C. (1989) "DNA Sequence-Specific Adducts of Adriamycin and Mitomycin C", FEBS Letters 246, 233-240.
23. Herrera, J. E. and Chaires J. B. (1989) "A Premelting Conformational Transition in Poly(dA)-Poly(dT) Coupled to Daunomycin Binding", Biochemistry 28, 1993-2000.
24. Hammer, B. C., Russel R. A., Warrener, R. N., and Collins, J. G. (1989) " A ^{19}F-NMR Study of 2-Fluoro-4-demethoxydaunomycin Intercalation Complexes with the Decanucleotides d(G-C)$_5$ and d(A-T)$_5$", Eur. J. Biochem. 178, 683-688.
25. Wang, A. H.-J., Ughetto G., Quigley G. J., and Rich, A. (1987) "Interactions between an Anthracycline Antibiotic and DNA: Molecular Structure of Daunomycin Complexed to d(CpGpTpApCpG) at 1.2-A Resolution", Biochemistry 26, 1152-1163.
26. Rizzo, V., Battistini, C., Vigevani, A., Sacchi, N., Razzano, G., Arcamone, F., Garbesi, A., Colonna, F. P., Capobianco, M., and Tondelli, L. (1989) "Association of Anthracyclines and Synthetic Hexanucleotides. Structural Factors Influencing Sequence Specificity", J. Mol. Recognition 2, 132-141.
27. Moore, M. H., Hunter, W. N., Langlois d'Estainot, B., and Kennard, O. (1989) "DNA-Drug Interactions. The Crystal Structure of d(CGATCG) Complexed with Daunomycin" J. Mol. Biol. 206, 693-705.
28. Gresh, N., Pullman, B., Arcamone, F., Menozzi, M., and Tonani, R. (1989) "Joint Experimental and Theoretical Investigation of the Comparative DNA Binding Affinities of Intercalating Anthracycline Derivatives", Mol. Pharmacol. 35, 251-256.
29. Fox, K. R. and Kunimoto, S. (1989) "Sequence Selective Binding of Ditrisarubicin B to DNA: Comparison with Daunomycin", FEBS Letters 250, 323-327.
30. Glisson, B. S. and Ross, W. E. (1987) "DNA Topoisomerase II: a Primer on the Enzyme and its Unique Role as a Multidrug Target in Cancer Chemotherapy", Pharmac. Ther. 32, 89-106.
31. Chaires, J. B. (1986) "Allosteric Conversion of Z DNA to an Intercalated Right-handed Conformation by Daunomycin", J. Biol. Chem. 261, 8899-8907.
32. Arcamone, F. (1986) "Design and Synthesis of Anticancer Drugs with Selective DNA Binding Properties", in H.C. van der Plas, M. Simonyi, F.C. Aldeweireldt, J.A. Lepoivre (Eds.) Bio-Organic Heterocycles 1986 - Synthesis, Mechanisms and Bioactivity. Proceedings of

the 4th FECHEM Conference on Heterocycles in Bio-Organic Chemistry, Elsevier, Amsterdam, pp. 119-136.

33. Bialer, M., (1984) "Distamycin and Derivatives: the Pyrrole Amidine Antiviral Antibiotics" in Y. Becker (Ed.), Antiviral Drugs and Interferon. The Molecular Basis of Their Activity, Martinus Nijoff, Boston, pp. 143-156.

34. Orlowski, S., Belehradek, J. Jr., Paoletti, C., and Mir, L. M. (1988) "Transient Electropermeabilization of Cells in Culture. Increase of the Cytotoxicity of Anticancer Drugs" Biochem. Pharmacol. 37, 4727-4733.

35. Arcamone, F., Migliacci, A., Morvillo E., Nicolella, V., Sanfilippo, A., and Schioppacassi, G. (1875) "On the Mechanism of Inhibition of Enzyme Induction in Escherichia coli by Distamycin A", Il Farmaco, Ed. Sci. 30, 859-869.

36. Arcamone, F., Lazzari, E., Menozzi, M., Soranzo, C., and Verini, M. A. (1986) "Synthesis, DNA Binding and Antiviral Activity of Distamycin Analogs Containing Different Heterocyclic Moieties", Anti-Cancer Drug Design, 1, 235-244.

37. Arcamone, F. M., Animati F., Barbieri, B., Configliacchi, E., D'Alessio, C., Geroni, C., Giuliani, F. C., Lazzari, E., Menozzi, M., Mongelli, N., Penco, S., and Verini, M. A. (1989) "Synthesis, DNA-Binding Properties and Antitumor Activity of Novel Distamycin Derivatives", J. Med. Chem. 32, 774-778.

38. Fesen, M. and Pommier, Y. (1989) "Mammalian Topoisomerase II Activity is Modulated by thev DNA Minor Groove Binder Distamycin in Simian Virus 40 DNA", J. Biol. Chem. 264, 11354-11359.

39. Woynarowski, J. M., Mchugh, M., Sigmund, R. D., and Beerman, T. A. (1989) "Modulation of Topoisomerase II Catalytic Acxtivity by DNA Minor Groove Binding Agents Distamycin, Hoeghst 33258, and 4',6-Diamidine-2-phenylindole", Mol/ Pharmacol. 35, 177-182.

40. Bruzik J. P., Auble, D. T., and deHaseth, P. L. (1987) "Specific Activation of Transcription Initiation by the Sequence Specific DNA Binding Agents Distamycin A and Netropsin", Biochemistry 26, 950-956.

41. Martello, P. A., Bruzik, J. P., de Haseth P., Youngquist, R. S., and Dervan, P. B. (1989) "Specific Activation of Open Complex Formation at an Escherichia coli Promoter by Oligo(N-methylpyrrolecarboxamide)s: Effect of Peptide Length and Identification of DNA Target Sites", Biochemistry 28, 4455-4461.

42. Levy, A., Weisman-Shomer, P.,, and Fry, M. (1989) "Distamycin Paradoxically Stimulates the Copying of Oligo(dA).Poly(dT) by DNA Polymerases", Biochemistry 28, 7262-7267.

43. Broggini, M., Ponti, M., Ottolenghi, S., D'Incalci, M., Mongelli, N., and Mantovani, R. (1989) "Distamycin Inhibits the Binding of OTF-1 and NFE-1 transfactors to the conserved DNA Elements", Nucleic Acids Res. 17, 1051-1059.
44. Gambari, R., Barbieri, R., Nastruzzi, C., Chiorboli, V., Feriotto, G., Natali, P. G., Giacomini, P., and Arcamone, F. (1990) "Distamycin Inhibits the Binding of a Nuclear Factor to the -278/-256 Upstream Sequence of the Human HLA-DRα Gene" Manuscript in preparation.
45. Weil, R., et al. (1990), Manuscript in preparation.

CATIONIC PORPHYRIN-DNA COMPLEXES: SPECIFICITY OF BINDING MODES

ROBERT J. FIEL, BRUCE G. JENKINS* and JAMES L. ALDERFER
Department of Biophysics
Roswell Park Memorial Institute
Buffalo, New York 14263

^{1}H-imino NMR is used to resolve the interaction of the cationic porphyrin meso-tetra(4-N-methylpyridyl)porphine (T4MPyP) with poly[d(G-C)$_2$]. At 0.1 M Na$^+$ several slow exchange intercalation sites were identified, with multiple peaks located upfield from the GH-1 resonance of the free polynucleotide. Coexisting "outside" binding sites were identified with a small downfield peak. Under the same conditions only line broadening of the TH3 resonance was produced in the spectrum of poly[d(A-T)$_2$], although a downfield shift (+0.34 ppm) of the AH2 resonance suggests that T4MPyP binds to the minor groove of this polynucleotide. At 0.5 M salt T4MPyP induces a transition in poly[d(A-T)$_2$], characterized by ^{31}P spectroscopy, that correlates with a "Z-like form" described earlier by circular dichroism. Results from electrophoresis experiments indicate that this form is an aggregate complex. The ^{1}H-imino spectra of the T4MPyP/calf thymus DNA complex is characteristic of intercalation at GC sites; however, the exact nature of the binding at AT sites, although not intercalation, has not been determined.

Introduction

The recognition that certain cationic porphyrins bind to DNA by intercalation evolved from a series of spectroscopic and hydrodynamic observations (1). It was noted that upon mixing with calf thymus DNA, meso-tetra(4-N-methylpyridyl)porphine (T4MPyP), Figure 1, undergoes a marked hypochromicity and bathochromic shift in its visible absorption spectrum. Induction of circular dichroism in the Soret band of the porphyrin was also observed. In addition, T4MPyP was found to bind to DNA with a large binding constant, and to increase its melting temperature and its relative viscosity.The induced circular dichroism was of particular interest in that the concentration and ionic strength dependence of the spectra indicated that T4MPyP is targeted to at least two binding sites on the DNA. At high ionic strength (ca. 1.0 M Na$^+$), a single positive band is observed at 425 nm. Initially, this band was identified (incorrectly) as representing intercalative binding; however, it was later shown that the induction of a positive CD band in a porphyrin/DNA complex is generally characteristic of outside binding and identified with AT binding sites (2-4).

At low ionic strength (ca. 16 mM Na$^+$), a negative band predominates, but the spectrum tends toward a conservative profile as the ratio of porphyrin to DNA is increased. This negative band is now generally considered to be characteristic of intercalative binding and is identified with GC binding sites (2-4).

* Current address:

Massachusetts General Hospital
NMR Center
Building 149, 13th St.
Charlestown, Massachusetts 02129

B. Pullman and J. Jortner (eds.), Molecular Basis of Specificity in Nucleic Acid-Drug Interactions, 385–399.
© 1990 Kluwer Academic Publishers. Printed in the Netherlands.

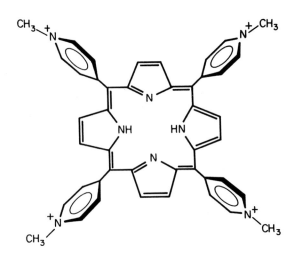

Figure 1 - Structure of meso-tetra(4-N-methylpyridyl)porphine (T4MPyP).

At intermediate ionic strength (ca. 0.2 M Na+), the CD spectrum is conservative with a cross-over at 427 nm, close to the absorption maximum at 424 nm for the free porphyrin. The CD profile at this intermediate range appears to be a composite of the spectra noted for the high and low extremes of ionic strength. Therefore, the conservative profile can be attributed to the occupation by T4MPyP of two binding sites (AT and GC) via two binding modes: outside binding and intercalation, respectively (2-4).

Taken together, the spectroscopic and hydrodynamic data provided strong evidence that T4MPyP binds to DNA by intercalation. Although the seemingly prohibitive stereochemical barrier presented by the meso-substituents (see Figure 1) argued against this possibility, the subsequent demonstration that T4MPyP unwinds (and rewinds) supercoiled DNA provided decisive evidence in favor of intercalation (5). This result was confirmed by two independent studies employing similar assay systems (6,7). In addition, many details as to the nature of the T4MPyP/DNA intercalation complex have been obtained from recent NMR experiments (6,8-11) and theorectical modelling studies (12,13), and the concept is now generally accepted (for review see [14])

Following the initial demonstration of T4MPyP-induced DNA unwinding, the assay was employed to investigate the ionic strength dependence of T4MPyP intercalation (2). Using a viscosimetric technique, it was shown that the endpoint marking the T4MPyP-induced conversion of supercoiled PM-2 DNA to the relaxed (open circular) form shifts to larger values of R as ionic strength increases. This shift corresponds to a decrease in the unwinding angle of T4MPyP from 20° to approximately 2.5° as the concentration of Na+ is increased from 0 to 0.3 M. In accordance with the circular dichroism studies, these data were interpreted as resulting from a shift from GC specificity at low ionic strength to AT specifity at high ionic strength, (i.e., a transition from intercalative binding which unwinds DNA, to outside binding which does not).

The DNA-unwinding assay was also used to monitor the binding characteristics of a related series of porphyrins and metalloporphyrins. It was shown that T3MPyP (where the number 3 designates the meta position of the N-methylpyridyl group) is an intercalator (2). Similarly, the Ni and Cu derivatives of T4MPyP were also identified as intercalators, whereas the Fe, Mn, Co, and Zn derivatives were classified as outside binding porphyrins (15). T2MPyP and TMAP (meso-tetra[4-N-trimethylanilinium]porphine), neither of which unwind supercoiled DNA, were classified as outside binding porphyrins (2,3). Interestingly, DNA-bound TMAP was found to stabilize itself by self-stacking in a parallel array, perpendicular to the axis of the double helix (16). Thus, outside binding with self-stacking represents a third binding mode or, as discussed below, a subgroup of a broadly defined outside binding mode.

By combining data from the unwinding assay with spectroscopic results on DNA and synthetic polynucleotides, a working hypothesis was formulated to address the question of the base specificity of these three binding modes (2). Briefly stated, at small R (the ratio of porphyrin to base pairs) and low ionic strength, T4MPyP, its corresponding Ni and Cu derivatives, and T3MPyP bind by intercalation with some degree of GC specificity. As R increases, the GC selectivity decreases with an apparent shift to AT specificity. The nonintercalating metal analogs of T4MPyP (Fe,Mn,Co,and Zn), T2MPyP, TMAP, and the intercalating porphyrins at large R and/or medium to high ionic strength, bind by a broadly defined outside mode that appears to express some degree of AT selectivity. The outside binding mode can be classified into three subgroups: (A) nonspecfic electrostatic, as illustrated by T2MPyP (B) self-stacking, as illustrated by TMAP (C) groove binding, as illustrated by nonintercalating metalloporphyrins and, at specific conditions of R and ionic strength, porphyrin intercalators. Groove binding, probably the most complex of the subgroups, may include "partial intercalation" among its variations.

The present work further investigates the nature the binding of T4MPyP to AT and GC base pairs using corresponding synthetic polydeoxyribonucleotides and calf thymus DNA as substrates.

Materials and Methods

Meso-tetra(4-N-methylpyridyl)porphine (T4MPyP) was purchased as the tetraiodide salt from Strem Chemicals Inc. and used without further purification. Poly[d(A-T)$_2$] and poly[d(G-C)$_2$] were purchased from Boehringer-Mannheim Biochemicals. Calf thymus DNA was purchased from Worthington Biochemicals. The polymers were sonicated under nitrogen and their lengths estimated by electrophoresis against a standard prepared from the Hae III restriction products of plasmid pAL III, kindly provided by Dr. B. R. Munson. Analysis of the electrophoresis data indcated the following lengths, given in base pairs: poly[d(A-T)$_2$]=100±20, poly[d(G-C)$_2$]=90±15, calf thymus DNA=130±25.

NMR spectra were collected in the Fourier transform mode on either a Bruker WP-200 operating at 200 MHz for ^{1}H and 81 MHz for ^{31}P or a Bruker AM-400 operating at 400 MHz for [1]H.

Results

Poly[d(G-C)$_2$]: Many recent investigations employing ^{1}H and ^{31}P NMR have provided important information about the base specificity of ligand binding to DNA. Feigon et al.(17), utilized ^{1}H-imino NMR to probe the effects of over 70 clinical and experimental antitumor drugs on the Watson Crick base-pairs in DNA. Application of this technique to studies of porphyrin/DNA complexes has also proven to be useful (8-11). In order to develop a basis for the analysis of effects on DNA, it is helpful to first determine effects on synthetic polynucleotides.

It is instructive to first examine how T4MPyP influences the ^{1}H imino spectrum of poly[d(G-C)$_2$]. As shown in Figure 2, the addition of T4MPyP produces a complicated pattern for the imino protons of guanine (GH1). Significant line broadening is apparent with increasing R. The position of the GH1 peak is shifted upfield and at least four new peaks are resolved in the upfield region.A small downfield peak is also observed. The appearance of a downfield peak is generally thought to be typical of groove binding (17-19). It is also noted that this peak has a strong electrostatic component, shown by its enhanced resolution at conditions of very low ionic strength (Figure 3).

The multiple upfield peaks shown in Figure 2 indicate intercalative binding and can be explained as arising from different intercalation sites under conditions of slow exchange (17). These effects are similar to those shown by Marzilli, et al. (9). Figure 4 is a simple schematic

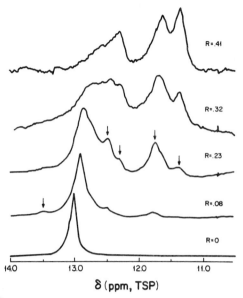

<u>Figure 2</u> - 400 MHz ^{1}H imino spectrum of 14 mM poly[d(G-C)$_2$] in 0.1 M NaCl, 10 mM sodium phosphate pH 7.2, 1 mM EDTA at 30°C as a function of R.

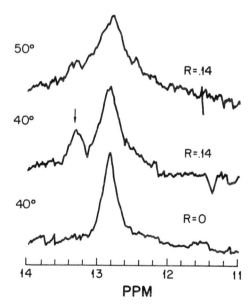

<u>Figure 3</u> - 200 MHz ^{1}H imino spectrum of poly[d(G-C)$_2$] in 10 mM sodium phosphate buffer pH 7.2, with no added salt. R = 0.1.

depicting possible intercalation sites for R=0.23, i.e., when there are approximately three molecules of T4MPyP for every 13 base-pairs. The site labeled Type 0 is unoccupied and therefore a relatively unperturbed struture with a "normal" imino resonance. The Type I site is defined as a "nearest neighbor", being one base-pair away from the intercalated ligand. Type II is defined as being one and two base-pairs away from two molecules of T4MPyP. Type III produces resonance from imino protons directly adjoining an intercalation site and

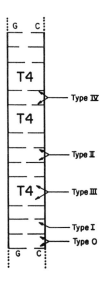

Figure 4 - Schematic for some of the various imino types possible upon slow intercalation of T4MPyP (T4) into poly[d(G-C)$_2$] when R = 0.23.

Type IV corresponds to resonances from imino protons in an "excluded" site between two intercalated molecules of T4MPyP. Although the spectral assignments of these types of binding sites are somewhat arbitrary (Figure 5), they are generally supported by NOE and ring current calculations (20).

Additional evidence in support of slow exchange for the intercalation model of T4MPyP binding to poly[d(G-C)$_2$] is shown in Figure 5. As the temperature is raised the peaks broaden and shift, indicating that the T4MPyP/GC exchange rate is increasing. The T4MPyP/GC exchange rate exhibits a strong correlation with base-pair opening rates determined for poly[d(G-C)$_2$] in 0.1 M NaCl (21), suggesting that the rate of intercalation of T4MPyP may be a base-pair opening limited event.

Poly[d(A-T)$_2$]: T4MPyP causes very little effect on the ^{1}H imino spectrum of poly[d(A-T)$_2$] relative to that seen for the GC polymer. As shown in Figure 6, T4MPyP does not affect the position of the peak for the imino proton from thymine (TH3) up to R=0.33, but does induce some line broadening, probably due to an increased correlation time for the complex. Although some interaction is apparent, the effect is not indicative of intercalation. Generally, the outside binding mode of T4MPyP and the other members of subgroup (C) to AT base pairs is not well understood. Our original contention that the binding of T4MPyP to AT sites increases with increasing ionic strength (2) was confirmed by Pasternack et al. (22). Although their conclusions were also based on CD measurements, they extended the work by determining the distribution of T4MPyP between AT and GC sites in calf thymus DNA as

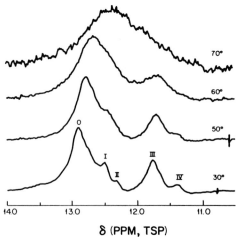

<u>Figure 5</u> - 400 MHz ^{1}H imino spectrum of poly[d(G-C)$_2$] + T4MPyP in buffer (R = 0.23) as a function of temperature (°C).

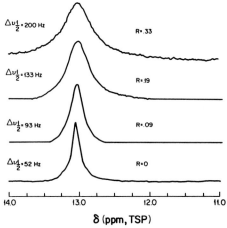

<u>Figure 6</u> - 400 MHz ^{1}H imino spectrum of 12.5 mM poly[d(A-T)$_2$] in the above buffer at 30°C as a function of the R value (R = [T4MPyP]/[b.p.]).

a function of ionic strength. The total number (GC+AT) of occupied binding sites was shown to decrease with increasing ionic strength, but the fraction of GC sites occupied decreases while that of the AT sites increases over a range of ionic strength of 0.004 to 2.0. In a detailed investigation of the equilibrium binding of T4MPyP to synthetic polynucleotides, Strickland et al. (23) concluded that there was no significant preference for either poly[d(A-T)$_2$] or poly[d(G-C)$_2$] between 0.115 and 0.515 M Na$^+$, in apparent contradiction to the CD results. However, a close examination of Strickland's data reveals that the ratio of K_{GC}/K_{AT} is 1.075 at 0.115 M; 0.682 at 0.215 M; 0.657 at 0.315 M; and 0.838 at 0.515 M Na$^+$. We cannot determine whether these differences are significant, but they suggest that further studies will be required to fully resolve the point.

Differences in the nature of the binding of T4MPyP to the AT and GC synthetic polymers is further revealed in the NOE difference spectra of the T4MPyP/poly[d(A-T)$_2$] (not shown). The primary observation is that the AH2 and TH6 resonances which overlap at 400 MHz in the absence of T4MPyP are resolved as porphyrin is added. Moreover, the AH2 resonance shifts downfield (+0.34 at R=0.33) while all others (AH8,TH6,H1',H2',H2",TCH$_3$) shift upfield by relatively small amounts, the largest being -0.16 at R=0.33 for TCH$_3$. This along with other NOE data indicating a dipolar interaction between protons of T4MPyP and AH2 identify the porphyrin as bound to the minor groove, and suggest that the B-type conformation of the polymer is generally unperturbed (24,25).

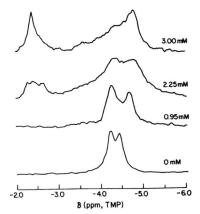

Figure 7 - 81 MHz ^{31}P spectrum of 15 mM poly[d(A-T)$_2$] in 0.5 M NaCl, 10 mM sodium phosphate pH 7.2 at 30°C as a function of T4MPyP concentration.

By contrast, it is interesting that at high ionic strength, T4MPyP induces a dramatic change in the apparent conformation of poly[d(A-T)$_2$]. This change may have some bearing on the mechanism by which T4MPyP binds to AT sites in natural DNA. First observed in CD studies by Carvlin, et al. (26), it was described as a Z-like transition. This transition is also readily demonstrated using ^{31}P NMR spectroscopy as shown in Figure 7, where increasing concentrations of T4MPyP induce significant changes in the spectrum of poly[d(A-T)$_2$] in 0.5 M NaCl. The spectra closely resemble the B to Z transition in the poly[d(G-C)$_2$] and poly[d(G-meC)$_2$] systems (27,28). There is a gradual separation of the two peaks, a feature generally characteristic of binding by Na$^+$ and Cs$^+$ (29,30), followed by an apparently cooperative transition from B to a Z-like form. The transition appears to be complete at a T4MPyP concentration of 3 mM. Note that the peak at -4.35 ppm is shifted downfield by approximately 2.0 ppm to -2.42 ppm, while the peak at -4.80 ppm remains unchanged. This effect is similar to that observed in the B to Z transition of poly[d(G-meC)$_2$] (downfield shift of the downfield component of the doublet) and implies that the high concentration of T4MPyP induces a gauche-trans conformation for the dTpA phosphodiester linkage of poly[d(A-T)$_2$] (31).

It should be noted that the concentration of T4MPyP has been expressed in terms of molarity in these examples rather than as R, since the transition is dependent upon the concentration of the polynucleotide rather than the ratio of T4MPyP to poly[d(A-T)$_2$]. This is demonstrated by the results shown in Figure 8 (top spectrum). With the concentration of the polynucleotide at 7.5 mM and the concentration of T4MPyP at 2.25 mM (R=0.60), the ratio of the area of the downfield peak (Z-type) to the B-type peak is 0.33. Compare this to Figure 7, where the polynucleotide concentration is 15 mM and T4MPyP is 3 mM (R=0.40), ratio of the peaks is 1. The two bottom spectra of Figure 8 show the effect of dilution while holding R and ionic strength constant. Diluting the polynucleotide from 15 mM to 3 mM (R=0.4) restores the original (B-type) spectrum.

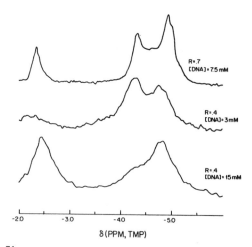

Figure 8 - 81 MHz ^{31}P spectrum of poly[d(A-T)$_2$] in 0.5 M NaCl buffer at 30°C. Bottom - 15 mM poly[d(A-T)$_2$] + 3 mM T4MPyP (R = 0.4). Middle - same exact sample as above diluted five fold (R = 0.4). Top - 7.5 mM poly[d(A-T)$_2$] in 0.5 M NaCl buffer with 2.6 mM T4MPyP (R = 0.7).

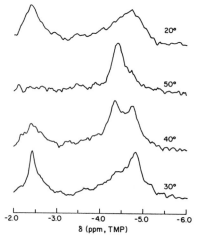

Figure 9 - 81 MHz ^{31}P spectrum of 15 mM poly[d(A-T)$_2$] + 3 mM T4MPyP as a function of temperature. The top spectrum at 20°C is cooled from over 50°C demonstrating the transition is reversible.

The intensity of the characteristic downfield peak of the T4MPyP/polynucleotide complex is diminished on raising the temperature from 30° to 40° and is absent at 50°, well below the T_m of approximately 80° (Figure 9). Also note that the characteristic doublet of the B-form is lost at elevated temperatures, indicating that a conformational transition to an intermediate state may have occurred since cooling to 20° restores the Z-type spectrum.

Another feature of the stability of this complex is that the addition of ethanol to only 10% (v:v) reverses the B to Z-like transition (Figure 10). This is not simply an effect of dilution since it does not occur with the addition of an equivalent volume of water. By comparison, it is known that ethanol facilitates the B to Z transition in poly[d(G-C)$_2$] (32,33). Therefore, a major and characteristic difference exists between these two systems.

Additional information on the nature of the T4MPyP/poly[d(A-T)$_2$] complex can be obtained from the proton spectra of the aromatic region of the polynucleotide upon addition

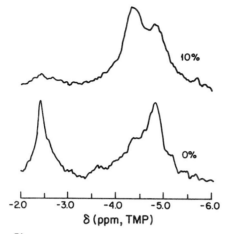

Figure 10 - 81 MHz ^{31}P spectrum of Z-type poly[d(A-T)$_2$] from above as a function of ethanol (% v:v).

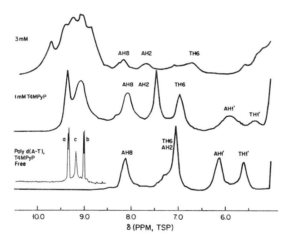

Figure 11 - ^{1}H NMR spectrum of 15 mM poly[d(A-T)$_2$] in 0.5 M NaCl buffer at 30°C as a function of T4MPyP concentration. The control ^{1}H spectra (free T4MPyP and free poly[d(A-T)$_2$]) are shown at the bottom.

of the porphyrin. Figure 11 (bottom) is a spectrum of free T4MPyP superimposed onto a spectrum of free poly[d(A-T)$_2$]. The addition of T4MPyP to a concentration of 1 mM induces a downfield shift of AH2 and upfield shifts of TH6, AH1'and TH1'. At 3 mM T4MPyP, the concentration at which the Z-like ^{31}P spectrum occurs, signicant increases in linewidths and shifts of the resonance peaks are noted in the spectrum for the porphyrin and the polynucleotide. Except for TH6, the base protons now undergo a downfield shift while the H1'protons undergo large upfield shifts. Although specific conformational changes cannot be assigned, these data clearly indicate that a change in the secondary structure has occurred.

¹H NOE difference studies have also been carried out (Figure 12) and, although an unequivocal assignment of the conformation of the complex cannot be made from these spectra, they provide useful information. Pre-saturation at AH8 enhanced the resonance at AH1'to a greater degree than at TH1', i.e., an effect opposite to that seen for B-DNA (24,25). Enhancement is absent in the H2',H2" region, but some cross relaxation is noted in the H4',H5',H5" region at about 4 ppm. In addition, strong enhancement is seen in the H3' region around the HDO peak. This effect, combined with the lack of intensity in the H2',H2" region, indicates that the conformation of the adenine sugar is C3' endo.

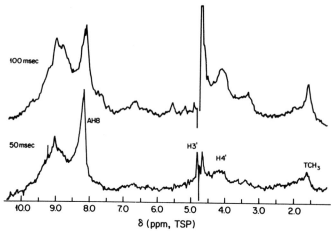

Figure 12 - ¹H NOE difference spectra of the aggregated poly[d(A-T)₂] 3 mM T4MPyP from above as a function of pre-saturation time of AH8.

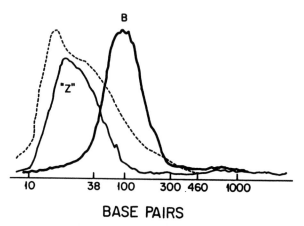

Figure 13 - Densitometer tracings of electrophoretic migration of free poly[d(A-T)₂] in the aggregated form with 3 mM T4MPyP; and with 1 mM T4MPyP in 2.5% agarose gels. Numbers at the bottom correspond to mobilities of known base-pair markers of Hae III restricted DNA, with the exception of the marker labeled 10 bp which is from a sample of d(pGpT)₅.d(pApC)₅ which was co-loaded on the gel.

Even though these data defining the nucleic acid conformational form of the complex are limited, they clearly illustrate that it must be substantially different from the B-form. In reviewing all the data at hand, it seems likely that the complex formed by T4MPyP and poly[d(A-T)$_2$] at high concentrations of Na$^+$ is an aggregate. This conclusion is supported by results obtained using gel electrophoresis to study the complex, as shown in the densitometer tracings of Figure 13. The abscissa represents the mobility in base-pairs of the digestion products from the plasmid pAL-III and the restriction enzyme Hae III (34). Both the fully (3 mM T4MPyP) and partially (1 mM T4MPyP) converted Z-like forms migrate as if they were either shorter fragments than the control poly[d(A-T)$_2$] or more negatively charged. However, the starting materials all have the same length and T4MPyP is a cation (+4), so the complex must have a lower net negative charge than the control polynucleotide. The results of the gel electrophoresis can be explained if the T4MPyP were to induce aggregation in the polynucleotide such that the compact structure would migrate more easily through the pores of the gel.

Calf Thymus DNA: Measurements of the effect of T4MPyP on the imino protons of calf thymus DNA (TH3,GH1) are shown in Figure 14. The appearance of the spectrum for the porphyrin/DNA complex is that generally expected for intercalation (17). A nearly equivalent upfield shift is noted for both peaks along with some line broadening. However, the GC sites appear to have undergone a somewhat greater change, as shown by the reduction in the intensity of the GH1 resonance relative to that of TH3 and the resolution of one or two small peaks in the upfield region. Similar spectra have been obtained by Banville, et al. (8) using salmon sperm DNA. The overall effect on GH1 may be best described as resulting from intercalation at an intermediate exchange rate, at least relative to the slow exchange rate apparent for T4MPyP/poly[d(G-C)$_2$]. Interpretation of the effect on TH3 is intractable except to note that it is inconsistent with groove binding, such as exhibited by distamycin, which characteristically induces downfield shifts of the imino protons (17,19). The result is consistent, however, with our earlier data and subsequent hypothesis that T4MPyP targets two sites on DNA, one at AT base pairs and the other at GC base pairs.

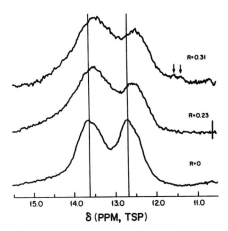

Figure 14 - 400 MHz ^{1}H imino spectrum of 13 mM calf thymus DNA in 0.1 M NaCl, 10 mM sodium phosphate pH 7.2, 1 mM EDTA at 30°C as a function of the R value of added T4MPyP. Arrows indicate slow exchange imino resonances similar to those seen in the poly[d(G-C$_2$] spectrum (Figure 2).

Discussion

The evidence supporting the conclusion that several cationic porphyrins bind to poly[d(G-C)$_2$] by intercalation appears to be uneqivocal. Many of the details of the binding have only recently become known. In addition to the hydrodynamic evidence (8), the ^{1}H imino NMR data presented here and by others (8) clearly describe an intercalative complex for the interaction of T4MPyP with this polynucleotide. Several intercalation sites have been resolved upfield of the unperturbed GH1 resonance and preliminary assignments have been made. In addition, a downfield peak has been observed that is consistent with the coexistence of an outside binding mode for T4MPyP for solutions that are approximately 100 mM Na$^+$ and below.

The intercalative binding mode of several cationic porphyrins and metalloporphyrins to DNA has also been well established by supercoiled DNA unwinding and NMR studies (5-11,15). More recent NMR investigations of the interaction of porphyrins with synthetic oligomers led to the identification of 5'CpG 3' as a predominant sequence for intercalation (9,11). This pivotal result has been confirmed by footprinting experiments using DNA fragments of known sequence (35-37), and is supported by theoretical studies (12,13). Bütje, et al. (38), in an interesting application of resonance Raman specroscopy to study the interaction of cationic porphyrins with a series of hexadeoxyribonucleotides, also conclude that 5'CpG 3' is a primary sequence for intercalation. In addition, these authors note that the 5'GpC 3' site and the mismatched hexamer d(TGTGCA)$_2$ are capable of intercalation. Complicating this picture is the fact that those cationic porphyrins which have been identified as intercalators are also thought to bind to DNA via at least one other mode. Some of the experimental evidence for a second mode is consistent with outside binding at AT base pairs. Interestingly, footprinting experiments have identified the sequence 5'TpA 3' as another binding site for intercalative porphyrins. Although this technique cannot discriminate between binding modes, theoretical studies indicate that intercalation into this sequence is energetically unfavorable and that the porphyrin probably occupies the site by outside groove binding (12,13). Therefore, it seems reasonable to view the T4MPyP/DNA complex as a dynamic process in which porphyrin molecules bind to various sites with a spectrum of exchange rates determined by the nature of the site. Binding at some of these sites will induce substantial distortions of the native structure of DNA and thereby modify the nature of neighboring sites, while little effect will be noted at other sites.

The nature of the complex at equilibrium will depend upon many factors; however, since one component is a multicharged cation (+4) and the other a polyanion, the ionic strength of the solution must be a critical parameter. It may in fact limit intercalation to solution conditions with Na$^+$ below 300 mM, as shown by PM-2 DNA unwinding studies. Also, the ionic strength dependence of T4MPyP-induced unwinding of supercoiled DNA runs counter to that of more conventional intercalators. For example, the apparent unwinding angle for chloroquine and 9-amino-1,2,3,4-tetrahydroacridine increases with increasing ionic strength (39). For T4MPyP however, a low concentration of counter ion (Na$^+$) reduces screening between the phosphate groups of DNA and the cationic N-methylpyridyl groups of the porphyrin. Maximizing the electrostatic interaction between these groups apparently provides an essential stabilization force to the intercalation complex. This point was also apparent in earlier model building experiments in which close contact between the phosphate and N-methylpyridyl groups was seen to be a critical factor (5).

Although intercalative binding of T4MPyP to DNA may be limited to an Na$^+$ concentration below ca. 300 mM, outside binding appears to extend well beyond this limit (1-3). Clearly this cannot be described as the type of nonspecific electrostatic binding characterized by T2MPyP (subgroup A), since binding of this porphyrin to DNA is not detected above 200 mM Na$^+$ (2). The term 'groove binding' has been used; however, the ^{1}H imino NMR spectra shown here (Figure 14) and by others (8) does not typify the characteristic spectra obtained for known groove binding ligands such as distamycin (17,19). Although a direct link has not been made, we believe that the unique Z-like transformation induced in poly[d(A-T)$_2$] by T4MPyP at 0.5

M Na$^+$ may be associated with the mechanism by which this porphyrin binds to DNA at high salt conditions. This may also relate more generally, with the exception of subgroup (A), to outside binding porphyrins and could involve partial intercalation.

It is noteworthy that the DNA interactive porphyrins and metalloporphyrins are of general interest because of their unexpected ability to intercalate; however, at this junction it appears that the mechanism by which these porphyrins bind to the "outside" of DNA may be of even greater interest, and even greater complexity.

REFERENCES

1. Fiel, R.J., Howard, J.C., Mark, E.H. and Datta-Gupta, N. (1979) 'Interaction of DNA with a porphyrin ligand: evidence for intercalation ', Nucleic Acids Res. 6, 3093-3118.
2. Carvlin, M.J. and Fiel, R.J. (1983) 'Intercalative and nonintercalative binding of large cationic porphyrin ligands to calf thymus DNA', Nucleic Acids Res. 11, 6121-6139.
3. Carvlin, M.J., Mark, E., Fiel, R.J. and Howard, J.C. (1983) 'Intercalative and nonintercalative binding of large cationic porphyrin ligands to polynucleotides', Nucleic Acids Res. 11, 6141-6154.
4. Pasternack, R.F., Gibbs, E.J. and Villafranca, J.J. (1983) 'Interactions of porphyrins with nucleic acids', Biochem. 22, 2406-2414.
5. Fiel, R.J. and Munson, B.R. (1980) 'Binding of meso-tetra(4-N-methylpyridyl)porphine to DNA', Nucleic Acids Res. 8, 2835-2842.
6. Banville, D.L., Marzilli, L.G. and Wilson, W.D. (1983) '^{31}P NMR and viscometric studies of the interaction of meso-tetra-(4-N-methylpyridyl)porphine and its Ni(II) and Zn(II) Derivatives with DNA', Biochem. Biophys. Res. Commun. 113, 148-154.
7. Kelly, J.M., Murphy, M.J., McConnell, D.J. and OhUigin, C. (1985) 'A comparative study of the interaction of 5, 10, 15, 20-tetrakis(N-methylpyridinium-4-yl)porphyrin and its zinc complex with DNA using fluorescence spectroscopy and topoisomerisation', Nucleic Acids Res. 13, 167-184.
8. Banville, D.L., Marzilli, L.G. and Strickland, J.A. (1986) 'Comparision of the effects of cationic porphyrins on DNA properties: influence of GC content of native and synthetic polymers', Biopolymers 25, 1837-1858.
9. Marzilli, L.G., Banville, D.L., Zon, G. and Wilson, W.D. (1986) 'Pronounced ^{1}H and ^{31}P NMR spectral changes on meso-tetra(4-N-methylpyridyl)porphyrin binding to poly d(G-C).poly d(G-C) and to three tetradecaoligodeoxyribonucleotides: evidence for symmetric, selective binding to 5'CG3'sequences', J. Am. Chem. Soc. 108, 4188-4192.
10. Strickland, J.A., Banville, D.L., Wilson, W.D. and Marzilli, L.G. (1987) 'Metalloporphyrin effects on properties of DNA polymers', Inorg. Chem. 26, 3398-3406.
11. Strickland, J.A., Marzilli, L.G., Wilson, W.D. and Zon, G. (1989) 'Metalloporphyrin DNA interactions: insights from NMR studies of oligodeoxyribonucleotides', Inorg. Chem. 28, 4191-4206.
12. Ford, K.G., Pearl, L.H. and Neidle, S. (1987) 'Molecular modeling of the interactions of tetra-(4-N-methylpyridyl)-porphin with TA and CG sites on DNA', Nucleic Acids Res. 15, 6553-6562.
13. Hui, X., Gresh, N. and Pullman, B. (1990) 'Modelling of the binding specificity in the interactions of cationic porphyrins with DNA', Nucleic Acids Res. 18, 1109-1114.
14. Fiel, R.J. (1989) 'Porphyrin-nucleic acid interactions: a review', J. Biomol. Struct. Dynam. 6, 1259-1274.
15. Fiel, R.J., Carvlin, M.J., Byrnes, R.W. and Mark, E.H.   (1984) 'DNA interactive porphyrins: metalloporphyrin derivatives', in R. Rein (ed.), Molecular Basis of Cancer, Part B: Macromolecular Recognition, Chemotherapy, and Immunology, Alan R. Liss, New York, pp. 215-226.
16. Carvlin, M.J., Datta-Gupta, N. and Fiel, R.J. (1982) 'Circular dichroism spectroscopy of a cationic porphyrin bound to DNA', Biochem. Biophys. Res. Commun. 108, 66-73.

17. Feigon, J., Denny, W.A., Leupin, W. and Kearns, D.R. (1984) 'Interactions of antitumor drugs with natural DNA: ^{1}H NMR study of binding mode and kinetics', J. Med. Chem. 27, 450-465.
18. Patel, D.J., Pardi, A. and Itakura, K. (1982) 'DNA conformation dynamics and interactions in solution',Science 216, 581-590.
19. Patel, D.J. (1982) 'Antibiotic-DNA interactions: intermolecular nuclear Overhauser effects in the netropsin-d(C-G-C-G-A-A-T-T-C-G-C-G) complex in solution', Proc. Natl. Acad. Sci. (USA) 79, 6424-6428.
20. Jenkins, B.G. (1986) 'The effect of base sequence upon the properties of naturally occuring DNA polymers: a biophysical investigation', Ph.D. dissertation, RPMI Division, SUNY/Buffalo.
21. Mirau, P.A. and Kearns, D.R. (1984) 'Effect of environment, conformation, sequence and base substituents on the imino proton exchange rates in guanine and inosine-containing DNA, RNA, and DNA-RNA duplexes', J. Mol. Biol. 177, 207-   227.
22. Pasternack, R.F., Garrity, P., Ehrlich, B., Davis, C.B., ibbs, E.J., Orloff, G., Giartosio, A. and Turano, C. (1986) 'The influence of ionic strength on the binding of   a water soluble porphyrin to nucleic acids', Nucleic Acids Res. 14, 5919-5931.
23. Strickland, J.A., Marzilli, L.G., Gay, K.M. and Wilson, W.D.   (1988) 'Porphyrin and metalloporphyrin binding to DNA polymers: rate and equilibrium binding studies', Biochem. 27, 8870-8878.
24. Assa-Munt, N. and Kearns, D.R. (1984) 'Poly(dA-dT) has a right-handed B conformation in solution: a two-dimensional NMR study', Biochem. 23, 791-796.
25. Borah, B., Cohen, J.S. and Bax, A. (1985) 'Conformation of double-stranded polydeoxynucleotides in solution by proton two-dimensional nuclear Overhauser enhancement spectroscopy', Biopolymers 24, 747-765.
26. Carvlin, M.J., Alderfer, J.L. and Fiel, R.J. (1983) 'DNA complexes on intercalative and outside binding porphyrins', Third Conversation in Biomolecular Stereodynamics, symposium report, p. 127.
27. Patel, D.J., Canuel, L.L. and Pohl, F.M. (1979) "Alternating B-DNA' conformation for the oligo(dG-dC) duplex in high-salt solution', Proc. Natl. Acad. Sci. (USA) 76, 2508-2511.
28. Patel, D.J., Kozlowski, S.A., Nordheim, H. and Rich, A. (1982) 'Right-handed and left-handed DNA: studies of B-DNA and Z-DNA by using proton nuclear Overhauser effect and P NMR', Proc. Natl. Acad. Sci. (USA) 79, 1413-1417.
29. Patel, D.J., Kozlowski, S.A., Suggs, J.W. and Cox, S.D. (1981) 'Right-handed alternating DNA conformation: poly(dA-dT) adopts the same dinucleotide repeat with cesium, tetraalkylammonium, and 3 alpha, 5 beta, 17 beta-dipyrrolidinium, and steroid dimethodide cations in aqueous solution', Proc. Natl. Acad. Sci. (USA) 78, 4063-4067.
30. Jenkins, B.G., Wartell, R.M. and Alderfer, J.L. (1986) 'Conformational properties of poly[d(G-T)].poly[d(C-A)] and poly[d(A-T)] in low- and high-salt solutions: NMR and laser Raman analysis', Biopolymers 25, 823-849.
31. Chen, C.-W., Cohen, J.S. and Behe, M. (1983) 'B to Z transition of double-stranded poly[deoxyguanylyl(3'-5')-5-methyldeoxycytidine] in solution by phosphorus-31 and carbon-13 nuclear magnetic resonance spectroscopy', Biochem. 22, 2136-2142.
32. Pohl, F.M. (1976) 'Polymorphism of a synthetic DNA in solution', Nature 260, 365-366.
33. Behe, M. and Felsenfeld, G. (1981) 'Effects of methylation of a synthetic polynucleotide: the B-Z transition in poly(dG-m5dC).poly(dG-m5dC)' Proc. Natl. Acad. Sci. (USA) 78, 1619-23.
34. Lenard, A., Weinberger, M., Munson, B. and Helmstetter, C. (1980) 'The effects of oriC-containing plasmids on host cell growth', in B. Alberts (ed.), Mechanistic Studies of DNA, Replication and Genetic Recombination, Academic Press, Inc., New York, pp. 171-179.

35. Ward, B., Skorobogaty, A. and Dabrowiak, J.C. (1986) 'DNA binding specificity of a series of cationic metalloporphyrin complexes', Biochem. 25, 7827-7833.
36. Ford, K., Fox, K.R., Neidle, S. and Waring, M.J. (1987) 'DNA sequence preferences for intercalating porphyrin compound revealed by footprinting', Nucleic Acids Res. 15, 2221-2234.
37. Kuroda, R., Takahashi, E., Austin, C.A. and Fisher, L.M. (1990) 'DNA binding and intercalation by novel porphyrins: role of charge and substituents probed by DNase I footprinting and topoisomerase I unwinding', FEBS 262, 293-298.
38. Bütje, K., Schneider, J.H., Kim, J.-J.P., Wang, Y., Ikuta, S. and Nakamoto, K. (1989) 'Interactions of water-soluble porphyrins with hexadeoxyribonucleotides: resonance Raman, UV-visible and ^{1}H NMR studies', J. Inorg. Biochem. 37,119-134.
39. Jones, R.L., Lanier, A.C., Keel, R.A. and Wilson, W.D. (1980) 'The effect of ionic strength on DNA-ligand unwinding angles for acridine and quinoline derivatives', Nucleic Acids Res. 8, 1613-1624.

Acknowledgments

The authors thank Drs. Ben Munson and Patricia Johnson for their helpful suggestions and James J. Alletto, Joia Di Stefano, and Ester Mark for their help in preparing the manuscript.

COMPLEMENTARY STUDIES ON SEQUENCE SPECIFICITY IN DNA-ANTITUMOR DRUGS INTERACTIONS(*).

BERNARD PULLMAN
Institut de Biologie Physico-Chimique, Fondation
Edmond de Rothschild, 13, rue Pierre et Marie
Curie, PARIS 75005, France.

ABSTRACT

The author presents results of recent computations on the DNA sequence specificity of four types of antitumor drugs which have been relatively little explored theoretically till now : 1) analogues of the classical groove binding ligand distamycin A in which the heteroaromatic rings are replaced by hydrocarbon units (benzene rings or saturated β-alanine moieties), 2) tetracationic porphyrins, 3) dicationic steroid diamines and 4) aurelic acid derived antibiotics. Some of these groups have been actively investigated experimentally by authors present at this meeting.

The somewhat ambiguous title of my contribution (complementary to what ?) obviously needs an explanation. The complementarity is with respect to the large recent review of our studies on "Molecular Mechanisms of Specificity in DNA-Antitumor Drug Interactions" which I have presented in the last published issue of Advances in Drug Research [1]. In fact, I shall also omit from this paper the latest developments on the problems of lexitropsins [2, 3], isolexins [4] and vinylexins [5], a review of which was presented in two recent publications [6, 7]. This paper will thus deal only with our current exploration of newer, practically untouched as yet by theory, groups of antitumor drugs. There are four of them, three of which are by a "lucky" coincidence also

---

(*) This paper is dedicated to Professor G.B. Marini Bettolo of the Universita Degli Studi di Roma, President of the Pontifical Academy of Sciences, at the occasion of his 75th birthday.

B. Pullman and J. Jortner (eds.), Molecular Basis of Specificity in Nucleic Acid-Drug Interactions, 401–422.
© 1990 Kluwer Academic Publishers. Printed in the Netherlands.

investigated experimentally by authors present in this meeting.

I - ANALOGUES OF THE CLASSICAL GROOVE BINDING LIGANDS NETROPSIN AND DISTAMYCIN A IN -WHICH THE HETEROAROMATIC PYRROLE RINGS ARE REPLACED BY HYDROCARBON UNITS.

Netropsin (I) and distamycin A (II) are fundamental DNA groove-binding ligands with the characteristic specificity of these type of antitumor drugs for the minor groove of A-T sequences of B-DNA. A number of theories have been proposed to explain the origin of this specificity which according to some of them would be due to such local factors as hydrogen bonding between the NH groups of the amide linkages of these compounds with the N_3 and/or O_2 receptor sites at the AT base pairs [8], or van der Waals contacts between the CH groups of the pyrrole rings of the drugs and the adenine C-2 hydrogens [9]. We have been able to demonstrate the insufficiency or inadequacy of such proposals and have shown the fundamental role played in the origin of the specificity by the localisation of the deepest electrostatic molecular potential of DNA in the minor groove of its AT sequences [see 1 and the references therein].

The recent preparation, essentially by Sasisekharan and his collaborators [10-13], of synthetic analogues of distamycin A in which the pyrroles are replaced by benzene rings or by a saturated β-alanine moiety, yielding the representative compounds III and IV, offers the possibility of exploring the effect of the changes on the specificity of the parent compound and simultaneously of reexamining the validity of the different proposals concerning the origin of this specificity.

Computations have been carried out for this sake in our Laboratory on the comparative interaction energies of distamycin A and compounds III and IV with the representative DNA sequences $d(ATATATA)_2$ and $d(GCGCGCG)_2$, using the JUMNA procedure which was developed in our Laboratory [14, 15] and used in several investigations on ligand-oligonucleotide complexations (reviewed in ref. 1).

Table I presents the main results obtained [16]. In this table E_{inter} represents the intermolecular interaction energy between the drug and the minor groove of the two alternating oligonucleotides, ΔE_{lig} the deformation energy of the ligand and ΔE_{DNA} that of the oligonucleotide upon binding, ΔE the overall complexation energy and δ the complexation energy of all compounds studied with respect to their most stable association taken as energy zero.

The results contained in Table I clearly indicate :
1) the persistence of the preference of the two analogues III and IV of distamycin A for the minor groove of the AT

I

II

III

IV

TABLE I

Interaction energies (kcal/mole) of distamycin A and its analogues III and IV with the minor groove of AT and GC oligomers (see text for the significance of the symbols).

	DISTAMYCIN A		ANALOGUE III		ANALOGUE IV	
	$d(ATATATA)_2$	$d(GCGCGCG)_2$	$d(ATATATA)_2$	$d(GCGCGCG)_2$	$d(ATATATA)_2$	$d(GCGCGCG)_2$
E_{inter}	-136.3	-110.4	-107.8	- 99.0	-121.1	- 98.1
ΔE_{lig}	3.2	5.6	7.1	4.2	2.0	3.6
ΔE_{DNA}	12.9	13.4	6.1	10.0	5.8	9.0
ΔE	-120.2	- 91.4	- 94.6	- 84.8	-113.3	- 85.5
δ	0.0	28.8	25.6	35.4	6.9	34.7

sequences ; 2) the decrease of the AT versus GC specificity in the two analogues with respect to distamycin A. ; 3) the decrease of the value of the complexation energy of the most stable association of the analogues III and IV with respect to that of distamycin A. The decrease of both the AT selectivities and the affinities is the most pronounced with analogue III. One of the obvious conclusions of these results is that pyrrole rings are in no way necessary for producing AT specificity in these type of compounds and that, consequently, this specificity cannot be attributed to the van der Waals contacts between their CH groups and the adenine C-2 hydrogens as postulated in [9].

The exploration of the ligand-oligonucleotide H-bonding interactions that intervene in the stabilization of the investigated complexes is interesting [16]. In the interaction with the AT sequence the ligand-nucleotide hydrogen bonds involve essentially the NH groups of the amide linkages of the ligands with the O_2 and N_3 sites on the bases of the oligonucleotides and, to a lesser extent, of the hydrogens of the propionamidinium end of the ligand with the $O_{1'}$ atom of the deoxyriboses and $O_{3'}$ or $O_{5'}$ atoms of the backbone. The persistence of the binding to $O_2(T)$ and $N_3(A)$ in the three compounds is consistent with the situation observed in the crystal structure of the distamycin A d(CGCAAATTTGCG)$_2$ complex [17] and, as shown both theoretically and experimentally (see 1 and references therein), is a general feature of minor groove binders which possess such a hydrogen bonding capability. In the interaction with the GC sequence the number of hydrogen bonds with the O_2/N_3 sites is significantly smaller. As demonstrated, however, with compounds devoid of such capability and showing nevertheless the same specificity

[1] the hydrogen bonds, although they strengthen the binding and the specificity, are not a necessary condition for it. On the other hand the "contacts" which exist in distamycin A between the H atoms of the pyrroles and the C-2 atoms of adenines are practically lost in III and IV, the related "contacts" which would involve the H atoms of the benzene ring of III or of the β-alanine moiety of IV being much weaker. These molecules conserve nevertheless the same AT specificity.

The high numerical values of the E_{inter} component of the binding energies and their greater values in the associations with the AT sequence than with the GC one confirm the fundamental significance for the specificity of the maximal concentration of the molecular electrostatic potential in the minor groove of the former sequence. On the other hand, a detailed analysis of the numerical data of Table I indicates an interesting phenomenon : it shows that while the smaller AT specificity of III with respect to that of II stems essentially from the smaller value of its E_{inter}, the smaller specificity of IV is due rather to the greater values of its deformation energies ΔE_{DNA} and ΔE_{lig}. This situation confirms the necessity, strongly advocated by us in our previous publications and underlined as one of the major conclusions in ref. 1, for computing explicitly the intermolecular complexation energies between the ligands and DNA and all their components in any serious attempt to understand the factors responsible for the specificity of the association. Invaluable as may be X-ray crystallographic or NMR (or in fact any other physicochemical) determinations of the structural features of the associations, they do not provide unambigous information on these factors.

The persistence but the simultaneous decrease of AT specificity of III and IV in comparison to that of distamycin A are in agreement with the essential experimental findings obtained (using UV and CD spectroscopy) of Sasisekharan and his colleagues [10-13]. They provide a theoretical basis for their explanation.

II- CATIONIC PORPHYRINS : INTERCALATORS OR GROOVE BINDERS ?

Drugs which bind non covalently to DNA are generally considered to do so by two distinct mechanisms : intercalation between base pairs or binding in a groove and are thus classified as intercalators or groove binders. It was occasionally considered that "outside" binding may represent a transient state for intercalation or a concomitant but much weaker mode of interaction and it was also recognized that bulky substituents may possibly transform an intercalator into a groove binder (see e.g. 18). Essentially, however, the division of "physically"

bound drugs into intercalators or groove binders is an
established practice.
    Recently a number of antitumor drugs have been
shown to exhibit the unaccustomed feature of being able to
behave both as intercalators and groove binders in their
interaction with synthetic polynucleotides and with DNA.
Prominent among these compounds are a series of large
cationic porphyrins (for a recent review see e.g. 19) the
fundamental, parent molecule of which is the meso-tetra-(4-
N-methylpyrydil)-porphyrin (T4MPyP, V). Studies with
synthetic poly- and oligonucleotides and with DNA have
indicated in their great majority [19-25] a sequence
selective interaction, intercalative in GC sequences but
non intercalative ("outside binding") in AT sequences. The
dual interaction is also observed with cationic
metalloporphyrins [19, 23-28], the sequence and mode of
binding preference depending then strongly, if not
decisively, on the nature of the bound metal ion and the
porphyrin structure.

V

    A molecular modelling of the interaction of T4MPyP
with the dinucleotide monophosphates d(CpG) and d(TpA) [29]
presented evidence in favour of the feasibility of a full
intercalation into the d(CpG) sequence and the
impossibility of such an intercalation into the d(TpA)
sequence, without excluding, however, the possibility of a

"partial" intercalation in the latter case. While obviously significant with respect to the prominent difference of the two sequences towards the acceptance of T4MPyP as an intercalator, this study appears nevertheless as incomplete in the sense that, because of the very short dimensions of the receptor sites, it could not permit a comparative investigation of the competitive, also sequence dependent, groove binding ability of the drug. Such an investigation can only be achieved by considering oligonucleotide receptors of dimensions sufficient to shape the major features of the groove properties. Moreover the use of such longer segments enables also to investigate the influence of flanking base pairs on the reactivity of the intercalation site, a problem which raised recently some questions [30-32]. For these major reasons we have carried out modelling studies using deoxyhexanucleotides as drug receptors. Although our investigations are being presently extended to cationic porphyrins carrying a variable number of charges, we shall present here only the results referring to the fundamental T4MPyP compound [33].

The computational procedure used in this study is again the JUMNA method [14, 15].

The major part of the computations bears on the interaction of T4MPyP with the alternating sequences $d(CGCGCG)_2$ and $d(TATATA)_2$, the intercalative complexes being considered to occur between the central pyrimidine-purine base pairs. They were investigated for both sequences in three distinct orientational possibilities : a symmetrical one with two N-methylpyridinium rings lying in the major groove and two in the minor groove and two asymmetrical ones, one with three N-methylpyridinium rings in the major groove and the fourth in the minor groove and the other with three N-methylpyridinium rings in the minor groove and the fourth in the major groove. These configurations will be denoted as 2M-2m, 3M-1m, and 1M-3m, respectively.

Because of the proposal in [19, 32] that in natural DNAs, as contrasted to poly(dA-dT), intercalation of T4MPyP could take place at a d(TpA) step (whereas in the synthetic polymer only groove-binding occurs), we were led to investigate theoretically the intercalative binding of T4MPyP to the "mixed" sequence $d(CGTACG)_2$ as well.

The non intercalative complexes of T4MPyP with $d(CGCGCG)_2$ have been investigated both for the major and minor groove binding. Similar binding to $d(TATATA)_2$ was explored solely in the minor groove, on account of the well-known preference of groove binders for this groove in this sequence (for a review see e.g. 1), due principally, as demonstrated abundantly in our Laboratory to the significantly more attractive molecular electrostatic potential in this groove of this sequence. Outside binding

of T4MPyP, by which we mean binding essentially to the
sugar-phosphate backbone, was investigated only with
d(TATATA)$_2$, as no great difference in this type of binding
is expected as a function of base sequence.

The results of the computations are reported in
Table II.

TABLE II
Values of the binding energies of T4MPyP to d(CGCGCG)$_2$, d(TATATA)$_2$ and
d(CGTACG)$_2$. Energies in kcal/mole.

	d(CGCGCG)$_2$			d(TATATA)$_2$			d(CGTACG)$_2$	
Interca-lation	2M-2m	3M-1m	1M-3m	2M-2m	3M-1m	1M-3m	2M-2m	3M-1m
E_{inter}	-159.9	-162.1	-129.7	-141.7	-157.3	-109.8	-144.1	-174.0
ΔE_{DNA}	33.5	57.4	33.2	32.2	44.2	32.9	37.2	67.5
ΔE_{lig}	2.3	1.7	0.3	2.4	2.5	1.1	2.0	2.5
ΔE	-124.1	-103.0	- 96.2	-107.1	-110.6	- 75.8	-104.9	-104.3
δ	0.0	21.1	27.9	17.0	13.5	48.3	19.2	19.8

	d(CGCGCG)$_2$		d(TATATA)$_2$	
Groove Binding	major groove	minor groove	minor groove	sugar phosphate backbone
E_{inter}	-133.1	-116.5	-133.5	- 74.9
ΔE_{DNA}	31.1	20.7	15.6	14.4
ΔE_{lig}	2.0	1.9	1.3	1.0
ΔE	-100.0	- 93.9	-116.6	- 59.5
δ	24.1	30.2	7.5	64.6

Let us consider successively the results for the
intercalative and the groove binding complexes.
1) Intercalated complexes.
The most stable intercalative complex occurs with
the d(CGCGCG)$_2$ oligonucleotide demonstrating thus the
preference of the drug for CG sites. Moreover the binding
of the porphyrin to this oligonucleotide occurs in the
following order of relative configurational preferences :

$$2M-2m > 3M-1m > 1M-3m$$

indicating a marked specificity for the "symmetrical" mode of intercalation. This marked preference for 2M-2m over 3M-1m is dictated by the highly unfavourable ΔE_{DNA} term in the latter configuration. The least favourable energy balance in the 1M-3m configuration, on the other hand, stems from the low intermolecular energy term.

Experimental evidence, confirmatory of the preference for the symmetrical selective intercalative binding to CG sequences was provided recently in ^{1}H and ^{31}P NMR exploration of the T4MPyP-poly(dG-dC).poly(dG-dC) complex [34].

The intercalation of T4MPyP into the d(TATATA)$_2$ oligonucleotide is at obvious disadvantage with respect to intercalation into the regular CG hexamer. Moreover, the binding configurations to d(TATATA)$_2$ are ranked in the following order of decreasing stabilities :

$$3M-1m > 2M-2m < 1M-3m$$

indicating that, if feasible, such an intercalation would be asymmetric. The preference for 3M-1m over 2M-2m is due in this case to a distinctly more favourable E_{inter} term (by 15.6 kcal/mole) overcompensating for the less favourable ΔE_{DNA} term.

A detailed examination of the structure of the T4MPyP-d(TATATA)$_2$ 3M-1m complex shows moreover [33] that it is associated with a dramatic distortion of the DNA backbone so that, at most, this mode of interaction of T4MPyP with d(TATATA)$_2$ could be considered as representing a distorted partial intercalation. (Detailed information on the geometries of all these intercalative and also the further discussed groove binding complexes may be found in ref. 33).

The binding energetics of T4MPyP to the mixed sequence d(CGTACG)$_2$, in the two representative arrangements 2M-2m and 3M-1m, do not differ markedly from those in the corresponding complex with the regular sequence d(TATATA)$_2$. These results do not support thus the hypothesis [31] that the inherently unfavourable intercalation of T4MPyP at a d(TpA) step could be facilitated by adjacent GC pairs.

2) Groove binding complexes.

In clear distinction to the situation with the intercalative complexes, groove-binding occurs preferentially with the d(TATATA)$_2$ sequence, showing thus a net specificity for AT base pairs. Moreover, as stated above, numerous previous studies (see ref. 1) on groove binders do not leave any doubt that this sequence preference is associated with a parallel preference for its minor groove.

The binding of T4MPyP to the sole sugar-phosphate backbone of d(TATATA)$_2$ inolves a relatively reduced number of stabilizing interactions concerning solely two adjacent rings of the ligand. Hence the particularly small complexation energy of such an arrangement.

3) Energetics of intercalation versus groove-binding interactions.

This was the central problem investigated in this particular exploration. The results of the computations lead to the straightforward indication that while the intercalative mode of binding is the dominating one for the interaction of T4MPyP with the d(CGCGCG)$_2$ sequence, in which it is favoured by 24.1 kcal/mole over the most efficient groove binding association, the reverse is true in the association of this drug with the d(TATATA)$_2$ sequence, in which the groove binding interaction is by 9.5 kcal/mole more stable than intercalation.

To our knowledge, this is the first theoretical demonstration of a differential selection of binding modes, intercalation in d(CGCGCG)$_2$ and minor groove binding to d(TATATA)$_2$, elicited by a DNA-binding ligand and is fully consistent with available experimental data for T4MPyP binding. These data [21, 22, 35] confirm also explicitly that the AT polymer forms a more stable external complex than does the GC polymer and also that the GC intercalated complex is more stable than the complexes formed on the exterior of this polymer.

This situation indicates that, as elegantly stated in [36], dealing with the interaction of a series of unfused tricyclic aromatic cations with DNA, "intercalation and groove binding modes should be viewed as two potential wells on a continuous energy surface".

III - DICATIONIC STEROID DIAMINES.

The binding of steroid diamines to DNA has been the object of studies by a variety of physico-chemical methods : high-resolution ^{1}H [37-40] and ^{31}P NMR [41], thermal denaturation monitored by U.V. spectroscopy [40, 42], hydrodynamic and electric dichroïsm [43], and topological and viscosimetry investigations [44]. These studies seem to rule out an intercalative mode of binding and converge towards the conclusion that the binding of steroid diamines induces a kink in the structure of DNA, enabling the drug to insert between partially unstacked base pairs [45]. The precise structural characteristics of the complex remain, however, largely unknown. Thus unanswered questions concern, e.g., the amplitude of the kink induced by the drug, a possible preference for its occurence at a py-(3', 5')-pu or at a pu-(3', 5')-py step, its possible base-pair dependence, its orientation to face preferentially a precise segment of the steroid diamine, the possible repuckering of the sugars produced at the bound nucleotides, etc.

In an attempt to contribute to the solution of these questions, we have undertaken a theoretical exploration by the JUMNA method of the structural characteristics of the complexes of a representative steroid diamine, dipyrandium VI, with double-stranded hexanucleotides d(TATATA)$_2$, d(ATATAT)$_2$ and d(CGCGCG)$_2$ [46].

VI

Various distinct modes of binding were investigated :

a) groove binding to a native B-DNA structure ;

b) intercalative binding with varying amplitudes of base pair separation Δz and unwinding angles ;

c) interaction with a kinked model, using as starting point for energy minimization the structure put forward by Sobell et al. [45] for the complex of irehdiamine with d(GCGCGC)$_2$;

d) interaction with kinked conformations, in which different kinks are induced, at a well-defined dinucleoside step, with varying amplitudes, along an axis that is either perpendicular or parallel to the dyad axis, or along a combination of the two.

The results of the computations [46] pertaining to the first three of the above-mentioned binding modes are reported in Table III.

For groove binding to the sequences d(TATATA)$_2$ and d(ATATAT)$_2$ only minor groove complexes were investigated, whereas for groove binding to d(CGCGCG)$_2$, both minor and major groove complexes were studied. As already mentioned above, such a choice was justified on the basis of experience acquired in our many previous studies on this

TABLE III
Values of the binding energies of dipyrandium upon groove binding,
intercalation and complexation with the kinked structure of ref. [45].

	GROOVE BINDING				INTERCA-LATION	COMPLEXATION WITH THE KINKED STRUCTURE OF Ref. [45]
Oligonu-cleotide	$d(TATATA)_2$	$d(ATATAT)_2$	$d(CGCGCG)_2$	$d(CGCGCG)_2$	$d(TATATA)_2$	$d(TATATA)_2$
	Minor groove	Major groove	Minor groove	Major groove		
ΔE	- 84.9	- 77.1	- 71.4	- 60.5	- 77.2	- 72.3
δlig	0.0	0.0	0.0	0.0	0.1	0.1
δDNA	3.0	3.6	14.8	5.4	24.7	5.7
δE	- 81.9	- 73.5	- 56.6	- 55.1	- 52.3	- 66.5
$\delta *$	8.0	16.5	33.3	34.9	37.6	23.5

* Exceptionally (for reasons see text) these $\delta 'S$ are indicated with
respect to the most stable association of Table IV.

mode of interaction and understandable from the
distribution of the molecular electrostatic potential in B
DNA.
        Intercalative binding of dipyrandium was
investigated with the representative sequence $d(TATATA)_2$
only, intercalation taking place between the central bases
T_3 and A_4. The optimized value of Δz is 6.8 Å, with an
unwinding angle of 15°. The two cationic ends of the ligand
are in distinct grooves.
        The main results of the computations indicate
that :
        a) Groove binding is preferred over intercalation
by a considerable amount of energy (nearly 30 kcal/mole).
This energy difference stems from both the intermolecular
interaction term ($\simeq$ 8.5 kcal/mole) and the conformational
deformation of the oligonucleotide, Δ_{DNA} ($\simeq$ 23 kcal/mole).
        b) In groove binding both AT sequences are
preferred (by more than 15 kcal/mole) over the GC one.
        c) A distinct preference (of $\simeq$ 7 kcal/mole) is
computed for binding to the $d(TATATA)_2$ sequence rather than

to the isomeric d(ATATAT)$_2$ one. The former has a central d(TpA) step facing the tricyclohexane moiety of the ligand. This preference is imposed essentially by the intermolecular interaction term ΔE.

d) Binding in the kinked structure of ref. [45], although significantly favoured over intercalation (by $\simeq$ 15 kcal/mole) is nevertheless less favourable (again by $\simeq$ 15 kcal/mole) than groove binding to the B-DNA conformation. This last result is imposed essentialy by the intermolecular interaction term ΔE, which favours groove binding by about 12 kcal/mole and by a smaller DNA deformation (by 3 kcal/mole).

On account of the last disappointing result obtained with the kinked structure of ref. [45], we have undertaken to search for other theoretically generated kinked structures susceptible to bind dipyrandium in a more efficient way. This can be achieved by the JUMNA algorithm by imposing a "kink" on the helicoïdal axis at a given dinucleoside step, undergoing a rotation of varying amplitudes along an axis that may be either perpendicular to the dyad axis of the upstream base pair, or parallel to it. In keeping with the notations of ref. [15], we will denote the first rotations by "P" (kink propeller) and the second one by "T" (kink tilt). Mixtures of the two rotations may and have also been considered. Varying amplitudes of the kink were introduced in order to select appropriate starting structures for energy minimization.

Details of all the complexes explored may be found in ref. 46. Among those two complexes with a kinked structure are significantly more stable than the corresponding unkinked ones of Table III for all three oligonucleotide sequences.

TABLE IV

Values of the binding energies of dipyrandium in the kinked structures generated theoretically and more stable than those of Table III.

Oligonu-cleotide	d(TATATA)$_2$ kinked T$_3$A$_4$	d(ATATAT)$_2$ kinked T$_2$A$_3$
ΔE	- 98.9	- 91.1
δlig	0.0	0.2
δDNA	9.0	6.5
δE	- 89.9	- 84.4
δ	0.0	5.5

TABLE V
Structural information on the two best energy-minimized kinked complexes of dipyrandium with d(TATATA)$_2$ and d(ATATAT)$_2$ at their kink sites.

		d(TATATA)2 : kink site T$_3$A$_4$		d(ATATAT)$_2$: kink site T$_2$A$_3$	
		First strand	Second strand	First strand	Second strand
	ϕ'	- 157.5	- 156.5	160.0	164.5
Conforma-tional angles	ω'	- 76.7	- 146.8	- 85.4	- 132.7
	ω	114.4	- 79.8	118.1	- 65.5
	ϕ	- 177.8	175.5	180.0	177.8
	ψ	69.5	51.9	61.0	54.4
	χ_T	41.5	78.1	58.7	67.0
	χ_A	50.4	41.3	57.0	76.6
kink parameters					
kink propeller		- 39.4			- 20.0
kink twist		3.2			- 10.0
kink dislocation along dyad axis		0.7			0.2
kink dislocation perpendicular to dyad axis		- 0.7			- 1.9

The best complex of dipyrandium occurs with d(TATATA)$_2$, in which a kink is induced at the central T$_3$A$_4$ step facing the tricyclohexane moiety of the ligand. The second best (δ = 5.5 kcal/mole) is that with the isomeric sequence d(ATATAT)$_2$, with a kink induced now in the upstream step T$_2$A$_3$ facing the cationic pyrolidine ring linked to ring A. The details of the energetics of these two complexes are indicated in Table IV. Table V reports the details of their structures.

IV AUREOLIC ACID DERIVED ANTIBIOTICS.

The most representative of this type of ligands are olivomycin, mithramycin and chromomycin A$_3$. They possess very similar structural features consisting of an aglycone moiety and five attached hexopyranoses. The structure of one of them, olivomycin, is illustrated in VII. Although footprinting patterns obtained with the three drugs are not identical they point to a definite specificity of all of them for G-C rich sequences of DNA [47, 48], a result

VII

confirmed by other physicochemical techniques (see e.g. 49, 50).

Another important feature of the interaction is the association of the specificity essentially with the aglycone moiety : thus GC specificity is maintained upon successive elimination of the sugars down to the derivative consisting of the aglycone and sugar D only [49]. The sugars do affect, however, the strength of binding which they increase significantly, in particular by decreasing the rate of dissociation of the DNA-drug complex [49].

While there exists a general consensus on the above features of specificity, no understanding is available on the nature of the factors responsible for it. In fact, strikingly divergent opinions are held even on the fundamental characteristics of the interaction of these antibiotics with their nucleic acid receptor. Thus while some authors [51, 52] proposed originally an intercalative mode of association, they consider more recently a groove binding mode [53, 54], the latter advocated also by others [50, 55-59]. In view of the data presented in particular in [56] groove binding appears indeed as the most probable mode of interaction. A difference of opinion persists, however, among the protagonists of the groove binding

mechanism of whether the interacttion involves the major
[53, 54, 60] or minor [50, 55-59] groove of DNA. Moreover,
quite different architectures of the minor groove
interaction are considered in the last references,
involving e.g. the interaction of the side-chain keto
oxygen of the aglycone [50] or of its O_{11} hydroxyl [56, 57]
with the NH_2 group of guanine.

        Significant divergences of opinion prevail also on
the role of the Mg^{2+} ions in the association. While it is
frequently considered that their presence constitutes an
absolute requirement for the binding to occur [61, 62],
with, however, different views on their exact role and
positioning [49, 50, 56], a recent publication [63]
indicates that at pH 4.5 the binding of the drug, which is
present then in its neutral monomeric form, does not
require divalent cations (although the strength of binding
is greatly enhanced in their presence). At pH > 7,0 and low
DNA/drug ratio (<20) metal cations are necessary, but at
high DNA/drug ratio an appreciable proportion of the drug
is bound even in the absence of the metal. These authors
consider that the insertion of Mg^{2+} into the drug-DNA
complex is accompanied by deprotonation of the drug.
Following [56, 57] the divalent cation is implied in the
interaction of a dimeric form of the drug in an anionic
form with the DNA receptor.

VIII

        In view in this situation and as a first step
towards elucidating the essential factors governing the
groove and base sequence preferences of this type of
antibiotics we have investigated [64], by the Jumna method
again, the binding to three heptanucleotides, d(CGCGCGC)2,
d(TATATAA)2 and d(CICICIC)$_2$, of a shortened model of
olivomycin composed of its aglycone and its D sugar (VIII).

This shortened form of the natural antibiotic (symbol ASD) was considered in its neutral monomeric form and the interaction was carried out in the absence of any divalent cation. Although we are well aware of the fact that such a simplified model is far from corresponding to the usual conditions and the mode of interaction of the natural aureolic acid derived antibiotics with DNA, it seems nevertheless plausible, on the basis of the literature reviewed above, that this investigation may nevertheless bring into evidence the <u>basic features</u> responsible for the <u>specificity</u> of this type of drugs for the GC sequences of DNA and provide information on their <u>groove preference</u>.

Two binding configurations were explored and energy-minimized for the interaction of ASD with the grooves of the $d(CGCGCGC)_2$ and $d(TATATAT)_2$ sequences following the approach of the chromophore to the oligonucleotides through its hydrophilic or hydrophobic side. The corresponding results are presented in columns (a) and (b) of table VI.

TABLE VI
Values of the binding energies of ASD to $d(CGCGCGC)_2$, $d(TATATAT)_2$ and $d(CICICIC)_2$

	$d(CGCGCGC)_2$ (a)	(b)	$d(TATATAT)_2$ (a)	(b)	$d(CICICIC)_2$ (b)
(1) Minor groove binding					
ΔE_{inter}	- 63.2	- 56.9	- 60.2	- 52.2	- 55.5
ΔE_{lig}	3.6	1.1	6.3	1.5	3.1
ΔE_{DNA}	7.4	5.0	8.9	4.8	6.5
δE	- 52.2	- 50.7	- 45.1	- 45.9	- 45.9
δ	0.0	1.5	7.2	6.3	6.3
(2) Major groove binding					
ΔE_{inter}	- 52.6	- 54.0	- 43.4	47.8	- 52.0
ΔE_{lig}	0.5	4.4	0.0	1.7	4.5
ΔE_{DNA}	8.6	8.5	5.4	5.2	6.9
δE	- 43.4	- 41.1	- 38.0	- 40.8	- 40.7
δ	8.8	11.1	14.0	11.4	11.6

Its examination shows that the most stable association occurs in the minor groove of the GC oligonucleotide and it involves the hydrophilic side of the chromophore (column a in table V). Structurally, this preferential stabilization is due essentially to hydrogen bond interactions of the hydroxyl oxygens of O_{11} and O_{12} of the chromophore with the 2-amino group of guanine G4 and of the hydrogen of OH_{12} with $O1'$ of $S5'$. It may nevertheless be noted that the complex formed through the interaction of the hydrophobic side of the drug with the same oligonucleotide (column b of table VI) is only 1.5 kcal/mole less stable than the previous one. Its stabilizing interactions involve hydrogen bonds of the hydroxyl groups of the side chain of the aglycone with the amino group of G3' and with O_1' of S4' and also of the hydroxyl hydrogen of O_{21} of the chromophore with $O1'$ of S6'.

The similar interactions with the minor groove of the AT oligonucleotide are significantly weaker (by 7.2 and 6.3 kcal/mole) than the corresponding ones with the GC oligonucleotide. Although they involve also a series of hydrogen bonds between the drug and the nucleic acid receptors, these bonds occur essentially between the proton donor hydroxyl groups of the aglycone side chain and different acceptor oxygen atoms on the oligonucleotide. They are obviously unable to produce as strong an association as the bonds formed with the GC oligonucleotide by the interactions involving the NH_2 group of guanine. The interactions with the IC oligomer are essentially similar to those found with the AT oligomer.

Complexes with the major groove of the oligonucleotides, whether involving the hydrophilic or hydrophobic side of the drug, are constantly significantly weaker then those formed in the minor groove. The binding to the AT oligonucleotide is particularly disfavoured in this case.

In conclusion, it seems that the ASD model of the aureolic acid derived antibiotics correctly reflects the major aspects of the specificity in the interaction of these drugs with DNA. It confirms the specificity of this drug system for GC sequences, and moreover, supports its preference for the minor groove of these sequences. The major structural features which appear to be responsible for this specificity are the hydrogen bonded interactions between the hydroxyl groups O_{11} and O_{12} of the chromophore and the 2-amino group of guanine G4.

As it has been abundantly stated and shown above groove binding antibiotics generally show a marked specificity for the minor groove of AT sequences. The particular behaviour of the aureolic acid derived antibiotics is thus worth stressing. It may be observed that in the few cases in which binding of drugs is observed to the minor groove of GC sequences, it always seems to involve a hydrogen bonding interaction between an oxygen atom of the drug and the 2-amino group of guanine on the nucleic acid receptor. (For a discussion see ref. 1).

ACKNOWLEDGEMENT.

This work was supported by the Association for International Cancer Research (Brunel and Saint-Andrews Universities, United Kingdom) to which the authors wish to express their deep thanks.

BIBLIOGRAPHY.

[1]   Pullman, B. : Advances in Drug Research, $\underline{18}$, 1 (1989).
[2]   Zakrzewska, K., Lavery, R. and Pullman, B. : J. Biomol. Struct. Dynam., $\underline{4}$, 833 (1987).
[3]   Randrianarivelo, M., Zakrzewska, K. and Pullman, B. : J. Biomol. Struct. Dynam., $\underline{6}$, 769 (1989).
[4]   Zakrzewska, K. and Pullman, B. : J. Biomol. Struct. Dynam., $\underline{5}$, 1043 (1988).
[5]   Zakrzewska, K., Randrianarivelo, M. and Pullman, B. : J. Biomol. Struct. Dynam., $\underline{6}$, 331 (1988).
[6]   Pullman, B. in : Perspectives in Quantum Chemistry (J. Jortner and B. Pullman Ed) Kluwer Academic Press, Dordrecht, Holland, 1989, p. 123.
[7]   Pullman, B. in : Modelling of Molecular Structure and Properties in Physical Chemistry and Biophysics, (J.L. Rivail Ed) Elsevier Science Publishers, Amsterdam, in press.
[8]   Krylov, A.J., Grokhovsky, S-L, Zasedatelev, A.S., Shuze, A.L., Gursky, G.V. and Gottikh, B.P. : Nucleic Acid Research, $\underline{6}$, 289 (1979).
[9]   Kopka, M.L., Yoon, Ch., Goodsell, D., Pjura, P. and Dickerson, R.E. : Proc. Natl. Acad. Sci. U.S.A., $\underline{82}$, 1376 (1985).
[10]   Dasgupta, D., Rajagopalan, M. and Sasisekharan V. : Biochem. Biophys. Res. Comm., $\underline{140}$, 626 (1986).
[11]   Parrack, P., Dasgupta, D., Ayyer, J. and Sasisekharan, V. : FEBS. Let. $\underline{212}$, 297 (1987).
[12]   Dasgupta, D., Parrack, P. and Sasisekharan, V. : Biochemistry, $\underline{26}$, 6381 (1987).
[13]   Rao, K.E., Dasgupta, D. and Sasisekharan, V. : Biochemistry, $\underline{27}$, 3018 (1988).
[14]   Lavery, R. in : DNA Bending and Curvature (Olson, W., Sundaralingam, M., Sarma, M.H. and Sarma, R.H. Eds Adenine Press, New York), 191 (1988).
[15]   Lavery, R. in : Unusual DNA Structures (Wells, R. and Harvey, S. Eds Springer Verlag, New York) 189 (1988).
[16]   Pullman, B., Hui, X. and Gresh, N. in : New Trends in Biological Chemistry (T. Ozawa Ed.) Japan Scientific Press, Tokyo, in press.
[17]   Coll, M., Frederick, C., Wang, A-H and Rich, A. : Proc. Natl. Acad. Sci. U.S.A., $\underline{84}$, 8385 (1987).

[18]   Müller, W., Crothers, D.M. and Waring, M.J. : Eur. J. Biochem., 39, 223 (1973).

[19]   Fiel, R.J. : J. Biomol. Struct. Dynam., 6, 1259 (1989).

[20]   Wilson, W.D., Tanious, F.A., Barton, H.J., Strekowski, L. and Boykin, O.W. : J. Am. Chem. Soc., 111, 5008 (1989).

[21]   Pasternak, R., Gibbs, E.I. and Villafranca, J.J. : Biochemistry, 22, 2406 (1983).

[22]   Pasternak, R., Gibbs, E.I. and Villafrance : Biochemistry, 22, 5409 (1983).

[23]   Kelly, J.M., Murphy, M.J., McConnell, D.J. and Ohlligin, C. : Nucleic Acids Res., 13, 167 (1985).

[24]   Banville, D.L., Marzilli, L.G. and Wilson, W.D. : Biochim. Biophys. Res. Comm., 113, 148 (1983).

[25]   Strickland, J.A., Marzilli, L.G., Gay, K.M., and Wilson, W.D. : Biochemistry, 27, 8870 (1988).

[26]   Bromley, S.D., Ward, B.W. and Dabrowiak, J.C. : Nucleic Acid Res, 14, 9133 (1986).

[27]   Ward, B., Skorobogaty, A. and Dabrowiak, J.C. : Biochemistry, 25, 6875 (1986).

[28]   Ward, B., Skorobogaty, A. and Dabrowiak, J.C. : Biochemistry, 25, 7827 (1986).

[29]   Ford, K.G., Pearl, L.H. and Neidle, S. : Nucleic Acid Res., 15, 6553 (1987).

[30]   Banville, D.L., Marzilli, G.G., Strickland, J.A. and Wilson, W.D. : Biopolymers, 25, 1837 (1986).

[31]   Jenkins, B.G. : Dissertation, Ph. D., RPMI Division SUNY/Buffalo, quoted in [19] (1986).

[32]   Ford, K., Fox, K.R., Neidle, S. and Waring, M.J. : Nucleic Acid Res., 15, 2221 (1987).

[33]   Hui, X., Gresh, N. and Pullman, B. : Nucleic Acid Res., 18, 1109 (1990).

[34]   Marzilli, G., Banville, D.L., Zon, G. and Wilson, W.D. : J. Am. Chem. Soc., 108, 4188 (1986).

[35]   Pasternak, R.F. and Bibbs, E.J. in : Metal-DNA Chemistry, ACS Symposium series N° 402 (T.D. Tullins Ed.). American Chem. Soc., 1989, p. 59.

[36]   Wilson, W.D., Tanious, F., Watson, R., Barton, M., Strekowska, A., Harden, D. and Strekowski, L. : Biochemistry, 28, 1984 (1989).

[37]   Patel, D. and Canuel, L. : Proc. Natl. Acad. Sci. U.S.A., 76, 24 (1979).

[38]   Patel, D., Kozlowski, S., Suggs, J. and Cox, S. : Proc. Natl. Acad. Sci. U.S.A., 78, 4063 (1981).

[39]   Patel, D. : Accounts of Chemical Research, 2, 118 (1979).

[40]   Gourevitch, M. and Puigdomenech, P. : Int. J. Biol. Macromol., 8, 97 (1986).

[41]   Suggs, J. and Taylor, D. : FEBS Letters, 189, 77 (1985).

[42] Waring, M. and Henley, S. : Nucleic Acids Res., 2, 567 (1975).

[43] Dattagupta, N., Hogan, M. and Crothers, D. : Proc. Natl. Acad. Sci. U.S.A., 75, 4286 (1978).

[44] Waring, M. and Chisholm, J. : Biochim. Biophys. Acta, 262, 18 (1972).

[45] Sobell, H., Tsai, C-C, Jain, S. and Gilbert, S. : J. Mol. Biol., 114, 335 (1977).

[46] Hui, X., Gresh, N. and Pullman, B. : Nucleic Acid Res., 17, 4177 (1989).

[47] Van Dyke, M. and Dervan, P. : Biochemistry, 22, 2373 (1983).

[48] Fox, K. and Howarth, N. : Nucleic Acids Res., 13, 8695 (1985).

[49] Behr, W., Honikel, K. and Hartmann, G. : Eur. J. Biochem., 9, 82 (1969).

[50] Brikenstein, V., Pitina, L., Barenboim, G. and Gurskii, G. : Molekularnaya Biologiya, 18, 1606 (1985).

[51] Harwitz, K.B. and McGuire, W.L., : J. Biol. Chem., 253, 6319 (1978).

[52] Berman, E., Brown, S.C., James, T.L. and Shafer, R.H. : Biochemistry, 24, 6887 (1985).

[53] Keniry, M., Brown, S., Berman, E. and Shafer, R. : Biochemistry, 26, 1058 (1987).

[54] Kam, M., Shafer, R.H. and Berman, E. : Biochemistry, 27, 3581 (1988).

[55] Waring, M., : J. Mol. Biol., 54, 247 (1970).

[56] Gao, X. and Patel, D.J. : Biochemistry, 28, 751 (1989).

[57] Gao, X. and Patel, D.J. : Quart. Rev. Biophys, 22, 93 (1989).

[58] Cons, B.M.G. and Fox, K.R. : Nucleic Acid Res., 17, 5447 (1989).

[59] Sarker, M. and Chen, F-M : Biochemistry, 28, 6651 (1989).

[60] Berman, E. and Kaw, M. in : Computer-Assisted Modelling of Receptor-Ligand Interactions : Theoretical Aspects and Applications to Drug Design (R. Rein and A. Golourbek, Eds) Alan R. Riss, 1989, p. 217.

[61] Ward, O., Reich, F. and Goldberg, I. : Science, 149, 1259 (1965).

[62] Nayak, R., Sirsi, M. and Podder, S.K. : Biochim. Biophys. Acta, 378, 195 (1975).

[63] Weinberger, S., Shafer, R. and Berman, E. : Biopolymers, 27, 831 (1988).

[64] Chen, K.X., Gresh, N., Hui, X., Pullman, B. and Zakrzewska, K. : FEBS Letters, 245, 145 (1989).

[65] Stockert, J.C. : J. Theor. Biol., 137, 107 (1989).

[66]   Kuwahara, J. and Sugiura, J. : Proc. Natl. Acad.
         Sci., 85, 2459 (1988).

URANYL PHOTOFOOTPRINTING. DNA STRUCTURAL CHANGES UPON BINDING OF MITHRAMYCIN.

Peter E. Nielsen, Benjamin M.G. Cons (a), Keith R. Fox (a) and Vibeke Beck Sommer (b),
Research Center for Medical Biotechnology, Department of Biochemistry B, The Panum Institute, University of Copenhagen, Blegdamsvej 3c, DK-2200 Copenhagen N, DENMARK; (a) Department of Physiology & Pharmacology, University of Southampton, Bassett Crescent East, Southampton SO9 3TU, U.K.; (b) Chemical Laboratory II, The H.C. Ørsted Institute, University of Copenhagen, Universitetsparken 5, DK-2100 Copenhagen Ø, DENMARK

ABSTRACT. The mithramycin DNA complex was analysed by uranyl photofootprinting. A decrease in photocleavage was observed at the GC-rich mithramycin binding sites, but more noteworthy the cleavage of the regions surrounding the mithramycin binding sites were significantly enchanced. Furthermore, these off-binding site regions were also cleaved by the uranyl-citrate complex which does usually not photocleave DNA. These results show that binding of mithramycin to DNA results in an altered DNA conformation in the vicinity of the binding sites and that this conformation has a higher affinity for uranyl than ordinary B-DNA.

1. INTRODUCTION

We have recently shown that the uranyl(VI) ion induces cleavage of the DNA backbone in a photochemical reaction, and that this property of the uranyl ion makes it a useful and promising photochemical probe for determining contacts between phosphates of the DNA backbone and DNA bound proteins. Specifically we have analysed the complexes between λ-repressor and the O_R1 operator (Nielsen et al., 1988), E.coli RNA polymerase and the deoP1 promoter (Jeppesen & Nielsen, 1989), Xenopus transcription factor IIIA and the internal control region (ICR) of the 5S RNA gene (Nielsen & Jeppesen, 1990) as well as DNA complexes with catabolic regulator protein (CRP) and deoR repressor (Jeppesen, Hammer & Nielsen, unpublished). Furthermore, we have shown that uranyl mediated photocleavage of double stranded DNA (in acetate buffer) is reflecting the conformation of the DNA in terms of minor groove width and/or

423

B. Pullman and J. Jortner (eds.), Molecular Basis of Specificity in Nucleic Acid-Drug Interactions, 423–431.
© 1990 Kluwer Academic Publishers. Printed in the Netherlands.

electronegative potential. In particular, we have found that
1) uranyl photocleavage of the A-tracts of bent kinetoplast
DNA is most efficient where the minor groove according to
hydroxyl radical probing is more narrow 2) AT-rich regions
known to bind distamycin are cleaved in accordance with
preferred uranyl binding in the center of the minor groove
and 3) the uranyl photocleavage modulation pattern of the
ICR region of the Xenopus 5S RNA gene shows a 5.5 base pair
repeat which is compatible with the cleavage pattern previously obtained with DNase I; the DNA cleavage of which is
known to sense DNA minor groove width (Nielsen et al.,
1990a).

The antitumour antibiotic mithramycin (Figure 1) interacts with GC-rich regions of DNA via the minor groove
(Gao & Patel, 1989; Cons & Fox, 1989a). This interaction is
unusual in that it requires the presence of a divalent metal
ion, especially magnesium (Cons & Fox, 1989b). NMR studies
have demonstrated that the drug binds as a dimer with the
magnesium coordinating the two mithramycin molecules (Gao &
Patel, 1989). In this structure the DNA minor groove is
forced open to accommodate the bulky dimer, rendering the
structure more like A-DNA. Previous footprinting studies
have shown that the drug specifically recognises the
dinucleotide step GpG (Van Dyke & Dervan, 1983; Fox &
Howarth, 1985) though the ability to interact with this sequence is profoundly affected by other local factors. These
footprinting studies have also suggested that mithramycin
causes changes in DNA structure.

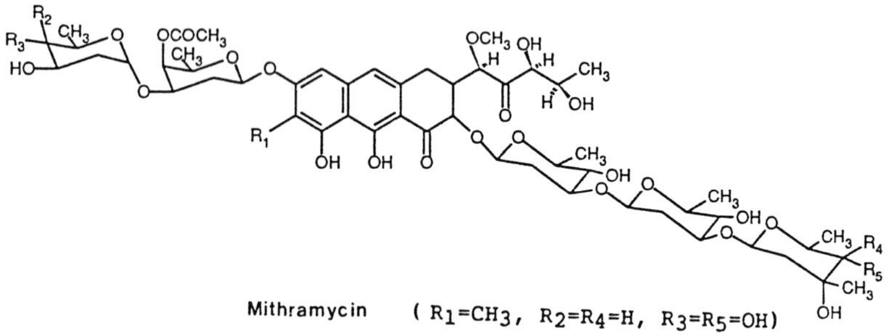

Mithramycin   (R_1=CH_3, R_2=R_4=H, R_3=R_5=OH)

Figure 1. Structure of mithramycin.

A number of other DNA binding drugs have been shown to
induce DNA structural changes in regions surrounding their
binding sites. These changes are detected by increased susceptibility to nucleases such as DNase I and DNase II (Fox &
Waring, 1984; Low et al., 1984) and increased reactivity
towards diethyl pyrocarbonate osmium tetroxide and potassium

permanganate (McLean & Waring, 1989; Jeppesen & Nielsen, 1988). The molecular details of these changes are still obscure, though they probably arise from some combination of changes in groove width and helix unwinding.

2. MATERIALS AND METHODS

Preparation and $3'-{}^{32}P$-labeling of the EcoRI site of the 135 base pair fragment of pXb1 was performed as described (Cons & Fox, 1989a). $5'-{}^{32}P$-labeling of the EcoRI site was performed by linearizing the plasmid with EcoRI, treating this with bacterial alkaline phosphatase, purifying the DNA on low-melting agarose, labeling with $\gamma^{32}P$-ATP and T_4-kinase and finally restricting the plasmid with AvaI. Mithramycin was obtained from Pfizer Inc., U.S.A. A typical reaction mixture contained 0.25 μg calf thymus DNA, 500 cps ${}^{32}P$-labeled "135-fragment" and the desired amount of mithramycin in 50 μl buffer (100 mM Tris-HCl, pH 6.8, 10 mM $MgCl_2$). To this was added 50 μl 2 mM uranyl nitrate solution (eventually containing 4 mM Na-citrate) freshly diluted from a 100 mM stock solution. The samples were irradiated for 30 min at room temperature using a Philips TL 40W/03 fluorescent light tube ($\lambda \sim 420$ nm, $20 \ J \cdot S^{-1} \cdot m^{-2}$). Subsequently, 20 μl 0.5 M Na-acetate pH 4.5 was added (this avoids co-precipitating of the uranyl) and the DNA was precipitated with 250 μl ethanol. Gel electrophoresis was performed on 10% polyacrylamide, 7 M urea sequencing gels in TBE buffer (90 mM Tris-borate, 1 mM EDTA pH 8.7).
   Oligonucleotides containing methylphosphonates were prepared by standard methods (Agrawal & Goodchild, 1987). Uranyl photocleavage of these were performed as above, except that 100 mM Tris pH 6.8 was used as buffer.

3. RESULTS AND DISCUSSION

The effect of mithramycin (Figure 1) on the uranyl photocleavage of the 135bp fragment of the pXbs1 plasmid (Cons & Fox, 1989a) is shown in Figure 2. Mithramycin causes distinct changes of the cleavage pattern. These are characterized by a relative decrease in the cleavage of the "GC-regions" (Figure 3), which have previously by DNase I and EDTA/FeII footprinting been shown to bind mithramycin (Cons & Fox, 1989) (Figure 5c). It is noteworthy, however, that mithramycin seems to enhance the uranyl mediated DNA photocleavage proximal to its binding sites. This could indicate an increased affinity for binding of the uranyl cation to the DNA phosphates.
   We have previously suggested that the photocleavage of DNA by uranyl is caused by photooxidation of a deoxyribose

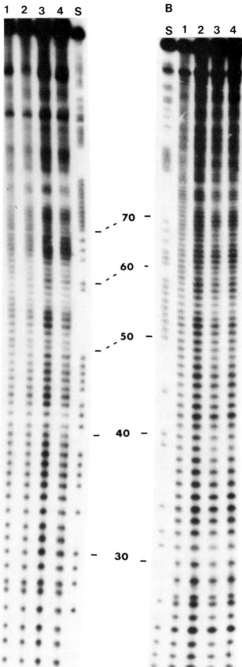

Figure 2. Uranyl photofootprinting of mithramycin binding sites. The following concentrations of mithramycin were used: Lanes 1-4: 0, 5, 50 & 100 μM, respectively. Lane S is a A+G sequence reaction. a: $5'-^{32}P$-labeled DNA fragment, b: $3'-^{32}P$-labeled DNA fragment. The numbers refer to the DNA sequence shown in Figure 3.

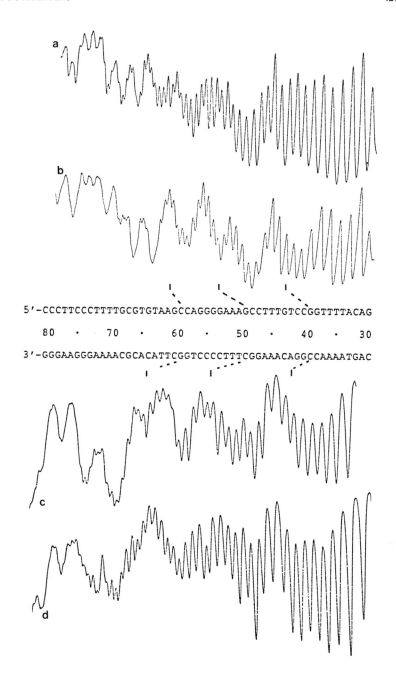

5'-CCCTTCCCTTTTGCGTGTAAGCCAGGGGAAAGCCTTTGTCCGGTTTTACAG

80   ·   70   ·   60   ·   50   ·   40   ·   30

3'-GGGAAGGGAAAACGCACATTCGGTCCCCTTTCGGAAACAGGCCAAAATGAC

Figure 3. Densitometric scanning of the autoradiographs presented in Figure 2. Traces a–d correspond to lanes 1b, 4b, 1a & 4a, respectively.

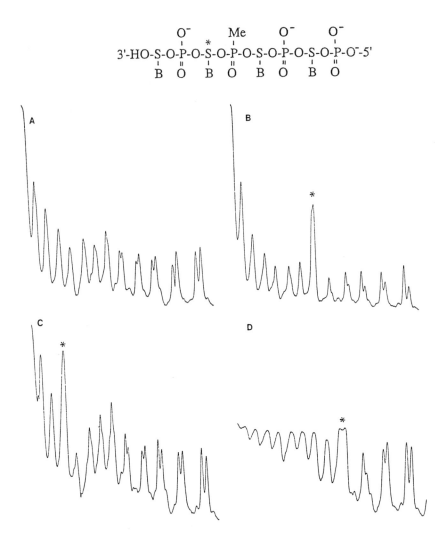

Figure 4. Uranyl photocleavage of oligonucleotides containing methylphosphonates. The sequence is 3'-GATGGA$_c$TCCA$_b$GA$_a$TGGCACGG where A$_x$ denotes the position of the methylphosphonate in oligonucleotide x. Oligonucleotide "a" was a control without methylphosphonate. The asterix denotes the 5'-deoxyribose of the methylphosphonate and also the position of the products of methylphosphonate hydrolysis. Occurance of "double bands" were observed in some experiments (as the one shown), but we have not yet been able to define the conditions under which they form. They are most likely due to incomplete removal of the deoxyribose oxidation product from the 3'-phosphate, <u>e.g.</u>, as found in phosphoglycolates.

proximal to a phosphate where to the uranyl ion is complexed. In order to substantiate this suggestion, we have
analysed the uranyl mediated photocleavage of
oligonucleotides in which a specific phosphate has been replaced by a methylphosphonate. The methylphosphonate carries
no negative charge and thus is expected to complex less efficiently with the uranyl. Accordingly, photocleavage of the
deoxyribose attached to the methylphosphonate via the 3'-
hydroxyl group is significantly reduced (~ 50%) (Figure 4).

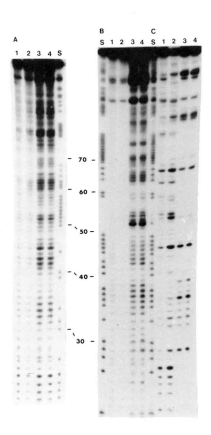

Figure 5. Uranyl-citrate photofootprinting of mithramycin
DNA complexes. The following concentrations of mithramycin
were used: lanes 1-4: 0, 5, 50 & 100 µM. Lanes S are A+G
sequence reactions. a & b are 3' and 5' labeled DNA fragment, respectively. c: DNase I footprint of mithramycin-5'-
labeled DNA fragment complex. Lanes 1-4 as in a & b.

We can not estimate the amount of cleavage at the 5'-
deoxyribose since the products from hydrolysis of the
methylphosphonate – which is quite labile – migrate at the
same position in the gel.

    Citrate forms a strong complex with uranyl and this
complex does not photocleave DNA (Jeppesen & Nielsen, 1989)
(Figure 5, lanes 1). If, however, the DNA structure is perturbed, e.g., as found in the E.coli RNA polymerase promoter
open complex (Jeppesen & Nielsen, 1989) strong uranyl binding sites may be created and the sites are photocleaved by
uranyl citrate, presumably due to transfer of uranyl from
citrate to the DNA. Since binding of mithramycin to DNA appears to increase uranyl photocleavage at specific sites
(Figure 2), we also analysed the mithramycin DNA complex by
uranyl–citrate photocleavage. These results (Figure 5) show
that when mithramycin is complexed to the DNA, this is
cleaved by uranyl–citrate. These new sites correlate with –
but do not exactly correspond to – the regions of enhanced
uranyl photocleavage in the absence of citrate and they are
found in the vicinity of mithramycin binding sites.

    Two mechanisms may account for the formation of strong
uranyl binding sites in the DNA mithramycin complex. Either
the DNA structural changes induced outside the mithramycin
binding sites favour uranyl binding or uranyl coordination
sites are created by the DNA and the mithramycin in combination.

    Although we are not at present able to interpret these
results in molecular conformational terms, they show that
mithramycin when bound to DNA induces pronounced changes in
the DNA structure and that these changes are sensitively
probed by uranyl mediated DNA photocleavage. In particular,
we wish to draw attention to the experiment with the uranyl
citrate complex which we believe shows that the DNA conformational changes induced by mithramycin are basically
different from those induced by e.g. echinomycin, distamycin
or nogalamycin, since we did not observe any changes in the
uranyl photocleavage pattern with these drugs (results not
shown), unless very high drug concentrations are used, in
which case bis–intercalators provoke enhanced photocleavage
at G–residues (Nielsen et al., 1990b).

    Further studies on the mechanism of uranyl mediated
DNA photocleavage should eventually enable us to interpret
results like the ones presented in this paper in more
rigorous structural terms.

ACKNOWLEDGEMENTS

The financial support of "Forskerakademiet" (a visiting fel-
lowship to BMGC) and the NOVO Foundation (PEN is a Hallas-
Møller fellow) is gratefully acknowledged.

REFERENCES

Agrawal,S. & Goodchild,J. (1987) Tetrahedren Lett. **28**, 3539–3542.

Cons, B.M.G. & Fox,K.R. (1989a) Nucleic Acids Res. **17**, 5447–5459.

Cons,B.M.G. & Fox,K.R. (1989b) Biochem. Biophys. Res. Commun. **160**, 517–524.

Fox,K.R. & Howarth,N.R. (1985) Nucleic Acids Res. **13**, 8695–8714.

Fox,K.R. & Waring,M.J. (1984) Nucleic Acids Res. **12**, 9271–9285.

Gao,X. & Patel,D.J. (1989) Biochemistry **28**, 751–762.

Jeppesen,C. & Nielsen,P.E. (1988) FEBS Letters. **231**, 172–176.

Jeppesen,C. & Nielsen,P.E. (1989) Nucleic Acids Res. **17**, 4947–4956.

Low,L., Drew,M.R. & Waring,M.J. (1984) Nucleic Acids Res. **12**, 4865–4879.

McLean,M. & Waring, M. (1988) J. Mol. Recog. 138–151.

Nielsen,P.E., Jeppesen,C. & Buchardt,O. (1988) FEBS Letters, **235**, 122–124.

Nielsen,P.E.& Jeppesen,C. (1990) Trends Photochem. Photobiol. (in press.).

Nielsen,P.E., Møllegaard,N.E. & Jeppesen,C. (1990a) Nucleic Acids Res. (submitted).

Nielsen,P.E., Møllegaard,N.E. & Jeppesen,C. (1990b) Anti Cancer Drug Res. **5**, 105–111.

Van Dyke,M.W. & Dervan,P.B. (1983) Biochemistry **22**, 2373–2377.

CHARACTERISTICS OF NONCOVALENT AND COVALENT INTERACTIONS OF (+) AND
(-) *ANTI*-BENZO[a]PYRENE DIOL EPOXIDE STEREOISOMERS OF DIFFERENT
BIOLOGICAL ACTIVITIES WITH DNA.

N.E. GEACINTOV, M. COSMAN, V. IBANEZ, S.S. BIRKE and C.E.
SWENBERG[1]
*Chemistry Department, New York University, New York, NY,
10003, and [1]Radiation Biochemistry Department, Armed Forces
Radiobiology Research Institute, Bethesda, MD 20814*

ABSTRACT. The (+) and (-) enantiomers of *anti*-7,8-diol-9,10-epoxy-
7,8,9,10-tetrahydrobenzo[a]pyrene (BPDE) are characterized by striking
differences in their tumorigenic and mutagenic activities. The covalent binding of these isomers leads to DNA adducts with different distributions of conformations. The adducts derived from (+)-BPDE cause
significant unwinding of supercoiled DNA, while those derived from
(-)-BPDE do not. The conformations of these covalent lesions are
different from those of classical intercalation complexes and include
external binding modes and adducts which exhibit some though not all
characteristics of intercalative binding. New insights into these
different adducts are obtained from studies of oligonucleotides of defined base composition and sequence modified covalently at the exocyclic amino group of guanine by *cis* and *trans* addition of (+)-BPDE
and (-)-BPDE.

1. INTRODUCTION

Polycyclic aromatic hydrocarbons (PAH) are characterized by varying
degrees of tumorigenic and mutagenic activities. It is well established that these compounds are metabolically activated to epoxide and
diol epoxide derivatives, which are the ultimate mutagenic and tumorigenic forms of these molecules [1,2]. Mutations probably constitute
the initial first steps in the complex, multi-stage phenomenon of
chemical carcinogenesis [3]. The detailed molecular mechanisms by
which PAH diol epoxides and other chemical carcinogens induce mutations are not yet well understood. Subtle structural differences,
such as the stereochemical properties of the PAH diol epoxide
molecules, can manifest themselves in terms of dramatic differences in
their biological activities [1]. The formation of covalent adducts
with cellular DNA in specific genomic sequences is critical for the
expression of the mutagenic potentials of PAH diol epoxides [4-8].
Experimental [9-13] and theoretical [14,15] studies have shown that
base composition and sequence effects can play an important role in
determining the chemical reactivities of PAH diol epoxides and the
characteristics of the DNA adducts formed *in vitro*.

433

B. Pullman and J. Jortner (eds.), Molecular Basis of Specificity in Nucleic Acid-Drug Interactions, 433–450.
© 1990 *Kluwer Academic Publishers. Printed in the Netherlands.*

The most biologically active metabolites of the ubiquitous environmental pollutant benzo[a]pyrene (BP) are the bay-region diol epoxide
derivatives in which an epoxide ring bridges the 9,10-positions of BP
with two -OH groups located at the 7 and 8 positions. [1].   There
are four different stereoisomers of this metabolite.   The most
tumorigenic isomer is the (+) enantiomer of the diastereomer trans-
7,8-dihydroxy-anti-9,10-epoxy-7,8,9,10-tetrahydrobenzo[a]pyrene,   or
BPDE (known also as anti-BPDE, BPDE 2, or BPDE I).   Unlike (+)-BPDE,
the (-) enantiomer is not tumorigenic to any significant extent [1].
Both enantiomers are mutagenic, but their activities are different
from one another in mammalian [16] and in bacterial [17] cell
systems.   These pairs of anti-BPDE enantiomers, denoted here by (+)-
BPDE and (-)-BPDE, constitute a fascinating example of structure-bio-
logical activity relationships. Brookes and Osborne [16]   proposed
that the conformational properties of covalent adducts derived from
the chemical binding of these two isomers to DNA are different from
one another, and that these differences may be of critical importance
in the processing of these adducts by the cellular machineries of repair and replication.   The mechanisms of interaction of these BPDE
stereoisomers with DNA in aqueous model systems, and the properties of
the noncovalent complexes and covalent adducts formed, have been studied extensively (reviewed in 18-20].
    The relationships between biological activities, chemical structure,
steric characteristics, and reactivities of chemical carcinogens and
their metabolites with DNA, have long been subjects of intense interest to researchers in the field of chemical carcinogenesis [2].   In
this article we discuss and review the   differences in the interactions of (+)-BPDE and (-)-BPDE with nucleic acids, and the
characteristics of the covalent adducts formed. Recent results obtained with supercoiled DNA and with oligodeoxynucleotides of defined
base composition and sequence have provided new insights into the
characteristics of the covalent BPDE-DNA lesions.

2. Characteristics   of Reaction and Binding of BPDE with Nucleic Acids.

2.1 REACTION PATHWAYS IN AQUEOUS DNA SOLUTIONS

The hydrophobic BPDE stereoisomers and similar PAH diol epoxides are
known to form noncovalent complexes with DNA. Such physical BPDE-DNA
complexes are formed on time scales of milliseconds, while the covalent binding reactions occur on time scales of minutes [21].   The
dominant reaction of the BPDE enantiomers is hydrolysis to the
tetraols BPT (7,8,9,10-tetrahydrotetrahydroxybenzo[a]pyrene).   This
reaction is catalyzed by DNA under conditions of relatively low ionic
strengths [18, and references quoted therein].   Only 16-20% of (+)-
BPDE and 4-5% of the (-)-BPDE molecules form covalent addition products with native double-stranded DNA.   The reaction pathways of BPDE
in aqueous solutions are summarized in Fig. 1.

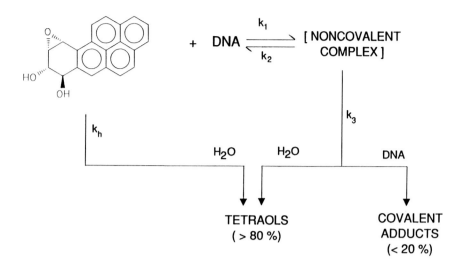

Fig. 1

The characteristics of the non-covalent BPDE-DNA complexes are consistent with those of classical intercalation complexes because of the following observations: (1) unwinding of supercoiled DNA [22,37], and (2) the pyrenyl residues are oriented parallel to the planes of the DNA bases [23]. The association constants K (M^{-1}) characterizing the formation of these noncovalent complexes appear to be approximately equal for (+)-BPDE and (-)-BPDE [24,25]. The formation of BPT is by far the dominant reaction pathway of BPDE in aqueous DNA solutions.

Most recently, the chemical adduct distribution of the two enantiomers of BPDE with calf thymus DNA has been reinvestigated by Cheng et al. [26]; in the case of (+)-BPDE, the major adduct (~94%) involves trans opening of the BPDE epoxide ring and covalent adduct formation between the C-10 position of BPDE and the exocyclic amino group of guanine in DNA. In the case of (-)-BPDE, a similar trans-dG adduct is also formed (63%), with lesser amounts of a cis-adduct (22%) and a trans -N6-adenine adduct (15%) [26].

2.2. BASE-SEQUENCE SELECTIVITIES

2.2.1. *Noncovalent Complex Formation and Reaction Kinetics.* Based on the reaction scheme in Fig. 1, it can be shown [27] that the overall reaction rate constant (k) of BPDE molecules in aqueous DNA solutions can be approximated by the following equation:

$$k = \frac{k_h + k_3 K [DNA]}{1 + K [DNA]} \qquad (1)$$

where k_h is the reaction rate constant in the absence of DNA, k_3 is the rate constant for reaction of BPDE molecules at DNA binding sites, and $K = k_1/k_2$ is the apparent association constant averaged over all types of different noncovalent binding sites [12]. The quantity [DNA] denotes the concentration of binding sites expressed in units of DNA nucleotides.

The reactivity of BPDE with DNA can be expressed in terms of f_{cov}, the fraction of BPDE molecules which react by forming covalent adducts with DNA rather then by reacting with water to form tetraols.

A comparison of the base sequence dependences of the parameters $k_3 (s^{-1})$, f_{cov}, and $K(M^{-1})$ obtained with synthetic polynucleotides is summarized in Fig. 2.

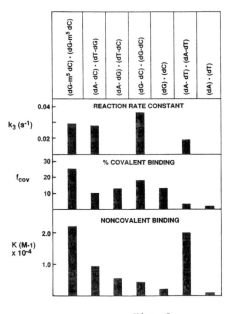

Fig. 2

In these studies [12], racemic mixtures of BPDE $((\pm)$-BPDE), rather then the resolved enantiomers were used because of the greater availability of $(\pm)$-BPDE. Roche has shown that f_{cov} for $(+)$-BPDE is larger than for $(-)$-BPDE by factors of ~5 in poly(dG-dC)·(dG-dC) and in poly(dG)·(dC). For the adenine-containing polymers, the values of f_{cov} are comparable for $(+)$-BPDE and $(-)$-BPDE [28].

The noncovalent association constant K is highest in poly(dG-m^5dC)·(dG-m^5dC) and in poly(dA-dT)·(dA-dT) sequences, and is lowest in the non-alternating polymers poly(dG)·(dC) and poly(dA)·(dT). The reaction rate constant of physically bound BPDE, k_3, is uniformly higher for all dG-containing polymers, reflecting the preferred reactions of BPDE with guanine [10]. Values of k_3 are not listed for any of the non-alternating polynucleotides; the reactions at these sequences do not appear to involve the formation of pre-reaction physical complexes [12], and thus Eq. (1) is not appropriate in these cases. The values of the covalent binding efficiency factor f_{cov} are significantly higher for all polymers which contain guanine than for those which do not (Fig. 2). Again this fact reflects the preference for the covalent binding of both BPDE enantiomers to guanine rather than to adenine [26].

2.2.3. *Relationship Between Noncovalent Complex and Covalent Adduct Formation*. It is evident from the data shown in Fig. 2, that there is no obvious and direct correlation between noncovalent complex formation (which is presumed to be intercalative in nature [12]) and the efficiency of covalent adduct formation. It is often assumed that intercalation is an important pre-requisite for covalent adduct formation, and thus for the manifestation of the biological activities of PAH diol epoxides [19,27,29]. While this hypothesis is not supported by the data in Fig. 2, it is likely that in a complex cellular environment the value of K may be important. Large values of K for any particular PAH diol epoxide tend to keep these reactive molecules associated with the DNA where they can ultimately undergo chemical reactions to form mutagenic adducts with guanine or adenine [27]. If the K values are small, a greater fraction of the PAH diol epoxide molecules will be located in complexes with other cellular macromolecules and in the aqueous phases of the cell. In these other environments they can be deactivated by reacting with glutathione or other agents which serve to detoxify these mutagenic compounds. Thus, the ability of these compounds to form noncovalent complexes with DNA, whether intercalative or not, may be important because high values of K increase the probability of reaction with DNA.

In the case of native DNA, which contains all possible sequence, the results summarized in Fig. 2 suggest that a higher fraction of BPDE molecules which have not yet reacted can be found at alternating dA-dT sequences. However, covalent binding reactions at dG can still occur efficiently; the residence time of any given BPDE molecule at a particular noncovalent binding site is expected to be quite short, of the order of several milliseconds, or less. Since the covalent binding and hydrolysis reactions occur on time scale of minutes, any particular BPDE molecule has the opportunity to sample many different DNA binding sites during its lifetime by physically diffusing from site to site and, finally, reacting with guanine [12].

2.3. CHARACTERISTICS OF COVALENT ADDUCTS.

2.3.1. *Types of Adduct Conformations*. Two different types of BPDE-DNA binding sites have been identified based on linear dichroism and other spectroscopic techniques [18-20].

Site I-type conformations are characterized by mean tilts of the long axis of the pyrenyl residues of 25-30° with respect to the aver-

age orientations of the planes of the DNA bases; these site I conformations are characterized by broadened and red shifted absorption
spectra (~10 nm with respect to the absorption spectra of BPDE
molecules in aqueous solutions). Such red shifts are usually attributed to extensive π - electron stacking interactions between the
pyrenyl residues and DNA bases; however, other causes for these effects cannot be excluded. In linear dichroism experiments in which
the DNA molecules are oriented in hydrodynamic flow gradients, any
BPDE molecules bound to the DNA either noncovalently or covalently,
contribute to the linear dichroism (LD) signal [18,20]. The LD spectra above 300 nm resemble the absorption spectra of the pyrenyl
residues, but are either negative or positive in sign, depending on
the mean orientation of the pyrenyl long axis with respect to the
planes of the DNA bases. Site I adducts contribute *negative* LD signals because, on the average, the long axes of the pyrenyl residues
are tilted close to the planes of the DNA bases.

Site II adducts, in contrast to Site I adducts, are considerably
more exposed to the aqueous solvent environment [30-32]. The red
shift in the absorption spectrum is only 2-3 nm, and the LD spectrum
again resembles the absorption spectrum, but is positive in sign.
Taken together, these characteristics suggest that the long axes of
the planar pyrenyl residues, on the average, are tilted away from
the DNA base planes at relatively large angles (> 65°).

2.3.2. *Conformations of Noncovalent Complexes and Covalent Adducts Derived from (+)-BPDE- and (-)-BPDE.* The complexes formed when either
enantiomer binds to native DNA are of the Site I-type, and intercalative in nature [23]. Linear dichroism measurements indicate that
there is a pronounced change in orientation of (+)-BPDE as a result
of the chemical binding reaction, whereas the apparent changes in
conformations of the (-)-BPDE pyrenyl residues appear to be less pronounced [25].

Examples of LD spectra obtained from the covalent binding of (+)-
BPDE or (-)-BPDE to poly(dG)·(dC) are shown in Fig. 3.

The adducts derived from (-)-BPDE are characterized by an absorption
spectrum with maxima at 338 and 353 nm (Fig. 3A), which constitutes a
9-10 nm red shift with respect to the absorption spectrum of free BPDE
in aqueous solutions. The LD spectrum is negative in sign with the
minima occurring at the same wavelength as the absorption maxima (Fig.
3C). The (-)-BPDE-poly(dG)·(dC) adducts are clearly of the site I-
type.

The adducts derived from (+)-BPDE are characterized by a positive LD
spectrum (Fig. 3D) which resembles the absorption spectrum (Fig. 3B),
with maxima at 329 and 345 nm in both cases. These maxima are red-
shifted by only 1-2 nm with respect to free BPDE, suggesting that π-
stacking interactions are not occurring to any significant extent; a
small negative Site I LD signal with a minimum at 260 nm is also
observed (Fig. 3D). Overall, the LD spectrum is attributed mainly to
Site II adducts with a small contribution from site I adducts. Similar effects are observed with the alternating poly(dG-dC)·(dG-dC).
In the case of(+)-BPDE bound covalently to calf thymus DNA, site II

adducts are even more dominant than in the case of poly(dG)·(dC), whereas in (-)-BPDE-DNA adducts the Site I:Site II distribution is about 2:1 [32].

2.3.3   *Adduct Conformations and Biological Activities.*   The conforma-
tions of covalent native DNA adducts derived from nearly 20 different PAH diol epoxides have been examined by linear dichroism techniques [18,33].   All of the highly tumorigenic isomers give rise to adducts which are   predominantly of the Site II-type, whereas all less active or inactive isomers give rise predominantly to Site I adducts.   This empirical correlation between biological activity and adduct conformation has been observed for PAH diol epoxide molecules which are known to bind predominantly to N2 of guanine.   The conformations of DNA adducts derived from other PAH diol epoxides which bind extensively to adenine [7,34] have not yet been studied, and may well be different.

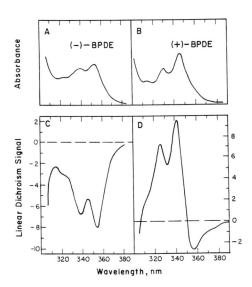

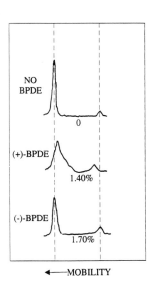

Fig. 3 (left). Characteristic absorption and linear dichroism spectra
        of BPDE enantiomers covalently bound to poly(dG)·(dC).

Fig. 4 (right). Electrophoretic agarose gel patterns of PIBI 30
        supercoiled DNA covalently modified with (+)-BPDE or (-)-BPDE
        with 1.4% and 1.7% of bases modified, respectively.

3. Unwinding of Supercoiled DNA Induced by (+)-BPDE and (-)-BPDE.

The unwinding of supercoiled DNA by drugs and carcinogens can provide information on the mechanism of complex formation and DNA damage, *e.g.* the formation of single strand breaks. The unwinding of supercoiled DNA by racemic BPDE via noncovalent complex formation [22] and covalent adduct formation [35-37], has been previously investigated. It is well established that adducts derived from the covalent binding of (±)-BPDE to SV40 DNA [35] and ϕX174 DNA [36] cause the removal of left-handed supercoils and a decreased electrophoretic mobility on agarose gels. Based on such results, Drinkwater *et al.* [35] proposed a covalent intercalative binding model for racemic BPDE. Gamper and co-workers [36] suggested that the unwinding may be due to a disruption of base pairing at BPDE binding sites, with the aromatic hydrocarbon residues lying in the minor or the major groove.

We have investigated the unwinding of supercoiled DNA using (+)-BPDE and (-)-BPDE enantiomers instead of (±)-BPDE, and supercoiled PIBI 30 DNA (2926 base pairs, superhelical density $\sigma \approx$ -0.04). Utilizing a kinetic flow linear dichroism method [37], we have observed that both enantiomers when bound *noncovalently* to DNA cause reversible unwinding (data not shown); this is in agreement with the earlier observations of Meehan *et al.* [22]. However, the effects of *covalently* bound (+)-BPDE and (-)-BPDE residues are dramatically different from one another. Typical densitometer traces of gel electrophoresis patterns obtained with unmodified PIBI 30, and the same DNA sample modified covalently by either (+)-BPDE or (-)-BPDE are shown in Fig. 4.

The unmodified DNA is characterized by two electrophoretic bands of different mobilities. The higher mobility, more intense band is due to the supercoiled form, while the slower and smaller band is attributed to the relaxed nicked form (σ = 0). The shape of the highly supercoiled DNA is more compact than that of the relaxed DNA, and thus its electrophoretic mobility is higher than that of the nicked form. The covalently bound (+)-BPDE causes a significant decrease and broadening of the electrophoretic mobilities of PIBI30, while adducts arising from the (-) enantiomer clearly are much less effective, since the mobility distribution of the modified DNA is somewhat broadened but otherwise unaffected even at a high level of binding (1.7% of all PIBI 30 nucleotides modified).

Based on the results shown in Fig. 4, we conclude that the previously observed unwinding of supercoiled DNA induced by covalently bound residues derived from (±)-BPDE [35,36] was predominantly caused by the (+) enantiomer.

The striking difference in unwinding effects produced by (+) and (-) BPDE indicates that the covalently bound (+)-BPDE residues cause significant changes in the tertiary structure of DNA, while those arising from (-)-BPDE do not. It is interesting to note that simi-

lar effects are observed with the anti-tumor agent cis-dichloro-
diammineplatinum (II) (cis-DPP) and its inactive isomer trans-DPP
[38]. While cis-DPP causes significant unwinding of supercoiled DNA,
trans-DPP does not.

Based on results of linear dichroism measurements on supercoiled DNA
[37] covalently modified with either (+)-BPDE or (-)-BPDE, the adduct
conformations appear to be similar in linear and in supercoiled DNA
(data not shown). Unexpectedly, the external Site II adducts derived
from (+)-BPDE cause extensive unwinding, while the Site I adducts derived from (-)-BPDE do not, even though the conformations of these
Site I adducts resemble those of classical intercalation complexes in
several respects. These results are in accord with previous conclusions that covalent adduct conformations other than intercalative ones
can cause significant unwinding of supercoiled DNA [37]. Indeed other
causes of duplex unwinding not involving the formation of intercalation complexes have been documented [39].

It is known that racemic BPDE, upon covalent binding to native DNA,
gives rise to a bend or kink at the binding site [40]. Eriksson et
al. [41] have provided evidence that such effects are caused by (+)-
BPDE, but not by (-)-BPDE. Thus, the observed unwinding effects may
be due to the formation of such kinks, and/or flexible hinge joints.
The lack of unwinding of supercoiled DNA by the covalent binding of
(-)-BPDE is difficult to rationalize since the pyrenyl residues appear
to be at least partially stacked with the DNA bases. A more thorough
understanding of these complex effects must await a detailed elucidation of the structures of these covalent BPDE-DNA adducts.

4.   Synthesis and Characterization of Defined Covalent BPDE-Oligodeoxynucleotide Adducts.

The characteristics of covalent adducts derived from the reactions of
the tumorigenic (+)-BPDE and non-tumorigenic (-)-BPDE with native linear DNA, synthetic polynucleotides, and supercoiled DNA, are strikingly different from one another. However, in order to achieve a
deeper understanding of these characteristics on a more detailed
molecular level, it is necessary to extend these investigations to DNA
of well defined base composition and sequence. We have therefore initiated a series of studies on the characteristics of covalent adducts
derived from the binding of (+)-BPDE and (-)-BPDE to oligonucleotides
of defined base composition and sequence. Our initial studies have
focused on single strands containing a single dG residue, the primary
site of attack of both enantiomers, and the characteristics of the duplexes formed with modified strands.

4.1.   SYNTHESIS AND IDENTIFICATION

Initial studies in our laboratory have shown that pyrimidine-guanine-
pyrimidine sequences in oligonucleotides 9-11 bases long   can be
efficiently modified with either of the two BPDE enantiomers at the

single dG residues; the details of the synthesis and chemical characterization of the adducts are described more fully elsewhere [42].
Only a brief summary is provided here.

4.1.1.   *Reaction Yields.*   Both (+)-BPDE and (-)-BPDE form predominantly *cis* and *trans* addition products with the exocyclic amino
group of guanine (defined in Fig. 5).

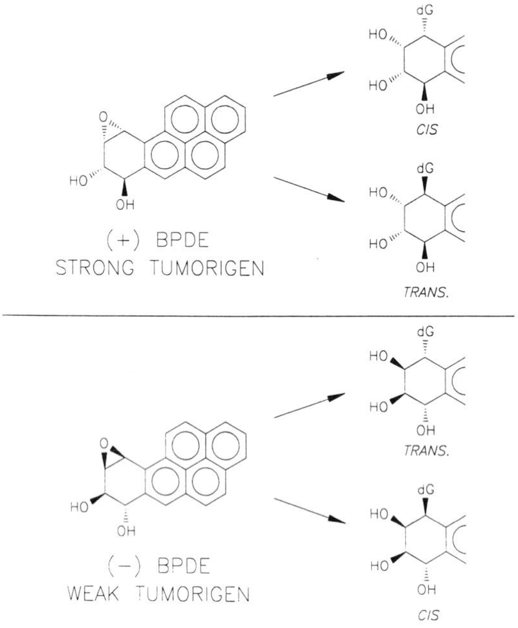

Fig. 5

Good levels of covalent binding were achieved with the 9-mer sequence
d(ATATGTATA) and 11-mer sequence d(CACATGTACAC). The reaction yields,
or fraction of BPDE molecules which react by covalently binding to the
oligonucleotides, were 34 - 45%   in the case of (+)-BPDE and 15% -
20% in the case of (-)-BPDE.

4.1.2.   *Isolation and Characterization of Addition Products.*   The
adducts were characterized by basically following methods similar to
those of Cheng *et al.* [26].   The BPDE-oligonucleotide adducts were
separated from the unmodified oligonucleotides by HPLC methods utilizing an ODS-Hypersil preparatory column and a 0-90% methanol/20 mM
sodium phosphate (pH = 7.0) buffer gradient in 60 minutes (24 ± 1 C°)
with a 3.0 ml/min flow rate.   Typical elution profiles of the

oligonucleotide d(ATATGTATA) modified with (+)-BPDE or (-)-BPDE are shown in Fig. 6. Three major elution peaks corresponding to the

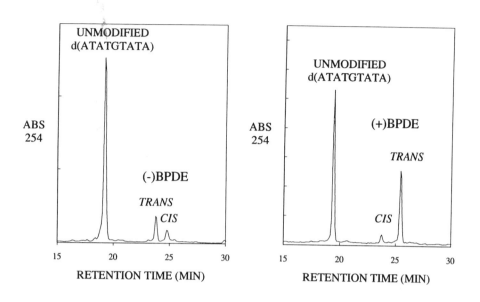

Fig. 6

unmodified oligonucleotide, and two elution peaks corresponding to BPDE-modified oligonucleotides are observed (minor elution peaks were also observed at later elution times and are not shown in Fig. 6).

Each of the elution peaks were collected and subjected to enzyme hydrolysis [42] for adduct analysis and identification. The enzyme hydrolysates were separated by HPLC using an ODS-Hypersil analytical column with a linear 0-90% methanol/buffer gradient in 60 min and a flow rate of 1.5 ml/min. The retention times of the BPDE-nucleoside adducts were compared with those of BPDE-N2-dG and BPDE-N6-dA standards prepared by the methods of Cheng et al. [26]. The cis and trans assignments shown in Fig. 6 were further confirmed by comparing the CD spectra (data not shown) of each of the eluates with those published by Cheng et al. [26]. With both enantiomers, the yield of N2-dG adducts was dominant over the yield of BPDE-dA adducts. We shall therefore focus our attention here only on the cis and trans BPDE-N2-dG adducts. With the oligonucleotide d(ATATGTATA) and (+)-BPDE, the relative yield of trans/cis adducts was about 7:1, whereas in the case of (-)-BPDE the trans/cis adduct ratio was only 2:1. Thus, in each case, trans addition dominates, although this effect is much less pronounced in the case of (-)-BPDE. Out of these four adducts, only the (-)-cis-BPDE-N2-dG adduct was found to be chemically unstable upon storage in the dark.

4.2.   PROPERTIES OF BPDE-OLIGONUCLEOTIDE ADDUCTS

4.2.1.   *Absorption Spectra of Single-Stranded BPDE-Oligonucleotide Adducts.*   The absorption spectra of the (+)-*trans*-N2-dG   and (+)-*cis*-N2-dG BPDE-d(ATATGTATA) adducts are shown in Fig. 7.

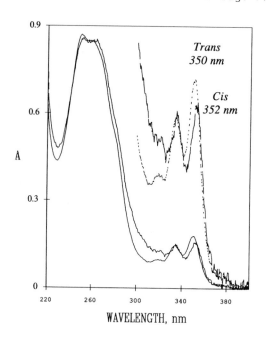

Fig. 7

The longest wavelength absorption band of the (+)-*trans* adduct displays a maximum at 350 nm, while the corresponding maximum of the *cis* adduct is at 352 nm.   The (-)-*trans* and (-)-*cis* adducts display absorption maxima at 350 and 353 nm, respectively.   These 6-9 nm red shifts in the absorption spectra relative to free BPDE in buffer solution   suggests   that   there   are   significant   base   stacking interactions between the pyrenyl residues and the DNA bases.   Upon heating to 80 C^O, the absorbance maxima shift from 350 - 353 nm to 346 - 348 nm, suggesting that these stacking interactions are disrupted at the higher temperatures.

4.2.2.   *Changes in Absorption Spectra Upon Stoichiometric Addition of the Complementary Strands.*   Upon addition of the complementary strand d(TATACATAT) to any of the four N2-dG-BPDE-d(ATATGTATA) adducts described in the above paragraph, there were no perceptible changes within the absorption band of the pyrenyl residues above 300 nm. These results suggest that this 9-mer is not capable of forming good duplexes when the single dG is modified with BPDE.

It was therefore of interest to construct a longer oligonucleotide with the same central d(..TAT..) motif, but with a few dC residues instead of dT residues in the flanking regions in order to provide extra stability upon duplex formation with the complementary strand. We selected the 11-mer d(CACATGTACAC) which, upon reaction with either (+)-BPDE or (−)-BPDE gave rise to the sets of four adducts described for the sequence d(ATATGTATA) in the above paragraphs. Interestingly, upon addition of the complementary strand d(GTGTACATGTG), the absorption maxima of the pyrenyl residues of the (+)-*trans* and (−)-*trans*-adducts blue-shifted from 350 nm to 346 nm; these absorption characteristics are reminiscent of the external Site II adducts observed when (+)-BPDE binds to native DNA, poly(dG-dC)·(dG-dC), or poly(dG)·(dC) (see above). On the other hand, when the complementary strands are added to the (+)-*cis*- and (−)-*cis*-BPDE-N2-dG adducts of the modified sequence d(CACATGTACAC), the absorption maxima remain approximately at the same positions near 352-353 nm. These results suggest that there are considerable base stacking interactions in both the single-stranded and double-stranded DNA. The absorption characteristics of these adducts are reminiscent of the Site I-type conformations observed when (−)-BPDE binds to high molecular weight DNA or polynucleotides.

4.2.3. *Melting Curves - the Destabilization of Duplex DNA by BPDE.* The unmodified 9-mer d(ATATGTATA) and 11-mer d(CACATGTACAC) in the duplex form (titrated with their respective complements) exhibit classical two-state helix coil melting curves (Figs. 8A and 8B) with T_m values of 20 and 39 C, respectively, and hyperchromicities of 16 ± 1% in both cases.

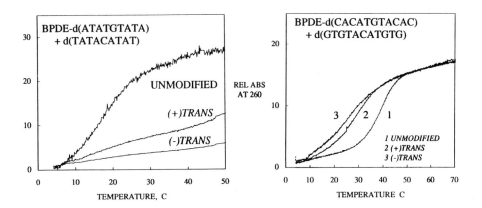

Fig. 8. Melting curves of modified oligonucleotides.

In the case of the modified 9-mer duplex, the presence of the covalently bound BPDE residues appears to disrupt base pairing and duplex
formation, as is evident from the melting curves obtained with the
(+)-*trans* and (-)-*trans* 9-mer adducts (Fig. 8A); neither of the *cis*
adducts displayed any hyperchromicity upon heating (data not shown).

The modified (+)-*trans* and (-)-*trans*-11-mer adducts are characterized by well defined helix coil transition curves with significantly
lower T_m values (27 and 25 C, respectively). In the case of the (+)-
trans 11-mer adduct, the hyperchromicity is the same as in the case of
the unmodified 11-mer; however, the large decrease in the T_m value
suggests a significant loss in hydrogen-bonding in the covalent
adduct. In the case of the (-)-*trans* 11-mer adduct, the hyperchromicity is lowered from 15% to 11% upon covalent adduct formation, suggesting a small decrease in base stacking interactions as well as a
loss in base pairing. The (+)-*cis* adduct exhibited a low T_m (27 C)
and the hyperchromicity was 12%.

It is evident that the covalent binding of (+)-BPDE or (-)-BPDE has
a tendency to destabilize duplex DNA, regardless of the *cis* or *trans*
conformation of the adducts. This effect can be minimized by introducing additional dC-dG base pairs on either side of the BPDE-modified
oligonucleotide residues.

5. Conclusions

The preparation of stereochemically defined oligonucleotides modified
at specific dG residues with either (+)-BPDE or (-)-BPDE has allowed,
for the first time, a comparison of the spectroscopic characteristics
of *cis* and *trans* BPDE-N2-dG adducts incorporated in sequences 9-11
bases long. The spectroscopic properties of these BPDE-oligonucleotide adducts suggest that the Site I and Site II conformations observed with higher molecular weight DNA modified with (+)-BPDE and
(-)-BPDE may, in part, correspond to such *cis* and *trans* adducts,
respectively. The characteristics of DNA adducts derived from the
covalent binding of the tumorigenic (+)-BPDE and non-tumorigenic (-)-
BPDE are quite different from one another in the relative proportions
of Site I and Site II adducts, and in the unwinding of supercoiled
DNA. How these differences manifest themselves biologically remains
to be elucidated.

6. ACKNOWLEDGEMENTS

This work was supported by the Department of Energy (Grants DEFGO2-
86ER60405 and DEFGO2-88ER60674) and in part by Grant CA 20851 from the
US Public Health Service, Department of Health and Human Resources,
awarded by the National Cancer Institute. We wish to thank Dr. D.
Patel for suggesting the sequence of the 11-mer oligonucleotide
described in this work.

7. REFERENCES

1. Conney, A.H. (1982) 'Induction of microsomal enzymes by foreign chemicals and carcinogenesis by polycyclic aromatic hydrocarbons', G.H.A. Clowes memorial lecture, Cancer Research 42, 4875-4917.
2. Singer, B. and Grunberger, D. (1983) Molecular Biology of Mutagens and Carcinogens, Plenum Press, New York.
3. Loeb, L.A. (1989) Endogenous carcinogenesis: molecular oncology in the twenty-first century-presidential address. Cancer Res. 49, 5489-5496.
4. Yang, J.-L, Maher, V.M., and McCormick, J.J. (1987) 'Kinds of mutations formed when a shuttle vector containing adducts of (+)-7b,8a-dihydroxy-9a,10a-epoxy-7,8,9,10-tetrahydrobenzo[a] pyrene replicates in human cells, Proc. Natl. Acad. Sci. USA 84, 3787-3791.
5. Kootstra, A., Lew, L.K., Nairn, R.S., and McLeod, M.C. (1989) 'Preferential modification of GC boxes by benzo[a]pyrene-7,8-diol-9,10-epoxide', Mol. Carcinogenesis 1, 239-244.
6. Boles, T.C. and Hogan, M.E. (1984) 'Site specific carcinogen binding to DNA, Proc. Natl. Acad. Sci. USA 81, 5623-5627.
7. Reardon, D.B., Bigger, C.A.H., Strandberg, J., Yagi, H., Jerina, D.M., and A. Dipple (1989) 'Sequence selectivity in the reaction of optically active hydrocarbon dihydrodiol epoxides with rat h-*ras* DNA', Chem. Res. Toxicol. 2, 12-14.
8. Warpehoski, M.A. and Hurley, L.H. (1988) 'Sequence selectivity of DNA covalent modification', Chem. Res. Toxicol. 1, 315-333.
9. Maher, V.M., Yang, J.-L., Ma, M.C.-M. and McCormick, J. (1989) 'Comparing the frequency and spectra of mutations induced when an SV-40 based shuttle vector containing covalently bound residues of structurally related carcinogens replicates in human cells', Mutation Res. 220, 83-92.
10. MacLeod, M.C. and Zachary, K.L. (1985) 'Involvement of the exocyclic amino group of deoxyguanine in DNA-catalyzed carcinogen detoxification', Carcinogenesis 6, 147-149.
11. Chen, F.M. (1986) 'Binding of enantiomers of *trans*-7,8-dihydroxy-*anti*-9,10-epoxy-7,8,9,10-tetrahydrobenzo[a]pyrene to polynucleotides, J. Biomol. Struct. and Dynamics 4, 401-418.
12. Geacintov, N.E., Shahbaz, M., Ibanez, V., Moussaoui, K. and Harvey, R.G. (1988) 'Base-sequence dependence of noncovalent complex formation and reactivity of benzo[a]pyrene diol epoxide with polynucleotides.' Biochemistry 27, 8380-8387.
13. Kim, S.-K., Geacintov, N.E., Zinger, D., and Sutherland, J.C. (1988) 'Fluorescence spectral characteristics and fluorescence decay profiles of covalent polycyclic aromatic carcinogen-DNA adducts', in R.M. Sweet and A.D. Woodhead (eds), Synchrotron Radiation in Structural Biology, Basic Life Sciences, Vol, 51, Plenum Press, New York, pp. 187-205.

14. Miller, K.J., Taylor, E.R., and Dommen, J. (1985) A mechanism for the stereoselectivity and binding of benzo[a]pyrene diol epoxides, in R.G. Harvey (ed.) Polycyclic Hydrocarbons and Carcinogenesis, ACS Symposium Series 283, American Chemical Society, Washington, D.C., pp. 239-288

15. Zakrzewska, K. and Pullman, B. (1987) 'Sequence selectivity, a test of the nature of the covalent adduct formed between benzo[a]pyrene and DNA', J. Biomol. Struct. Dynamics 4, 845-858.

16. Brookes, P. and Osborne, M.R. (1982) Mutation in mammalian cells by stereoisomers of *anti*-benzo[a]pyrene diol epoxide in relation to the extent and nature of the DNA reaction products', Carcinogenesis 3, 1223-1226.

17. Stevens, C.W., Bouck, N., Burgess, J.A., and Fahl, W.E. (1985) Benzo[a]pyrene diol-epoxides: differential mutagenic efficiency in human and bacterial cells', Mutation Res. 152, 5-14.

18. Geacintov, N.E. (1988) 'Mechanisms of reaction of polycyclic aromatic epoxide derivatives with nucleic acids', in S.K. Yang and B.D. Silverman (eds.) Polycyclic Aromatic Hydrocarbon Carcinogenesis: Structure Activity Relationships, CRC Press, Boca Raton, FL, Vol. 2, pp. 181-206.

19. Harvey, R.G. and Geacintov, N.E. (1988) 'Intercalation and binding of carcinogenic hydrocarbon metabolites to nucleic acids', Accts. Chem. Res. 21, 66-73.

20. Gräslund, A. and Jernström, B. (1989) 'DNA-carcinogen interaction: covalent DNA-adducts of benzo[a]pyrene 7,8-dihydrodiol 9,10-epoxides studied by biochemical and biophysical techniques', Quart. Rev. Biophys. 22, 1-37.

21. Geacintov, N.E., Yoshida, H., Ibanez, V., and Harvey, R.G. (1981) Non-covalent intercalative binding of 7,8-dihydroxy-9,10-epoxybenzo[a]pyrene to DNA, Biochem. Biophys. Res. Comm. 100, 1569-1577.

22. Meehan, T., Gamper, H. and Becker, J.F. (1982) 'Characterization of reversible physical binding of benzo[a]pyrene derivatives to DNA', J. Biol. Chem. 257, 10479-10485.

23. Roche, C.J. Geacintov, N.E., Ibanez, V., and Harvey, R.G. (1989) 'Linear dichroism properties and orientations of different ultraviolet transition moments of benzo[a]pyrene derivatives bound noncovalently and covalently to DNA', Biophys. Chem. 33, 277-288.

24. MacLeod, M.C. and Zachary, K. (1985) 'Catalysis of carcinogendetoxification by DNA: comparison of enantiomeric diol epoxides', Chem. Biol. Interactions 54, 45-55.

25. Geacintov, N.E., Yoshida, H., Ibanez, V., Jacobs, S.A. and Harvey, R.G. (1984) Conformations of adducts and kinetics of binding to DNA of the optically pure enantiomers of anti-benzo[a]pyrene diol epoxide', Biochem. Biophys. Res. Comm. 122, 33-39.

26. Cheng, S.C., Hilton, B.D., Roman, J.M. and Dipple, A. (1989) 'DNA adducts from carcinogenic and noncarcinogenic enantiomers of benzo[a]pyrene dihydrodiol epoxide', Chem. Res. Toxicol. 2, 324-340.

27. Geacintov, N.E. (1986) 'Is intercalation a critical factor in the covalent binding of mutagenic and tumorigenic polycyclic aromatic diol epoxides to DNA? Carcinogenesis 7, 759-766.

28. Roche, C.J. (1987) The Physical and Covalent Interactions of Polycyclic Aromatic Hydrocarbon Epoxide Derivatives with Nucleic Acids. Ph.D. Dissertation, New York University.

29. LeBreton, P.R. (1985) 'The intercalation of benzo[a]pyrene and 7,12-dimethyl-benz[a]anthracene metabolites and metabolite model compounds in DNA', in R.G. Harvey (ed.) Polycyclic Hydrocarbons and Carcinogenesis, ACS Symposium Series 283, American Chemical Society, Washington, D.C., pp.64-84.

30. Kolubayev, V., Brenner, H.C., and Geacintov, N.E. (1987) Stereoselective covalent binding of enantiomers of anti-benzo[a]pyrene diol epoxide to DNA as probed by optical detection of magnetic resonance', Biochemistry 26, 2638-2641.

31. Kim, S.-K., Brenner, H.C., Soh, B.J., and Geacintov, N.E. (1989) 'Fluorescence spectroscopy of benzo[a]pyrene diol epoxide-DNA adducts. Conformation-specific emission spectra', Photochem. Photobiol. 50, 327-337.

32. Zinger, D., Geacintov, N.E., and Harvey, R.G. (1987) Conformations and selective photodissociation of heterogeneous benzo[a]pyrene diol epoxide enantiomer-DNA adducts', Biophys. Chem. 27, 131-138.

33. Carberry, S.E., Geacintov, N.E., and R.G. Harvey (1989) 'Reactions of stereoisomeric non-bay-region benz[a]anthracene diol epoxides with DNA and conformations of non-covalent complexes and covalent adducts', Carcinogenesis 10, 97-103.

34. Bigger, C.A.H., Strandberg, J., Yagi, H., Jerina, D.M. and Dipple, A. (1989) 'Mutagenic specificity of a potent carcinogen, benzo[c]phenanthrene (4R,3S)-dihydrodiol (2S,1R)-epoxide, which reacts with adenine and guanine in DNA', Proc. Natl. Acad. Sci. USA 86, 2291-2295.

35. Drinkwater, N.R., Miller, J.A., Miller, E.C., and N.-C. Yang (1978) Covalent intercalative binding to DNA in relation to the mutagenicity of hydrocarbon epoxides and N-acetone-2-acetylamine fluorene', Cancer Res. 38, 3247-3255.

36. Gamper, H.B., Straub, K., Calvin, M., and Bartholomew, J.C. (1980) DNA alkylation and unwinding induced by benzo[a]pyrene diol epoxide: Modulation by ionic strength and superhelicity', Proc. Natl. Acad. Sci. USA 77, 2000-2004.

37. Yoshida, H. and Swenberg, C.E. (1987) 'Kinetic flow dichroism study of conformational changes in supercoiled DNA induced by ethidium bromide and noncovalent and covalent binding of benzo[a]pyrene diol epoxide', Biochemistry 26, 1351-1358.

38. Scovell, W.M. and Collert, F. (1985) Unwinding of supercoiled DNA by cis- and trans-diamminedichloroplatinum(II): influence of the torsional strain on DNA unwinding', Nucl. Acids Res. 13, 2881-2895.

39. Bauer, W.R. (1978) 'Structures and reaction of closed duplex DNA' Ann. Rev. Biophys. Bioeng. 7, 287-313.

40. Hogan, M.E., Dattagupta, N., and Whitlock, J.P. Jr. (1981) 'Carcinogen-induced alteration of DNA structure', J. Biol. Chem. 256, 4504-4513.

41. Eriksson, M., Nordén, B., Jernström, B., and Gräslund, A. (1988) 'Binding geometries of benzo[a]pyrene diol epoxide isomers covalently bound to DNA. Orientational distribution', Biochemistry 27, 1213-1221.

42. Cosman, M., Ibanez, V., Geacintov, N.E. and Harvey, R.G. (1990) 'Preparation and isolation of adducts in high yield derived from the binding of two benzo[a]pyrene-7,8-dihydroxy-9,10-oxide stereoisomers to the oligonucleotide d(ATATGTATA)'. Submitted for publication.

43. McLaughlin, L.W. and Piel, N. (1984) 'Chromatographic purification of synthetic oligonucleotides', in M.J. Gait (ed.), Oligonucleotide Synthesis: A Practical Approach, IRL Press, Oxford, pp. 117-133.

AFLATOXIN-DNA BINDING AND THE CHARACTERIZATION OF AFLATOXIN B$_1$-OLIGODEOXYNUCLEOTIDE ADDUCTS BY ^{1}H NMR SPECTROSCOPY

MICHAEL P. STONE, S. GOPALAKRISHNAN, KEVIN D. RANEY, VERONICA M. RANEY, SUZANNE BYRD, AND THOMAS M. HARRIS
Department of Chemistry and Center in Molecular Toxicology
Vanderbilt University
Nashville, TN 37235
USA

ABSTRACT. Aflatoxins B$_1$ and B$_2$ intercalate with B-DNA, as demonstrated by NMR analysis of association with d(ATGCAT)$_2$ and d(GCATGC)$_2$, alteration of pBR322 electrophoretic mobility, and flow dichroism using linearly oriented calf thymus DNA. The less planar δ-lactone ring of aflatoxins G$_1$ and G$_2$ reduces DNA binding affinity by approximately one order of magnitude, but binding studies suggest that aflatoxins G$_1$ and G$_2$ also bind B-DNA by intercalation. At low DNA concentration, the number of adducts formed by either aflatoxin B$_1$-8,9-epoxide or aflatoxin G$_1$-9,10-epoxide is reduced with a concomitant increase in formation of the respective dihydrodiols, and the ratio of adducts formed by aflatoxin G$_1$-9,10-epoxide to those formed by an equivalent concentration of aflatoxin B$_1$-8,9-epoxide decreases. Reaction of aflatoxin B$_1$-8,9-epoxide with d(ATCGAT)$_2$ exhibits a limiting stoichiometry of 1:1 aflatoxin B$_1$:d(ATCGAT)$_2$. In contrast, reaction of aflatoxin B$_1$-8,9-epoxide with d(ATGCAT)$_2$ exhibits a limiting stoichiometry of 2:1 aflatoxin B$_1$:d(ATGCAT)$_2$. ^{1}H NOE experiments, non-selective ^{1}H T$_1$ relaxation measurements, and ^{1}H chemical shift perturbations demonstrate that in each case the aflatoxin moiety is intercalated above the 5' face of the modified guanine. These data are consistent with an intercalated transition state complex between aflatoxin B$_1$-8,9-epoxide and B-DNA.

1. Introduction

The aflatoxins are among the most potent mutagens implicated in human carcinogenesis. These compounds are coumarin derivatives elaborated by *Aspergillus flavus* and *parasiticus*. Aflatoxins are of world concern because of their widespread occurrence in corn, peanuts, cottonseed, and other agricultural commodities (for a review, see Busby & Wogan, 1984). Recent drought conditions have raised concerns regarding aflatoxin contamination of the corn crop in the United States. Widespread poisoning of poultry by contaminated feed led to the isolation of the four aflatoxins which will be the focus of this monograph, designated B$_1$, B$_2$, G$_1$, and G$_2$ on the basis of their fluorescence and thin layer chromatography elution characteristics.

The structural identification of aflatoxins was made by Büchi, Wogan, and coworkers (Asao et al., 1963; 1965); crystal structures were determined by Van Soest and Peerdeman (1970), and by Cheung and Sim (1964). Aflatoxin mutagenicity results as a consequence of oxidative activation *in vivo*, which requires the presence of the vinyl ether functionality in the terminal furan ring. Consequently, aflatoxins B$_2$ and G$_2$ do not form

451

B. Pullman and J. Jortner (eds.), Molecular Basis of Specificity in Nucleic Acid-Drug Interactions, 451–480.
© 1990 *Kluwer Academic Publishers. Printed in the Netherlands.*

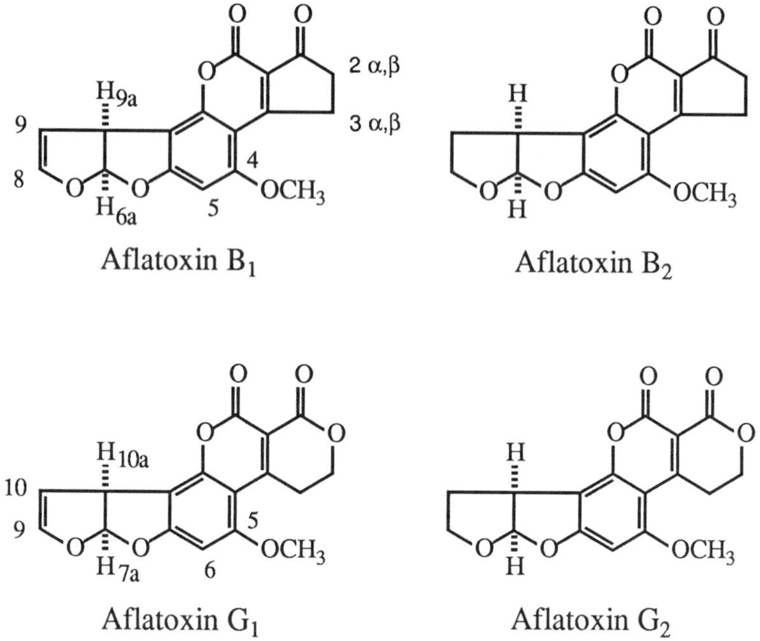

Figure 1. Structures of aflatoxins B_1, B_2, G_1, and G_2. The protons are numbered following the IUPAC convention and correspond with the NMR data. In earlier literature, the aflatoxin B_1 vinyl ether group is numbered as 2,3. The definitions of the enantiotopic protons at C2 and C3 of aflatoxin B_1 are based upon the Cahn, Ingold, and Prelog nomenclature. In this report, we have defined H2α to be the pro-R proton at C2; H2β is defined to be the pro-S proton at C2. H3α is defined to be the pro-S proton at C3, and H3β is defined to be the pro-R proton at C3. H2α and H3α lie on the same face of the cyclopentenone ring as do H6a and H9a; H2β and H3β lie on the other face of the cyclopentenone ring.

DNA adducts. Cytochrome P-450$_{NF}$, the nifedipine oxidase, is reported to be the principal human liver enzyme involved in the bioactivation of aflatoxin B_1 (Shimada & Guengerich, 1989). Aflatoxin B_1 can be activated *in vitro* by either chemical or enzymic methods (Swenson et al., 1977; Gorst-Allman et al., 1977; Croy et al., 1978; Martin & Garner, 1977). The ultimate mutagen of aflatoxin B_1 is postulated to be *exo* aflatoxin B_1-8,9-epoxide, formed by oxidation of the 8,9 double bond. This labile intermediate has never been detected *in vivo*. However, experimental evidence provides strong support that this epoxide is indeed the key intermediate. Essigmann et al. (1977) and Lin et al. (1977) showed that activated aflatoxin B_1 reacts with DNA exclusively at guanine N7; 8,9-dihydro-8-(N7-guanyl)-9-hydroxy-aflatoxin B_1 was identified as the primary adduct of duplex DNA. Nucleophilic attack by guanine N7 on C8 of activated aflatoxin B_1 occurs exclusively from the more hindered back side to give *trans* adduct (Hertzog et al., 1982). The successful preparation and characterization of *exo* aflatoxin B_1-8,9-epoxide was reported by Baertschi et al. (1988), who demonstrated that this epoxide reacts directly with calf thymus DNA to yield 8,9-dihydro-8-(N7-guanyl)-9-hydroxy-aflatoxin B_1.

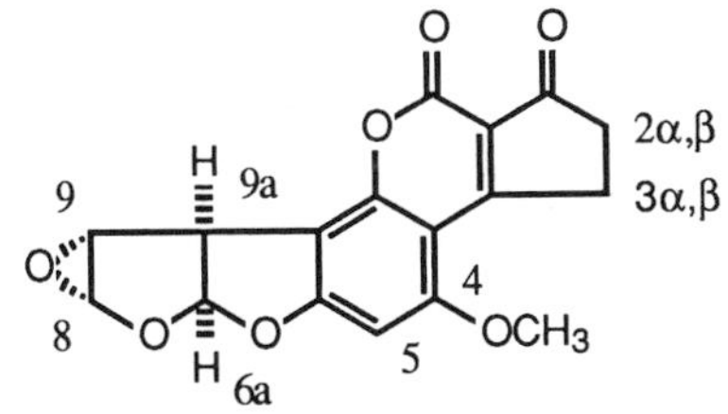

Aflatoxin B$_1$-8,9-epoxide

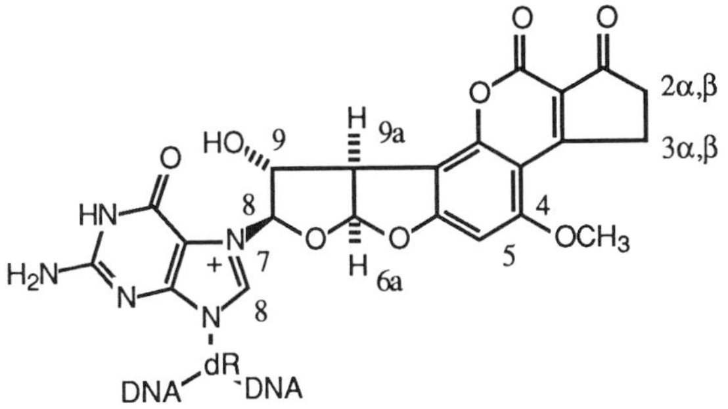

8,9-Dihydro-8-(N7-guanyl)-9-hydroxy-aflatoxin B$_1$ Adduct

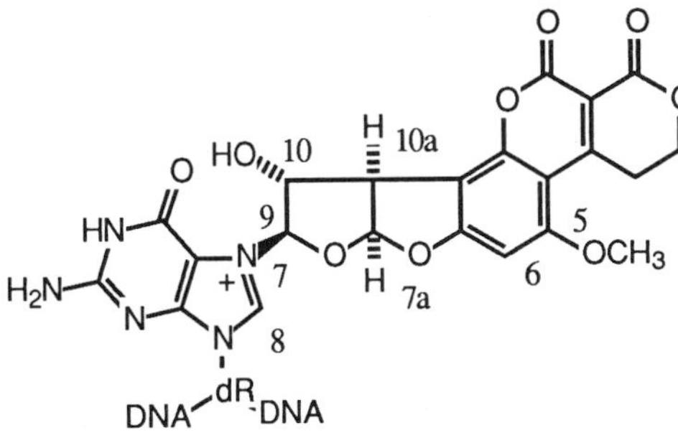

9,10-Dihydro-9-(N7-guanyl)-10-hydroxy-aflatoxin G$_1$ Adduct

Figure 2. Structures of aflatoxin B$_1$-8,9-epoxide, and the DNA adducts 8,9-dihydro-8-(N7-guanyl)-9-hydroxy-aflatoxin B$_1$ and 9,10-dihydro-9-(N7-guanyl)-10-hydroxyaflatoxin G$_1$.

Adduct formation by aflatoxin B$_1$ elicits a strong biological response, as measured by *S. typhimurium* reversion (McCann et al., 1975; Wong et al., 1977; Mori et al., 1986; Yourtee et al., 1987) or *umu* gene activation (Oda et al., 1985; Shimada et al., 1987; Baertschi et al., 1989; Shimada & Guengerich, 1989), and induces both frameshift and substitution mutations (Foster et al., 1983; Sambamurti et al., 1988). However, the instability of 8,9-dihydro-8-(N7-guanyl)-9-hydroxy-aflatoxin B_1 complicates adduct-directed mutagenesis studies. This instability is due to the presence of an alkyl group at guanine N7 which results in a positive charge localized on the guanine imidazole ring. The primary adduct is subject to cleavage of the glycosyl bond resulting in depurination. Under basic conditions, hydroxide attack at guanine C8 results in cleavage of the imidazole ring to give a formamidopyrimidine (FAPY) adduct which is relatively stable. A third pathway of degradation is hydrolysis of the aflatoxin B_1-guanine N7 bond to remove the adduct and release the hydrolysis product *trans* aflatoxin B_1-8,9-dihydrodiol.

We have sought to understand the properties which make aflatoxin B_1-8,9-epoxide uniquely well suited to forming DNA adducts. Although the half-life of this epoxide has not been measured, it is estimated to be on the order of seconds in aqueous solution. Nevertheless, it reacts readily with DNA, both regiospecifically and stereospecifically. This observation suggests that adduct formation is directed by binding[1] of aflatoxin B_1-8,9-epoxide to the DNA helix with a specific orientation which promotes formation of a favorable transition state (Misra et al., 1983; Benasutti et al., 1988; Gopalakrishnan et al., 1989a). The present report summarizes our current understanding of DNA adduct formation by aflatoxin B_1-8,9-epoxide, and the geometry of the primary DNA adduct, 8,9-dihydro-8-(N7-guanyl)-9-hydroxy-aflatoxin B_1.

2.   Materials and Methods

2.1   MATERIALS

Aflatoxins B_1, B_2, G_1, and G_2 were purchased from Sigma Chemicals, Inc., or Aldrich Chemicals. Calf thymus DNA Type I was purchased from Sigma Chemicals, Inc. Oligodeoxynucleotides were synthesized with an automated synthesizer, using solid-phase phosphoramidite chemistry. Plasmid pBR322 was purchased from Promega. Topoisomerase I was purchased from Bethesda Research Laboratories, Inc. Ethidium bromide and actinomycin D were purchased from Sigma Chemicals Inc.

2.2   METHODS

2.2.1. *Handling of aflatoxins and aflatoxin epoxides.* Crystalline aflatoxins are hazardous due to their electrostatic nature and should be handled using appropriate containment procedures and respiratory mask to prevent inhalation. Aflatoxins can be destroyed by treatment with NaOCl. It should be assumed that aflatoxin epoxides are highly toxic and carcinogenic. Manipulations should be carried out in a well-ventilated hood with suitable containment procedures.

2.2.2. *Equilibrium Binding Measurements.* (a) Equilibrium Dialysis. Calf thymus DNA was prepared in 10 mM sodium phosphate buffer at pH 7.2. The DNA concentration was 0.5 mM for aflatoxin B_1 experiments and 3 mM for aflatoxin G_1 experiments. Stock

---

[1]Throughout this paper, the term "binding" refers to non-covalent association with DNA, and the corresponding establishment of equilibrium between bound and free ligand.

solutions of aflatoxins B_1 and G_1 in dimethylsulfoxide (0.2 mM and 1.2 mM) were prepared. Aliquots of aflatoxins were added to 1 mL of buffer and to 1 mL of buffer containing DNA to give a range of aflatoxin concentrations from 0.4 μM to 36 μM. Dialysis cells were constructed from 14 mL polypropylene centrifuge tubes and caps (Horowitz & Barnes, 1983; MacLeod et al., 1987; Raney et al., 1990). A series of cells, with differing aflatoxin concentrations, were dialyzed at 4 °C for 60 hours. The aflatoxin concentration in each chamber of the cell was determined by reverse-phase HPLC.

(b) Measurement of excess ^{1}H NMR linewidth. Stock solutions (2.5 mM) of aflatoxins B_1, G_1, B_2, and G_2 were prepared in d_6-dimethylsulfoxide. 10 μL of each stock was added to 400 μL of D_2O buffer (0.1 M NaCl, 5 x 10^{-5} M Na_2EDTA, and 0.01 M sodium phosphate, pH 7.0) to make a 61 μM NMR sample. Stock solutions of 0.22 mM and 1.05 mM calf thymus DNA in D_2O buffer containing 61 μM aflatoxin were prepared by addition of aflatoxin stock solution. The aflatoxin NMR sample was titrated with the DNA-aflatoxin mixture, with gentle mixing after each addition of DNA. After each addition of DNA, a NMR spectrum was obtained at 27 °C. Measurement of DSS line width served as a check for proper shimming and changed by less than 0.5 Hz from sample to sample.

(c) Oligodeoxynucleotide Binding. A 70 μM sample of each of the aflatoxins in deuteriated NMR buffer containing 2.5% v/v d_6-dimethylsulfoxide was titrated with d(ATGCAT)$_2$ or d(GCATGC)$_2$ dissolved in deuteriated NMR buffer containing an equivalent concentration of the aflatoxin. The chemical shifts of the various aflatoxin protons were plotted as a function of oligodeoxynucleotide concentration.

2.2.3. *Conformational Studies on Bound Aflatoxins.* (a) Electrophoresis.

Closed circular plasmid pBR322 was electrophoresed on a series of cylindrical agarose gels which contained increasing amounts of aflatoxins (Espejo & Lebowitz, 1976; Gopalakrishnan et al., 1989a; Raney et al., 1990). 0.27 μg of pBR322 in a loading buffer was layered on each gel and electrophoresed at 60 V for 4 to 6 hours. The gels were stained with ethidium bromide, visualized by short-wave UV light, and photographed. The topoisomerase-treated DNA was incubated at 37 °C for 5 hrs with 10 units of topoisomerase I prior to electrophoresis.

(b) Flow Dichroism. Aflatoxins were added to aliqouts of a 2 mM DNA solution and equilibrated in the dark for two days at 5 °C. The total concentration of aflatoxin in solution was ~100 μM. A syringe pump forced the solution of each aflatoxin and calf thymus DNA through a modified CARY cell holder which contained a 0.025 cm pathlength quartz cell at 13 mL/minute, and induced partial linear orientation of the DNA. Absorbance of plane-polarized light was monitored at 260 nm and 365 nm as the aflatoxin-DNA solution flowed through the cell (Raney et al., 1990).

(c) Nuclear Magnetic Resonance. For preparation of oligodeoxynucleotide NMR samples in D_2O, 400 μL of NMR buffer (0.1 M NaCl, 5 x 10^{-5} M Na_2EDTA, and 0.01 M sodium phosphate, pH 7.0) was added to the amount of oligodeoxynucleotide required to prepare a 0.5 mM sample in a volume of 400 μL (Gopalakrishnan et al., 1989a; Raney et al., 1990). Samples were lyophilized three times from D_2O and then dissolved in 400 μL of 99.96% D_2O. 10 μL from a 40 mM stock solution of aflatoxin in d_6-dimethylsulfoxide was added to give a solution in which the oligodeoxynucleotide was in equilibrium with a saturated solution of each aflatoxin. Chemical shifts were referenced internally to DSS; sample temperature was maintained at 5 °C.

(d) <u>Competitive Binding Experiments</u>. NMR samples for competitive binding experiments were 0.5 mM in oligodeoxynucleotide and contained a saturated solution of aflatoxin. A parallel experiment was performed on a saturated sample of aflatoxin with no oligodeoxynucleotide present. As the competing ligand was titrated into the two samples, the aflatoxin chemical shifts in the sample containing the oligodeoxynucleotide were compared to those in the parallel sample containing no oligodeoxynucleotide (Gopalakrishnan et al., 1989a; Raney et al., 1990). The blank sample served to control any contribution in the observed chemical shift of the carcinogen derived from interaction of the aflatoxin with the competing ligand.

2.2.4. Preparation of Aflatoxin B_1 Adducts. Aflatoxin B_1-8,9-epoxide was prepared as described by Baertschi et al. (1988). Adducts were prepared by addition of excess aflatoxin B_1-8,9-epoxide in dichloromethane to double-stranded oligodeoxynucleotide at 5 °C. All reactions with aflatoxin B_1 were performed under subdued light to minimize potential formation of aflatoxin B_1 photoproducts or photo-decomposition of the resulting carcinogen-DNA adduct (Israel-Kalinsky et al., 1982; Misra et al., 1983). A typical reaction and the subsequent purification is described by Gopalakrishnan et al. (1989b).

2.2.5. Formation of Guanine Adducts by Aflatoxin Epoxides. Calf thymus DNA was prepared at three concentrations: 1.0 mg/mL (1.5×10^3 μM base pairs), 0.1 mg/mL (1.5×10^2 μM base pairs), and 0.01 mg/mL (15 μM base pairs). An aliquot of either aflatoxin B_1-8,9-epoxide or aflatoxin G_1-9,10-epoxide in acetone was added to a stirring 5 mL sample of DNA at a particular concentration to yield a 4 μM solution of epoxide. The solutions were then heated at 95 °C for 15 minutes to liberate guanine N7 adducts, which were quantitated by reverse-phase HPLC (Raney et al., 1990).

2.2.6. Optical Spectroscopy of Aflatoxin-Oligodeoxynucleotide Adducts. UV-vis spectroscopic measurements were made using a Varian CARY 2390 spectrophotometer interfaced with a Neslab ETP-3 temperature programmer unit for performing melting experiments. Optical measurements on the double-stranded species were made by dilution of concentrated solutions into 1 M NaCl buffer to favor maintenance of duplex conformation at low concentration.

2.2.7. Nuclear Magnetic Resonance of Aflatoxin-Oligodeoxynucleotide Adducts. The oligodeoxynucleotide duplex concentrations ranged from 0.8 mM to 2.0 mM. Samples were dissolved in 400 μL NMR buffer consisting of 0.1 M NaCl, 5×10^{-5} M disodium EDTA, and 0.01 M sodium phosphate, pH 7.0. For observation of the non-exchangeable protons, the buffered samples were dissolved in 99.96 % D_2O, and purged with dry N_2. For observation of the exchangeable ^1H resonances, samples were dissolved in H_2O containing 10% D_2O. 1D spectra in H_2O were acquired using the 1331 water suppression pulse sequence (Hore, 1983a,b). A NOESY spectra in H_2O was acquired using the 1:1 (jump-return) suppression pulse sequence (Plateu & Gueron, 1982) as developed by Sklenar & Bax (1987) and Sklenar et al. (1987). Non-selective spin-lattice relaxation measurements were performed using the standard inversion recovery pulse technique. NOESY spectra were acquired in the phase-sensitive mode at 6 °C (S. Gopalakrishnan, T. M. Harris, & M. P. Stone, manuscript in preparation).

3.   Results

3.1   EQUILIBRIUM BINDING OF AFLATOXINS WITH DNA

The association of aflatoxin epoxides with DNA cannot be examined directly, due to their reactivity. We have utilized aflatoxins B_1, B_2, G_1, and G_2 as surrogates for the labile aflatoxin epoxides in equilibrium binding studies. All four aflatoxins spontaneously associate with a variety of B-form polymeric nucleic acids, as well as double-stranded oligodeoxynucleotides.

3.1.1. *Measurement of DNA Association.* (a) <u>Scatchard Analysis</u>. Aflatoxins are sparingly soluble in aqueous buffer, which limits binding measurements to low values of r (i.e., the fraction of occupied binding sites on the DNA lattice is controlled by the concentration of unbound aflatoxin). Aflatoxins will associate with both A:T and G:C base pairs in B-DNA. Comparison of the binding of aflatoxins B_1 and B_2 to calf thymus DNA, poly(dGdC):poly(dGdC), and poly(dAdT):poly(dAdT) showed similar affinity for each polymer (Stone et al., 1988). The magnitude of the DNA association constant for aflatoxins B_1 and B_2 is measured by Scatchard analysis to be ~ 10^3 M^{-1}. A typical Scatchard plot of aflatoxin B_1 binding to calf thymus DNA at 4 °C derived from equilibrium dialysis is shown in Figure 3A (Scatchard, 1949; McGhee & von Hippel, 1974). Substitution of the cyclopentenone ring in aflatoxin B_1 for the δ-lactone ring in aflatoxin G_1 (Figure 1) reduces DNA binding affinity by approximately one order of magnitude, as evidenced by the Scatchard plot shown in Figure 3B.

(b) ^{1}<u>H NMR Linebroadening</u>. Chemical exchange between free and bound aflatoxins is rapid on the NMR time scale and a single ^{1}H resonance is observed for each proton (Gopalakrishnan et al., 1989a; Stone et al., 1988). The reduced binding affinity of aflatoxin G_1 is corroborated by monitoring NMR linewidth at half height ($v_{1/2}$) of the aflatoxin methoxy proton signal as a function of calf thymus DNA concentration (Figure 4). Addition of DNA results in broadening of the methoxy signal because bound aflatoxin undergoes more rapid spin-spin relaxation than free aflatoxin. An approximate six-fold

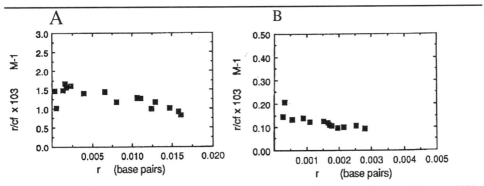

Figure 3. Scatchard plots for the binding of (A) aflatoxin B_1 and (B) aflatoxin G_1 to calf thymus DNA. r=[bound aflatoxin]/[DNA]$_{bp}$; C_f=[free aflatoxin].

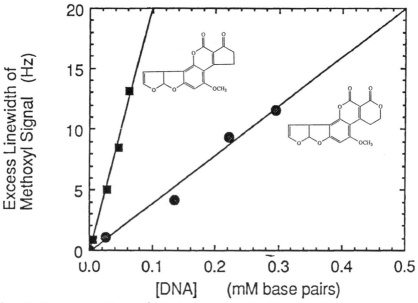

Figure 4. Measurement of excess ^{1}H NMR line width of the aflatoxin 4-OCH$_3$ resonance upon titration of aflatoxin B$_1$ and aflatoxin G$_1$ with calf thymus DNA.

excess concentration of calf thymus DNA is required to achieve comparable excess line width of the aflatoxin G$_1$ 4-OCH$_3$ signal as compared to the aflatoxin B$_1$ 4-OCH$_3$ signal.

(c) <u>Oligodeoxynucleotide Binding</u>. The binding of aflatoxins B$_1$, B$_2$, G$_1$, and G$_2$ to d(ATGCAT)$_2$ and of aflatoxins B$_1$, B$_2$, and G$_2$ to d(GCATGC)$_2$ was examined. To measure relative binding affinities to d(ATGCAT)$_2$ and d(GCATGC)$_2$, the chemical shift of the methoxy resonance from each aflatoxin was monitored as a function of oligodeoxynucleotide concentration. The results of a typical experiment are shown in Figure 5. For both d(ATGCAT)$_2$ and d(GCATGC)$_2$, greater perturbation of aflatoxin B$_1$ or B$_2$ methoxy resonances as compared to aflatoxin G$_1$ or G$_2$ methoxy resonances is observed for equal concentrations of oligodeoxynucleotide. In the limit of infinite oligodeoxynucleotide concentration, it is estimated that the methoxy resonances of all four aflatoxins have similar chemical shift.

3.1.2. *Conformational Studies of Bound Aflatoxins*. (a) <u>Alteration of Superhelical Density by Aflatoxin Binding</u>. Aflatoxins B$_1$ and B$_2$ relieve negative superhelical twists in supercoiled plasmid as well as introduce positive superhelical twists into relaxed closed circular plasmid; properties characteristic of intercalators (Figure 6 A,B). The equilibrium between free and bound aflatoxin is rapid compared to the time scale of electrophoresis, so the observed change in DNA results from the time-averaged population of bound aflatoxins. The critical free concentration (c'; the amount of free ligand required to completely relax supercoiled plasmid) is attained at a concentration of free aflatoxin B$_1$ approximately equal to its aqueous solubility. Aflatoxin B$_2$ is less effective in altering

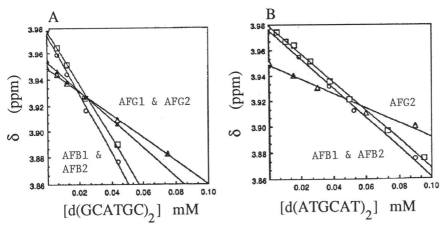

Figure 5. Measurement of ^{1}H NMR chemical shift of the 4-OCH$_3$ resonance upon titration of aflatoxin B$_1$, aflatoxin B$_2$, aflatoxin G$_1$, and aflatoxin G$_2$ with (A) d(GCATGC)$_2$ or (B) d(ATGCAT)$_2$.

electrophoretic mobility and c' is not attained in our experiments (Figure 6B), although unwinding and rewinding of plasmid in a manner similar to aflatoxin B$_1$ is observed.

Aflatoxins G$_1$ and G$_2$ are substantially less effective in altering the electrophoretic migration of supercoiled or relaxed plasmid, and even saturated aqueous solutions produce only small changes in plasmid migration (Figure 6 C, D). This is probably due to the lower binding affinity of aflatoxin G$_1$ and G$_2$ as compared to aflatoxins B$_1$ and B$_2$, since a sufficient fraction of pBR322 sites must be filled to create an observable perturbation in electrophoretic mobility. A greater concentration of free aflatoxin G$_1$ or G$_2$ is required to fill an equivalent fraction of sites. In these experiments, the attainable free concentration of aflatoxins G$_1$ and G$_2$ is limited by solubility.

(b) <u>Flow Dichroism of Aflatoxins</u>. Linearly oriented calf thymus DNA exhibits negative reduced dichroism at 260 nm which results from the approximately perpendicular orientation of the purine and pyrimidine bases relative to the helical axis. In the presence of this oriented DNA, the reduced dichroism of aflatoxins B$_1$ and B$_2$ at 360 nm is negative (Table 1). Limitations in sensitivity arise due to the short pathlength of the flow dichroism cell, and because only partial (20%) orientation of DNA is achieved. We could not observe flow dichroism for aflatoxins G$_1$ and G$_2$.

(c) ^{1}<u>H NMR</u>. The binding orientation of aflatoxins G$_1$ and G$_2$ to d(ATGCAT)$_2$ and d(GCATGC)$_2$ was probed by ^{1}H NMR. These data were compared with corresponding data obtained for aflatoxins B$_1$ and B$_2$. Figure 7 shows spectra of saturated solutions of aflatoxins B$_1$, B$_2$, G$_1$, and G$_2$ in the presence of 0.5 mM d(ATGCAT)$_2$ (3 mM base pairs). The ^{1}H spectra represent population-weighted averages of the various species in rapid exchange between free and bound states. These consist of unbound aflatoxin, free oligodeoxynucleotide, and all species in which one or more aflatoxin molecules are bound. Large effects are observed for aflatoxin protons while only small changes are observed for the oligodeoxynucleotide protons, because the fraction of aflatoxin which is bound is large while the fraction of binding sites which are occupied is small. The direction of the chemical shift changes is upfield for all four aflatoxins. These

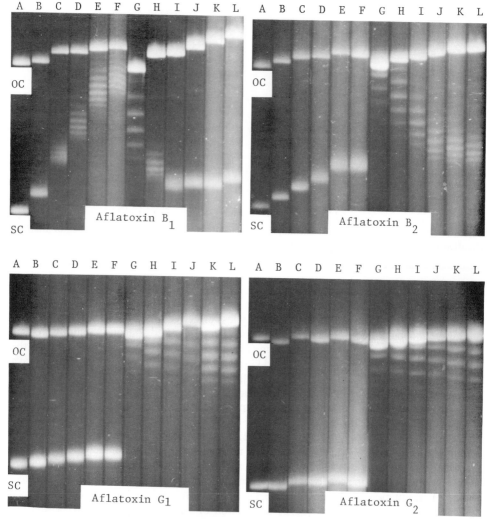

Figure 6. Conformational changes in pBR322 caused by binding of aflatoxin B_1 (AFB1), aflatoxin B_2 (AFB2), aflatoxin G_1 (AFG1), and aflatoxin G_2 (AFG2). The first six tube gels in each set show the migration of negatively supercoiled plasmid as a function of increasing aflatoxin concentration. The subsequent six gels show the migration of relaxed closed circular plasmid as a function of increasing aflatoxin concentration. The positions of supercoiled (SC) and open circular (OC) forms of plasmid are indicated. The amounts of aflatoxins B_1 and B_2 added to each lane are as follows: lanes A and G, no added aflatoxin; B and H, 15 μg/mL; C and I, 20 μg/mL; D and J, 45 μg/mL; E and K, 60 μg/mL; F and L, 75 μg/mL. The amounts of aflatoxins G_1 and G_2 added to each lane are as follows: lanes A and G, no added aflatoxin; B and H, 20 μg/mL; C and I, 40 μg/mL; D and J, 60 μg/mL; E and K, 80 μg/mL; F and L, 100 μg/mL. The amount of aflatoxin in each tube gel is presented as weight/volume added to the gel rather than molar concentration due to difficulty in measurement of aflatoxin concentration in the gels. The weight/volume of aflatoxin spanned a range of values which exceeded the aqueous solubility of aflatoxin, which ensured that the fraction of occupied binding sites reached a maximum when the concentration of free aflatoxin was equal to the aqueous solubility.

TABLE 1. Flow Dichroism Data for Aflatoxins B_1 and B_2.

Compound	λ (nm)	$red_D{}^a$
Aflatoxin B_1	260	-0.14
	365	-0.034
Aflatoxin B_2	260	-0.17
	365	-0.017

$^a red_D = [A_{0°} - A_{90°}/A]$

TABLE 2. Chemical shift changes ($\Delta\delta = \delta_{equilibrium} - \delta_{free}$; ppm) observed for aflatoxin B_1, aflatoxin B_2, aflatoxin G_1, and aflatoxin G_2 in equilibrium with 0.5 mM d(ATGCAT)$_2$ at 5 °C.

proton	$\Delta\delta$, Aflatoxin B_1	$\Delta\delta$, Aflatoxin B_2	$\Delta\delta$, Aflatoxin G_1	$\Delta\delta$, Aflatoxin G_2
H5	-0.72	-0.47		
H6a	-0.17	-0.11		
H8	-0.08			
H9	-0.17			
4-OCH$_3$	-0.26	-0.23		
H6			-0.34	-0.25
H7a			-0.11	-0.07
H9			-0.07	
H10			-0.13	
5-OCH$_3$			-0.16	-0.12

chemical shift changes are quantitated in terms of $\Delta\delta$ tabulated in Table 2, where $\Delta\delta = \delta_{free\ aflatoxin} - \delta_{observed}$. The largest increased shielding is observed in all cases for the aflatoxin H5 and OCH$_3$ protons, ranging from 0.72 ppm in aflatoxin B_1 to 0.25 ppm in aflatoxin G_2. At greater oligodeoxynucleotide concentrations, where the fraction of bound aflatoxin is increased, the magnitudes of the chemical shift changes for specific protons in aflatoxins G_1 and G_2 become similar to those of corresponding protons in aflatoxins B_1 and B_2.

(d)   Competitive Binding Experiments. Figure 8A illustrates the changes in the observed chemical shifts of various aflatoxin B_1 protons as the equilibrium between the carcinogen and d(ATGCAT)$_2$ is altered by the addition of ethidium bromide. The observed chemical shifts of the aflatoxin B_1 protons are plotted as a function of the ratio [ethidium bromide]:[d(ATGCAT)$_2$]. The chemical shifts of the aflatoxin B_1 protons approach the chemical shifts of free aflatoxin B_1 as the [ethidium]:[duplex] ratio reaches 4.0, indicating that at a ratio of 4 ethidiums per duplex, the aflatoxin B_1 molecules are displaced from the DNA. This experiment was repeated with actinomycin D; the results are shown in Figure 8B. When the molar ratio of [actinomycin D] to [d(ATGCAT)$_2$] is 1:1, the aflatoxin B_1 molecules are entirely displaced from the oligodeoxynucleotide. In contrast to the results observed for ethidium bromide and actinomycin D, when spermidine trihydrochloride was titrated into the aflatoxin B_1-oligodeoxynucleotide equilibrium mixture, aflatoxin B_1 was not displaced. Similar results are observed for aflatoxins G_1 and G_2 (Raney et al., 1990).

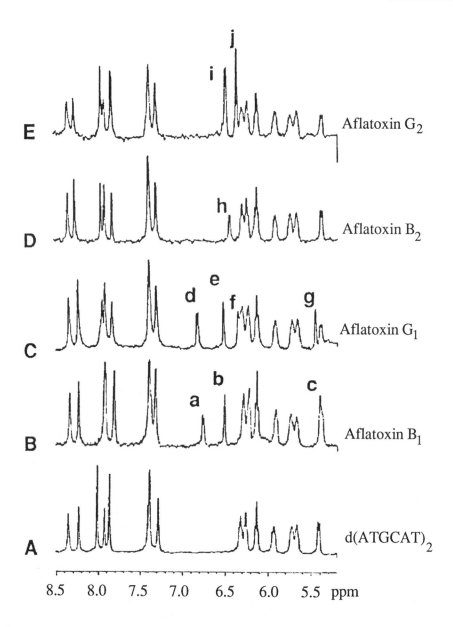

Figure 7. Expansions of ^{1}H NMR spectra of (A) 0.5 mM d(ATGCAT)$_2$ at 5°C, and 0.5 mM d(ATGCAT)$_2$ at 5 °C in equilibrium with saturated aqueous solutions of (B) aflatoxin B$_1$, (C) aflatoxin G$_1$, (D) aflatoxin B$_2$, and (E) aflatoxin G$_2$. In all cases, exchange between free and bound aflatoxins is rapid on the NMR time scale. The aflatoxin resonances are assigned as follows: (a) aflatoxin B$_1$ H6a, (b) aflatoxin B$_1$ H8, (c) aflatoxin B$_1$ H9, (d) aflatoxin G$_1$ H7a, (e) aflatoxin G$_1$ H9, (f) aflatoxin G$_1$ H6, (g) aflatoxin G$_1$ H10, (h) aflatoxin B$_2$ H6a, (i) aflatoxin G$_2$ H7a, and (j) aflatoxin G$_2$ H6.

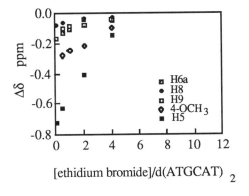 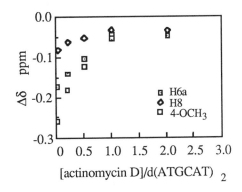

Figure 8. Chemical shift of the various aflatoxin B₁ proton resonances as a function of (A) [ethidium bromide]/[d(ATGCAT)₂] and (B) [actinomycin D]/[d(ATGCAT)₂]. As the competing ligand is titrated into the equilibrium mixture, aflatoxin B₁ is displaced from d(ATGCAT)₂.

3.2.   ADDUCT FORMATION BY AFLATOXIN B₁-8,9-EPOXIDE AND AFLATOXIN G₁-9,10-EPOXIDE

Aflatoxin epoxides react with regio- and stereospecificity at the N7 position of deoxyguanosine in double stranded B-DNA. Treatment of calf thymus DNA produces extensive covalent reaction. Thermal hydrolysis of the covalent linkages followed by reverse-phase HPLC purification yield an adduct identified as 8,9-dihydro-8-(N7-guanyl)-9-hydroxy-aflatoxin B₁ (Baertschi et al., 1988). The reactivity of guanine N7 with aflatoxin B₁-8,9-epoxide is strongly dependent upon maintenance of B-DNA conformation. Table 3 summarizes the results of a series of experiments in which the yield of 8,9-dihydro-8-(N7-guanyl)-9-hydroxy-aflatoxin B₁ was monitored as a function of DNA conformation.

The yields of 8,9-dihydro-8-(N7-guanyl)-9-hydroxy-aflatoxin B₁ and 9,10-dihydro-9-(N7-guanyl)-10-hydroxy-aflatoxin G₁ upon reaction of calf thymus DNA with aflatoxin B₁-8,9-epoxide and with aflatoxin G₁-9,10-epoxide were determined at three DNA concentrations (Table 4). As DNA concentration is decreased, the yield of both 8,9-dihydro-8-(N7-guanyl)-9-hydroxyaflatoxin B₁ and 9,10-dihydro-9-(N7-guanyl)-10-hydroxyaflatoxin G₁ falls, with a concomitant increase in formation of the respective dihydrodiols which are the products of epoxide hydrolysis. For example, in the presence of 1.5 mM DNA, a 78% conversion of aflatoxin B₁-8,9-epoxide to 8,9-dihydro-8-(N7-guanyl)-9-hydroxy-aflatoxin B₁ is observed, whereas if DNA concentration is lowered to 15 µM, adduct yield is reduced to 29%. This effect is even more striking in the case of aflatoxin G₁-9,10-epoxide, in which case adduct yield is reduced to <8% as DNA concentration is reduced to 15 µM.

TABLE 3. Comparative yield of 8,9-dihydro-8-(N7-guanyl)-9-hydroxy-aflatoxin B_1 as a function of DNA conformation. The concentration of guanine in each polynucleotide was 3 x 10^{-5} M. Monomers were each 1 mM. DNA conformation was monitored by CD spectroscopy.

Conformation	nucleotide	% yield of N7 adduct
B form	calf thymus DNA	15
	poly(dGdC):poly(dGdC)	13
	poly dG:poly dC	15
	calf thymus DNA	
	+ 30 µM Co(NH3)6Cl3	9
Z form	poly(dGdC):poly(dGdC)	0.2
A form	poly(rC):poly(dG)12-18	1.7
monomers	dG	4.5
	dA	---
	dC	---
	T	---

TABLE 4. Measurement of Aflatoxin N7-guanine adducts at varying DNA concentration.[a]

Epoxide	DNA concentration, µM		
	1.5 x 10^3	1.5 x 10^2	15
aflatoxin B_1	78 ± 5	64 ± 2	29 ± 5
aflatoxin G_1	62 ± 2	36 ± 3	8 ± 2

[a]Efficiency of guanine N7 adduct formation measured as % yield of total aflatoxin epoxide added to the reaction mixture.

3.3.   PREPARATION OF TWO OLIGODEOXYNUCLEOTIDE ADDUCTS

3.3.1. *Reaction Stoichiometry.* The reaction of aflatoxin B_1-8,9-epoxide with the oligodeoxynucleotide sequence isomers d(ATCGAT)2 and d(ATGCAT)2 is summarized in Scheme 1. Reaction between aflatoxin B_1-8,9-epoxide and d(ATCGAT)2 is characterized by a limiting stoichiometry of 1:1 aflatoxin B_1:d(ATCGAT)2 to yield d(ATC^{AFB}GAT):d(ATCGAT) (Gopalakrishnan et al., 1989b), where the modified guanine is designated as ^{AFB}G. Reaction of the sequence isomer d(ATGCAT)2 with aflatoxin B_1-8,9-epoxide results in a markedly different result: d(ATGCAT)2 reacts with two equivalents of epoxide to yield the bis adduct d(AT^{AFB}GCAT)2 in which both guanines are modified. Chromatographic analysis of the reaction mixtures obtained by addition of excess epoxide to each of the two oligodeoxynucleotides demonstrates that for d(ATCGAT)2, reaction is complete when 50% of the oligodeoxynucleotide [strands] is consumed, and for d(ATGCAT)2 proceeds until 100% of the oligodeoxynucleotide has undergone reaction.

Formation of d(ATC^{AFB}GAT):d(ATCGAT) from d(ATCGAT)2 results in a doubling of oligodeoxynucleotide ^{1}H NMR resonances due to loss of pseudo-dyad strand symmetry (Figure 9A). The magnitude of the aflatoxin-induced resolution of symmetry-related protons varies, such that for some protons (e.g., the thymine CH_3 protons) loss of symmetry is evidenced only by line broadening. In contrast, formation of d(AT^{AFB}GCAT)2 from d(ATGCAT)2 occurs with retention of pseudo-dyad symmetry, and no doubling of NMR resonances is observed (Figure 9B).

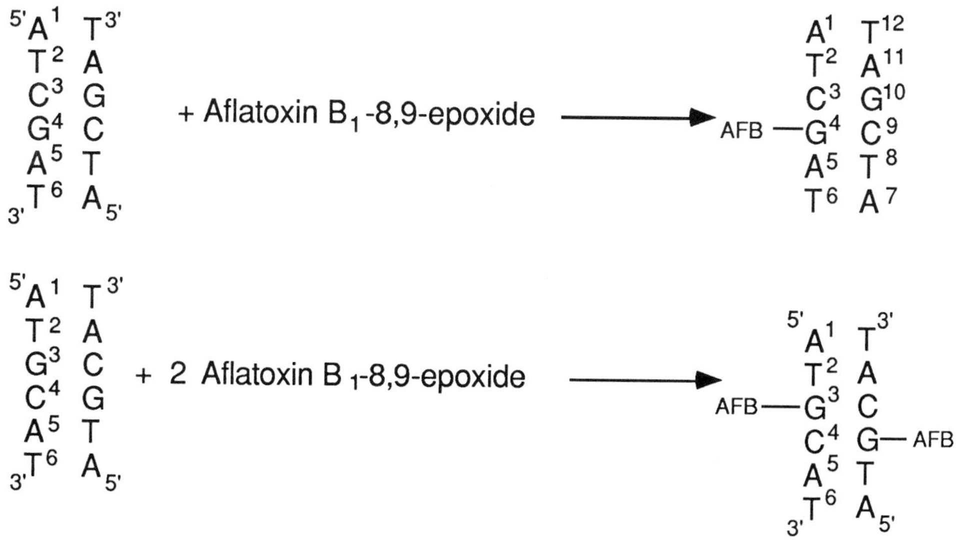

```
5'A¹ T³'                                              A¹  T¹²
  T² A                                                T²  A¹¹
  C³ G        + Aflatoxin B₁-8,9-epoxide  ──────►     C³  G¹⁰
  G⁴ C                                          AFB —G⁴  C⁹
  A⁵ T                                                A⁵  T⁸
 ₃'T⁶ A₅'                                             T⁶  A⁷


5'A¹ T³'                                                 5'A¹ T³'
  T² A                                                     T² A
  G³ C        + 2 Aflatoxin B₁-8,9-epoxide ──────►   AFB—G³ C
  C⁴ G                                                     C⁴ G—AFB
  A⁵ T                                                     A⁵ T
 ₃'T⁶ A₅'                                               ₃'T⁶ A₅'
```

Scheme 1. Comparative reaction stoichiometry in the reaction of aflatoxin B₁-8,9-epoxide with d(ATCGAT)₂ and d(ATGCAT)₂. The nucleotide numbering schemes for both adducts are as indicated and correspond with the NMR data.

3.3.2. *Adduct Stability.* Figure 10 shows UV melting curves for d(ATC^{AFB}GAT):d(ATCGAT) and d(AT^{AFB}GCAT)₂ as compared to the respective parent duplexes. The two unmodified hexamers each exhibit broad melting transitions for which the lower baseline is not attained even at 0 °C. In contrast, the appearance of lower baselines is clearly observed for the modified oligodeoxynucleotides. A shift in the midpoint of the melting transition to higher temperature and increased sigmoidal appearance of the plot is observed. It is concluded that both adducts have increased thermal stability as compared to the parent duplexes.

3.3.3. *¹H NMR Analysis of d(ATC^{AFB}GAT):d(ATCGAT) and d(AT^{AFB}GCAT)₂.* Incorporation of ^{AFB}G creates a characteristic set of perturbations to the ¹H NMR spectra of these adducts. Figure 11 shows the results of a series of 1D NOE experiments on the hydrogen bonded imino protons of d(ATC^{AFB}GAT):d(ATCGAT). Aflatoxin adduct formation interrupts the pattern of imino-imino NOEs on the 5' side of the adducted guanine. In Figure 11 no NOE is observed between the two guanine H1 protons; each guanine H1 shows connectivity only to an internal thymine H3 proton. A sharp signal which which integrates as one proton and which exchanges with solvent is observed at 9.75 ppm. This is the H8 proton of the adducted guanine, which has shifted downfield approximately 1.6 ppm. This resonance does not exhibit an NOE to the imino protons located between 12-14 ppm, but does show NOE connectivity to the deoxyribose protons of the adducted guanine, as well as to the H8 proton of the adducted aflatoxin moiety.

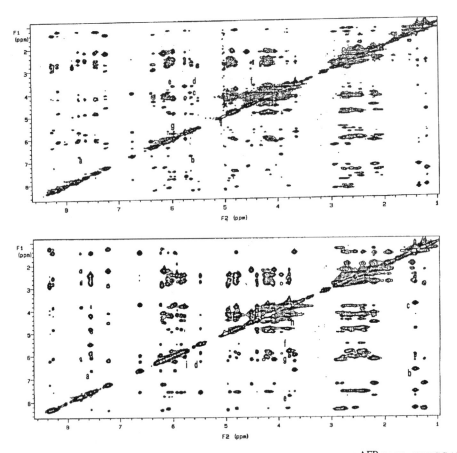

Figure 9. Typical NOESY spectra (top) the non-symmetrical adduct d(ATC^{AFB}GAT):d(ATCGAT) and (bottom) the symmetrical adduct d(AT^{AFB}GCAT)$_2$. A doubling of ^{1}H resonances occurs in d(ATC^{AFB}GAT):d(ATCGAT) which results from loss of pseudo-dyad symmetry. The labelled crosspeaks in the spectra correspond to specific aflatoxin B$_1$-DNA dipole-dipole interactions, and are cross-referenced in Tables 5 and 6. Key to crosspeaks for d(ATC^{AFB}GAT):d(ATCGAT)] (top): a, AFB H6a-- C^3 H6; b, AFB H6a --C^3 H5; c, AFB H9a--C^3 H5; d,e,f, AFB 4-OCH$_3$--^{AFB}G^4 H1', C^3 H1', ^{AFB}G^4 H4'; g, AFB H5-- C^3 H1'. Key to crosspeaks for d(AT^{AFB}GCAT)$_2$ (bottom): a, AFB H6a--T^2 H6; b AFB H6a--T^2 CH$_3$; c, AFB H9a--T^2 CH$_3$; d, AFB H8--C^4 H5; e,f,g,h AFB 4-OCH$_3$--A^5 H2; ^{AFB}G^3 H1', T^2 H1', ^{AFB}G^3 H4'; i, AFB H5--T^2 H1'. These spectra were obtained at 499 MHz in D$_2$O buffer.

Adduction by ^{AFB}G interrupts the sequential intrastrand NOE connectivities between DNA base protons and deoxyribose units (Feigon et al., 1983; Hare et al., 1983) in both strands of the oligodeoxynucleotide duplex on the 5' side of the modified guanine. Figure 12 shows expansions of NOESY crosspeaks arising from DNA base-deoxyribose H1' connectivities. In d(ATC^{AFB}GAT):d(ATCGAT) no NOE is observed between G^{10} H8 and C^9 deoxyribose. Likewise, in d(AT^{AFB}GCAT)$_2$, no NOE is observed between A^5

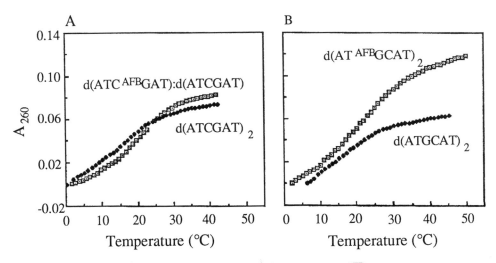

Figure 10. Formation of either d(ATCGAT):d(ATC^AFBGAT) or d(AT^AFBGCAT)₂ results increased thermodynamic stability as compared to the corresponding unmodified duplexes d(ATCGAT)₂ and d(ATGCAT)₂. (A) Thermal denaturation of d(ATCGAT):d(ATC^AFBGAT) followed by UV absorbance. (B) Thermal denaturation of d(AT^AFBGCAT)₂ followed by UV absorbance. In both cases, the open squares represent the modified oligodeoxynucleotide while the solid diamonds represent the unmodified oligodeoxynucleotide.

H8 and C⁴ deoxyribose. Guanine H8 of ^AFBG is not observed in Figure 12 due to exchange with solvent. A complementary experiment, performed in H₂O buffer, demonstrates that ^AFBG guanine H8 exhibits an NOE only to ^AFBG deoxyribose protons, and not to the deoxyribose protons of the 5'-neighbor base pair. This effect is particularly striking for d(AT^AFBGCAT)₂, in which case the pattern of sequential NOESY connectivities appears as if three right-handed dinucleotide units are present: ApT, ^AFBGpC, and ApT.

For d(ATC^AFBGAT):d(ATCGAT), intranucleotide NOEs are observed between the aflatoxin 4-OCH₃ and H5 protons and ^AFBG⁴. Internucleotide NOEs are observed between 4-OCH₃, H5, H6a, and H9a, and the 5'-neighbor C³:G¹⁰ base pair. The 4-OCH₃ and H5 protons exhibit NOEs to C³ H1' and ^AFBG⁴ H1'; 4-OCH₃ also shows an intranucleotide NOE to ^AFBG⁴ H4'. All of these oligodeoxynucleotide protons face into the minor groove of d(ATC^AFBGAT):d(ATCGAT). Aflatoxin H6a and H9a exhibit major groove internucleotide NOEs to C³ H5, and H6a also shows an NOE to C³ H6. A weak NOE is observed between aflatoxin H2β, located on the cyclopentenone ring, and C⁹ H1'. These NOEs are summarized in Table 5.

For d(AT^AFBGCAT)₂, intranucleotide NOEs are observed between the aflatoxin 4-OCH₃ and H5 protons and ^AFBG³, and (in H₂O buffer) between aflatoxin H8 and ^AFBG³ guanine H8. Internucleotide NOEs are observed between 4-OCH₃, H5, H6a, and H9a, and the 5'-neighbor T²:A⁵ base pair. The 4-OCH₃ protons exhibit NOEs to T² H1', ^AFBG³ H1' and ^AFBG³ H4'; 4-OCH₃ also shows an internucleotide NOE to A⁵ H2. All

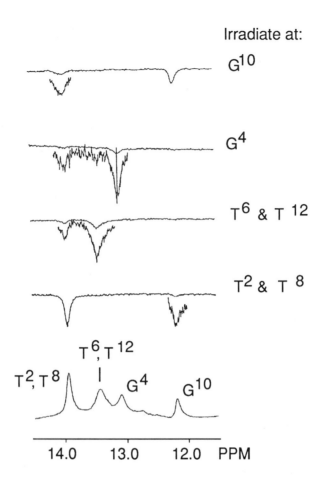

Figure 11. Irradiation of the broad resonance located at 13.4 ppm assigned as the imino protons of T^6 and T^{12}, shows an NOE only to the set of imino protons located at 14 ppm, which are assigned as the imino protons of T^2 and T^8. Irradiation of the 14 ppm signal fails to give the reciprocal NOE to the broad peak at 13.4 ppm, presumably because of the increased exchange rate of T^6 and T^{12} H3 with solvent. Irradiation at 14 ppm also gives an NOE to the signal located at 12 ppm, assigned as a guanine imino proton. The reverse experiment, in which the 12 ppm signal is irradiated, gives an NOE to the signal at 14 ppm. Irradiation of the remaining signal at 13.2 ppm, which is assigned as the other guanine imino resonance, also shows an NOE to the 14 ppm signal. Of the two guanine H1 resonances, the 13.2 ppm signal is assigned as ^{AFB}G H1. This resonance is shifted downfield and has increased linewidth as compared to the second guanine H1 observed at 12 ppm and the two guanine H1 protons of unmodified d(ATCGAT)$_2$, consistent with the rationale that formation of ^{AFB}G should result in increased acidity for guanine H1.

of these oligodeoxynucleotide protons face into the minor groove of d(AT^{AFB}GCAT)$_2$. Minor groove NOEs are also observed between aflatoxin H5 and T^2 H1', T^2 H2", and T^2 H2'. The latter crosspeak is likely a second order NOE resulting from spin diffusion in the 100 msec data. Additional minor groove interactions are observed between aflatoxin H3α,

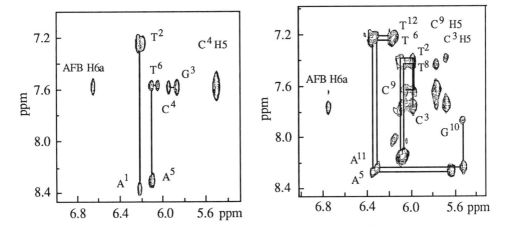

Figure 12. Sequential base-H1' NOE connectivities for (right) d(ATC^{AFB}GAT):d(ATCGAT) and (left) d(AT^{AFB}GCAT)$_2$. The solid lines in each NOESY panel represent intrastrand sequential connectivities between base protons and deoxyribose H1' protons. No crosspeaks are observed from ^{AFB}G guanine H8 due to exchange with deuteriated buffer. The sequential intrastrand NOEs are perturbed in the unmodified strand due to the presence of ^{AFB}G. Thus for d(ATC^{AFB}GAT):d(ATCGAT) there is no connectivity between G^{10} H8 and C^9 H1' (right panel) and for d(AT^{AFB}GCAT)$_2$ there is no connectivity between A^5 H8 and C^4 H1' (left panel). The phase-sensitive NOESY spectra were recorded at 499 MHz, 5 °C, with a mixing time of 100 msec.

located on the cyclopentenone ring, and A^5 H2, and between H2β and C^4 H1'. In the major groove of d(AT^{AFB}GCAT)$_2$, aflatoxin H6a and H9a exhibit internucleotide NOEs to T^2 CH$_3$, and H6a also shows an NOE to T^2 H6. Therefore, aflatoxin H6a and H9a are in close proximity to major groove protons of the 5'-neighbor T^2:A^5 base pair. Aflatoxin H8 shows a weak NOE in the 3' direction, to C^4 H5. These NOEs are summarized in Table 6.

Incorporation of ^{AFB}G results in a characteristic pattern of upfield chemical shifts for the protons of the aflatoxin moiety in both d(ATC^{AFB}GAT):d(ATCGAT) and d(AT^{AFB}GCAT)$_2$, as compared to unbound aflatoxin. The pattern and magnitude of these shifts is similar for each adduct, which suggests that the two adducts have a common geometry. Furthermore, comparison of Tables 2, 7, and 8 reveals that chemical shift changes of aflatoxin protons upon adduct formation are also similar to those observed for aflatoxin B$_1$ association with d(ATGCAT)$_2$. The latter observation suggests that the orientation of the covalently attached aflatoxin moiety remains similar to that of the complex between aflatoxin B$_1$ and the oligodeoxynucleotide.

Measurements of non-selective T$_1$ relaxation for d(ATC^{AFB}GAT):d(ATCGAT) and d(AT^{AFB}GCAT)$_2$ reveal that in addition to the adenine H2 protons, which are isolated in the minor groove, aflatoxin H8 and H9 also show slow spin-lattice relaxation in both oligodeoxynucleotide adducts. Figure 13 shows data for d(ATC^{AFB}GAT):d(ATCGAT) and d(AT^{AFB}GCAT)$_2$ at a τ value which is close to the T$_1$ null point for most of the oligodeoxynucleotide protons. The inverted signals represent those protons which

TABLE 5. Summary of NOEs observed between the aflatoxin moiety and the oligodeoxynucleotide in d(ATC^{AFB}GAT):d(ATCGAT).

Aflatoxin Proton of ^{AFB}G^4	Aflatoxin-DNA NOEs	Figure 9 reference
H2α		
H2β	C^9 H1'	
H3α		
H3β		
4-OCH$_3$	C^3 H1'	crosspeak e
	^{AFB}G^4 H1'	crosspeak d
	^{AFB}G^4 H4'	crosspeak f
H5	C^3 H1'	crosspeak g
	^{AFB}G^4 H1'	
H6a	C^3 H6	crosspeak a
	C^3 H5	crosspeak b
H8		
H9		
H9a	C^3 H5	crosspeak c

TABLE 6. Summary of NOEs observed between the aflatoxin moiety and the oligodeoxynucleotide in d(AT^{AFB}GCAT)$_2$.

Aflatoxin Proton of ^{AFB}G^3	Aflatoxin-DNA NOEs	Figure 9 reference
H2α		
H2β	C^4 H1'	
H3α	A^5 H	
H3β		
4-OCH$_3$	A^5 H2	crosspeak e
	^{AFB}G^3 H1'	crosspeak f
	^{AFB}G^3 H4'	crosspeak h
	T^2 H1'	crosspeak g
H5	T^2 H1'	crosspeak i
	T^2 H2'	
	T^2 H2"	
H6a	T^2 H6	crosspeak a
	T^2 CH$_3$	crosspeak b
H8	C^4 H5	crosspeak d
	^{AFB}G^3 guanine H8	observed in H$_2$O
H9		
H9a	T^2 CH$_3$	crosspeak c

undergo spin-lattice relaxation more slowly. Aflatoxin H6a also appears as an inverted signal at this τ value. Although the adenine H2 protons have slower spin-lattice relaxation than do the other oligodeoxynucleotide protons due to their location in the minor groove, adduct formation does result in increased relaxation rate for the adenine H2 protons as compared to the unmodified duplexes. The greatest perturbation of T_1 relaxation is observed for A^5 H2 of d(AT^{AFB}GCAT)$_2$.

TABLE 7. Chemical shifts (ppm from DSS) of aflatoxin B_1 protons in the d(ATC^{AFB}GAT):d(ATCGAT) adduct at 6 °C. Sample concentration was 0.8 mM duplex. δ_{free} values are derived from a saturated solution of aflatoxin B_1 in D_2O (Gopalakrishnan et al., 1989a). Negative ppm values of $\Delta\delta$ refer to increased shielding in the adduct as compared to free aflatoxin B_1.

proton	δ_{free}	δ_{adduct}	$\Delta\delta$
H6a	6.93	6.75	-0.18
H8	6.59	6.37	-0.22[a]
H9	5.57	6.07	+0.50[a]
H5	6.69	5.75	-0.94
H9a	4.80[b]	3.93	-0.87[b]
4-OCH$_3$	3.98	3.64	-0.34

[a]Change in hybridization at carbons C8 and C9 upon epoxidation and formation of adduct to guanine N7.
[b]In sample of free aflatoxin B_1, this proton resonance is obscured by residual HDO in the sample.

TABLE 8. Chemical shifts (ppm from DSS) of aflatoxin B_1 protons in the d(AT^{AFB}GCAT)$_2$ adduct at 6 °C. Sample concentration was 1.2 mM duplex. δ_{free} values are derived from a saturated solution of aflatoxin B_1 in D_2O (Gopalakrishnan et al., 1989a). Negative ppm values of $\Delta\delta$ refer to increased shielding in the adduct as compared to free aflatoxin B_1.

proton	δ_{free}	δ_{adduct}	$\Delta\delta$
H6a	6.93	6.65	-0.28
H8	6.59	6.26	-0.33[a]
H9	5.57	5.98	+0.41[a]
H5	6.69	5.80	-0.89
H9a	4.80[b]	3.71	-1.09[b]
4-OCH$_3$	3.98	3.81	-0.17

[a]Change in hybridization at carbons C8 and C9 upon epoxidation and formation of adduct to guanine N7.
[b]In sample of free aflatoxin B_1, this proton resonance is obscured by residual HDO in the sample.

4. Discussion

The striking characteristics of DNA adduct formation by aflatoxin B_1 are the regio- and stereoselectivity for reaction at guanine N7, and the efficiency of adduct formation. No adducts other than 8,9-dihydro-8-(N7-guanyl)-9-hydroxy aflatoxin B_1 or the corresponding formamidopyrimidine (FAPY) rearrangement products have been observed (Essigmann et al., 1977; Lin et al., 1977; Baertschi et al., 1988). Conversion of aflatoxin B_1-8,9-epoxide to 8,9-dihydro-8-(N7-guanyl)-9-hydroxy aflatoxin B_1 with yields of as high as 80% was observed in the presence of 1.5 mM DNA (Baertschi et al., 1989). These observations are remarkable considering the hydrolytic instability of aflatoxin B_1-8,9-epoxide. Aflatoxin B_1 has a relatively low affinity for DNA, with an association constant on the order of 10^3 (Figure 2A), and it is probable that aflatoxin B_1-8,9-epoxide has similar DNA binding affinity. The principle goal of the research presented here was to

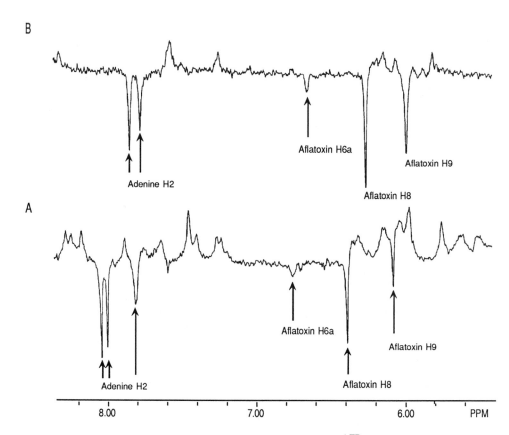

Figure 13. Non-selective spin-lattice relaxation data for (A) d(ATC^{AFB}GAT):d(ATCGAT), and (B) d(AT^{AFB}GCAT)$_2$. The spectra were obtained at 400.13 MHz, 5 °C, and represent τ values at which most of the oligodeoxynucleotide protons are close to the null point. The slowly relaxing adenine H2 protons are located in the minor groove of the modified oligodeoxynucleotide. The spin-lattice relaxation rate of adenine H2 protons in the modified oligodeoxynucleotides is increased but remains slower than for the other oligodeoxynucleotide protons. The aflatoxin protons H8 and H9 face into the major groove of the modified oligodeoxynucleotide. Aflatoxin H6a also faces into the major groove and exhibits a slightly slower T$_1$ relaxation rate.

explore the hypothesis that adduct formation is promoted by association of aflatoxin B$_1$-8,9-epoxide with the DNA helix in a specific orientation which leads to formation of a favorable transition state (Misra et al., 1983; Benasutti et al., 1988; Gopalakrishnan et al., 1989a).

4.1.   INTERCALATIVE ASSOCIATION OF AFLATOXINS WITH THE B-DNA HELIX

The coumarin moiety and attached cyclopentenone ring of aflatoxins B$_1$ and B$_2$ form a planar chromophore which can intercalate into DNA. Examination of molecular models reveals that intercalation of aflatoxin B$_1$-8,9-epoxide above the 5' face of guanine positions the *exo*-epoxide in an orientation to facilitate *endo* attack by the nucleophile followed by *trans* opening of the epoxide (Loechler et al., 1988; Bonnett & Taylor, 1989), which probably accounts for the observed efficiency of adduct formation despite the relatively low DNA binding affinity. The intercalation model also derives support from the observation that aflatoxin B_1 can form photoadducts with DNA (Israel-Kalinsky *et al.*, 1982; Misra *et al.*, 1983); these photoadducts may result from 2 + 2 cycloadditions between DNA bases and the coumarin chromophore of aflatoxin B_1, analogous to the well-characterized photoadducts formed between psoralen and DNA (Straub *et al.*, 1981; Kanne *et al.*, 1982) but their structures have not been investigated.

The observed chemical shift changes for aflatoxin protons upon interacting with d(ATGCAT)$_2$ are substantial (0.1-0.7 ppm increased shielding; Table 2) and clearly consistent with the intercalation model for the interaction. In the intercalated state, the chemical shifts for aflatoxin protons would be influenced by ring-current effects arising from interactions with the adjacent aromatic rings of the nucleotide bases. These ring-current shielding effects are expected to be as large as 1.5 ppm (Giessner-Prettre & Pullman, 1976). By varying the d(ATGCAT)$_2$ concentration to increase the fraction of aflatoxin in the bound state, we have observed upfield chemical shifts of as large as ~1.4 ppm for the aflatoxin B_1 H5 proton.

The ability of aflatoxins B_1 and B_2 to unwind and rewind the supercoiled plasmid is diagnostic of intercalation (Waring, 1970; Espejo & Lebowitz, 1976; Gopalakrishnan et al., 1989a; Raney et al., 1990). Unwinding results in the removal of supercoils from the closed circular plasmid DNA and is readily quantitated by techniques sensitive to the tertiary structure of the DNA molecule, such as gel electrophoresis. Non-intercalative binding could also effect electrophoretic mobility of DNA, if the binding involves charge neutralization or a change in the persistence length of the DNA molecule. However, non-intercalative binding would not be expected to introduce positive supercoils into relaxed closed circular DNA (Figure 6B). Furthermore, aflatoxins are neutral molecules which should not effect the negative charge on the DNA upon binding.

Competitive binding experiments involving actinomycin D and ethidium bromide provide evidence in support of the intercalation model. These experiments indicate that both actinomycin D and ethidium bromide displace aflatoxins from binding sites on d(ATGCAT)$_2$. Misra *et al.* (1983) examined the effect of ethidium bromide on aflatoxin adduct formation in pBR322 DNA. Their experiments showed that in the presence of ethidium bromide, the aflatoxin B_1 adduct formation was substantially reduced.

The structural change from the cyclopentenone ring of aflatoxin B_1 to the less planar δ-lactone ring of aflatoxin G_1 was expected to decrease intercalative binding to DNA. Scatchard analysis of aflatoxin binding to calf thymus DNA demonstrates that aflatoxin G_1 does in fact have lower DNA binding affinity. Under conditions in which the association constant for the binding of aflatoxin B_1 to calf thymus DNA is ~1.7 x 10^3 M^{-1}, the association constant of aflatoxin G_1 is ~1.5 x 10^2 M^{-1} (Figure 3). Lower binding affinity for aflatoxin G_1 relative to aflatoxin B_1 is also observed upon titration with d(ATGCAT)$_2$ or d(GCATGC)$_2$ (Figure 5). Utilization of oligodeoxynucleotides makes possible the examination of both the magnitude and orientation of aflatoxin binding to specific DNA sequences. The chemical shift changes for aflatoxins G_1 and G_2 which are observed in the presence of d(ATGCAT)$_2$ or d(GCATGC)$_2$ are similar to those observed for aflatoxins B_1 and B_2, although a higher concentration of oligodeoxynucleotide is required to achieve comparable chemical shift perturbations. As for aflatoxins B_1 and B_2, binding of aflatoxin G_2 to the oligodeoxynucleotides results in more rapid T_1 relaxation for adenine H2 (not shown), and aflatoxin G_2 is displaced from DNA by ethidium bromide. These results lead to the conclusion that the binding of aflatoxins G_1 and G_2 to B-DNA is probably also intercalative and similar to that of aflatoxins B_1 and B_2.

4.2.   CONFORMATION OF ADDUCTED AFLATOXIN B_1

The preparation of d(ATC^{AFB}GAT):d(ATCGAT) and d(AT^{AFB}GCAT)$_2$ provides a striking example of the ability of DNA sequence to mediate aflatoxin B_1 adduct formation at guanine N7. Whereas only one molecule of aflatoxin B_1-8,9-epoxide will react with d(ATCGAT)$_2$, two molecules of aflatoxin B_1-8,9-epoxide will react with d(ATGCAT)$_2$ (Scheme 1).

Analysis of the ^{1}H NMR data leads to the conclusion that the aflatoxin B_1 moiety is intercalated above the 5' face of guanine in ^{AFB}G for both d(ATC^{AFB}GAT):d(ATCGAT) and d(AT^{AFB}GCAT)$_2$. In this orientation, the overall conformation of the DNA remains right handed, and the principal perturbation of the B-DNA structure occurs adjacent to the binding site at N7 of ^{AFB}G. The aflatoxin methoxy and cyclopentenone ring protons face into the minor groove of the DNA, whereas the furofuran ring protons face into the major groove of the DNA. Evidence in support of this conclusion accrues from inspection of DNA-DNA NOE connectivities, aflatoxin B_1-DNA NOE connectivities, spin-lattice relaxation measurements, and chemical shift perturbations of aflatoxin B_1 protons.

The presence of ^{AFB}G in either d(ATC^{AFB}GAT):d(ATCGAT)   or d(AT^{AFB}GCAT)$_2$ perturbs intrastrand DNA base-deoxyribose connectivities on the 5' side of ^{AFB}G. The presence of ^{AFB}G also suppresses the imino-imino NOE between ^{AFB}G and the 5'-neighbor base pair, as demonstrated for d(ATC^{AFB}GAT):d(ATCGAT) in Figure 11.   The suppression of this NOE supports the conclusion that the aflatoxin moiety lies on the 5' face of the modified base pair. Aflatoxin B_1-DNA NOE connectivities are localized to the immediate 5' neighbor base pair relative to ^{AFB}G, and involve both the major and minor groove of the modified oligodeoxynucleotide. NOEs are observed between the aflatoxin H6a and H9a protons and major groove protons of the 5'-neighbor base (Tables 5 and 6). These NOEs demonstrate that aflatoxin H6a and H9a must be located in the major groove, and oriented in the 5'-direction from the site of modification. The aflatoxin methoxy and H5 proton show NOEs to the minor groove of the modified oligodeoxynucleotides. This pattern of NOEs also suggests that the aflatoxin moiety must be oriented 5' to the modified guanine residue. In d(AT^{AFB}GCAT)$_2$, an NOE is observed between 4-OCH$_3$ and A^5 H2 which provides strong evidence that the aflatoxin methoxy group is located in the minor groove of the modified oligodeoxynucleotide. Only one NOE was observed between the aflatoxin moiety and a 3'-neighbor base pair to the site of adduction: a weak NOE between aflatoxin H8 and C^4 H5 in d(AT^{AFB}GCAT)$_2$. This NOE is consistent with intercalation of aflatoxin above the 5'-face of guanine, which would orient aflatoxin H8 in the major groove facing toward the 3'-neighbor base pair. Spin lattice relaxation measurements support the notion that the aflatoxin moiety interacts with both the minor and major grooves of d(ATC^{AFB}GAT):d(ATCGAT) and d(AT^{AFB}GCAT)$_2$. In the minor groove, the aflatoxin adduct produces an increased spin-lattice relaxation rate for adenine H2 protons, which suggests that the presence of the aflatoxin protons in the minor groove provides a more efficient relaxation path for these normally isolated protons. This observation confirms the proximity of aflatoxin protons to adenine H2 in the minor groove, and is consistent with the observation of NOEs between A^5 H2 in d(AT^{AFB}GCAT)$_2$ and aflatoxin H3α and 4-OCH$_3$ protons (Table 6). The T_1 experiments also show that aflatoxin H8 and H9 have substantially slower T_1 relaxation (Figure 13). Analysis of the aflatoxin-oligodeoxynucleotide NOE data, above, reveals that aflatoxin H8 and H9 must face into the major groove of both adducts. Furthermore, the observation of only one NOE between aflatoxin H8 and the 3' neighbor base pair in d(AT^{AFB}GCAT)$_2$ suggests that these two aflatoxin protons are relatively isolated from spin-lattice relaxation pathways. Aflatoxin B_1 protons exhibit substantial upfield chemical shifts in d(ATC^{AFB}GAT):d(ATCGAT) and d(AT^{AFB}GCAT)$_2$ as compared to free aflatoxins. The pattern and magnitude of these chemical shift perturbations is similar in both modified oligodeoxynucleotides, which suggests a similar orientation of the aflatoxin moiety in each instance. These chemical shift perturbations of as large as 0.9 ppm for aflatoxin H5 are consistent with the notion that the aflatoxin moiety is intercalated, and presumably arise from ring current effects due to stacking interactions between the aflatoxin moiety and the DNA base pairs. The chemical shifts of the DNA protons are less perturbed by the presence of the adduct, and are localized to the immediate vicinity of the modified base pair. Comparison of chemical shift changes and spin-lattice relaxation rates for corresponding protons in either the associative or the covalent adduct suggests that the orientation of aflatoxin in d(ATC^{AFB}GAT):d(ATCGAT) and d(AT^{AFB}GCAT)$_2$ is similar to that in the corresponding non-covalent complexes.

4.3.   NATURE OF THE TRANSITION STATE

The orientation of bound aflatoxin B_1-8,9-epoxide or aflatoxin G_1-9,10-epoxide in B-DNA is of biological significance in its relationship to the transition state leading to adduct (Loechler et al., 1988). We believe intercalative binding of aflatoxins B_1 and G_1 is relevant to adduct formation at guanine N7 by the respective epoxides. If the rate-limiting step in adduct formation proceeds from a precovalent complex in equilibrium with free epoxide, shifting the equilibrium to favor free epoxide should promote increased hydrolysis of the epoxide and formation of fewer adducts. This is observed when the bound fraction of either aflatoxin B_1-8,9-epoxide or aflatoxin G_1-9,10-epoxide is reduced by lowering DNA concentration. Structural changes which hinder aflatoxin epoxide intercalation should decrease adduct formation if the precovalent complex between the epoxide and DNA is intercalative. At low DNA concentration, the amount of guanine N7 adduct formed by aflatoxin G_1-9,10-epoxide as compared to aflatoxin B_1-8,9-epoxide correlates directly with the difference in binding constants for these two compounds (Table 4).

Determination of conformation for d(ATC^{AFB}GAT):d(ATCGAT) and d(AT^{AFB}GCAT)$_2$ does not provide direct information regarding the conformation of the transition state for the reaction between aflatoxin B_1-8,9-epoxide and DNA since rearrangement to the most thermodynamically stable conformation must occur following bond formation between guanine N7 and aflatoxin C8. Adduct conformation could in principle differ from that of the transition state complex, as has been proposed in the case of reaction between benzo[a]pyrene diol epoxide and guanine 2-NH$_2$ (Geacintov, 1985). However, a transition state complex involving epoxide intercalation successfully predicts the stoichiometry of adduct formation which is observed for these two oligodeoxynucleotides (Scheme 2). Insertion of aflatoxin B_1-8,9-epoxide above the 5'-

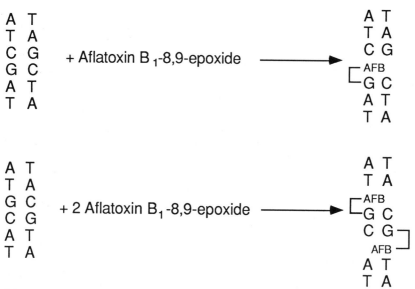

Scheme 2. Insertion of aflatoxin B_1-8,9-epoxide above the 5'-face of a guanine in d(ATCGAT)$_2$, followed by formation of the N7 guanyl adduct would prevent binding of a second molecule of aflatoxin B_1-8,9-epoxide. In contrast, two intercalation sites would be available for reaction with the sequence isomer d(ATGCAT)$_2$.

face of a guanine in d(ATCGAT)$_2$, followed by formation of the N7 guanyl adduct would prevent binding of a second molecule of aflatoxin B_1-8,9-epoxide. In contrast, two intercalation sites would be available for reaction with the sequence isomer d(ATGCAT)$_2$.

Loechler et al. (1988) argue that the structure of the transition state in the reaction between aflatoxin B_1-8,9-epoxide and guanine N7 and that of the resulting guanine N7 adduct are expected to be similar. Their argument is based upon the rationale that for the transition state in reactions involving oxocarbonium ions, bond formation is very incomplete whereas bond breaking is nearly complete (i.e., low β_{nuc} and high β_{lg}) (Koehler & Kordes, 1970; Cordes & Bull, 1974; Young & Jencks, 1977). Therefore, the transition state should be relatively insensitive to nucleophilic strength but quite sensitive to binding orientation. Consideration of the orientation of the available bonding orbital at guanine N7 suggests that intercalation of aflatoxin B_1-8,9-epoxide above the 5' face of the reactive guanine provides excellent positioning for nucleophilic attack on the *exo*-epoxide.

Our results are inconsistent with an alternative model advanced by Loechler and coworkers on the basis of molecular modeling studies, which proposes that both the transition state and the aflatoxin B_1 adduct are located in the major groove (Loechler et al., 1988). That model was formulated to explain data (Benasutti et al., 1988) which monitored the yield of reaction at specific guanines as a function of DNA sequence. Similar data have been collected in other laboratories (Misra et al., 1983; Muench et al., 1983; Marien et al., 1987; 1989). Loechler et al. (1988) argue that molecular modeling in combination with molecular mechanics can identify a transition state which rationalizes relative reactivity data. However, the inability of empirical rules which are based upon local sequence to unambiguously predict guanine reactivity (Misra et al., 1983; Marien et al., 1987) suggests that additional factors such as DNA conformational heterogeneity could exist, or that an incorrect transition state is being modeled. Molecular mechanics calculations locate conformations which represent local rather than global energy minima, and have additional limitations including uncertainty in accounting for solvent and counterion distribution. Loechler et al. (1988) acknowledge that their transition state does not account for the experimental observation that the most reactive guanines are those with a 5'-neighbor guanine, and have proposed that DNA conformation at these sites is altered.

Our conformational studies of adducted oligodeoxynucleotides demonstrate that the aflatoxin B_1 moiety interacts with both grooves of the modified oligodeoxynucleotide. The major groove model does not predict perturbation of minor groove protons such as adenine H2, which exhibits increased spin-lattice relaxation as well as dipolar coupling to aflatoxin H3α and 4-OCH$_3$ protons. The characteristic B-DNA pattern of internucleotide NOEs between the base and deoxyribose protons and between the hydrogen bonded imino protons would not be expected to be altered in both the modified and unmodified strands of the oligodeoxynucleotide were the aflatoxin moiety in the major groove. The substantially increased shielding of the aflatoxin protons, is also not consistent with the major groove model, which would predict instead some downfield shifts for aflatoxin protons oriented approximately perpendicular to the planes of the DNA base pairs. Finally, the difference in reaction stoichiometry for adduct formation with d(ATCGAT)$_2$ and d(ATGCAT)$_2$ is also difficult to reconcile with the major groove model, in which case one might expect that the reaction with d(ATCGAT)$_2$ could be forced to go beyond the observed stoichiometry of 1:2 aflatoxin:d(ATGCAT)$_2$.

4.4. SUMMARY

The present data lead to the conclusion that the interaction between aflatoxin B_1-8,9-epoxide and B-DNA is intercalative and leads to efficient formation via an intercalative transition state of the guanine adduct 8,9-dihydro-8-(N7-guanyl)-9-hydroxy-aflatoxin B_1. Upon adduct formation, the aflatoxin moiety remains intercalated in d(ATC^{AFB}GAT):d(ATCGAT) and d(AT^{AFB}GCAT)$_2$. Many questions remain to be answered. A more precise understanding of the solution conformation of these intercalated aflatoxin adducts remains to be determined. The principle experimental difficulty in working with these adducts is their tendency to undergo slow depurination, which hinders collection of quantitative NOE data for use in distance geometry calculations (Nerdal et al., 1989), or restrained molecular dynamics calculations (Gronenborn & Clore, 1989). Calculational studies using molecular mechanics and molecular dynamics simulations have been initiated in an effort to develop a better understanding of adduct conformation. Of future interest are the structures of the corresponding DNA adducts with aflatoxin G_1 and other aflatoxins, as well as the FAPY adducts which are the rearrangement products of these primary cationic adducts. Finally, it is hoped that the availability of specifically modified oligodeoxynucleotides will benefit future adduct-directed mutagenesis studies.

5. Acknowledgements

This research was funded by the National Institutes of Health, grants ES-03755 and ES-00267. We are indebted to a number of our colleagues who provided both experimental expertise and insightful comments. Dr. Elliot Gruskin and Professor Stephen Lloyd assisted with gel electrophoresis experiments. Professor W. David Wilson and Ms. Farial A. Tanious (Georgia State University) assisted in obtaining flow dichroism data. Professor Thomas R. Krugh and Suzanne Freeman (The University of Rochester) assisted in obtaining 500 MHz NOE data on aflatoxin B_1-oligodeoxynucleotide adducts. Professor Fu-Ming Chen (Tennessee State University) assisted in obtaining CD spectra. Dr. Steven W. Baertschi assisted in the preparation of dimethyldioxirane and aflatoxin B_1-8,9-epoxide. We also thank Professor Edward L. Loechler (Boston University) and Professor F. P. Guengerich for helpful discussions.

6. References

Asao, T., Büchi, G., Abdel-Kader, M. M., Chang, S. B., Wick, E. L., & Wogan, G. N. (1963) *J. Am. Chem. Soc.* 85, 1705-.

Asao, T. Büchi, G., Abdel-Kader, M. M., Chang, S. B., Wick, E. L., & Wogan, G. N. (1965) *J. Am. Chem. Soc.* 87, 882-.

Baertschi, S.W., Raney, K.D., Stone, M.P., & Harris, T.M. (1988) *J. Am. Chem. Soc.* 110, 7929-7931.

Baertschi, S. W., Raney, K. D., Shimada, T., Harris, T. M., & Guengerich, F. P. (1989) *Chem. Res. Toxicol.* 2, 114-122.

Benasutti, M., Ejadi, S., Whitlow, M. D., & Loechler, E. L. (1988) *Biochemistry* 27, 472-481.

Bonnett, M., & Taylor, E. R. (1989) *J. Biomol. Struct. Dyn.* 7, 127-149.

Busby, W. F., Jr., & Wogan, G. N. (1984) In *Chemical Carcinogens*, 2nd ed.; Searle, C., Ed.; American Chemical Society Series, Vol. 182, pp 945-1136.

Cheung, K. K., and Sim, G. A. (1964) *Nature (London) 201*, 1185-1188.

Cordes, E. H., & Bull, H. G. (1974) *Chem. Rev. 74*, 581-603.

Croy, R. G., Essigmann, J. M., Reinhold, V. M., & Wogan, G. N. (1978) *Proc. Natl. Acad. Sci. U.S.A.* 75, 1745-1749.

Espejo, R. T., & Lebowitz, J. (1976) *Anal. Biochemistry 72*, 95-103.

Essigmann, J. M., Croy, R. G., Nadzan, A. M., Busby, W. F., Jr., Reinhold, V. N., Büchi, G., & Wogan, G. N. (1977) *Proc. Natl. Acad. Sci. U.S.A. 74*, 1870-1874.

Feigon, J., Denny, W. A., Leupin, W., & Kearns, D. R. (1983) Biochemistry 22, 5943-5951.

Foster, P. L., Eisenstadt, E., & Miller, J. A. (1983) *Proc. Natl. Acad. Sci. USA 80*, 2695-2698.

Geacintov, N. E. (1985) in Polycyclic Hydrocarbons and Carcinogenesis, Harvey, R. G., Ed., ACS Symposium Monograph #283, American Chemical Society, Washington, D. C., pp. 107-124.

Giessner-Prettre, C. & Pullman, B. (1976) *Biochem. Biophys. Res. Commun. 70*, 578-581.

Gopalakrishnan, S., Byrd, S., Stone M.P., & Harris, T.M. (1989a) *Biochemistry 28*, 726-734.

Gopalakrishnan, S., Stone, M.P., & Harris, T.M. (1989b) *J. Am. Chem. Soc. 111*, 7232-7239.

Gorst-Allman, C. P., Steyn, P. S., & Wessels, P. L. (1977) *J. Chem. Soc. Perkin Trans. 1*, 1360-1364.

Gronenborn, A. M., & Clore, G. M. (1989) *Biochemistry 28*, 5978-5984.

Hare, D. R., Wemmer, D. E., Chou, S. H., Drobny, G., & Reid, B. R. (1983) *J. Mol. Biol. 171*, 319-336.

Hertzog, P. J., Lindsay Smith, J. R., & Garner, R. C. (1982) Carcinogenesis 3, 825-828.

Hore, P. J. (1983a) *J. Magn. Reson. 54*, 539-542.

Hore, P. J. (1983b) *J. Magn. Reson. 55*, 283-300.

Horowitz, P.M., & Barnes, L.D. (1983) *Anal. Biochem. 128*, 478-480.

Israel-Kalinsky, H., Tuch, J., Roitelaman, J., & Stark, A. A. (1982) *Carcinogenesis (London) 3*, 423-429.

Kanne, D., Straub, K., Hearst, J. E., & Rapoport, H. (1982) *J. Am. Chem. Soc. 104*, 6754-6764.

Koehler, K., & Cordes, E. H. (1970) *J. Am. Chem. Soc. 92*, 1576-1582.

Lin, J. K., Miller, J. A., & Miller, E. C. (1977) *Cancer Res. 37*, 4430-4438.

Loechler, E. L., Teeter, M. M., & Whitlow, M. D. (1988) *J. Biomol. Struct. Dyn. 5*, 1237-1257.

MacLeod, M., Smith, B., & McClay, J. (1987) *J.Biol.Chem. 262*, 1081-1087.

McCann, J., Spingarn, N. E., Kobori, J., & Ames, B. N. (1975) *Proc. Natl. Acad. Sci. USA 72*, 979-983.

McGhee, J. D., & von Hippel, P. H. (1974) *J. Mol. Biol. 86*, 469-489.

Marien, K., Moyer, R., Loveland, P., Van Holde, K., & Bailey, G. (1987) *J. Biol. Chem. 262*, 7455-7462.

Marien, K., Mathews, K., van Holde, K., & Bailey, G. (1989) *J. Biol. Chem. 264*, 13226-13232.

Martin, C.N., & Garner, R. C. (1977) *Nature (London) 267*, 863-865.

Misra, R. P., Muench, K. F., & Humayun, M. Z. (1983) *Biochemistry 22*, 3351-3359.

Mori, H., Sugie, S., Yoshimi, N., Kitamura, J., Niwa, M., Hamasaki, T., & Kawai, K. (1986) *Mutat. Res. 173*, 217-222.

Moyer, R., Marien, K., van Holde, K., & Bailey, G. (1989) *J. Biol. Chem. 264*, 12226-12231.

Muench, K. F., Misra, R. P., & Humayun, M. Z. (1983) *Proc. Natl. Acad. Sci. USA 80*, 6-10.

Nerdal, W., Hare, D.R., & Reid, B. R. (1989) *Biochemistry 28*, 10008-10021.

Oda, Y., Nakamura, S., Oki, I., Kato, T., & Shinagawa, H. (1985) *Mutat. Res. 147*, 219-229.

Plateu, P., & Gueron, M. (1982) *J. Am. Chem. Soc. 104*, 7310-7311.

Raney, K. D., Gopalakrishnan, S., Byrd, S., Stone, M. P., & Harris, T. M. (1990) *Chem. Res. in Toxicol. 3*, in press.

Sambamurti, K., Callahan, J., Luo, X., Perkins, C. P., Jacobsen, J. S., & Humayun, M. Z. (1988) *Genetics 120*, 863-873.

Scatchard, G. (1949) *Anal. N.Y. Acad. Sci. 51*, 660-672.

Shimada, T., Nakamura, S. -I., Imaoka, S., & Funae, Y. (1987) *Toxicol. Appl. Pharmacol. 91*, 13-21.

Shimada, T., & Guengerich, F. P. (1989) *Proc. Natl. Acad. Sci. USA 86*, 462-465.

Sklenar, V., & Bax, A. (1987) *J. Magn. Reson. 74*, 469-479.

Sklenar, V., Brooks, B. R., Zon, G., & Bax, A. (1987) *FEBS Lett. 216,* 249-252.

Stone, M. P., Gopalakrishnan, S., Harris, T. M., & Graves, D. E. (1988) *J. Biomol. Struct. Dyn. 5*, 1025-1041.

Straub, K., Kanne, D., Hearst, J. E., & Rapoport, H. (1981) *J. Am. Chem. Soc. 103*, 2347-2355.

Swenson, D. H., Lin, J.-K., Miller, E. C., & Miller, J. A. (1977) Cancer Res. 37, 172-181.

Van Soest, T. C., & Peerdeman, A. F. (1970) *Acta Cryst. B26*, 1940-1955.

Waring, M. J. (1970) *J. Mol. Biol. 54*, 247-279.

Wong, J. J., Singh, R., & Hsieh, D. P. H. (1977) *Mutat. Res. 44*, 447-450.

Young, P. R., & Jencks, W. P. (1977) *J. Am. Chem. Soc. 99*, 8238-8248.

Yourtee, D. M., Kirk-Yourtee, C. L., & Searles, S. (1987) *Life Sci. 41*, 1975-1803.

SEQUENCE SPECIFIC ISOTOPE EFFECTS ON THE CLEAVAGE OF DNA BY RADICAL-GENERATING DRUGS

JOHN W. KOZARICH
Department of Chemistry & Biochemistry
and Center for Agricultural Biotechnology
University of Maryland
College Park, MD 20742

ABSTRACT. We have recently developed a new technique which utilizes specifically deuteriated ^{32}P end-labelled DNAs in combination with high resolution gel electrophoresis to detect and quantitate potentially rate-limiting carbon-hydrogen bond cleavages by DNA-cleaving drugs at individual sequence site. Iron·bleomycin and neocarzinostatin have been studied using this technique. For bleomycin DNA cleavage appears to be associated exclusively with a rate-limiting 4'-hydrogen abstraction. For neocarzinostatin a more complex cleavage pattern emerges. Abstraction of the 1'-hydrogen is a major pathway for the cleavage at C residues in d(AGC) sequences. However, at d(GT) sequences a partitioning between a rate-limiting abstraction of hydrogen at either the 4'- or 5'-carbons. The applicability of this approach to other DNA cleavers is also discussed.

Introduction

The mechanistic elucidation of DNA cleavage by bleomycin (BLM) [1], neocarzinostain (NCS) [2], calicheamicin [3], esperamicin [4], and related compounds [5] has been an area of increased interest in recent years. High sensitivity and precision are required to evaluate the mechanistic changes that may accompany alterations in local DNA conformation or modifications in drug structure or both. We have recently developed a new and powerful technique that makes use of specifically deuteriated ^{32}P end-labeled DNAs in combination with gel electrophoresis to detect and quantitate potentially rate-limiting carbon-hydrogen bond cleavages by DNA-cleaving drugs at individual sequence sites. Our recent findings have verified the generality of this approach for the analysis of the cleavage chemistry of bleomycin [6] and neocarzinostatin [7]. In this account, I will discuss some of the observations which led to the development of this method and some of the issues which impinge upon the manifestation of kinetic isotope effects in DNA cleavage reactions.

Background

TRITIUM SELECTION EFFECTS ON THE CLEAVAGE OF DNA BY IRON·BLM

The basis of the technique rests on our previous work on the chemistry of DNA cleavage using iron·BLM. The in vitro activity of BLM depends upon Fe(II) and O_2 or Fe(III) and H_2O_2 [1]. The initial BLM·Fe(II)·O_2 complex (Figure 1) undergoes one-electron reduction to ultimately yield "activated BLM" which can initiate DNA damage. Two types of DNA damage are observed with "activated BLM" (Figure 2). Pathway A results in the

B. Pullman and J. Jortner (eds.), Molecular Basis of Specificity in Nucleic Acid-Drug Interactions, 481–493.
© 1990 *Kluwer Academic Publishers. Printed in the Netherlands.*

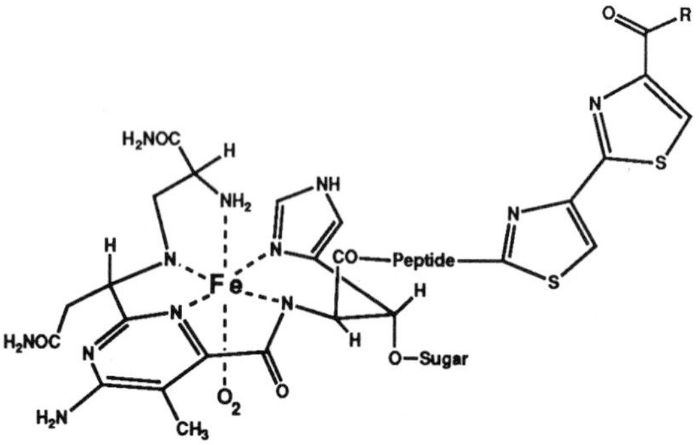

Figure 1. Proposed structure of the BLM·Fe.O₂ complex.

formation of nucleic acid base propenal and DNA strand scission that yields 3'-phosphoglycolate terminii and 5'-phosphate terminii. Pathway B results in the liberation of nucleic acid base plus an alkali-labile site that cleaves at pH 12 with piperidine to afford a 3'-phosphate terminus.

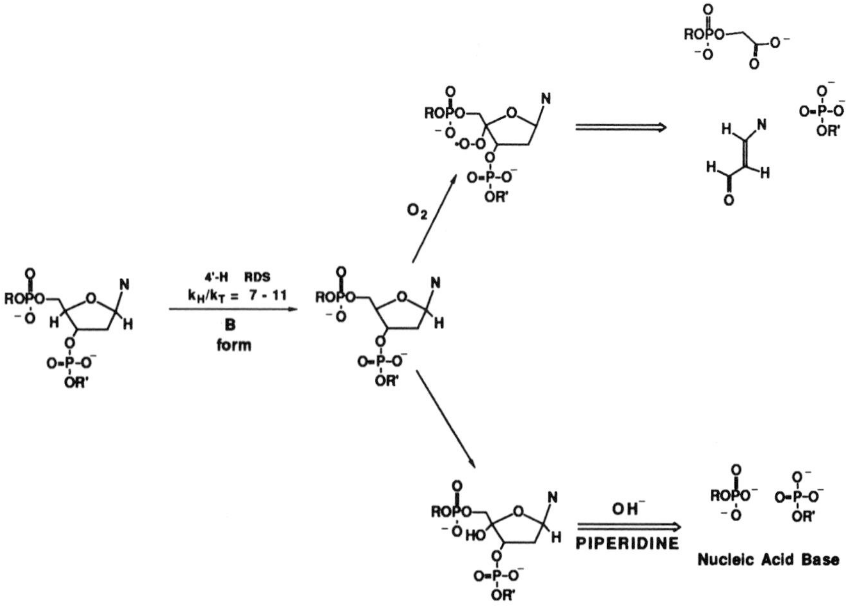

Figure 2. Proposed unified mechanism for the cleavage of DNA by "activated BLM".

We proposed the first unified mechanism for BLM action (Figure 2) based on our studies of the reaction of BLM with simple DNA models [such as poly(dA.dU)] which was tritiated at specific positions (i.e., at the 1'-, 2'-, 3'-, 4'- and 5'-carbons) in the deoxyribose ring [8]. The principle behind this approach is that the detailed analysis of the fate of the tritium at each position would serve as an indicator of the chemistry which was occurring at that position. Thus, if "activated BLM" abstracted a hydrogen (tritium) from the sugar and generated a radical intermediate, then we would anticipate that the tritium would be liberated as ^3H_2O. The amount of tritiated water can be quantitated and used as an indicator for chemistry at a particular carbon of the deoxyribose. Tritium may also be found and quantitated in one or more of the carbon-containing products generated by the reaction, thereby, suggesting chemical stability at that position. By performing this procedure for each specifically tritiated molecule, one can generate a "map" of the chemical reactivity of each carbon and propose a mechanistic hypothesis to explain the observations. Of course, there are several potential pitfalls in this type of analysis. The most critical problem is distinguishing between a tritium abstraction which is a drug-mediated homolytic carbon-tritium bond cleavage yielding tritiated water and a labilization of tritium to water due to a slow chemical exchange of the tritium via an intermediate formed after the crucial drug-mediated abstraction. For instance, the 2'-H_R proton is completely lost to solvent during the formation of base propenal (Figure 2, pathway A) due to the anti elimination to yield the trans propenal. Using a combination of chemical conditions and reductive trapping procedures, we were able to determine which hydrogens were labilized by chemical exchange processes [8]. Since a discussion of this data lies beyond the scope of this account, the reader should refer to our published work for a detailed description of the procedures.

The results of these studies suggested that "activated BLM" effected a 4'-carbon-hydrogen bond cleavage that was subject to a suprisingly large tritium selection effect (k_H/k_T = 7 to 11). The selection effect was determined by two independent procedures that are illustrated in Figure 3.

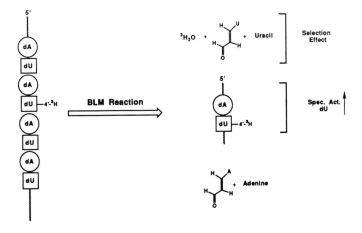

Figure 3. Tritium selection effects on the cleavage of poly(dA·[4'-^{3}H]dU) by activated bleomycin. Magnitude of the effect may be measured by quantitation of products (ambiguous) or by specific activity of reisolated starting material (unambiguous).

The poly(dA·[4'-³H]dU) contained tritium specifically at the 4'-carbon as a tracer isotope. By determining the amount of uracil and uracil propenal generated by "activated BLM", the total amount of deoxyuridine residues damaged could be determined. The total amount of tritiated deoxyuridine could also be determined by quantitating the amount of tritiated water produced (no tritium was found in any of the damage products). The consistent observation made under a variety of conditions was that the flux of tritiated material was less than that measured for the protio material - a strong indication of a selection effect against tritium. However, an ambiguity remained in the analysis. Since two protio-containing products were generated, it was not clear whether the tritium release should be correlated to one protio product or to both. In other words, did the rate-limiting tritium abstraction lead to a single product or to both? This question is not resolvable by this experiment. A converse approach was then taken (Figure 3). If the cleavage reaction is allowed to proceed to a determined extent of reaction and the remaining deoxyuridine in the damaged polymer is isolated, the specific activity of the deoxyuridine will increase. The increase in specific activity will be a function of the observed tritium selection effect and of the extent of reaction. The extent of reaction can calculated in one of three ways - by the amount of uracil generated, the amount of uracil propenal or by the sum of both products. Our analysis permitted us to establish that only when the extent of reaction was calculated on the basis of the sum of the two products did the increases in specific activity agree with a unique theoretical curve for a calculated isotope effect of 7 to 11. The result is unambiguous and provided the best evidence at the time for a single BLM-mediated event leading to all of the observed damage.

THEORETICAL DEVELOPMENT OF DEUTERIUM ISOTOPE EFFECTS

Our mechanistic proposal has received some criticism based on the experimental design and the lack of sensitivity of the approach. Several factors limit the overall sensitivity that one can obtain using tritiated nucleosides in this type of analysis. First, tritium is used as a tracer isotope. Roughly only one deoxyuridine residue in 10^8 or 10^9 is tritiated due to the synthetic procedures required to prepare this material. This requires, then, that extensive DNA damage be effected in order to generate enough tritiated water or to significantly increase the specific activity of the residual deoxyuridine. In a typical experiment, upwards of 50% of the deoxyuridine residues were damaged by "activated BLM" to meet these requirements. This, of course, required a very low DNA base to drug ratio of approximately 1. It is clear that these harsh conditions can not remotely be considered physiological; BLM is known to effect DNA damage and possibly cell kill under "one-hit" conditions. Second, such large tritium selection isotope effects (7 - 11) are usually subject to relatively large errors. This is due to a compression of the theoretical curves for the magnitude of the isotope effect generated by the analysis of the increase of the specific activity of the starting deoxyuridine. The compression is particularly severe for isotope effects greater than 4. In these cases, small inaccuracies in the determination of the extent of reaction can be amplified into large errors in the isotope effect. Third, the isotope effects determined are global in nature without specific sequence information. This is due to the tracer amount of tritium which creates a positional ambiguity with regard to which residues are tritiated and which are not. The problem is rendered moot in an alternating copolymer in which all deoxyuridine sites are essentially identical but is clearly a major obstacle when a heterogeneous DNA is used. In the latter case, the sequence heterogeneity suggests that specific BLM damage sites might differ in the magnitude of the tritium selection effect due to conformational differences in the DNA-drug complex that are due to subtle changes in sugar pucker, groove width, propeller twist etc. However, the uncertainty in the position of the tritium in any given molecule would make this type of detailed analysis impossible. Fourth, the analysis of the chemistry at minor damage sites is hampered by the lack of sensitivity. Poly(dA·dU) was chosen because deoxyuridine (a synthetically practical thymidine analog) constitutes a very good BLM damage site when it is 3'-downstream of a purine. However, minor damage site abound for BLM and other cleavers. The analysis of the chemistry at these sites is not feasible with this technology.

A potential solution to these problems rests in a simple hypothesis. We noted that the tritium selection effect of 7 - 11 predicts a deuterium kinetic isotope effect of 4 to 5. This prediction is based on the Swain-Schaad equation which relates both effects [9]:

$$k_H/k_D = \sqrt[1.44]{k_H/k_T} = 4 \text{ to } 5$$

This is a fairly substantial kinetic isotope effect which could in principle be detected at individual sequence sites in a heterogeneous DNA under "one hit" conditions. The basis of the experiment is outlined in Figure 4.

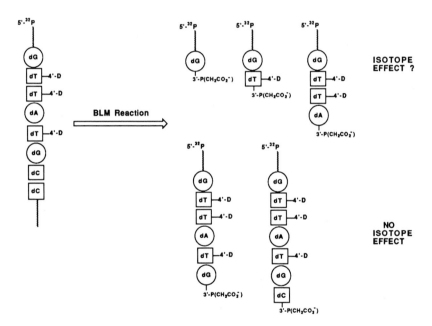

Figure 4. Deuterium kinetic isotope effects on the cleavage of a heterogeneous DNA by activated bleomycin. Each thymidine contains a 4'-deuterium (>95% ²H) and the reaction is carried out under a high DNA base to drug ration (i.E., "one-hit" conditions). The DNA fragment also contains a remote ³²P label at the 5'-terminus. Kinetic isotope effects are measured by the suppression of damage at deuteriated sites as determined by gel electrophoresis and autoradiography.

The concept behind the experiment is based on the ability to prepare 4'-deuteriated thymidine of a high isotopic content (>95% ^{2}H). The availability of deuteriated reagents of high isotopic purity makes this feasible. This immediately circumvents the problem of tracer isotope analysis associated with the use of tritium. The labeled thymidine may then be incorprated into a double stranded DNA fragment using standard molecular biology techniques (vide infra). The strand of the fragment which contains the deuteriated thymidine is remote labeled with a ^{32}P at the 5'-terminus (Figure 4). The ^{32}P serves as a high sensitivity radiolabel that will serve as the basis for the detection and quantitation of the damage sites. In a typical DNA cleavage reaction, the sequence specificity of the drug-mediated damage is usually assessed by reacting a defined 5'-end labeled DNA fragment with a limiting amount of drug in order to generate a nested set of damage fragments which represent a statistical distribution (biased for drug sequence specificity) of all possible damage sites. Since the reaction is performed under limiting drug concentrations, approximately one drug-hit per DNA molecule is obtained. This type of limiting chemistry is also the basis of DNA sequence analysis methods. The key feature of such conditions is that they may be viewed as constituting an initial velocity reaction. Since the determination of a kinetic deuterium isotope effect is ideally determined under initial velocity conditions, the principles of DNA sequence analysis should be adaptable to the measurement of these effects on a DNA cleavage reaction.

The nested set of fragments generated by drug-mediated cleavage can be separated on the basis on size and charge by high resolution polyacrylamide gel electrophoresis and detected by autoradiography. Comparison of the mobility of the damage fragments to known standards generated by the chemistry of DNA sequence analysis establishes the sequence specificity of the drug and the intensities of the autoradiographic bands are a valid indication of the amount of damage at a given site. The damage may also be quantitated by densitometry scanning of the autoradiogram or by very recently introduced phosphorimager technology. This general methodology has been used with many DNA cleavers to evaluate which nucleoside residues constitute major and which are minor damage sites.

If the unified mechanism for bleomycin action is correct, then the introduction of a 4'-deuterium at thymidine residues should result in a suppression of damage at those thymidine positions which are cleaved by a rate-limiting 4'-carbon-deuterium bond cleavage. That is, under the initial velocity conditions, deuteriated sites will be cleaved slower than protiated sites and, therefore, give less overall cleavage at that site vis a vis the undeuteriated sites. Using the remote ^{32}P label as the probe, this will manifest itself as a decrease in the autoradiographic intensity of those bands which can be quantitated by densitometry. If the amount of damage at a specific deuteriated site is compared to the amount at the same site in an undeuteriated DNA (after normalization), then the ratio is a direct measure of the effect of 4'-deuteriation at that position - a deuterium isotope effect.

The potential power of the method is obvious. One should in principle be able to measure isotope effects at individual sequence sites, many of which may be relatively minor. The damage at undeuteriated sites, such as cytidine residues (Figure 4), serves as an internal control since no isotope effect is expected at these positions. Finally, the partitioning of a 4'-radical intermediate (Figure 2) leading to both damage pathways (A & B) suggests that the same isotope effect should be identical on both pathways. Since pathway A results in the formation of a 5'-^{32}P end-labeled DNA containing a 3'-phosphoglycolate terminus and pathway B results in a 3'-phosphate terminus after alkali treatment, it is conceivable that the effect can be individually measured on both pathways. Under conditions of high resolution polyacrylamide (20%) gel electrophoresis, fragments smaller than ~40 bp can be sufficiently resolved to separate these fragments which differ slightly in charge and size.

Deuterium Isotope Effects on the Bleomycin-Mediated Cleavage of DNA

In these experiments we have routinely used the Eco RI-Bam HI (375 bp) or the Hin DIII-Bam HI (346 bp) fragment from pBR322 as the test sequence. Either fragment was cloned into M13mp19 and the single-stranded DNA was isolated from infected *Escherichia coli*, JM101. Synthesis of the complementary (+)-strand was accomplished by annealing kinase-treated primer to the single-stranded template and by performing the polymerization and with Sequenase (U.S. Biochemicals) and with dGTP, dATP, dCTP and either dTTP or [4'-^{2}H]dTTP. The polymerization and terminal ligation was monitored by comparison to replicative form on agarose gels. Samples were purified on Sephadex G-50. The replicative form containing either [4'-^{1}H]- or [4'-^{2}H]T was treated with Eco RI or Hin DIII(1 h; 37°), depending upon which fragment is desired, and 5' end-labeled by standard procedures. The desired fragment was produced by Bam HI digestion, resolved by electrophoresis on a 5% acrylamide gel and purified on an NACS Prepac (BRL) column.

By this procedure both the protiated (control) and the deuteriated fragments are prepared for cleavage by the drug. A typical DNA cleavage reaction is shown in Figure 5.

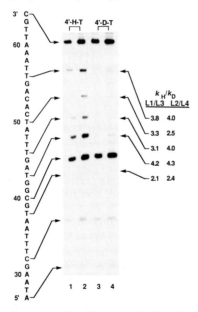

Figure 5. Autoradiogram of a high-resolution denaturing polyacrylamide gel of the reaction of "activated BLM" and the Eco RI-Bam HI pBR322 fragment containing [4'-^{1}H]T (lanes 1 and 3) or [4'-^{2}H]T (lanes 2 and 4). Each reaction (80 µL) contained 10 mM Na$_2$HPO$_4$, pH 7.6, 10 mM β-mercaptoethanol, salmon sperm DNA (0.2 µg/µL) and 70,000 cpm of fragment 5' end-labeled with ^{32}P. The BLM.Fe(II) (1:3) was added to a final concentration of 3.5 µM (~100 bp per BLM). Reactions were incubated for 10 min at 25°C and terminated by addition of 0.1 mM EDTA, 2.5 M sodium acetate and salmon sperm DNA (0.2 µg/µL) (100 µL final volume). Samples were precipitated with ethanol and subjected to gel electrophoresis. Samples in lanes 1 and 3 received no alkali treatment (pathway A, Figure 2). Samples in lanes 2 and 4 were treated with 1 M piperidine for 15 min at 90°C and repelleted before electrophoresis (pathways A and B, Figure 2).

The direct observation of an isotope effect on 4'-C-H bond cleavage is clearly demonstrated as is the known preference of BLM for cleavage at GC and GC sequences. The suppression of [^{32}P]DNA fragments resulting exclusively from damage at [4'-^{2}H]T sites is strong evidence for the kinetic discrimination by "activated BLM." In other experiments (data not shown) we have established that deuteriation at the 1'- or 5'-positions (additional potential hit sites from the minor groove) do not result in any observed isotope effect suggesting that no drug-mediated abstraction occurs at these positions. Since undeuteriated nucleosides such as C and A serve as internal controls, quantitation of the isotope effects may be performed by scanning densitometry of the autoradiogram (Figure 6).

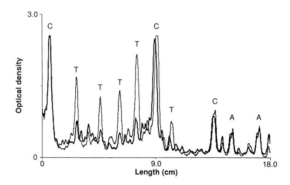

Figure 6. Densitometry scan of the autoradiogram shown in Figure 5. Total strand damage of [4'-^{1}H]T-containing DNA fragment (lane 2, light line) by "activated BLM" is compared to total strand damage of [4'-^{2}H]T-containing DNA fragment (lane 4, heavy line). The ratio of the integrated peak areas is a direct measure of the kinetic isotope effect at that position (L2/L4, Figure 5). Similar scans were performed on lanes 1 and 3 (L1/L3, Figure 5). The scans shown are the raw data without normalization.

The calculated isotope effects exhibit a range from ~2 to 4.5 (Figure 5). The differences in the magnitude of the effect at different sites is reproducible and suggests that local sequence variability may be important. Dissociation of the "activated BLM" from DNA must also be faster than bond cleavage to permit discrimination between labeled and unlabeled cleavage sites. The isotope effects on pathways A and B (Figure 5) are essentially the same for a particular damage site. The isotope effects on pathway A (L1/L3) were determined by quantitation of neutral strand scission, whereas those on pathways A plus B (L2/L4) were determined by quantitation of total alkali-induced scission. The effect on pathway B is similar to that on pathway A, which affirms the partitioning of a common intermediate at individual damage sites. Moreover, preliminary experiments varying the O_2 concentration to alter the partition ratio corroborate this proposal. Recent experiments have permitted us to resolve fragments containing 3'-phosphate termini from those containing 3'-phosphoglycolate termini. Quantitation of the isotope effect on the discreet bands has also corroborated that the effects on both pathways are virtually identical. From these results it is clear that the initial hypothesis using tritiated DNAs has been greatly strengthened by these experiments. The chemistry of DNA cleavage by BLM has been verified under "one-hit" conditions and the quality of the data suggests that the methodology may be applicable to the analysis of other DNA cleavers.

Deuterium Isotope Effects on the Cleavage of DNA by Neocarzinostatin

The findings with BLM have provided a strong impetus to explore the potential of the method, especially for the analysis of the chemistry at minor lesions. Our studies of the chemistry of cleavage of deoxycytidine residues in d(AGC) sequences by neocarzinostatin provide an excellent example of a minor lesion analysis [7].

Neocarzinostatin (NCS) binds to double-stranded DNA via intercalation of its napthoate moiety and interaction of its unusual [7.3.0]dodecadien-diyne epoxide with the minor groove (Figure 7) [2]. A putative diradical species of NCS, generated by nucleophilic addition of a thiol at C-12, abstracts hydrogen atoms from the deoxyribose residues of DNA to produce DNA strand breaks in the presence of O_2.

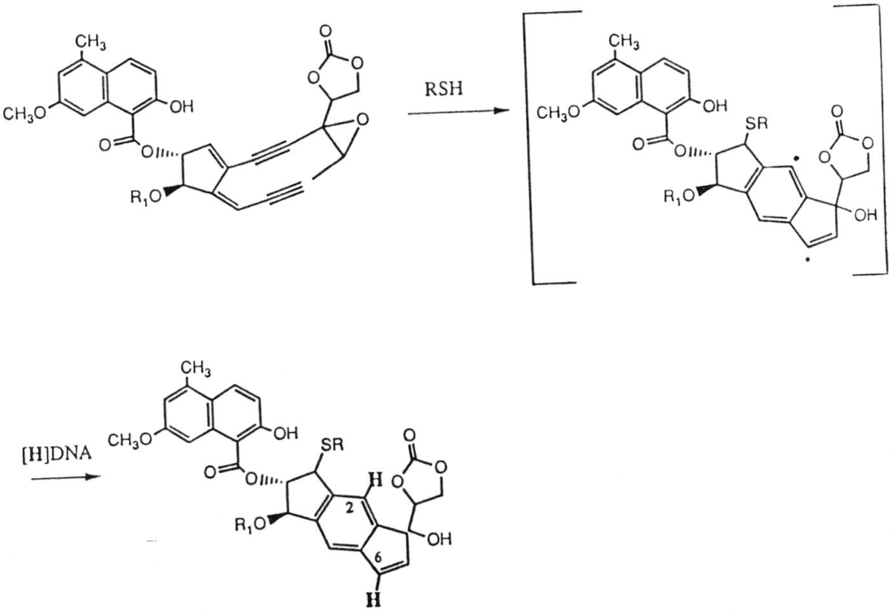

Figure 7. Proposed structure for NCS and its mechanism of action. One of the putative biradicals is postulated to be responsible for hydrogen atom abstraction of the C-1' hydrogen of dC_4 and dC_8 from $d(GAGC_4GAGC_8G)$ (vide infra).

The predominant DNA damage is direct strand breaks at thymidylate and deoxyadenylate residues resulting in thymidine and deoxyadenosine 5'-aldehyde at the 5'-terminus and phosphate at the 3'-terminus [2]. Recent research has uncovered a less prevalent lesion thought to be involved in NCS-induced mutagenesis in *E. coli*: alkali-labile sites at deoxycytidylate residues in d(AGC) sequences [10]. Two approaches using oligomers have been previously undertaken to understand the mechanism of formation of this minor lesion. The first utilized $d(AGCGAGC_7G)$ containing $[1',2',5-^3H]dC_7$. Incubation with NCS resulted in the release of cytosine and 0.04 eq of tritiated water suggesting an isotope effect of 25 on abstraction from the 1'-position. The distribution of the label was not rigorously established, however. The second approach involved quantitating the amount of deoxyribonolactone formed during the reaction (0.6 eq) [4].

In light of the observation of a potentially large tritium selection effect, it was clear that this new technique could permit a direct observation of the putative isotope effect on 1'-hydrogen abstraction solidifying earlier mechanistic proposals. At the same time we could test the premise that our new approach might provide a powerful method for the detection of the chemistry of minor lesions. The preparation of the oligomer follows standard procedures [7]. $[1'-^2H]dCTP$ or $[1'-^1H]dCTP$ was incorporated into the penultimate residue of the nanomer $d(GAGC_4GAGC_8G)$[10]. The nanomer containing unlabeled C_4 and $[1'-^1H]C_8$ in the $d(AGC)$ sequence was end-labeled, annealed to $d(CGCT)_3$, and incubated with limiting concentrations of NCS, and the degradation products were analyzed by polyacrylamide gel electrophoresis (Figure 8).

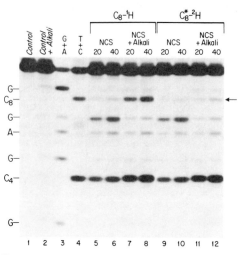

Figure 8. Isotope effects on alkali-dependent strand scissions by NCS at dC residues in $d(AGC)$ sequences. $5'-^{32}P$ end-labeled $d(GAGC_4GAGC_8G)$ annealed to $d(CGCT)_3$ was incubated with NCS (20 and 40 µM) and 5.0 mM glutathione in a standard reaction. Lanes 5-8, both C_4 and C_8 contain $1'-^1H$, and lanes 9-12, C_8 contains $1'-^2H$. Lanes 1 and 2 are minus NCS controls without and with alkali, respectively, for the nondeuteriated nanomer. Controls for the $[1'-^2H]C_8$ nanomer gave similar results (data not shown). Lanes 3 and 4 are Maxam-Gilbert markers and are shown as G + A and T + C. Lanes 5 and 9 and 6 and 10 are 20 and 40 µM NCS, respectively. Lanes 7 and 11 and 8 and 12 are 20 and 40 µM NCS + alkali, respectively. The arrow indicates cleavage at the C_8 position of the nanomer.

It can be seen in Figure 8 that there is alkali dependent cleavage at both C_4 and C_8 (compare lanes 5, 7, and 6, 8). Identical experiments were performed on the same oligomer containing $[1'-^2H]C_8$. The direct observation of an isotope effect on NCS mediated oligomer damage is apparent on the alkali-labile reaction by comparison of lanes 7 with 11 and 8 with 12 (see arrow). Quantitation of the isotope effect at C_8 ranges from 3.3 to 4.2 depending upon conditions. A similar analysis for cleavage at C_4, the internal control, gave the expected value of 1.0. The non-alkali dependent cleavage observed in lanes 5 and 6 with migration slightly faster than G_7 is presumably an intermediate $d(GAG)$ moiety attached to a 5'-phosphorylated α,β-unsaturated lactone that is the precursor to C_4 damage. Since the 1'-hydrogen of C_4 is unlabeled, there is no isotope effect on the formation of this intermediate (compare lanes 5,6 with lanes 9,10). These results provide the first direct evidence that NCS can effect removal of a 1'-hydrogen from a deoxycytidine residue in the d(AGC) sequence and demonstrate that a substantial isotope effect occurs on this reaction.

When the protio and deuterio nanomers were annealed to d(CICT)3 instead of d(CGCT)3, no significant isotope effect was observed (data not shown). Previous studies have shown that inosine (I), base-paired to C, enhances NCS-mediated cleavage 5-fold relative to G. If this enhancement is due to an increase in the relative rates of hydrogen abstraction by the activated drug versus dissociation from DNA of the activated drug, then the suppression of the isotope effect is explainable.

These studies have recently been extended to the Eco RI-Bam HI DNA fragment from pBR322 (375 bp) in which [1'-²H]dC or [1'-¹H]dC has been incorporated. Two d(AGC) sequences are present in this restriction fragment : ³²P-GACAGCTTATCATCGATAAGCT-. Results from densitometry scans (data not shown) indicate an alkali-labile isotope effect of 3.7 on the cleavage of dC closest to the 5'-end. Little damage was observed at the second d(AGC) sequence. The extent of damage at dC in the first d(AGC) sequence was less than 0.1% of the total damage. Such pathways, while difficult to investigate from a mechanistic point of view, may be the pathway of major biological significance. The direct observation of this 1'-carbon-hydrogen bond cleavage thus demonstrates the power of this method in establishing the chemistry involved in minor pathways.

We have recently turned our attention to the chemistry of cleavage of thymidine residues by NCS. This major damage mode is strand scission caused by hydrogen abstraction from the 5'-position of thymidylate and deoxyadenylate residues. The newly generated 5'- and 3'-termini contain a nucleoside 5'-aldehyde and a 3'-phosphate, respectively. A recent report by Saito and coworkers has suggested that the NCS-mediated cleavage of the self-complemementary hexanucleotide d(CGTACG) results in the formation of products that imply a partitioning between 4'- and 5'-hydrogen abstraction from the thymidylate [11]. They detected a very low amount of the phosphoglycolate terminus (<3%), a product diagnostic for 4'-chemistry in DNA cleavage by iron·bleomycin. We have recently demonstrated that in certain sequences thymidine is cleaved by activated NCS via a rate-limiting abstraction of either 4'- or 5'-hydrogen and that this partitioning can be modulated by deuteriation at either position. Moreover, the 4'-radical intermediate also partitions between a modified abasic carbohydrate terminus and a 3'-phosphoglycolate terminus. The relative ratio of these two termini is, in turn, modulated by the structure of the thiol activator/reductant (Figure 9).

Figure 9 provides a reasonable explanation for the observed chemistry. After abstraction of the 4'- or 5'-hydrogen by rate-limiting processes (that may, but not necessarily, occur through a common DNA·NCS*complex), the carbon-based radical is trapped by molecular oxygen to ultimately afford the corresponding peroxides. The 5'-peroxide is reduced to the hydroxyl species which spontaneously cleaves to yield the 5'-aldehyde terminus and the 3'-phosphate terminus (B). The 4'-peroxide may undergo thiol reduction to the 4'-hydroxy species and, ultimately, to the release of nucleic acid base, β-elimination and A. However, a competing fragmentation like that observed for iron·bleomycin would lead to the formation of the 3'-phosphoglycolate terminus (C).

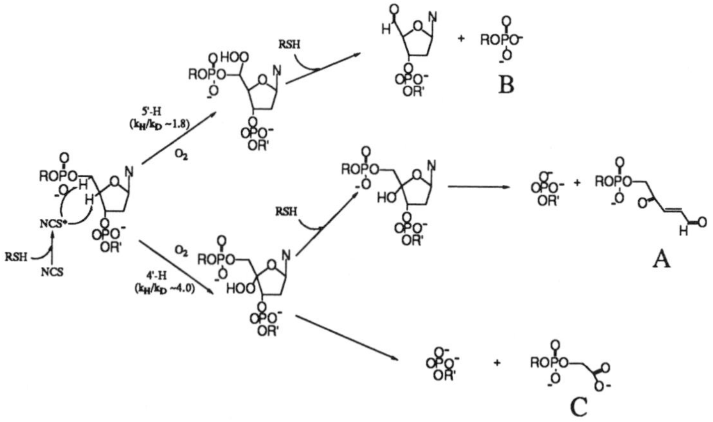

Figure 9. Proposed scheme for the cleavage of thymidine residues by NCS via partitioning between 4'- and 5'-hydrogen abstraction.

Conclusions

The results discussed here are illustrative of the power of using deuteriated DNAs as probes for mechanism of DNA cleavers. The high sensitivity and unambiguous visualization of specific damage chemistry is clearly advantageous for the simultaneuos analysis at multiple hit sites, some of which might constitute minor lesions. When coupled to other analytical methods, a complete reaction profile at a specific site is attainable. Such information is crucial to the understanding of the mode of action of these agents and serves to establish critical contacts points between DNA and drug that will be invaluable for the molecular modeling of these interactions.

Acknowledgments

This work is part of an ongoing collaboration with Professor JoAnne Stubbe of MIT. I thank my associates at Maryland, Leroy Worth, Jr., Bruce Frank and Donna Christner, who have played the major roles in the development of this research. Our recent collaboration with Professor Irving H. Goldberg and Dr. Lizzy S. Kappen on neocarzinostatin is gratefully acknowledged. This research was supported by the National Institutes of Health (GM 34454).

References

[1]   Stubbe, J. and Kozarich, J. W. (1987) 'Mechanism of bleomycin-induced DNA degradation', Chem. Rev. 87, 1107-1136.
[2]   Goldberg, I. H. (1987) 'Free radical mechanisms in neocarzinostatin-induced DNA damage', Free Radical Biol. & Med. 3, 41-54.

[3]   Zein, N. , Sinha , A. M. , McGahren, W. J. , and Ellestad , G. A. (1988) 'Calicheamicin γ_1^I : an antitumor antibiotic that cleaves double-stranded DNA site specifically', Science 240 , 1198-1201; Zein , N. , Poncin, M., Nilakantan, R., and Ellestad , G. A. (1988) 'Calicheamicin γ_1^I and DNA: molecular recognition process responsible for site-specificity', Science 244, 697-699.

[4]   Long , B.H. , Golik, J. , Forenza , S., Ward, B., Rehfuss, R., Dabroiak, J. C., Catino , J. J., Musial , S. T. , Brookshire , K. W., and Doyle , T. W. (1989) 'Esperamicins , a class of potent antitumor antibiotics : mechanism of action', Proc. Natl. Acad. Sci. U.S.A. 86, 2-6.

[5]   Dervan , P. B. , (1986) 'Design of sequence-specific DNA-binding molecules' , Science 232 , 464-471; Sigman, D. S. (1986) 'Nuclease activity of 1,10-phenanthroline-copper ion' , Acc. Chem. Res. 19, 180-186; Barton, J. K. (1986) 'Metals and DNA: Moleculat left-handed complements', Science 233, 727-734.

[6]   Kozarich, J. W., Worth, L. Jr., Frank, B. L., Christner, D. F., Vanderwall, D. E., and Stubbe, J. (1989) 'Sequence-specific isotope effects on the cleavage of DNA by bleomycin', Science, 245, 1396-1398.

[7]   Kappen, L. , Goldberg , I. H. , Wu, S. H. , Stubbe, J., Worth, L. Jr., and Kozarich, J. W. (1990) 'Isotope effects on the sequence-specific cleavage of dC in d(AGC) sequences by neocarzinostatin: elucidation of chemistry of minor lesions', J. Am. Chem. Soc., 112, 2797-2798.

[8]   Wu, J. C., Kozarich, J. W., and Stubbe, J. (1983) 'The mechanism of free base formation from DNA by bleomycin. A proposal based on site specific tritium release from poly(dA·dU)' J. Biol. Chem., 258, 4694-4697; Wu , J. C., Kozarich, J. W., and Stubbe, J. (1985) Biochemistry, 24, 7562-7568.

[9]   Swain, C. G., Stivers, E. C., Reuwer, J. F., Jr., and Schaad, L. J. (1958) 'Use of hydrogen isotope effects to identify the attacking nucleophile in the enolization of ketones catalyzed by acetic acid', J. Am. Chem. Soc., 80, 5885-5889.

[10]   Povirk , L. F., and Goldberg , I. H. (1985) 'Endonuclease-resistant apyrimidinic sites formed by neocarzinostatin at cytosine residues in DNA : evidence for a possible role in mutagenesis' , Proc. Natl. Acad. Sci. U.S.A., 82, 3182-3186; Povirk, L. F., and Goldberg, I. H. (1986) 'Base substitution mutations induced in the cI gene of lambda phage by neocarzinostatin chromophore: correlation with depyrimidination hot spots at the seguence AGC', Nucleic Acids Res., 14, 1417-1426; Kappen, L. S., Chen, C. -G., and Goldberg, I. H. (1988) 'Atypical abasic sites generated by neocarzinostatin at sequence-specific cytidylate residues in oligodeoxynucleotides', Biochemistry, 27, 4331-4340.

[11]   Saito, I., Kawabata, H., Fujiwara, T., Sugiyama, H., and Matsuura, T. (1989) 'A novel ribose C-4' hydroxylation pathway in neocarzinostatin-mediated degradation of oligonucleotides', J. Am. Chem. Soc., 111, 8302-8303.

QUINOLONE-DNA INTERACTION: HOW A SMALL DRUG MOLECULE ACQUIRES HIGH DNA BINDING AFFINITY AND SPECIFICITY

L. L. SHEN, M. G. BURES[*], D. T. W. CHU AND J. J. PLATTNER
Anti-infective Research Division
Computer-Assisted Molecular Design[*]
Abbott Laboratories
Abbott Park, Illinois 60064

ABSTRACT. Quinolones are a group of low molecular-weight, synthetic and extremely potent antibacterial agents. The functional target of these drugs have been shown to be the bacteria-specific DNA gyrase, a type II DNA topoisomerase. Quinolones may be classified as DNA-targeted drugs in a broader sense, since evidence has been provided that the direct binding target of the drug is the DNA substrate, but not the enzyme. There are two levels of drug binding specificity to pure DNA: (i) at the structure level it binds preferentially to the single-stranded DNA rather than the double-stranded, (ii) at the unpaired nucleotide level the drug binds better to guanine than other bases. At enzyme inhibition level, however, the binding specificity is controlled by the bound enzyme. Although double-stranded relaxed DNA substrate possesses no specific binding site for the drug, the binding of DNA gyrase to the DNA substrate in the presence of ATP has been shown to induce a saturable drug binding site, presumably a partially denatured "bubble" created during the intermediate gate-opening step. These small drug molecules acquire high binding affinity to this specific DNA site through a cooperative binding mechanism that involves self-associations among the drugs as shown by our model.

1. Introduction

Quinolones are a group of synthetic antibacterial agents. These drugs have recently become available for clinical uses and are believed to have a bright future due to their exceedingly high potency, low frequency of resistance development and rapid bactericidal effect (reviewed by Hooper and Wolfson, 1985; Hooper, 1986; Neu, 1987; Fernandes, 1988; Wolfson and Hooper, 1989).

The first member of the quinolone family of antibacterial agents, nalidixic acid, was synthesized more than a decade ago (Lesher et al., 1962). This compound has limited clinical value mainly because of its low potency and narrow antibacterial spectrum, and was used primarily for treating uncomplicated urinary tract infections. The discovery of DNA gyrase (a type II DNA topoisomerase) from bacteria (Gellert et al., 1976; Liu and Wang, 1978) and the subsequent finding that the enzyme is the target of nalidixic acid analogues greatly stimulated the development of the fluoroquinolones (Fig. 1). The new quinolones are fifty- to several hundred-fold more potent than nalidixic acid, have a broader antibacterial spectrum, and demonstrate much improved pharma-cokinetic properties.

Quinolones are specific inhibitors of DNA gyrase. This bacteria-specific enzyme is a type II DNA topoisomerase which negatively supercoils DNA in a reaction driven by ATP hydrolysis (Cozzarelli, 1980; Gellert 1981; Wang, 1985). The enzyme is a tetramer composed of two

B. Pullman and J. Jortner (eds.), Molecular Basis of Specificity in Nucleic Acid-Drug Interactions, 495–512.
© 1990 Kluwer Academic Publishers. Printed in the Netherlands.

A-subunits, the 105 kilodalton proteins encoded by the *gyrA* (formerly *nalA*) gene, and two B-subunits, the 95 kilodalton proteins encoded by the *gyrB* (formerly *cou*) gene. DNA gyrase catalyzes the supercoiling reaction via a reduction of the DNA linking number of covalently closed circular DNA by a concerted strand breaking-passing-resealing process (Brown and Cozzarelli, 1979). The end result of the enzyme reactions is the formation of an underwound DNA molecule which spontaneously adopts a negatively supercoiled form when the DNA strains resume the most energy-stable double-stranded configuration. Quinolones inhibit the supercoiling reaction by uncoupling the concerted DNA breakage and reunion process, and lead to the formation of a cleavable complex. On addition of a protein denaturant the complex yields a double-stranded break in DNA with subunit A attached covalently to the revealed 5'-ends (Gellert et al., 1977; Sugino et al., 1977). The 4-basepair staggered cleavage site induced by quinolones is in the center of the region bound by DNA gyrase (Morrison and Cozzarelli, 1981). This observation provided key evidence that the target of the drug is the subunit A of DNA gyrase. The notion that *gyrA* is the exclusive target of quinolones is being challenged by the studies on quinolone resistance mutants that give somewhat contradicting results. Although the majority of the results indicated that mutations leading to high-level drug resistance are in *gyrA* (Higgins et al. 1978), some mutations are also found in the *gyrB* (Yamagishi et al., 1981, 1986). We started our investigation on the mechanism of quinolone inhibition by having ^{3}H-norfloxacin synthesized, and found that ^{3}H-norfloxacin does not bind to DNA gyrase, but instead binds to pure DNA (Shen and Pernet, 1985).

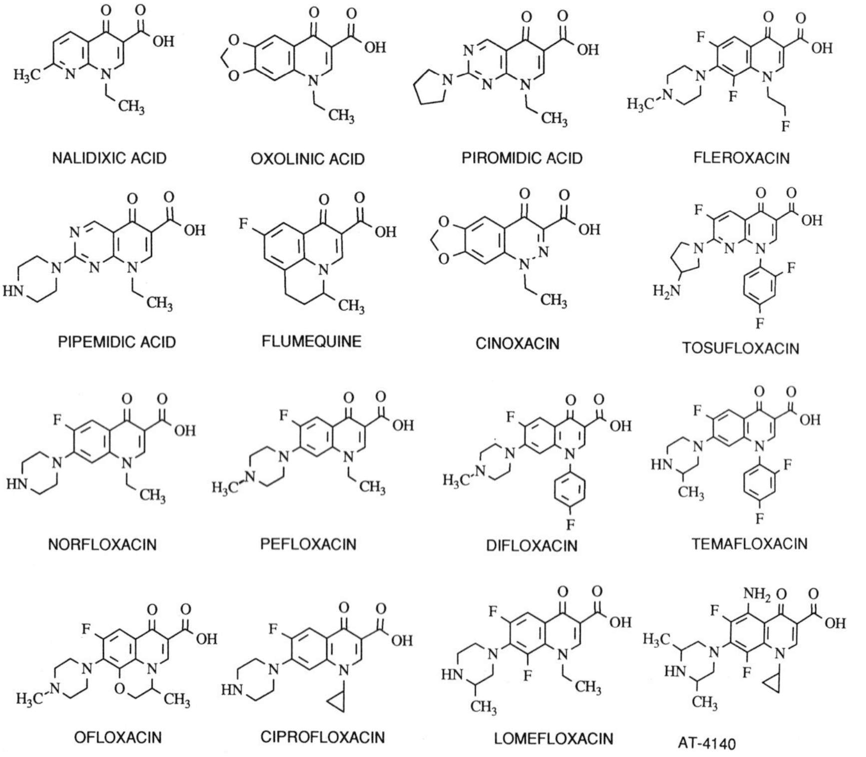

Figure 1. Structures of some major quinolones

The phenomenon of quinolone binding to the negatively charged DNA molecule was somewhat surprising. A few controversial observations were reported showing that quinolones do not bind to DNA (LeGoffic 1985; Palu' et al. 1988), though most of these negative results are accountable after a detailed examination (Shen, 1989). In addition, our finding was supported by the results of Tonaletti and Pedrini (1988) who demonstrated that quinolones in the presence of magnesium ions unwind DNA. Moreover, some direct effects of certain selected quinolones on DNA were also observed, including the quenching of the drug's intrinsic fluorescence, fluorine-19 NMR spectroscopic changes, as well as direct DNA cleavage effects (Shen et al. 1990).

Our finding on the binding of quinolone to DNA has led us to a more detailed investigation on the role of drug-DNA interaction in the enzyme inhibition, and a model was proposed (Shen et al., 1989c). The model offers an explanation of the puzzling observations with drug-resistance mutants, and provides a general guideline for the chemical synthesis of more potent compounds. In this paper we will review the mode of action of quinolones in terms of the proposed model, and in more detail we will present a theoretical consideration using the model to interpret the large activity difference between the two enantiomers of a well-known quinolone, ofloxacin.

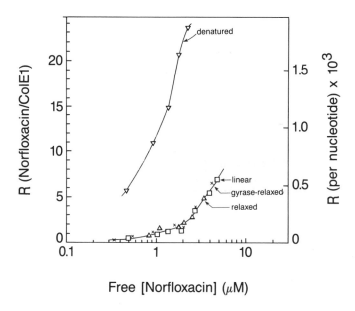

Figure 2. Binding of ^{3}H-norfloxacin to linear (square), relaxed (triangle), complex formed between gyrase and relaxed DNA (cross), and heat-denatured single-stranded ColE1 DNA (inverse triangle). Reaction mixtures contained 3.5 pmol of DNA. Denatured DNA was prepared by heating the linearized DNA to 98°C for 20 min and then rapidly cooling it to 25°C. The thermal denaturation process gave 25% increase in absorbance at 260 nm. The gyrase-relaxed DNA complex was prepared by incubating 3.5 pmol of relaxed DNA with 10 pmol of gyrase holoenzyme in the absence of ATP in 100 μl of binding buffer for 60 min before addition of ^{3}H-norfloxacin. R = molar binding ratio. Data retrieved from Shen and Pernet (1985).

2. Drug Binding Specificity

We have demonstrated that drug-DNA binding affinity is proportional to DNA gyrase inhibition potency, and thus concluded that binding affinity may be an important determinant of the drug potency (Shen and Pernet, 1985). In order to obtain a better understanding of the role of drug-DNA interaction in DNA gyrase inhibition and the molecular mode of the drug binding, we have extensively investigated the binding specificity of quinolone to pure DNA (Shen et al., 1989b). The binding affinity and specificity may be viewed at two levels. The first level of binding specificity is controlled by the topological structure of DNA. As our earlier studies indicated, ^{3}H-norfloxacin binds preferentially to single-stranded DNA rather than the double-stranded (Shen and Pernet 1985). This fact is illustrated in Fig. 2 which shows the binding of ^{3}H-norfloxacin to ColE1 DNA of different structural forms in a drug concentration range near its supercoiling inhibition constant. It is clearly demonstrated that the binding affinity is greatly favored by the heat-denatured single-stranded DNA rather than the double-stranded DNA of either linear or covalently closed relaxed form. The amount of the drug binding to the complex formed between relaxed DNA substrate and DNA gyrase in the absence of ATP, is indistinguishable from that to the DNA alone, again demonstrating the absence of the drug binding to DNA gyrase at this drug concentration range.

We further studied the possible base binding preference to three native double-stranded DNA of varying GC contents, although the amount of the binding was expected to be low. The results showed that there was no correlation between the amount of drug binding and the GC content of the native DNA (Shen et al., 1989b). These results suggest that when DNA strands are intact, the binding is limited and no trace of base preference can be demonstrated.

In case DNA strands are separated, the amount of drug binding increases and the second level of binding specificity, i.e. the base binding preference, is revealed. The emerging base preference on single-stranded DNA is illustrated in Fig. 3, which shows a sequence of preference to poly(dG), poly(dA), poly(dT) and poly(dC) in decreasing order, while the binding to the double-stranded poly(dA)-poly(dT) is virtually non-detectable. The binding to poly(dG) is distinctively greater than that to the other three polydeoxyribonucleotides. One unique structural feature of the guanine base is that it has two common hydrogen-bond donors while the other bases have only one. These results thus suggest that hydrogen bonds are involved with the drug binding, possibly between the hydrogen-bond donor groups on the base and the 4-keto and/or the 3-carboxyl group on the quinolone ring. Also shown in Fig. 3A is the drug binding to poly(dI) which shows a level of drug binding roughly equal to 20% of that to poly(dG). The structural difference between guanosine and inosine is that the former has an extra amino group on the pyrimidine ring (Fig. 3A insert) which is an important hydrogen-bond donor while pairing complementary strands. The result again suggests that the drug binds to unpaired DNA bases via hydrogen-bonding. Similar results supporting this conclusion were also obtained with synthetic polyribonucleotides (Fig. 3B). These results suggest that quinolones bind preferentially to an unpaired guanine base in a single-stranded DNA region, possibly through hydrogen bonds.

3. Two Types of Binding

We have observed various forms of drug binding with different forms of DNA throughout a wide range of drug concentration. These forms of binding, governed by the two levels of binding specificity described above, may be classified into the following two types according to their binding saturation pattern, binding affinity relative to the biological activity and the degree of cooperativity.

3.1 THE NON-SPECIFIC BINDING

The binding of ³H-norfloxacin to single-stranded heat-denatured DNA and to relaxed forms of
DNA illustrated in Fig. 2 are typical examples of this type of binding, since they lack binding
cooperativity and demonstrate no trace of binding saturation. The binding of ³H-norfloxacin to
supercoiled ColE1 DNA at high concentrations (up to the drug's solubility of 1 mM) also fits to
this type of binding (Shen et al. 1989b).

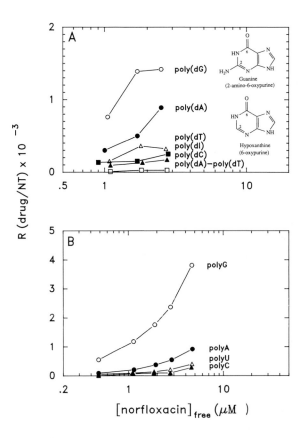

Figure 3. Binding of ³H-norfloxacin to synthetic model homopolymers. Panel A, membrane
filtration method was used to measure the drug binding to polydeoxyribonucleotides: poly(dG)
(open circle), poly(dA) (closed circle), poly(dT) (open triangle), poly(dI) (closed square), poly(dC)
(closed triangle) and poly(dA)-poly(dT) duplex (open square) at three drug concentrations. Panel
B, same membrane filtration method was used to determine the drug binding to polyribonucleotides:
poly(G) (open circle), poly(A) (closed circle), poly(U) (open triangle) and poly(C) (closed triangle)
at five drug concentrations. Amount of these synthetic homopolymers used was 20 μg in 400 μl.
Inserts in panel A show chemical structures of the bases making up poly(dG) and poly(dI), i.e.
guanine and hypoxanthine, respectively. Data retrieved from Shen et al. (1989b)

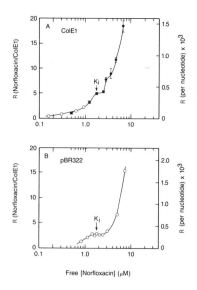

Figure 4. Binding of norfloxacin to supercoiled plasmid DNA. Binding mixtures contained 4.7 pmol of ColE1 (A) or pBR322 (B) DNA and the indicated amounts of ^{3}H-norfloxacin. r denotes the molar binding ratio. Vertical bars represent S.D.s. Different symbols indicate results obtained from different experiments. The value of supercoiling K_i (inhibition constant against *E. coli* DNA gyrase) are marked with arrows for comparison. Reprinted from Shen and Pernet (1985).

3.2 THE SPECIFIC BINDING

The binding of ^{3}H-norfloxacin to supercoiled ColE1 DNA (as illustrated in Fig. 4) represents a specific type of drug binding based on the following three characteristics: (i) the binding is saturable and takes place at the drug's supercoiling inhibitory concentration, (ii) the binding affinities of some selected quinolones to this DNA site are directly proportional to their supercoiling inhibition constants (Shen and Pernet 1985) and (iii) the binding to this saturable site is highly cooperative (Shen et al. 1989b)

Such a specific binding pattern was observed only with the supercoiled form of covalently closed circular (ccc) DNA (such as ColE1 DNA), but not with the relaxed or the linear form. It is evident that there exists in the supercoiled DNA a unique structural feature which serves as the drug's cooperative binding site. We know that the drug binds preferentially to single-stranded DNA. The question is therefore: does supercoiled DNA exhibit such a preferred structural feature? The answer is yes. It is known that supercoiled DNA is the stabilized form of a underwound ccc DNA with reduced linking numbers (Wang 1982). It is possible that a remnant single-stranded "bubble" may be retained in the supercoil and it is this denatured DNA pocket that the drug binds securely. Another possible candidate is the cruciform (or hairpin) structure that forms in the region with palindromic sequence, and this structure is known to be promoted by negative supercoiling (Gellert et al. 1978; Lilly 1980; Mizuuchi et al. 1982). The cruciform is characterized with single-stranded DNA bubbles that have been demonstrated by the cleavage with S1 nuclease which has specificity to single-stranded DNA (Panayotatos and Wells 1981; Singleton and Wells 1982). Further studies with DNA having such structural features are needed to verify these possibilities.

The next key question that needed to be answered before we could propose an inhibition model was: how the specific binding observed with supercoiled DNA, the product of the catalytic reaction, can be extrapolated to the drug inhibition of DNA gyrase that requires relaxed DNA as the substrate. The answer to this crucial question was obtained from our studies on the binding to the enzyme-DNA complexes (Shen et al., 1989a). We found that the binding of the drug to relaxed forms of DNA may be induced by the participation of DNA gyrase, and that the amount of the binding is far more than the total amount of the binding to the DNA and to the enzyme when the experiments were performed separately. The possibility that the increased drug binding was due to the conversion of the substrate DNA to the supercoiled form was ruled out by the use of a non-hydrolyzable triphosphate nucleotide in the binding mixture. The induced binding to the gyrase-DNA complex showed a saturable phase that highly resembles the binding pattern to the supercoiled DNA in terms of the amount of the binding and the binding cooperativity (Shen et al., 1989b). It was interesting to see that a similar cooperative drug binding was induced by the enzyme without the need of a nucleotide energy source when linearized DNA was used. This is presumably due to the open DNA ends that do not restrict the conformational change or DNA unwinding during the enzyme-wrapping around and gate-opening step following the DNA cleavage process. Moreover, there was a close relation between the appearance of such a characteristic binding and the drug-induced DNA breakage (Shen et al., 1989a) which is considered to be the central event responsible for the killing effect of the drug on bacterial cells (Kreuzer and Cozzarelli, 1979). These results suggested that the binding of DNA gyrase to the DNA substrate creates a site that allows the drug to bind in a cooperative manner.

4. Model of Quinolone Inhibition of DNA Gyrase

From the above observations, a quinolone-DNA cooperative binding model for the inhibition of DNA gyrase has been proposed (Shen 1986; Shen et al. 1986; 1989c). All experimental evidence favors the notion that the bound enzyme induces a drug binding site on the relaxed DNA substrate. We propose that the binding site is formed during the gate-opening step which requires the binding of ATP. The separated short DNA strands between the 4-basepair staggered cuts form a simulated denatured DNA bubble that is an ideal site for the drug to bind (Fig. 5). As the high binding affinity and cooperativity is concerned, we envision this to occur as follows: drug molecules acquire high binding affinity through a cooperative binding mechanism achieved via self-association of the drug molecules. Two types of interactions, supported by observations in the structure of nalidixic acid crystal, are feasible: the π-π stacking between the quinolone rings and the tail-to-tail hydrophobic interactions between the N1 substitution groups (Fig. 6). Such interactions result in the formation of a complex with multiple sets of hydrogen bond acceptors in a consolidated unit, thus extending the binding beyond a uni-dimensional domain, i.e. the assembled drug molecules can act together to occupy a site on DNA with binding groups distributed in a multi-dimensional space.

This working model has two unique features: (i) the unique configuration of a drug binding site on DNA, and (ii) the unique ability of the drug molecules to occupy such a site. Both aspects are

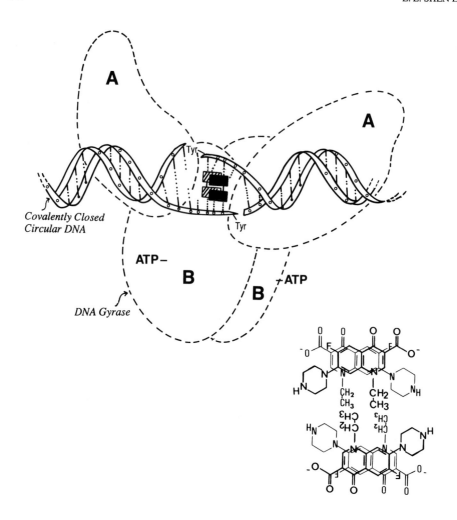

Figure 5. Model of the proposed quinolone-DNA cooperative binding in the inhibition of DNA gyrase. Quinolone molecules (filled and slashed rectangles that represent the mode of drug self-association shown at lower-right) bind to a gyrase-induced DNA site during the intermediate gate-opening step of DNA supercoiling process via hydrogen-bonds (indicated by dotted lines) to the unpaired bases. Gyrase A-subunits form covalent bonds between tyrosine 122 and the 5'-end of the DNA chain (Horowitz and Wang, 1987), and the subsequent opening of the DNA chains along the four-basepair staggered cuts results in a locally denatured DNA bubble which is an ideal site for the drug to bind. When relaxed DNA substrate (represented by the short DNA segment in the diagram) is used, ATP is required for the induction of the drug binding site. Dashed curves mimic the shape of the DNA gyrase, a tetramer composed of two A-subunits and two B-subunits, revealed by the electron microscopic image of the *M. Luteus* enzyme (Kirchhausen et al., 1985). Reprinted partly from Shen et al. (1989c).

equally important to drug binding specificity. It is evident that the DNA binding specificity at the enzyme inhibition level is controlled by the binding of the enzyme that creates a favorable DNA site for the drug. Thus, a third level of drug binding specificity is revealed.

This binding model offers an answer to questions concerning quinolone-resistance mutants. The model predicts that any mutations, either in gyrA of in gyrB, which affect the supercoiling process and which alters the configuration of the drug binding pocket on DNA could lead to drug resistance. Such a mutation may be more likely to occur in gyrA since the subunit A is directly involved in creating such a binding site and in stabilizing the drug bound to DNA. Gyrase subunit B is also important to the formation of such a site, since our binding and DNA cleavage experiments have clearly demonstrated the importance of ATP that has its receptor sites located on the subunit B.

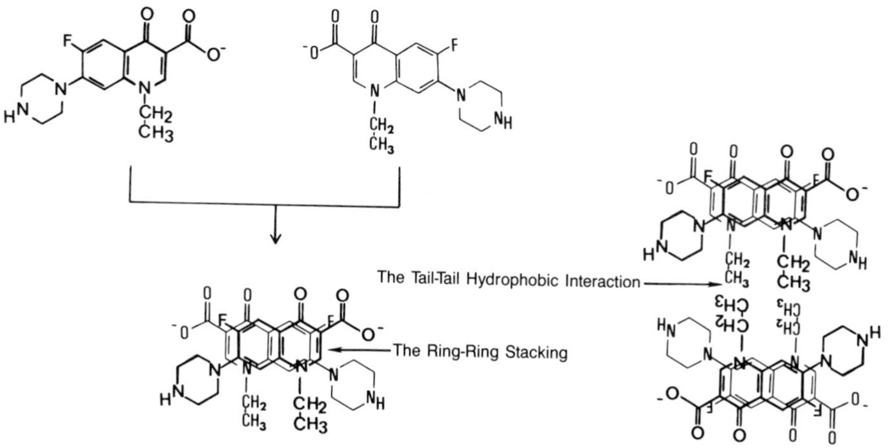

Figure 6. Details illustrating the proposed drug self-association mode at the quinolone binding site which is a single-stranded DNA pocket as shown in Fig. 5. Norfloxacin molecules (drawn separately by thick and thin lines) are used for illustrating the two types of proposed interactions: the π-π ring stacking of the quinolone rings and the tail-to-tail hydrophobic interaction between the N-ethyl groups. Only four drug molecules are shown in the diagram, the number of drug molecules involved may be higher depending on the size and the configuration of the binding site. Also the angle between the two molecules interacted through N1 hydrophobic tails is not necessarily fixed at 180° so as to provide flexibility in bond pairing. The result of such a self-association is the formation of a complex that mimics phospholipid micelles in solution, having a hydrophobic core and hydrophilic groups located outside of the complex.

5. Implications of the Model

The model has provided a guideline to the synthetic efforts for active compounds. Our model suggests that there are three functional domains on the quinolone molecule (Fig. 7). Accordingly, the following implications in drug design relative to the model may be summarized, and these features are consistent with the structure-activity relationship of quinolones (Mitscher et al. 1989, 1990):

(1) Hydrophobic substitutions at N-1 position are essential for activity. Though the strength of these interactions is not expected to be the sole determinant for maximum activity, a minimum level of interaction between these hydrophobic groups is necessary for optimal potency. It is evident from Fig. 1 that substituents at the N-1 positions of these active drug molecules are indeed hydrophobic; namely, ethyl, cyclopropyl, and phenyl groups.

(2) Highly potent compounds are usually very insoluble (Shen et al. 1989c). For increasing the solubility of the final product, pharmaceutical formulation method or a pro-drug approach is recommended.

(3) Extensive substitutions at C-7 are allowed. The function of this domain is less well defined by the proposed model, and it deserves more exploration. We speculate that this may be the area responsible for the drug-DNA interaction; this is the only domain that is spatially possible to accommodate large functional substituents. Indeed, some very potent inhibitors have been synthesized with a variety of large or small substituents at this position.

(4) C-4 keto group is absolutely essential for activity.

(5) C-3 carboxylic group and the C-4 keto group should be co-planar.

(6) Substitutions at C-5 with hydrogen-bond donors or acceptors are allowed. This is consistent with the good potency shown by AT-4140 (Fig. 1).

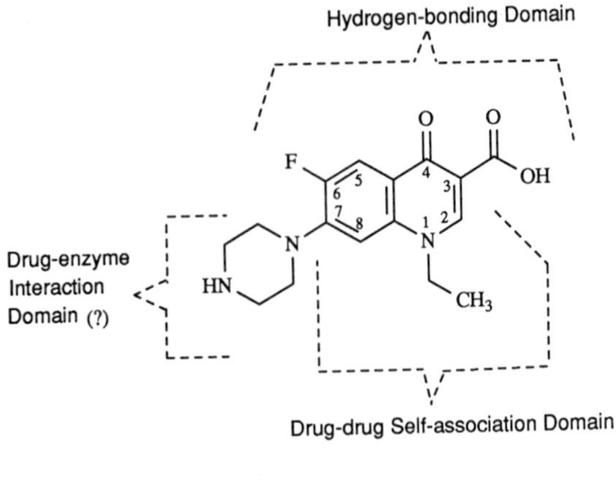

Figure 7. Functional domains of a quinolone antibacterial agent. Norfloxacin molecule is used as an example.

6. The Drug-stacking Model Explains the Large Activity Difference between Ofloxacin Enantiomers

Recently there has been an increasing interest in the investigation and synthesis of enantiomers of effective drugs that exert their biological action toward specific receptors or enzymes. This type of investigation has provided important information leading to the development of compounds with enhanced selectivity, higher potency, and less side effects (Ariens, 1986).

For quinolone antibacterials, large differential activities were observed between the enantiomers having chiral substituents attached to N-1 and to C-7 (Mitscher et al., 1989). As shown in Fig. 7, C-7 substituents are believed to interact with the enzyme sites. We may thus speculate that any difference in activity could result from the interaction between the chiral group attached to C-7 and a highly asymmetric enzyme site. Based upon our cooperative binding model, the same argument, however, cannot be applied to enantiomers with chiral substituents at N-1, since the hydrophobic substituents at this position are believed to involve only the drug self-assembly process at the DNA site. Yet, using ofloxacin enantiomers as an example, our proposed stacking model can also be used to explain the large activity difference between the enantiomers with chiral groups attached to N-1.

Ofloxacin was synthesized in the early 1980s at Daiichi Seiyaku Co.. The drug has an antibacterial potency roughly equal to that of norfloxacin against *E. coli* (Une et al. 1988) and has an supercoiling inhibition activity slightly less than that of norfloxacin (Shen, et al. 1989c). The compound has a unique tricyclic structure with a methyl group at C-3 position in the oxazine ring, thus providing an asymmetric center at this position (Fig. 8). Studies with selected laboratory bacterial strains revealed that the [S-(-)-ofloxacin] isomer is approximately 8 to 128 times more potent than the [R-(+)-ofloxacin (Hayakawa et al. 1986). Evidently, the sterochemical effect is on the enzyme rather than other factors such as the drug transport mechanism, since the S-isomer has also shown about 30 and 3 times more potent than the R-isomer for inhibiting DNA gyrases isolated from *E. coli* and from *M. leuteus*, respectively.

That the stereochemical dependence of an enzyme inhibition results from a small non-functional methyl group is fascinating. In the following discussion we will describe our attempts to explain the large activity difference between the two ofloxacin enantiomers using our proposed ring-stacking model.

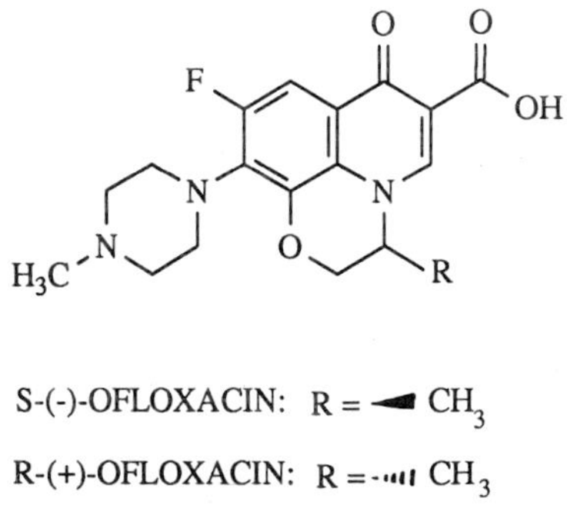

S-(-)-OFLOXACIN: R = ◄ CH₃

R-(+)-OFLOXACIN: R = ◄◄◄ CH₃

Figure 8. Structure of Ofloxacin Enantiomers

As our model indicates, the π-π stacking mechanism shown in Fig. 6 is an important determinant for the drug binding affinity to the DNA site. The ability of the molecule to stack properly to fit the binding site is therefore crucial in determining the inhibitory potency. We have used Abbott Laboratories' interactive molecular graphics system, SWAMI (Martin et al., 1988) to investigate several possible modes of stacking of ofloxacin enantiomers, as well as to examine different conformations of ofloxacin. This work demonstrated that the enantiomers and their conformers can not stack, with equivalent intermolecular interactions, in the same fashion.

Before presenting the results, we first need to evaluate the conformational flexibility of the ofloxacin molecule. As shown in Fig. 9 using (S)-ofloxacin as an example, the six-membered oxazine ring is not planar. Instead, the ring is puckered in two different ways, namely the *endo* (front) pucker and *exo* (back) pucker conformation (defined in Fig. 9). Molecular orbital calcu-lations using MOPAC with the MNDO force field (Stewart, 1989) indicate that the two ring-puckered forms are of approximately equivalent energy. The ring puckering, however, affects the projection of the methyl group from the plane of the quinolone ring system. In the *endo*-puckered form (Fig. 9, top), the methyl group is in an equatorial position and thus its protrusion from the plane of the quinolone ring in not pronounced. In the *exo*-puckered form (Fig. 9, bottom), however, the methyl group occupies a nearly axial position, and therefore is oriented almost perpendicular to the plane of the quinolone system. The steric relationship of the puckered forms to the quinolone ring will be seen to affect the manner of stacking of both enantiomers of ofloxacin.

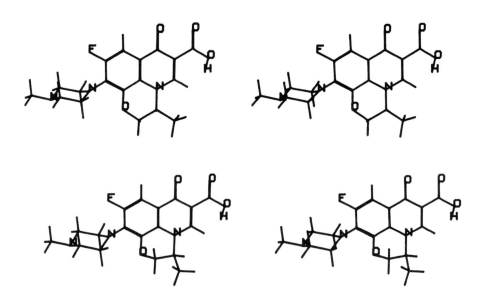

Figure 9. Stereo view of the conformation of (S)-ofloxacin with different oxazine ring puckers: the *endo*-form (top) and the *exo*-form (bottom). The terms, *endo* and *exo*, refer to the projection in space of the methyl group on the oxazine ring as a result of ring-puckering.

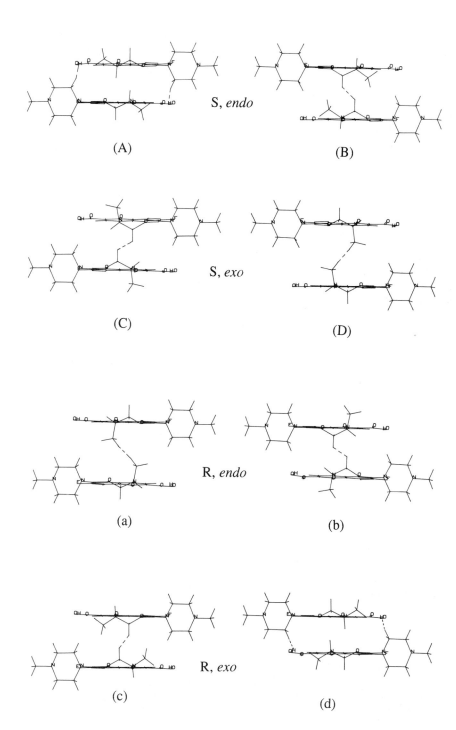

S, endo

(A)

(B)

S, exo

(C)

(D)

R, endo

(a)

(b)

R, exo

(c)

(d)

Figure 10 (preceding page).   The eight stacking orientations of ofloxacin enantiomers (S)-ofloxacin, panels A-D; (R)-ofloxacin, panels a-d and their conformers investigated in this model. The stacked molecular pairs were viewed from 4-keto to N-1 direction. For each enantiomer, two conformers (*endo, exo*) were studied, differing in the pucker of the oxazine ring (see Fig. 9). Then, for each puckered form, two orientations were possible: the rings stacked with the puckers facing outside the complex, and stacked with the puckers facing inside the complex. The close contacts between the two stacked rings determine the distance between the rings, which, in turn, relates to the π-π stacking energy present. Thus, the orientations that place both the pucker and methyl group outside the complex result in the shortest ring-ring distance, about 3.5 Å for orientations (A) and (D). The largest ring-ring distance, about 5.4 Å for orientations (D) and (a), was observed when the methyl groups were facing inside and the puckers were facing outside. A key finding of this exercise is that the most favorable orientation of (S)-ofloxacin (panel A) is stacked in a different manner than the best orientation of (R)-ofloxacin (panel d). Specifically, the S-isomer prefers to stack with the key hydrogen bond acceptors (3-carboxyl and 4-keto) located to the "northwest" and "southeast", while the best R-isomer orientation has these functional groups located to the "southwest" and "northeast". This difference in orientation could result in a higher binding affinity to DNA for one stacked orientation relative to the other, and is offered as a rationale for the greater activity of the (S)-isomer of ofloxacin. Geometric data for the different stacking orientations is reported in Table 1.

Molecular graphics was used to analyze the possible stacking orientations for the enantiomers of ofloxacin. As shown in Fig. 10, for each enantiomer, four orientations were investigated: (A) the *endo* form stacked with the ring puckers facing outside, (B) the *endo* form with the ring puckers facing inside, (C) the *exo* form with the ring puckers facing inside, and (C) the *exo* form with the ring puckers facing outside. Analogous stacking orientations for (R)-ofloxacin are depicted in Fig. 10, panels a-d. Additional requirements of the model are that in each orientation the rings are stacked nearly directly on top of each other and that the planes of the quinolone rings are essentially parallel. Therefore, the distance between the quinolone ring systems in each pair can be compared directly and used as a measure of the strength of the intermolecular π-π stacking interaction that, along with drug-DNA hydrogen-bonding and tail-to-tail hydrophobic interactions, promotes the cooperative mode of drug binding to DNA (Shen, et al., 1989c). In this regard, two parameters were considered especially important in our modeling of the stacking interaction. These parameters are the distance between the quinolone rings and the distance of any intermolecular close contacts present in each pair of ofloxacin molecules. In this preliminary model, we attempted to stack each pair to a minimum distance, while at the same time not introducing more than two intermolecular close contacts of less than ca. 1.6 Å. Currently we are working to optimize the distance between the ring systems and the conformation of ofloxacin necessary to achieve maximum interaction without introducing prohibitive steric hindrance. However, this empirical modeling used in this paper is instructional in evaluating the stacking mode and binding affinity of ofloxacin enantiomers to DNA, as discussed below.

As indicated in Fig. 10, the most favorable stacking orientation for either enantiomer is the pair with both the ring pucker and methyl group located on the outside of the stacked complex, i.e., Fig. 10A and Fig. 10d for (S) and (R)-ofloxacin, respectively. As expected for both enantiomers, it is this form that allows the closest distance between the quinolone rings. These two stacked pairs give a distance between the two quinolone rings of about 3.5 Å, and the closest intermolecular contact (ca. 1.6 Å) is between the hydrogen of the carboxyl group and a hydrogen on the piperazine ring (see Fig. 10 and Table 1). All other stacking orientations result in a distance between the ofloxacin rings of greater than 3.5 Å, and therefore these pairs would be expected to have less ring-ring stacking interaction. For example, with the oxazine ring puckered inward with either the *endo* (B, b) or *exo* (C, c)-forms, the distance between the quinolone rings is greater, at about 4.5 Å (see Fig. 10 and Table 1). In these pairs, the intermolecular close contacts are between hydrogens on the methylene groups in the oxazine rings. The least favorable (longest ring-ring distance) stacking orientations are the pairs with the oxazine rings puckered outward and the methyl groups positioned inside the complex (Fig. 10D, a), with a ring-ring distance of about 5.4 Å.

The key information obtained is that the most favorable pair (A) and pair (d) are in a mirror-image configuration, resulting in two molecular complexes with their functional groups (C=O) located in chiral positions. In our proposed cooperative binding model, the carbonyl groups are in the hydrogen-bonding domain interacting with the DNA. Hence, it is clear that the two ofloxacin enantiomers cannot stack in the same way to fit an asymmetric DNA site.

TABLE 1. Geometric data for the examined stacking orientations of ofloxacin enantiomers. For each orientation (see Fig. 10), the distance between the quinolone ring systems is reported, along with a description of the atoms in ofloxacin responsible for the closest intermolecular contact (ca. 1.6 Å).

Orientation (from Fig. 10)	Ring-Ring Distance (Å)	Atoms in Closest Contact (ca. 1.6 Å)
A	3.5	H of carboxyl - H of piperazine
B	4.5	H of oxazine - H of oxazine
C	4.4	H of oxazine - H of oxazine
D	5.4	H of methyl - H of methyl
a	5.4	H of methyl - H of methyl
b	4.5	H of oxazine - H of oxazine
c	4.6	H of oxazine - H of oxazine
d	3.5	H of carboxyl - H of piperazine

7. Concluding remarks

Quinolone antibacterial agents may be classified as DNA-targeted drugs in a broader sense, since the direct binding target of the drug is the DNA substrate, but not the enzyme. The drug binding specificity may be described at three different levels: the nucleotide level, the structure level, and the enzyme inhibition level. The drug shows a binding preference to guanine base rather than the other bases in unpaired condition, but the formation of double helix reduces the quantity of the drug binding and eliminates the binding specificity as well. At enzyme inhibition level, the binding is controlled by the bound enzyme. Although double-stranded relaxed DNA substrate possesses no specific binding site for the drug, the binding of DNA gyrase to the DNA substrate in the presence of ATP induces a saturable drug binding site, presumably a partially denatured DNA pocket created during the intermediate gate-opening step. These small drug molecules acquire high binding affinity to this specific DNA site through a cooperative binding mechanism that involves a tail-to-tail hydrophobic interaction and ring-ring stacking among the drugs as shown by our model. The mode of drug stacking may be used to elucidate the large activity difference of ofloxacin enantiomers, showing how an unimportant chiral group (the methyl group attached to the oxazine ring) affects the assembly of drug bound to an asymmetrical DNA site.

8. References

Ariens, E. J. (1986) 'Stereochemistry: a source of problems in medicinal chemistry', Med. Res. Rev. 6, 451-466.

Brown, P. O. & Cozzarelli, N. R. (1979) 'A sign inversion mechanism for enzymatic supercoiling of DNA', Science 206, 1081-1083.

Cozzarelli N. R. (1980) 'DNA gyrase and the supercoiling of DNA', Science 207, 953-960.

Fernandes, P. B. (1988) 'Mode of action, and in vitro and in vivo activities of the fluoroquinolones', J. Clin. Pharmacol. 28, 156-168.

Gellert, M. (1981) 'DNA topoisomerases', Ann. Rev. Biochem. 50, 879-910.

Gellert, M., Mizuuchi, K., O'Dea, M. H. & Nash, H. A. (1976) 'DNA gyrase: an enzyme that introduces superhelical turns into DNA', Proc. Natl. Acad. Sci. USA 73, 3872-3876.

Gellert, M., Mizuuchi, K., O'Dea, M. H., Itoh, T. and Tomizawa, J. (1977) 'Nalidixic acid resistance: a second genetic character involved in DNA gyrase activity', Proc. Natl. Acad. Sci. USA 74, 4772-4776.

Gellert, M., Mizuuchi, K., O'Dea, M. H., Ohmori, H. & Tomizawa, J. (1978) 'DNA gyrase and DNA supercoiling' Cold Spring Harbor Symp. Quant. Biol. 43, 35-40.

Hayakawa, I., Atarashi, S., Yokohama, S., Imamura, M., Sakano, K.-I. and Furukawa, M. (1986) 'Synthesis and antibacterial activities of optically active ofloxacin', Antimicrob. Agents Chemother. 29, 163-164.

Higgins, N. P., Peebles, C. L., Sugino, A., & Cozzarelli, N. R. (1978) 'Purification of subunits of *Escherichia coli* DNA gyrase and reconstitution of enzymatic activity', Proc. Natl. Acad. Sci. USA 75, 1773-1777.

Hooper, D. C. & Wolfson, J. S. (1985) 'The fluoroquinolones: pharmacology, clinical uses, and toxicities in humans', Antimicrob. Agents Chemother. 28, 716- 721.

Hooper, D. C. (1986) 'New quinolone antibacterial agents' in: L. Leive (ed.), Microbiology-1986, Amer. Soc. of Microbiol., Washington, D.C. pp 217-230.

Horowits, D. S., And Wang, J. C. (1987) 'Mapping the active site tyrosine of *Escherichia coli* DNA gyrase' J. Biol. Chem. 262, 5339-5344.

Kirchhausen, T., Wang, J. C. & Harrison, S. C. (1985) 'DNA gyrase and its complexes with DNA: direct observation by electron microscopy', Cell 41, 933-943.

Kreuzer, K. N., And Cozzarelli, N. R. (1979) '*Escherichia coli* mutants thermosensitive for deoxyribonucleic acid gyrase subunit A: effect on deoxyribonucleic acid replication, transcription, and bacteriophage growth', J. Bacteriol. 140, 424-435.

Le Goffic F (1985) Les quinolones, mecanisme d'action. In: Pocidalo J. J., Vachon F and Regnier B (eds) Les nouvelles quinolones. Editiones Arnette, Paris pp. 15-23.

Lesher, G. Y., Froelich, E. J., Gruett, M. D., Bailey, J. H. & Brundage, R. P. (1962) '1,8-Naph-thyridine derivatives. a new class of chemotherapeutic agents', J. Med. Chem. 5, 1063-1065.

Lilley, D. M. J. (1980) 'The inverted repeat as a recognizable structural feature in supercoiled DNA molecules', Proc. Natl. Acad. Sci. USA 77, 6468- 6472.

Liu, L. F. And Wang, J. C. (1978) 'Micrococcus luteus DNA gyrase: active components and a model for its supercoiling of DNA', Proc. Natl. Acad. Sci. USA 75, 2098-2102.

Martin, Y. C., Danaher, E. B., May, C. S., And Weininger, D. (1988) 'MENTHOR, a database system for the storage and retrieval of three-dimensional molecular structures and associated data searchable by substructural, biologic, physical, or geometric properties', J. Computer-Aided Molecular Design 2, 15-29.

Mitscher, L. A., Zavod, R. M., And Sharma, P. N. (1989) 'Structure-activity relationships of the newer quinolone antibacterial agents', in: P. B. Fernandes (ed.), International Telesymposium on Quinolones, J. R. Prous Science Publishers, Barcelona, Spain, pp. 3-20.

Mitscher, L. A., Sharma, P. N., And Zavod, R. M. (1989) 'The influence of the optical isomerism on the biological properties of quinolone antibacterials', in: P. B. Fernandes (ed.), Interna-tional Telesymposium on Quinolones, J. R. Prous Science Publishers, Barcelona, Spain, pp. 73-83.

Mitscher, L. A., Devasthale, P. V., And Zavod, R. M. (1990) 'Structure-activity relationships of fluoro-4-quinolones', in G. Crumplin (ed.), The 4-quinolones: antibacterial in vitro, Springer-Verlag London Limited, in press

Mizuuchi, K., Mizuuchi, M. & Gellert, M. (1982) 'Cruciform structures in palindromic DNA are favored by DNA supercoiling', J. Mol. Biol. 156, 229-243.

Morrison, A. And Cozzarelli, N. R. (1981) 'Contacts between DNA gyrase and its binding site on DNA: features of symmetry revealed by protection from nucleases', Proc. Nat. Acad. Sci. U.S.A 78, 1416-1420

Neu, H. C. (1987) 'Quinolones revisited: where are we?', Antimicrob. Newsletter 4, 9-14.

Palu' G, Valisena S, Peracchi M, Palumbo M (1988) 'Do quinolones bind to DNA?', Biochem Pharmacol 37,1887-1888.

Panayotatos, N. & Wells, R. D. (1981) 'Cruciform structures in supercoiled DNA', Nature 289, 466-470.

Shen L. L., And Pernet A. G. (1985) 'Mechanism of inhibition of DNA gyrase by analogues of nalidixic acid: the target of the drugs is DNA', Proc. Natl. Acad. Sci. USA 82, 307-311.

Shen, L. L. (1986) 'Quinolone antibacterials: structure, activity and mechanism of inhibition of DNA gyrase' (abstract No. 47) Cold Spring Harbor Meetings on the Biological Effect of DNA Topology 1986.

Shen, L. L., Baranowski, J., And Wai T. (1986) 'Mechanism of inhibition of DNA gyrase by quinolone antibacterials: a cooperative drug-DNA binding model' J. Cellular Biochemistry 1986; Suppl. 10B: abstract of UCLA Symposium on DNA replication and recombination, Park City, Utah

Shen L. L. (1989) 'A reply: "do quinolones bind to DNA?" - yes', Biochem. Pharmacol. 38, 2042-2044

Shen L. L., Kohlbrenner, W. E., Weigl, D., And Baranowski, J. (1989a) 'Mechanism of quinolone inhibition of DNA gyrase. Appearance of unique norfloxacin binding sites in enzyme-DNA complexes', J. Biol. Chem. 264, 2973-2978

Shen, L. L., Baranowski, J., And Pernet A. G. (1989b) 'Mechanism of inhibition of DNA gyrase by quinolone antibacterials. Specificity and cooperativity of drug binding to DNA', Biochemistry 28, 2879-2885.

Shen, L. L., Mitscher, L. A., Sharma, P. N., O'Donnell, T. J., Chu, D. W. T., Cooper, C. S., Rosen, T., And Pernet A. G. (1989c) 'Mechanism of inhibition of DNA gyrase by quinolone antibacterials. A cooperative drug-DNA binding model', Biochemistry 28, 2886-2894 .

Shen L. L., Baranowski, J., Nuss, M., Tadanier, J., Lee, C., Chu, D. T. W. and Plattner, J. J. (1990) 'Aspects of quinolone-DNA interactions', in G. Crumplin (ed.), The 4-quinolones: antibacterial in vitro, Springer-Verlag London Limited, in press

Stewart, J. J. P. (1989) 'Optimization of Parameters for Semiempirical Methods II. Applications', J. Computational Chem. 10, 221-264.

Singleton, C. K. & Wells, R. D. (1982) 'Relationship between superhelical density and cruciform formation in plasmid pVH51', J. Biol. Chem. 257, 6292-6295.

Sugino A, Peebles CL, Kreuzer KN, Cozzarelli N. R. (1977) 'Mechanism of action of nalidixic acid: purification of E. coli nalA gene product and its relationship to DNA gyrase and a novel nicking-closing enzyme', Proc. Natl. Acad. Sci. USA 74: 4767-4771.

Tornaletti, S. And Pedrini, A. M. (1988) 'Studies on the interaction of 4-quinolones with DNA by DNA unwinding experiments', Biochim, Biophy, Acta 949, 279-287.

Une, T.. Fujimoto, T., Sato, K., And Osada, Y. (1988) 'In vitro activity of DR-3355, an optically active ofloxacin', Antimicrob. Agents Chemother. 32, 1336-1340.

Wang, J. C. (1982) 'DNA topoisomerases' Sci. Am. 247, 94-109.

Wang, J. C. (1985) 'DNA topoisomerases', Ann. Rev. Biochem. 54, 665-697.

Wolfson, J. S. And Hooper, D. C. (1989) 'Fluoroquinolone Antibacterial agents', Clin. Microb. Rev. 2, 378-424.

Yamagishi, J., Furutani, Y., Inoue, S., Ohue, T., Nakamura, S., & Shimizu, M, (1981) 'New nalidixic acid resistance mutations related to deoxyribonucleic acid gyrase activity', J. Bacteriol. 148, 450-458.

Yamagishi, J., Yoshida, H., Yamayoshi, M., & Nakamura, S. (1986) 'Nalidixic acid-resistant mutations of the gyrB gene of Escherichia coli.', Mol. Gen. Genet. 204, 367-373.

MECHANISMS OF DNA SEQUENCE SELECTIVE MODIFICATIONS BY ALKYLATING AGENTS

J. A. HARTLEY

Department of Oncology
University College & Middlesex School of Medicine
91, Riding House Street
London, W1P 8BT, U.K.

ABSTRACT. This paper reviews our current knowledge of the DNA sequence selectivity of alkylating agents such as nitrogen mustards, chloroethylnitrosoureas and other chloroethylating agents, dimethane-sulphonates, and ethylating and methylating agents. In general, reaction at the guanine-N7 position for agents that can produce positively charged alkylating intermediates correlates well with the nearest neighbour base effect on the molecular electrostatic potential at the reaction site, resulting in a preferential reaction at runs of guanines. This suggests that the specific biological effects of such compounds may include preferential reaction at GC-rich genomic locations. In addition, the substituent attached to the reactive group can introduce a distinct sequence preference for reaction. In contrast, agents such as dimethyl sulphate and busulphan alkylate DNA with little overall sequence selectivity. The pattern of guanine-N7 alkylation can be quantitatively, and in some cases qualitatively, altered by increased ionic strength or the presence of cationic DNA affinity binders, and for agents with distinctly different patterns of alkylation *in vitro*, the sequence selectivities appear to be preserved to some extent in cells.

1. Introduction

DNA damage by alkylating agents is important in both the processes of chemical carcinogenesis and cancer chemotherapy. Even though agents such as the nitrogen mustards have been used clinically for over 40 years it is still not clear how such reactive compounds can produce their specific biological effects. Until relatively recently such agents were classed as being poorly sequence selective primarily because early studies were concerned with identifying the base to which the drug bound and the nature of the adduct rather than DNA primary base sequence selectivity. In the last five years, however, adaptation of DNA sequencing technology has enabled a detailed reexamination of these agents. This paper reviews our current knowledge of the sequence selectivity of alkylating agents and our understanding to date of the mechanisms underlying the selectivities observed.

2. Measurement of Sequence Selectivity by Alkylating Agents

An alkylation at the N7-position of guanine renders the guanine imidazole ring susceptible to ring opening at elevated pH (Kohn and Spears, 1967). Treatment with the secondary amine piperidine at 90°C quantitatively converts these modified base sites into strand breaks (Mattes et al., 1986a, figure 1). This is the basis of the dimethyl sulphate guanine-specific DNA sequencing reaction

513

B. Pullman and J. Jortner (eds.), Molecular Basis of Specificity in Nucleic Acid-Drug Interactions, 513–530.
© 1990 *Kluwer Academic Publishers. Printed in the Netherlands.*

(Maxam and Gilbert, 1980). Using DNA of known sequence, labeled at one end of one strand, the lengths of the fragments produced after treatment with an alkylating agent and subsequent piperidine treatment indicate the position of the original guanine-N7 monoalkylation. One nucleotide resolution can be achieved on denaturing polyacrylamide DNA sequencing gels, and provided the original drug treatment is restricted to at most one alkylation per DNA molecule the intensity of the autoradiographic image at each band gives an indication of the extent of guanine-N7 alkylation at that site. In addition, the overall extent of guanine-N7 alkylation for any dose of drug can be determined from the integrated area of the band corresponding to the full length fragment. (Boles and Hogan, 1984). This technique has now been used extensively to measure the sequence selectivity of guanine-N7 alkylation of a wide range of alkylating agents.

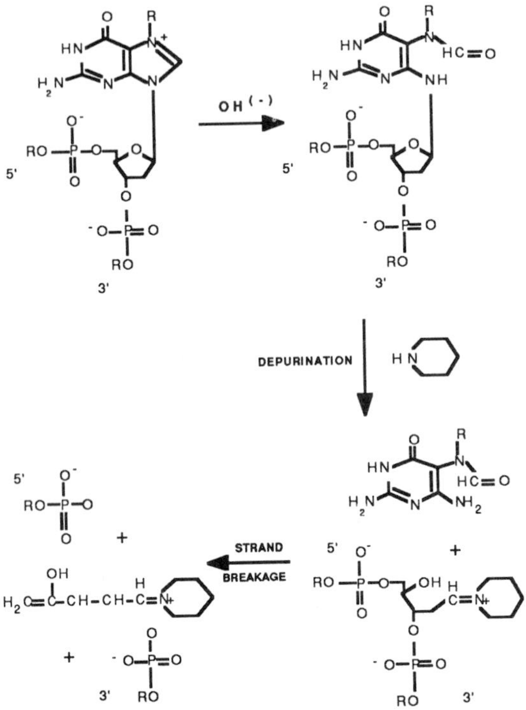

FIGURE 1. Mechanism by which piperidine creates DNA strand breaks at sites of guanine-N7 alkylations.

3. DNA Sequence Selectivity of Guanine-N7 Alkylation

3.1 NITROGEN MUSTARDS

Bis(2-chloroethyl)methylamine (mechlorethamine, figure 2) was the first clinically effective anticancer agent (Gilman and Philips, 1946) and derivatives such as L-phenylalanine mustard (melphalan, L-Pam, figure 2), cyclophosphamide and chlorambucil are still among the most useful clinical agents (Haskel, 1985), despite their apparently non-specific chemical reaction mechanisms. Covalent binding may occur at many nucleophilic sites within nucleic acids and proteins, but DNA is probably the most important target with reaction predominantly at the N7-position of guanine.

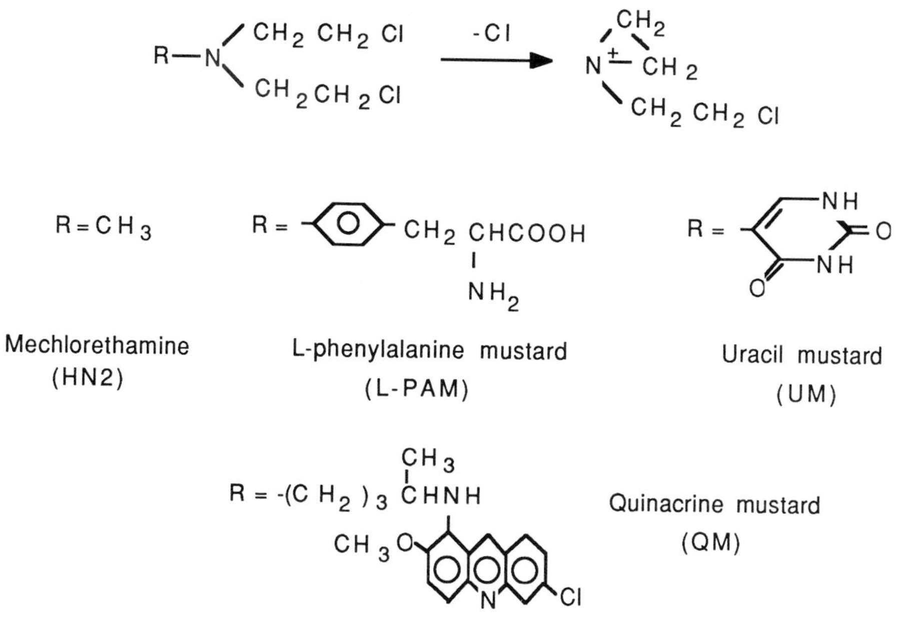

FIGURE 2. Some representative nitrogen mustards.

All the nitrogen mustards examined to date exhibit large variations in guanine-N7 alkylation intensities within a DNA sequence (Mattes et al, 1986b, figure 3). The reaction intensities of several mustards are closely correlated with that of the parent compound mechlorethamine in several different DNA segments. These include L-Pam, phosphoramide mustard, spiromustine, chlorambucil and mustamine (Kohn et al, 1987). With all these compounds preferential reaction is observed at guanines located in runs of guanines. In contrast, isolated guanines are generally alkylated weakly, particularly when followed by a cytosine on the 3' side. The requirement for antitumour activity of two alkylating groups within the mustard molecule suggests that this activity arises from the formation of crosslinks. DNA *inter*strand crosslinks and DNA-protein crosslinks have been observed in intact cells (Ewig and Kohn, 1977). The selective reaction of most mustards for guanines in runs of guanines would favour the formation of *intra*strand crosslinks, whereas reaction is weak at 5'-GC-3' sites which are the sites of potential *inter*strand crosslinking, although this has recently been disputed (Ojwang et al, 1989).

Two compounds, uracil and quinacrine mustard (UM and QM, figure 2) show a distinctly different alkylation pattern from other mustards (Mattes et al, 1986b, figure 3) and clearly

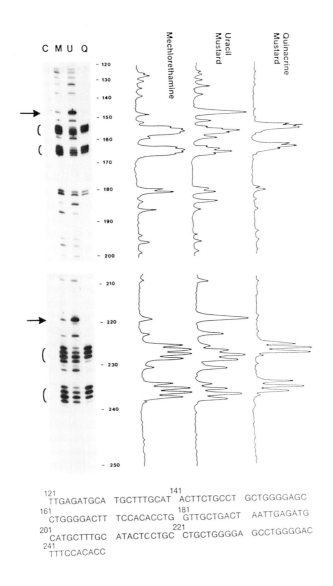

FIGURE 3.   Sites of guanine-N7 alkylation produced in a fragment of SV40 DNA following treatment with mechlorethamine (M), uracil mustard (U), or quinacrine mustard (Q).   C is control unalkylated DNA.   The dose of drug was chosen to give approximately the same level of overall alkylation in each case.   Brackets indicate the runs of contiguous guanines and the arrows sites of preferred alkylation by uracil mustard.   The corresponding densitometric scans are shown alligned with the gel, and the appropriate base sequence is given.

demonstrates that the substituent attached to the reactive group can impose a distinct sequence preference for reaction. In the case of UM the difference is due almost entirely to an enhanced reactivity for 5'-PyGC-3' (Py=pyrimidine) sequences which are only alkylated weakly by other mustards. The reaction intensities for quinacrine mustard show the greatest discrimination between guanines of any mustard. A detailed examination reveals that the sequence preference is dependent on the two bases on the 3'-side of the reacting guanine in the order of reactivity: GPu = TPu > GPy = TPy > AN = CN   (Pu = purine, Py = pyrimidine, N = any base).

3.2   CHLOROETHYLNITROSOUREAS (CNU's) AND OTHER CHLOROETHYLATING AGENTS

Chloroethylnitrosoureas (CNU's, figure 4) are among the most effective classes of compound so far tested in animal tumour systems but have proved disappointing in the clinic (Prestayko et al, 1981). The CNU's have a disproportionately strong reaction at runs of 3 or more guanines similar to the case with nitrogen mustards (Hartley et al,1986, figure 5). CNU's can produce chloroethyl and hydroxyethyl adducts in DNA; the latter are thought to have little importance in the expression of antitumour activity but may contribute to both the mutagenic and carcinogenic effects of CNU's. Four nitrosoureas known to differ in their relative production of hydroxyethylation and haloethylation (Tong et al, 1982) produce similar patterns of alkylation (Hartley et al, 1986) suggesting that the common alkyldiazohydroxide intermediate produced at neutral pH in aqueous solution (figure 4) gives rise to the observed sequence preference for runs of guanines. Furthermore both 7-chloroethylguanine and 7-hydroxyethylguanine are enhanced similarly by the presence of adjacent guanines in DNA (Hartley et al, 1986).

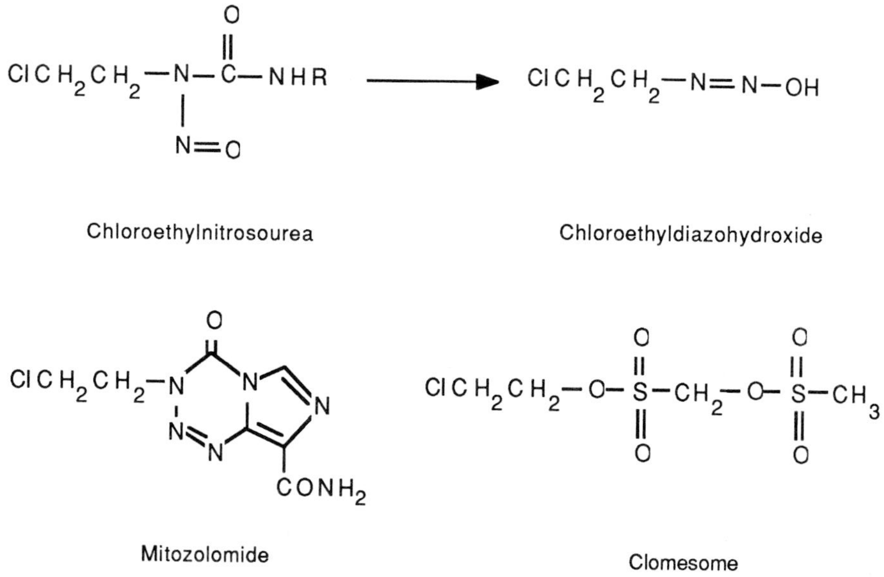

Chloroethylnitrosourea

Chloroethyldiazohydroxide

Mitozolomide

Clomesome

FIGURE 4.   Structure of CNU's, Mitozolomide and Clomesome.

Other chloroethylating antitumour agents capable of producing the same chloroethyldiazohydroxide species such as 8-carbamoyl-3-(2-chloroethyl)imidazo[5,1-d]-1,2,3,5-tetrazin-4(3H)one (mitozolomide, figure 4), 5-[3-(2-chloroethyl)-1-triazenyl]imidazole-4-carboxamide and 3-chloroethyl-1-(p-methoxycarbonylphenyl)triazene show the general preference for runs of guanines similar to, but not as strikingly as that observed for the CNU's (Hartley et al, 1986,1988). In contrast to the CNU's however the pattern of base sequence selectivity differs somewhat between agents suggesting that the nature of the non-alkylating portion of the molecule can again influence the ultimate alkylation preference.

The preference for runs of guanines is not however common to all chloroethylating agents. The agent 2-chloroethyl(methylsulphonyl)methanesulphonate (clomesome, figure 4) was developed as a more 'selective' antitumour agent than the CNU's because its simpler chemistry enables the production of chloroethyl- but not hydroxyethyl- adduct in DNA (Gibson et al., 1986). This compound produces 7-chloroethylguanine adducts in DNA with no discrimination between different guanines (Hartley et al., 1986). Thus although clomesome is more selective than the CNU's in its range of alkylation products, this is at the expense of DNA primary base sequence selectivity.

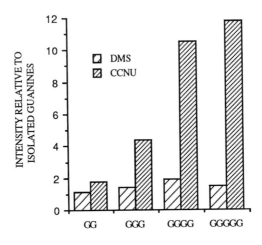

FIGURE 5. Average alkylation intensity at oligo-guanine sequences relative to isolated guanines for dimethyl sulphate (DMS) and a chloroethylnitrosourea (CCNU).

3.3 DABIS MALEATE

The guanine-N7 assay has also been used to help elucidate the mechanism of action of some alkylating agents. For example, the novel, positively charged antineoplastic agent 1,4-bis(2'-chloroethyl)-1,4-diazabicyclo-[2.2.1] heptane dimaleate (Dabis maleate, figure 6) although not expected to alkylate directly was assumed to yield an alkylating tertiary amine following loss of the bridging CH_2 as formaldehyde (Pettit et al., 1979). If this were the sole alkylating species generated then the reactivity patterns for the two compounds would be identical. The pattern of guanine-N7 alkylation however is distinctly different for the two compounds (Broggini et al.,

1990). As a result a mechanism was proposed in which loss of the bridging CH_2 occurs in two steps and that an intermediate with both a tertiary and quaternary 2-chloroethylamino group, possibly an aldehyde (figure 6) or hydroxymethyl derivative persists long enough to contribute significantly to the alkylation reaction and has a different sequence specificity to the compound with two tertiary amine groups.

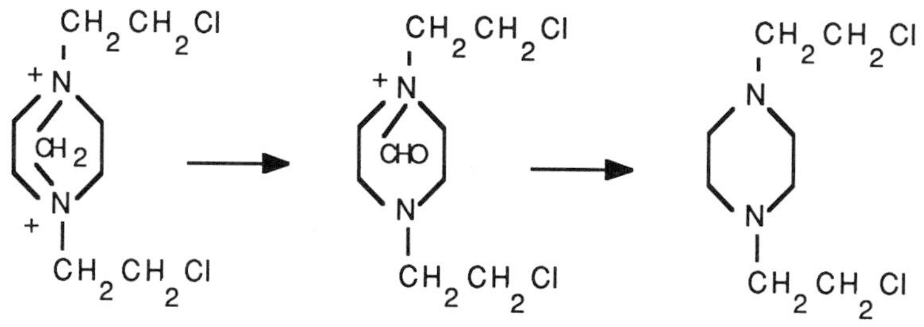

FIGURE 6. Structure of dabis maleate and a proposed mechanism for the loss of the bridging methylene to explain the sequence selectivity data.

3.4 DIMETHANESULPHONATES

The homologous series of dimethanesulphonic acid esters (figure 7) provides a valuable system for the study of bifunctional reactivity with key intracellular target sites. The best known member of the series, busulphan (n=4) is one of the drugs of choice in the treatment of chronic myeloid leukaemia. DNA interstrand crosslinking is found to be optimal with the n=6 member of the series (Bedford and Fox, 1983, Hartley et al, 1990a). In contrast, the extent of monoalkylation at guanine-N7 at an equimolar dose is greatest with busulphan although the extent of reaction is much lower (<1/100) compared to some nitrogen mustards, and shows little sequence selectivity. Interestingly, the 1 carbon member of the series MDMS is also able to crosslink DNA and at high doses shows a non-specific depurination of the DNA (as a result of generation of methanesulphonic acid on hydrolysis) but also a single strong site of guanine-N7 alkylation within a 276 base pair fragment of plasmid DNA in the sequence 5'-ATGG̲TGG-3' (figure 8, Hartley et al., 1990a). This unique reaction is not due to a strong sensitivity to acid depurination at this site since formic acid which is used to depurinate in the Maxam and Gilbert A and G specific sequencing reaction does not produce the same reactivity at this site.

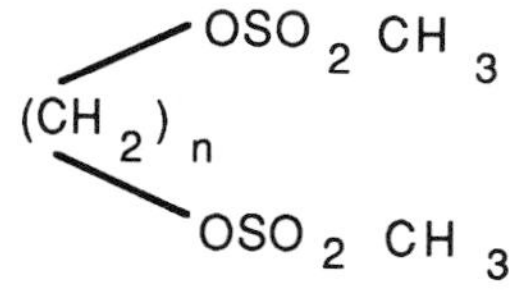

FIGURE 7. Structure of the dimethanesulphonates.

3.5 METHYLATING AGENTS

DNA methylation, particularly at the O^6-position of guanine is important in the processes of carcinogenesis and mutagenesis. Methylating agents appear to fall into two distinct classes with regard to their sequence selectivity for guanine-N7 alkylation. Dimethyl sulphate and methyl-

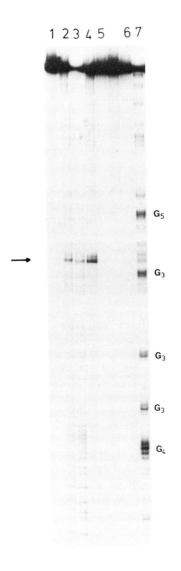

FIGURE 8. DNA sequence
selectivity of methylene dimethane-
sulphonate (MDMS).
Lane 1 5mM 10 mins
Lane 2 5mM 30 mins
Lane 3 5mM 1 hour
Lane 4 2.5mM 1 hour
Lane 5 0.5mM 1 hour
Lane 6 control
Lane 7 L-Pam 25μM 1 hour
showing alkylation at runs of
guanines.
The arrow shows the single strong
site of guanine-N7 alkylation in
this 276 base pair fragment in the
sequence 5'-ATGGTGG-3'

methanesulphonate (MMS) which act through a SN2 mechanism, although not sequence neutral, show little sequence selectivity (figure 5, Mattes et al, 1986, Latif et al, 1989) wheras the SN1 agents N-methyl-N-nitrosourea (MNU) and N-methyl-N'-nitro-N-nitrosoguanidine (MNNG) show the preference for runs of guanines (Wurdeman and Gold, 1988, Latif et al, 1989).   In addition the momomethyl analogues of the chloroethyltriazenes desribed above (section 3.2) also prefer runs of guanines (Hartley et al, 1988), although interestingly, monoethyl- analogues (and ethylnitrosourea) alkylate weakly with little sequence preference.
    A summary of the major sequence preferences for alkylating agents is given in table 1.

TABLE 1.   Summary of DNA sequence preferences for
alkylating agents.

Preference for runs of G's	Nitrogen mustards
	Chloroethylnitrosoureas
	Chloroethyltriazenes
	Dabis maleate
	Monomethyltriazenes
	MNU
	MNNG
Little sequence preference	Busulphan
	Clomesome
	Dimethyl sulphate
	MMS
	Ethyltriazenes
	ENU
Unique sequence preferences	UM (PyGC)
	QM (GTPu and GTPu)
	MDMS (ATGGTGG)

4.   Molecular Rationale for Sequence Selectivities of Alkylating Agents

4.1 DEPENDENCE OF REACTION ON MOLECULAR ELECTROSTATIC POTENTIAL

Why do many agents show a preference for runs of guanines? Highly electrophilic species would be expected to attack the most nucleophilic sites on DNA such as the N and O atoms of the bases. Such a reaction would be governed to a large extent, in particular in the absence of perturbing steric effects, by the molecular electrostatic potential (MEP) of the attacked site. The effects of the nearest neighbour base pairs on the MEP of a guanine-N7 position in B DNA have been calculated by Pullman and Pullman (1981). Figure 9 shows the good correlation between reaction intensity for mechlorethamine against MEP. For most nitrogen mustards the correlation coefficient was usually in the range 0.75-0.9 and the slope of the linear regression line in the range 0.085-0.1 mol/kcal at low ionic strength (Kohn et al., 1987).

The molecular rationale for the preference of nitrogen mustards for runs of guanines is therefore that the positively charged aziridinium group produced in aqueous solution (figure 2) is drawn towards the more electronegative guanine-N7 positions.   In the case of agents such as phosphoramide mustard and chlorambucil which bear negatively charged groups the aziridinium intermediates of these compounds will have no net charge but the -ve groups are some distance away from the aziridinium and do not appear to completely abolish the electrostatic interactions even though the slopes of the linear regression lines are somewhat lower than other mustards. In the case of the CNU's the preference for runs of guanines has led to the suggestion that a 'partial chloronium' ion may be involved (Hartley et al, 1986).   The lack of sequence selectivity for agents such as dimethyl sulphate, clomesome and busulphan is consistent with the inability of these agents to form charged or partially charged intermediates.

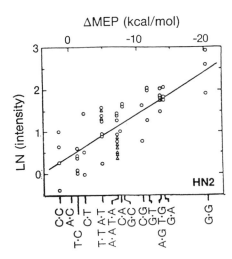

FIGURE 9. Dependence of reaction intensity of mechlorethamine at guanine-N7 positions on the molecular electrostatic potential.

Other factors may also contribute to the favoured reaction in runs of guanines. For example, three, or more, guanines in overall B DNA could possibly assume locally an A form. Theoretical calculations show that the MEP of guanine-N7 is significantly greater in A-form than in B-form DNA (Pullman et al, 1982). Also the accessibility of the N7 of a central guanine in GGG is greater than in other representative triplets for which its value has been computed eg. AGA and TGT (Lavery et al., 1981).

4.2 THE EXCEPTIONS TO THE RULE: URACIL AND QUINACRINE MUSTARDS

Unlike most nitrogen mustards the reaction intensities of uracil and quinacrine mustard did not correlate well with the MEP and models to account for their unique selectivities have been proposed (Kohn et al., 1987).

4.2.1 *Uracil Mustard.* In the case of UM computer modelling can help explain the observed preferences for 5'-Py<u>G</u>C-3' sequences. As the aziridinium ion approaches the guanine-N7 the uracil-O4 atom could interact with the N-H of the 3'-cytosine, thereby countering the positive 'suppressing' influence of the latter on the guanine-N7 position (figure 10). The geometry of the proposed interaction is improved if the guanine is displaced towards its sugar-phoshphate backbone. This can occur when the guanine is situated between 2 pyrimidines as predicted by the Calladine-Dickerson rules (Dickerson, 1983). The interaction between uracil-O4 and the cytosine N-H is also favoured by a rotation about the UM C5-N5 bond. A methyl group at the 6 position can block this rotation because of steric clash with the 2-chloroethyl group and as a result 6-meUM (the drug dopan) does not exhibit the same sequence preference as UM.

4.2.2 *Quinacrine Mustard.* QM is capable of intercalating in DNA prior to covalent reaction with guanine-N7.   A rapid initial non-covalent binding is indicated by the unusually low concentrations required for reaction (Mattes et al, 1986b).   The reaction preferences rely on the 2 bases 3' to the reactive guanine.   Models show that intercalation and guanine-N7 alkylation can only occur if the intercalation site is 3' to the reacting guanine in B DNA (Kohn et al, 1987).   It is also understandable why the immediate 3'-base must be a G or T.   If it were an A or C there there would be an amino group directly below the hydrocarbon side chain stretching from the intercalated quinacrine to the guanine-N7 reaction site.   This would be an unfavourable configuration compared to the case of G or T where there would be an oxygen atom in place of the amino group (figure 10).   As pointed out by Warpehoski and Hurley (1988) however, the model is not a simple case of preferred binding at these sequences, but rather represents a compromise between non-covalent binding (intercalation) and covalent bonding interactions since intercalation may be more favourable at other sites.

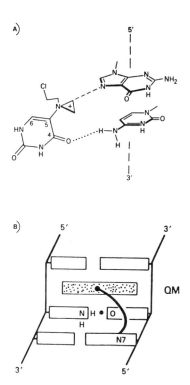

FIGURE 10.   Models for the unique sequence selectivities of uracil (A) and quinacrine (B) mustards as described in the text.

5.   Factors That Influence DNA Sequence Selective Alkylation

5.1   IONIC STRENGTH

Increased ionic strength produces the expected general reduction in reaction rate for most nitrogen mustards eg mechlorethamine, L-Pam (Kohn et al., 1987, table 2), consistent with a screening of the MEP, wheras alkylation by dimethyl sulphate is unaffected by salt (Wurdeman and Gold, 1988).   In addition Na and Mg do not substantially alter the rank order of reactivities for nitrogen mustards although the degree of selectivity (measured as the slope of the linear regression for reaction in the presence vs absence of Na) is altered in some cases.   For example, 100mM Na reduces the selectivity of mechlorethamine and L-Pam by 30% wheras phosphoramide mustard which bears a negatively charged group shows no reduction in selectivity and mustamine, which bears a positive ammonium group, shows a greater reduction.   The unique preferences of UM are less dependent upon long-range electrostatic interactions than other sites as shown by a lesser degree of suppression by Mg ions.   QM is essentially unaffected by the presence of 100mM Na (table 2) or 2mM Mg.

TABLE 2. Effect of ionic strength and cationic DNA affinity binders on total guanine-N7 alkylation by three nitrogen mustards

	% inhibition of guanine-N7 alkylation		
	L-Pam	UM	QM
sodium chloride (100mM)	91	93	8
ethidium bromide (10μM)	63	77	53
distamycin A (100μM)	72	66	0[a]
netropsin (100μM)	72	59	0[a]
spermine (10μM)	82	90	33

[a]Enhancement of overall alkylation.

5.2   CATIONIC AFFINITY BINDERS

Recently the effect of the intercalator ethidium bromide, the AT-specific minor groove binders distamycin A and netropsin, and the biological polyamine spermine on the guanine-N7 alkylation by L-Pam,UM and QM has been investigated in detail (Hartley et al., 1990b).   For L-Pam and UM the cationic DNA affinity binders dose-dependently inhibited the overall extent of alkylation (table 2).   This confirms theoretical evaluations which suggest that netropsin, distamycin A and spermine binding induce long-range modifications in the electrostatic properties of the DNA (weakening the negative MEP) which result in an altered reactivity of DNA toward alkylating agents (Zakrzewska and Pullman, 1983,1985).   QM alkylation is inhibited less by ethidium and spermine, and distamycin A and netropsin enhance overall QM alkylation (table 2) suggesting an increased accessibility for quinacrine intercalation.

More interestingly, however, the pattern of guanine-N7 alkylation is quantitatively altered by ethidium bromide, distamycin A and netropsin which differs with both the nitrogen mustard and the cationic agent (figure 11). The effect, which results in both suppression and enhancement of alkylation sites, was most striking in the case of netropsin and distamycin A, which differ from each other. DNA footprinting studies indicate that selective binding to AT sequences in the minor groove of DNA, in addition to affecting the electrostatic properties of the DNA, can produce long-range structural changes which produce an altered accessibility to alkylation at certain sites in the major groove. No sequence dependence was observed previously for the inhibition of MNU methylation by cationic affinity binders (Wurdeman and Gold, 1988), but this may be due to the small size and easy accessibility of the methyldiazonium ion.

6.   Genomic Targets for Alkylating Agents

6.1   GC-RICH SITES *IN VIVO*

The data suggests that GC-rich regions in genes could be preferred sites of damage to alkylating agents such as nitrogen mustards. The production of inter- and intrastrand crosslinks would be especially favoured. The human genome contains regions of unexpectedly high GC content (>80%) which accounts for approximately 1.3% of the genome and includes regions in a number of oncogenes (Zerial et al, 1986). The 5'-flank of the c-Ha-ras oncogene is particularly GC-rich and these sequences do indeed serve as preferred sites of alkylation *in vitro* by nitrogen mustards and CNU's (Mattes et al, 1988).

A functional role of GC rich sequences is suggested by their frequent occurrence in genes associated with proliferation. Certainly for some genes these runs are part of sequences known to be involved in the control of gene expression (eg SP1 transcription factor binding sites in the preferentially alkylated regions of the c-Ha-ras gene). A search of the Genbank DNA data base, as well as revealing a significant number of oncogenes (Mattes et al, 1988), also included a number of viral sequences, including the Epstein-Barr virus (EBV). In EBV large regions of extraordinary GC-richness are located within the 3kb repeats beginning about 3kb from the replication origin and have features suggesting an important control function (Karlin, 1986). It is interesting to note that the nuclei of tumour cells from the endemic African form of Burkitt's lymphoma contain multiple copies of the EBV genome, and this form of lymphoma is extraordinarily sensitive to chemotherapy: one or two doses of a nitrogen mustard (cyclophosphamide) can produce dramatic regression of the tumour (Ziegler, 1981).

6.2   ARE THE SEQUENCE SELECTIVITIES PRESERVED *IN VIVO* ?

Of course, the important question is to what extent the sequence selectivities observed are preserved in cells, and extreme caution must be used in extrapolating from *in vitro* findings to the *in vivo* situation. Nevertheless, recent data suggests that the different sequence selectivities observed for mechlorethamine, L-Pam, UM and QM are maintained to some extent in cells (Hartley et al, unpublished observations). Using the highly reiterated 340 base pair sequence of human alpha DNA, which accounts for about 1% of the human genome, extracted from cells treated with the nitrogen mustards it is possible to examine the sites of guanine-N7 alkylations

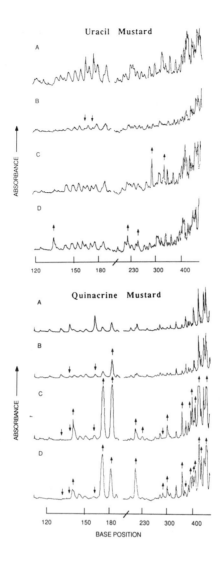

FIGURE 11. Densitometric traces of autoradiograms showing the patterns of alkylation for UM (25μM) and QM (125nM) either alone (A) or in the presence of 10μM ethidium bromide (B), 100μM distamycin A (C), or 100μM netropsin (D).Arrows correspond to major differences, down arrows show suppression, and up arrows enhancement of alkylation.

compared to the same DNA alkylated *in vitro*. Figure 12 shows the pattern observed in a portion of the alpha DNA from cells treated with mechlorethamine, UM and QM. Clear differences are observed between the alkylation patterns of the three compounds and the selectivities are qualitatively similar (but not identical) to those predicted and observed *in vitro*. This therefore suggests that if a DNA sequence is accessible to an alkylating agent the pattern of sequence dependent reactivity is not grossly affected by the nuclear milieu.

6.3 THE ROLE OF DNA REPAIR

In vivo the effectiveness of a particular alkylation will depend on both the initial damage and the persistence of that lesion. DNA repair processes may themselves be intrinsically sequence dependent in their efficiency. For example O^6-methylguanine adducts are repaired in different sequence contexts at different rates (Dolan et al, 1988), and are not converted to mutations in E.coli with equal efficiencies (Topal et al, 1986). In addition to repair at the primary sequence level preliminary data from the laboratory of Kohn suggests that although the same amount of damage to mechlorethamine was observed in both coding and non-coding sequences, the overall repair was much more efficient in the coding sequences (Wasserman et al, 1989).

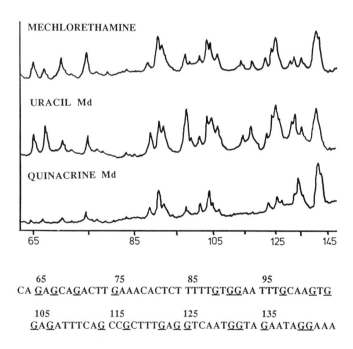

FIGURE 12. Sequence selectivity *in vivo*. Densitometric traces of the alkylation pattern of a sequence of human alpha DNA from cells treated with either mechlorethamine, UM or QM.

7. Future Prospects

It is clear that even simple alkylating agents can react with DNA in a sequence selective manner. It remains to be established, however, to what extent the sequence selectivity is related to the biological activities of these drugs. Since the specificity appears to be preserved to a degree in cells the next logical step would be to examine both the specificity of damage, and the specificity of the repair of that damage, in regions of the genome with defined functions, comparing for example coding versus non-coding sequences, oncogene versus normal genes, high GC-rich genes and so on. The newly developed technique of genomic sequencing may prove to be valuable and is presently being evaluated for this purpose. The piperidine cleavage assay has proved to be very useful in examining the sequence selectivity at the guanine-N7 position which is the major lesion produced by most alkylating agents. The most abundant lesion is not necessarily the most biologically significant, however, and assays are required to examine and quantitate lesions at other sites eg guanine-O6, interstrand crosslinks etc.

It is also clear that the substituent attached to the reactive group of the drug can impose a distinct sequence preference for reaction. This leads to the possibility of design of new alkylating agents with either an enhanced or altered sequence preference. One goal would be the production of more efficient and more selective interstrand crosslinking agents for use in cancer chemotherapy. It is also possible to direct the alkylation to sites other than the guanine-N7 by attaching the alkylating moiety to sequence selective vectors. For example the attachment of nitrogen mustard or chloroethylnitrosoureas to AT-specific minor groove binders such as distamycin A produces compounds that alkylate adenines in the minor groove of DNA with no detectable guanine-N7 alkylation in the major groove (D'Incalci et al, 1988, Church et al, 1989). A study of the biological effects of such compounds should give clues to the relative importance of different lesions. For example, the distamycin-nitrogen mustard compound is a very potent antitumour agent and is active in tumours resistant to conventional alkylating agents (D'Incalci et al, 1988). As our knowledge of the importance of different DNA lesions, the sequence specificities of these lesions, and the relative importance of different genomic locations increases we can hopefully approach the ultimate goal of targeting DNA damage to a particular gene sequence to obtain a specific chemotherapeutic result.

8. References

Bedford, P., and Fox, B.W. (1983) 'DNA-DNA interstrand crosslinking by dimethanesulphonic acid esters' Biochem. Pharm. 32, 2297.

Boles, T.C. and Hogan, M.E. (1984) 'Site-specific carcinogen binding to DNA', Proc.Natl.Acad. Sci. USA 81, 5623-5627.

Broggini, M., Hartley, J.A., Mattes, W.B., Ponti, M., Kohn, K.W., and D'Incalci, M. (1990) 'DNA damage and sequence selectivity of DNA binding of the new anti-cancer agent 1,4-bis(2'-chloroethyl)-1,4-diazabicyclo-[2.2.1]heptane dimaleate' Brit. J. Cancer, 61, 285-289.

Church, K.M., Wurdeman, R.L., and Gold, B. (1989) 'Control over the sequence specificity of DNA alkylation: synthesis and reactions of 2-chloroethylnitrosoureas linked to minor groove binding lexitropsins' Proc.Am.Ass.Cancer Res., 30, 626.

Dickerson, R.E. (1983) 'Base sequence and helix structure variations in B and A DNA' J. Mol.Biol. 166, 419.

D'Incalci, M., Broggini, M., Ponti, M., Erba, E., Mantovani, R., Capolongo, L., Geroni, C., Barbieri, B., Mongelli, N.and Giuliani, F.C. 'Studies on the mode of action of FCE24517, a new distamycin A derivative' Proc.Am.Ass.Cancer Res., 29, 329.

Dolan, M.E., Oplinger, M., and Pegg, A.E. (1988) 'Sequence specificity of guanine alkylation and repair' Carcinogenesis, 9, 2139-2143.

Ewig, R.A.G., and Kohn, K.W. (1977) 'DNA damage and repair in mouse leukaemia L1210 cells treated with nitrogen mustard, 1,3 bis(2-chloroethyl)-1-nitrosourea, and other nitrosoureas' Cancer Res., 37, 2114-2118.

Gilman A., and Phillips, F.S. (1946) 'The biological action and therapeutic applications of B-chloroethylamines and sulfides' Science, 103, 409.

Gibson, N.W., Hartley, J.A., Strong, J.M., and Kohn, K.W. (1986) '2-Chloroethyl-(methylsulfonyl)methanesulfonate (NSC-338947), a more selective alkylating agent than the chloroethylnitrosoureas' Cancer Res, 46, 553-557.

Hartley, J.A., Forrow, S.M., and Souhami, R.L. (1990b) 'Effect of ionic strength and cationic DNA affinity binders on the DNA sequence selective alkylation of guanine N7-positions by nitrogen mustards' Biochem., 29, 2985-2991.

Hartley, J.A., Gibson, N.W., Kohn, K.W., and Mattes, W.B. (1986) 'DNA sequence selectivity of guanine-N7 alkylation by three antitumor chloroethylating agents' Cancer Res., 46, 1943-1947.

Hartley, J.A., Mattes, W.B., Vaughan, K., and Gibson, N.W. (1988) 'DNA sequence specificity of guanine N7-alkylations for a series of structurally related triazenes' Carcinogenesis, 9, 669-674.

Hartley, J.A., Souhami, R.L., Fox, B.W., and Hartley, J.A. (1990a) 'DNA interstrand crosslinking and sequence selectivity of dimethanesulphonates' Brit. J. Cancer, submitted.

Haskel, C.M.(ed) (1985) 'Cancer Treatment' 2nd ed., W.B.Saunders, New York.

Karlin, S. (1986) 'Significant potential secondary structure in the Epstein-Barr virus genome' Proc. Natl. Acad. Sci. USA, 83, 6915.

Kohn, K.W., and Spears, C.L. (1967) 'Alkylated DNA: buoyant density changes and mode of decomposition' Biochim. Biophys. Acta, 145, 720-733.

Kohn, K.W., Hartley, J.A., and Mattes, W.B. (1987) 'Mechanisms of DNA sequence selective alkylation of guanine-N7 positions by nitrogen mustards' Nucleic Acids Res., 14, 10531-10549.

Latif, F., Bigger, C.A.H., Fishel, R., and Dipple, A. (1989) 'Effects of chemical reactivity on sequence selectivity in DNA methylation' Proc. Am.Assoc.Cancer Res., 30, 147.

Lavery, R., Pullman, A., and Pullman, B. (1981) 'Steric accessibility of reactive centers in B-DNA' Int. J. Quantum Chem., 20,49-62.

Mattes, W.B., Hartley, J.A., and Kohn, K.W. (1986a) 'Mechanism of DNA strand breakage by piperidine at sites of N7-alkylguanines' Biochim. Biophys. Acta, 868, 71-76.

Mattes, W.B., Hartley, J.A., and Kohn, K.W. (1986b) 'DNA sequence selectivity of guanine-N7 alkylation by nitrogen mustards' Nucleic Acids Res., 14, 2971-2987.

Mattes, W.B., Hartley, J.A., Kohn, K.W, and Matheson, D.W. (1988) 'GC-rich regions in genomes as targets for DNA alkylation' Carcinogenesis, 9, 2065-2072.

Maxam, A.M., and Gilbert, W. (1980) 'Sequencing end-labeled DNA with base-specific chemical cleavages' Methods Enzymol., 65, 499-560.

Ojwang, J.O., Gruenberg, D.A., and Loechler, E.L.(1989) 'Synthesis of a duplex oligonucleotide containing a nitrogen mustard interstrand crosslink' Cancer Res., 49, 6529-6537.

Pettit, G.R., Gieschen, D.P., and Pettit, W.E. (1979) 'Antineoplastic agents LXIV: 1,4-bis(2'chloroethyl)-1,4-diazabicyclo-[2.2.1]heptane dihydrogen dimaleate' J.Pharm. Sci., 68, 1539.

Prestayko, A.E., Crooke, S.T., Baker, L.H., Carter, S.K., and Shein, P.S.(eds.)(1981) 'Nitrosoureas', Academic Press, New York.

Pullman, B., Lavery, R., and Pullman, A. (1982) 'Two aspects of DNA polymorphism and microheterogeneity: molecular electrostatic potential and steric accessibility' Eur. J. Biochem., 124, 229-238.

Pullman, A., and Pullman, B. (1981) 'Molecular electrostatic potential of the nucleic acids' Q. Rev. Biophys., 14, 289-380.

Tong, W. P., Kohn, K.W., and Ludlum, D.B. (1982) 'Modifications of DNA by different haloethylnitrosoureas' Cancer Res., 42,4460-4464.

Topal, M.D., Eadie, J.S., and Conrad, M. (1986) 'O6-methylguanine mutation and repair is nonuniform' J.Biol. Chem., 261, 9879-9885.

Warpehoski, M.A., and Hurley, L.H. (1988) 'Sequence selectivity of DNA covalent modification' Chem. Res. Toxicol., 1, 315-333.

Wassermann, K., Kohn, K.W., and Bohr, V.A. (1989) 'Alkylation damage and repair of specific genes following exposure to nitrogen mustard' Proc.Am.Ass.Cancer Res., 30, 492.

Wurdeman, R.L., and Gold, B. 'The effect of DNA sequence, ionic strength, and cationic DNA affinity binders on the methylation of DNA by N-methyl-N-nitrosourea' Chem. Res. Toxicol., 1, 146-147.

Zakrzewska, K., and Pullman, B. (1983) 'A theoretical evaluation of the effect of netropsin binding on the reactivity of DNA towards alkylating agents' Nucleic Acids Res., 11, 8841-8845.

Zakrzewska, K., and Pullman, B. (1985) ' The effect of spermine binding on the reactivity of DNA towards carcinogenic alkylating agents' J. Biomol. Struct. Dynamics, 3, 437-444.

Zerial, M., Salinas, J., Filipski, J., and Bernardi, G. (1986) 'Gene distribution and nucleotide sequence organization in the human genome' Eur. J. Biochem., 160, 479.

Ziegler, J.L., (1981) 'Burkitt's lymphoma' N. Engl. J. Med., 305, 735-745.

ACKNOWLEDGEMENTS. The author acknowledges the important contribution to many of the studies reviewed here by William Mattes, Neil Gibson, Steve Forrow, John Bingham, Mauro Ponti,Robert Souhami, and in particular Kurt Kohn, in whose laboratory many of the initial experiments were performed and who continues to be a source of inspiration.

CONTRASTING MECHANISMS FOR THE SEQUENCE RECOGNITION OF DNA BY (+)- AND (-)-CC-1065

MARTHA A. WARPEHOSKI, PATRICK McGOVREN, and MARK A. MITCHELL, *Research Laboratories, The Upjohn Company, Kalamazoo, Michigan 49001*; LAURENCE H. HURLEY, *The Drug Dynamics Institute, College of Pharmacy, The University of Texas at Austin, Austin, Texas 78712*

ABSTRACT. The covalent reaction of (-)-CC-1065, the unnatural enantiomer of the potent and sequence-selective, DNA-reactive antibiotic (+)-CC-1065 with DNA was studied and compared to that of the natural product. Although (-)-CC-1065 also formed covalent adducts in which the cyclopropyl carbon was bonded to the N3 atom of adenine, and the thermal strand breakage that it produced paralleled that seen for (+)-CC-1065, it lay in the opposite direction along the minor groove and exhibited a markedly different sequence requirement for the covalently modified adenine. While (-)-CC-1065 and its full carbon framework analogue, (-)-AB'C', reacted readily at adenines near to, but generally distinct from, (+)-CC-1065-reactive adenines and exhibited potent cytotoxicity, their simpler analogues did not alkylate DNA under the conditions employed and were biologically nonpotent. An analysis of the reactivity patterns of (+)- and (-)-CC-1065 and their analogues with DNA restriction fragments supported the conclusion that the mode of sequence recognition for (-)-CC-1065 adduct formation is fundamentally different from that of (+)-CC-1065 and is primarily controlled by specific minor groove, AT-selective binding interactions, rather than by sequence requirements of the covalent step, as occurs for (+)-CC-1065 and the (+)-CPI analogues. Models are proposed comparing the interactions of the enantiomeric alkylating moieties variously oriented in the minor groove at potential reaction sites. The evolutionary significance of both the alkylating moiety and the minor groove binding segments of the natural product is discussed.

Introduction

The concept of DNA as a target of drug action has undergone considerable refinement in recent years, as information accrues on the range of structural and dynamic diversity available to this critical biomacromolecule.[1] In particular, drugs and carcinogens that react with nucleic acid bases in DNA to form covalent adducts have been found to exhibit sometimes remarkable discrimination toward the sequence contexts of the bases undergoing modification.[2-6] The mechanisms mediating this recognition of specific sequences of DNA by small molecules may include sequence-dependent molecular electrostatic potential,[3a] conformational flexibility[4b] catalytic functional group juxtapositions,[5] and precovalent binding affinities.[6]

Our previous work with CC-1065, an extremely potent antitumor antibiotic produced by *Streptomyces zelensis*,[7,8] revealed a DNA alkylating agent extraordinary both for the base-heteroatom specificity and the sequence selectivity of its covalent reaction with double-helical DNA.[9] The CC-1065 molecule consists of three repeating pyrroloindole subunits, one of which (the CPI, cyclopropylpyrrolindole subunit) contains a potentially reactive cyclopropyl function.[10] Naturally occurring CC-1065 has the 7bR,8aS stereochemical configuration of the cyclopropane

B. Pullman and J. Jortner (eds.), Molecular Basis of Specificity in Nucleic Acid-Drug Interactions, 531–550.

ring[11] and is dextrarotatory at the sodium D line.[12] (+)-CC-1065 binds strongly and reversibly in the minor groove of A-T regions in double-stranded DNA or duplex oligomers[13] and then bonds covalently to the N3 of adenine in target sequences, with opening of the cyclopropyl ring (Scheme I).[9]

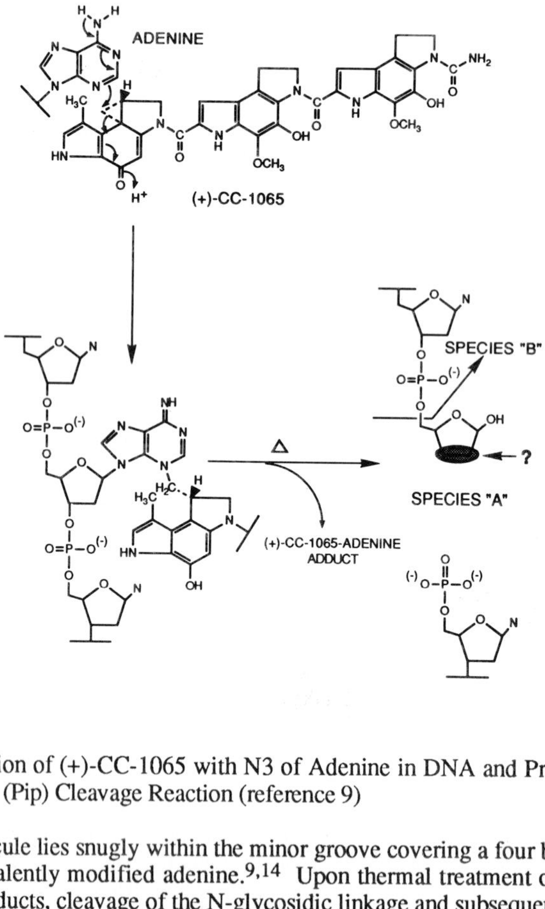

Scheme I. Reaction of (+)-CC-1065 with N3 of Adenine in DNA and Products from Thermal (Δ) and Piperidine (Pip) Cleavage Reaction (reference 9)

The reacted molecule lies snugly within the minor groove covering a four base pair region to the 5'-side of the covalently modified adenine.[9,14] Upon thermal treatment of (+)-CC-1065-(N3-adenine)-DNA adducts, cleavage of the N-glycosidic linkage and subsequent backbone breakage occurs to the 3'-side of the covalently modified adenine to leave a 5'-phosphate on the 3'-side of the break and, we presume, a modified deoxyribose on the 5'-side.[9a] Using a heat-induced strand breakage assay based on this observation, we determined that the most reactive adenines in a set of different DNA fragments were generally found in two sequences, 5'AAAAA* or 5'PuNTTA*, where * indicates the covalently modified adenine and N indicates any of the four bases in DNA.[9b] The construction of a site-directed (+)-CC-1065-(N3-adenine)-DNA adduct in a 117 base pair fragment[15] allowed us to determine the effect of covalent bonding on local DNA structure. Analysis of DNase I footprinting experiments and restriction enzyme cutting patterns using the DNA fragment containing the site-directed adduct demonstrated that (+)-CC-1065 adduct formation correlates with an asymmetric effect on DNA structure that extends more than one helix turn to the 5'-side of the covalent bonding site.[16]

Despite this novel mechanism of action of (+)-CC-1065, and its highly potent *in vitro*[17] and *in vivo*[18] antitumor activity, clinical development of the natural product was precluded by an unusual toxicity, which led to delayed death in mice receiving therapeutic doses.[19] Synthetic analogues have been prepared that are more efficacious, do not cause delayed death, and that, therefore, show considerable promise as chemotherapeutic agents.[20] Chirally resolved synthetic analogues are shown in Chart I.

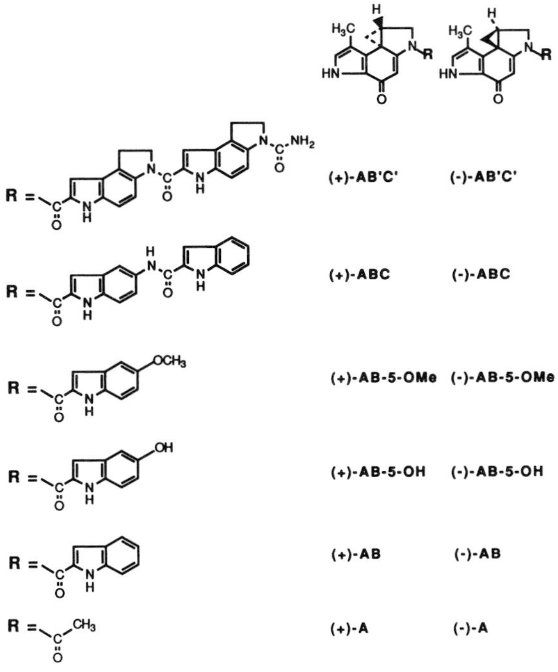

Chart I. Structures of (+)- and (-)-Cyclopropylpyrroloindole (CPI) Compounds Used in This Study

Like (+)-CC-1065, analogues of the 7b*R*,8a*S* cyclopropyl configuration [(+)-CPI agents] react with double-helical DNA to produce adducts that, on thermal treatment, cause strand breaks at alkylated adenines.[21] We recently described the unexpected finding that an acetyl-substituted (+)-CPI [(+)-A, Chart I], which shows little to no evidence of the noncovalent binding to DNA so characteristic of (+)-CC-1065, nevertheless has sufficient structural information to mediate the sequence specificity of the entire drug molecule, i.e., (+)-A preferentially alkylates the same sequences in DNA as does (+)-CC-1065.[22] This suggests that the sequence specificity of (+)-CC-1065 depends, not on the relative noncovalent binding affinity of the different sequences as originally thought,[9] but on sequence-dependent site reactivity (i.e., upon covalent bonding dynamics). It was suggested that the prime function of the middle and right-hand segments in the (+)-CPI series is to increase the equilibrium constant for non-specific reversible interaction with the minor groove of DNA where alkylation can occur, accounting for the difference in absolute reactivity among the analogues. Several sequences in an SV40 fragment that reacted with (+)-CC-1065 and (+)-AB'C' failed to be alkylated by (+)-ABC, indicating that specific reversible interactions between the inside edge of the B and C subunits and the floor of the minor groove of DNA can "modulate or fine tune" the sequence specificity of the (+)-CPI subunit.[22]

Materials and Methods

COMPOUNDS USED IN ALKYLATION SEQUENCE SPECIFICITY STUDIES
(+)-CC-1065 isolated from the fermentation broth of *S. zelensis*[7,18] was used in the sequencing experiments. All of the other compounds were synthetic analogues (Chart I) synthesized at The Upjohn Co., as described previously.[23]

DNA FRAGMENTS
The 117 base pair *Msp*I-*Bst*NI fragment of M13mpl bacteriophage DNA was isolated as described previously.[15] The 180 base pair fragment of M13mpl DNA was isolated by successive restriction endonuclease treatments with *Hinp*I and *Alu*I. Similar methods were used to isolate 118 and 276 base pair fragments from SV40 DNA using *Mbo*I and *Hinf*I.

3'- AND 5'-^{32}P-END LABELING.
3'- and 5'-^{32}P-end-labeling methods were as described previously.[15,22]

MODIFICATION OF DNA WITH CC-1065 AND ANALOGUES
Various 10-fold dilutions of a 280 mM stock solution of CC-1065 or analogues were added to aliquots of 3'- or 5'-^{32}P-labeled DNA fragments and incubated at 37 °C for 2 h, followed by ethanol precipitation to terminate the reaction.

MPE·Fe(II) FOOTPRINTING AND THERMAL BREAKAGE OF CC-1065-MODIFIED FRAGMENTS
MPE·Fe(II) footprinting and heat-induced breakage of DNA fragments containing CC-1065-DNA adducts were performed as described.[9b,15]

Results

The DNA base specificity and structure of the DNA adduct of (-)-CC-1065 has been shown to be the same as (+)-CC-1065. However, in contrast to (+)-CC-1065, the unnatural enantiomer was orientated to the 3'-side of the covalent bonding site,[23] which is in agreement with molecular modeling predictions.

COMPARISON OF (+)- AND (-)-CC-1065 SEQUENCE SELECTIVITY
The heat-induced strand breakage assay with (-)-CC-1065 was carried out on two DNA restriction fragments from SV40 (one strand each of a 276 base pair fragment and a 118 base pair fragment), and three restriction fragments from M13mpl (one strand of a 180 base pair fragment and both strands of a 117 base pair fragment). Exclusion of base pairs further from the end label than the largest cleavage fragment identified on each gel, and closer to the end label than the smallest identified cleavage fragment (i.e., experimentally unevaluable regions), leaves about 470 base pairs in these five restriction fragments that can be analyzed for sequence context of (-)-CC-1065 alkylation. This sampling of DNA is A-T rich, with 34% adenine, 32% thymine, and about 17% each of guanine and cytosine bases. The DNA was incubated with the drug at a range of concentrations. Figure I illustrates the gel results obtained with one of the DNA fragments. The appearance of each band on the polyacrylamide gel indicates that a detectable level of heat-induced cleavage has occurred at that adenine (Scheme I). Since relative band intensity is not sensitive to the heat-induced strand scission variable under the conditions employed, we infer that this intensity is proportional to the rate of alkylation of the adenine at which subsequent cleavage

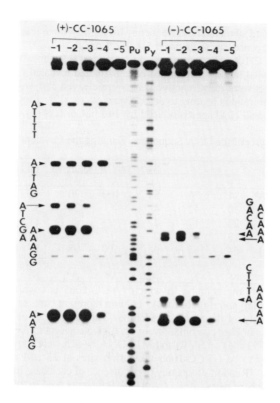

Figure I. Concentration dependence of alkylation by (+)-CC-1065 and (-)-CC-1065 of the 117-bp *Msp*I-*Bst*NI fragment of M13mpl DNA. Single 5'-^{31}P-labeled (+)-strand of the 117-bp DNA fragment was isolated and labeled as described in the Materials and Methods Section. Fragments were modified with 2.8 nM (lanes labeled -5), 28 nM (-4), 280 nM (-3), 2.8 mM (-2), or 28 mM (-1) of either (+)-CC-1065 (left-hand lanes) or (-)-CC-1065 (right-hand lanes). Thermal cleavage was as described in the Materials and Methods Section and samples were electrophoresed adjacent to Maxam-Gilbert purine- and pyrimidine-specific DNA cleavage reactions. The drug alkylation sequences are shown and arrows indicate adenines modified by either (+)- or (-)-CC-1065.

occurs. Under a given set of drug-DNA incubation conditions, the cleavage bands produced at the lowest concentration of a drug identify the most reactant adenines for that drug. Each reactive adenine, *when placed in the context of its consensus sequence* (vide infra) identifies a reactive *alkylation* site for the drug under consideration.

At 2.8-28 nM (-)-CC-1065, only eight heat-induced cleavage bands (and thus alkylation sites) were detected (unpublished results). At 20 nM (-)-CC-1065, twenty-four additional alkylation sites are produced. These thirty-two reactive sites include fourteen 5'AA*A, ten 5'TA*A, and seven 5'AA*T sequences, and a single 5'AA*C sequence (A* refers to the covalently modified adenine). Thus, while the alkylated adenine can be flanked by one thymine in this set of sequences, either on the 5'- or 3'-side, there is always at least one adjacent adenine. When the DNA is incubated with 2.8-28 mM (-)-CC-1065, twenty-one additional alkylation sites are generated, one-third of which now contain the 5'TA*T sequence not seen at lower drug concentrations. The composite tabulated base frequency for all fifty-three reaction sites is given in Table I, with the resulting consensus sequence[24] 5'A/T A* A/T.

Previous analysis of (+)-CC-1065 alkylation sites on restriction fragments from SV40 and T7 DNA representing a somewhat larger sampling of base pairs had revealed that the sequences 5'PuNTTA* and 5'AAAAA* were most reactive (i.e., were cleaved at 1.4-0.14 mM (+)-CC-1065, the lowest drug concentrations leading to detectable cleavages).[9a] They also represented the consensus sequences for all cleavages produced by 140 nM or less (+)-CC-1065 in these

Table I. Analysis of the DNA Sequences Flanking the Covalent Bonding Sites of (-)-CC-1065

		frequency of occurence,[a] (%)							
		-3	-2	-1	A*	+1	+2	+3	+4
A	(34)[b]	34	26	53	100	70	43	30	34
T	(32)	30	42	42		28	32	36	32
G	(17)	23	17	4			11	11	17
C	(17)	13	15	2		2	13	23	17
A/T	(66)	64	68	94	100	98	75	66	66
consensus[c]		N	N	A/T	A*	A/T	N	N	N

[a]A* represents the adenine to which (-)-CC-1065 is covalently bonded. Columns to the left and right of the A* column represent bases to the 5'- and 3'-sides, respectively, of the alkylated adenine. The frequencies of occurrence of individual bases or base combinations in the region adjacent to A* are computed from a total of 53 sites on SV40 and M13 DNA, which were shown by thermal cleavage assay to be (-)-CC-1065 alkylation sites at 28 μM or lower drug concentrations. [b]Random frequency of occurrence of each base in the evaluable region (ca. 470 base pairs) of the five DNA fragments used in this study. [c]The consensus sequences were determined by χ^2 analysis ($\alpha = 0.0005$) using the raw frequency values observed.

DNA fragments under the conditions employed. These conditions differed from those of the present study in that the DNA-drug incubations were carried out for 24 h at 4 °C rather than for 2 h at 37 °C. To allow a more direct comparison, the alkylation sites for (+)-CC-1065 in the present set of DNA fragments, under conditions identical with those used for (-)-CC-1065, were determined. Three triplets comprised the great majority of the thirty-eight most reactive sites: 5'TTA* (thirteen sites), TAA* (thirteen sites), and AAA* (seven sites). The high frequency of 5'TAA* in these experiments is a feature not found in earlier work[9b] and may possibly reflect the higher temperature of drug incubation used in the present study. At higher (+)-CC-1065 concentrations, seventeen additional alkylation sites are produced. Table II tabulates the frequency of occurrence of the bases at each position surrounding these fifty-five alkylated adenines, giving the consensus sequence 5'A/TA/TA*.

It is clear that (+)-CC-1065 and (-)-CC-1065 have distinct sequence requirements, even though both favor AT-rich DNA. The most striking difference is that (+)-CC-1065 strongly prefers two A/T base pairs in the 5'-direction from its adenine alkylation site, while (-)-CC-1065 requires A/T flanking on both sides of the reactive adenine. Comparison of these two drugs, not merely with respect to flanking sequence requirements, but with respect to the relative location of reaction sites along the DNA strands as illustrated in Figure I for one DNA fragment, leads to several intriguing observations. Chart II shows a 117 base pair fragment from M13mpl on which are indicated for both strands the adenines alkylated by (+)- and (-)-CC-1065. In this DNA

Table II. Analysis of the DNA Sequence Flanking the Covalent Bonding Sites of (+)-CC-1065

		frequency of occurrence,[a] %						
		-4	-3	-2	-1	A*	+1	+2
A	(34)[b]	27	36	31	44	100	36	42
T	(32)	45	27	58	55		27	24
G	(17)	18	29	4			9	24
C	(17)	9	7	7	2		27	11
A/T	(66)	73	64	89	98	100	64	65
consensus[c]		N	N	A/T	A/T	A	N	N

[a]A* represents the adenine to which (+)-CC-1065 is covalently bonded. Columns to the left and right of the A* column represent bases to the 5'- and 3'-sides, respectively, of the alkylated adenine. The frequencies of occurrence of individual bases or base combinations in the region adjacent to A* are computed from a total of 55 sites on SV40 and M13 DNA, which were shown by thermal cleavage assay to be (+)-CC-1065 alkylation sites at 28 μM or lower drug concentrations. [b]Random frequency of occurrence of each base in the evaluable region (ca. 470 base pairs) of the five DNA fragments used in this study. [c]The consensus sequences were determined by χ^2 analysis ($\alpha = 0.0005$) using the raw frequency values observed.

Chart II. Diagrammatic Presentation of the (+)- and (-)-CC-1065 Bonding Sequences on the 117 Base Pair *Msp*I-*Bst*NI Fragment of M13mp1[a]

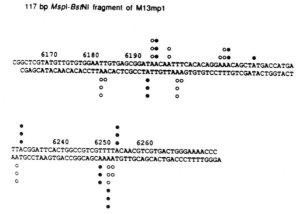

117 bp *Msp*I-*Bst*NI fragment of M13mp1

[a]The data are taken from concentration-dependency experiments in which 10-fold dilutions of a standard (+)- or (-)-CC-1065 stock solution (280 μM) were incubated for 2 h at 37 °C and then heated under standard conditions to produce DNA strand breakage. The O show the (-)-CC-1065 bonding sites and the ● show the (+)-CC-1065 bonding sites. The number of circles reflects the relative reactivity toward alkylation as indicated by the minimum concentration of drug producing detectable strand breakage at that site: O O O, 2.8-28 nM; O O, 280 nM; O, 2.8-28 μM.

fragment, one adenine reacts with both enantiomers of CC-1065. It must be emphasized that the sequence context (i.e., consensus sequence) for (+)- and (-)-CC-1065 is different, and hence, while the specific adenine that is alkylated by (+)- or (-)-CC-1065 is the same, the alkylation *sites* themselves are not identical. (We use the term "alkylation site" to include both the alkylated base and its sequence context.) All of three favorable 5'TTA* (+)-CC-1065 sites are accompanied by a favorable 5'TA*A (-)-CC-1065 reaction site directly opposite on the complementary strand. Another common motif, occurring five times in this sampling, is the occurrence of a (-)-CC-1065 alkylated adenine one or two base pairs in the 5'-direction from a (+)-CC-1065 reactive adenine on the same strand. Since the footprinting experiments discussed above show that (+)-CC-1065 extends in the 5'-direction from its modified adenine, while (-)-CC-1065 extends in the 3'-direction from its covalent link, this indicates that alkylation with either enantiomer leads to occupation of the same or overlapping spaces in the minor groove in many instances. A similar analysis of the set of (+)- and (-)-CC-1065 reaction sites on all of the five restriction fragments used in the (-)-CC-1065 consensus studies leads to some interesting observations. Of the fifty-three adenines that react with (-)-CC-1065, twenty-four can also react with (+)-CC-1065 [i.e., these adenines belong to both the set of (+)-CC-1065 alkylation sites and the set of (-)-CC-1065 alkylation sites]. Of the remaining adenines targeted by (-)-CC-1065, fifteen occur one or two base pairs 5' to a (+)-CC-1065 alkylated base. Of the fourteen (-)-CC-1065 reactive adenines that do not appear to be near (+)-CC-1065 alkylation sites (in some cases it is possible that such reaction would occur on the complementary strand), eight are in or adjacent to 5'AATT sequences. Four of these are represented in the most reactive (-)-CC-1065 sites. In this sequence the (-)-CC-1065 molecule prefers to react at the 5'-adenine (5'A*ATT) whenever the 5' flanking A/T requirement is honored. If not, it will react at the 3'-adenine (5'AA*TT).

Of the fifty-five (+)-CC-1065 alkylation sites identified in this DNA sample at 28 mM drug, nearly two-thirds occur at adenines that also react with (-)-CC-1065 or are one or two base pairs 3' to adenines that react with (-)-CC-1065. Of the nineteen exceptions, nine do not contain the consensus sequence (A/T A* A/T) for (-)-CC-1065. Six of the remaining ten are 5'TTA*T sequences, which are predicted to have reactive (-)-CC-1065 sites on the opposite strand, leaving only four (+)-CC-1065 sites that do not have an obvious reason for failing to react with (-)-CC-1065. One of these sites, 5'CTA*TG, reacts with (+)-CC-1065 only at relatively high concentrations. The other three sites contain 5'AAG sequences adjacent to, or incorporating, the target adenine.

In summary, within the constraints of their distinct consensus sequence requirements, (+)- and (-)-CC-1065 tend to react within the same A/T-rich sequences, such that much of the same space in the minor groove is accessible to either enantiomer. Nearly half of the alkylation sites for each of the enantiomers permit covalent bond formation with the same adenines. Another large fraction of sequences contains a (-)-CC-1065 alkylated adenine one or two base pairs 5' to an adenine targeted by (+)-CC-1065 on the same strand, or directly across on the complementary strand, such that the consensus sequence overlap. The relatively small number of exceptions to this appear to fall mainly into three categories: (1) 5'AATT sequences, which are excellent (-)-CC-1065 alkylation sites, but which react poorly or not at all with (+)-CC-0165; (2) sequences that can accommodate (+)-CC-0165, but not the stringent 5'A/T A* A/T (-)-CC-1065 consensus; and (3) nominally consensus (-)-CC-1065 sequences in which the target adenine is within or directly 5' to a 5'AAG sequence. While these are usually highly reactive sequences for (+)-CC-1065 when A/T base pairs lie in the 5'-direction, they are generally poorly reactive or unreactive with (-)-CC-1065. Indeed, three of the five 5'A*AG cleavage sites that appear at the higher concentrations of (-)-CC-1065 are highly susceptible (+)-CC-1065 sites and could conceivably arise at least partly from traces of enantiomeric contamination.

COMPARISON OF (+)- AND (-)-CPI ANALOGUES OF CC-1065

Sequence Specificity and Reactivity of DNA Alkylation. The (+)- and (-)-CPI analogues listed in Chart I were subjected to the thermally induced strand breakage assay using the 5'-labeled 117 base pair (+ strand) M13mpl DNA fragment, with a range of concentrations for each drug. This fragment contains five identified (+)-CC-1065 alkylation sites, and four identified (-)-CC-1065 sites (Figure II and Chart II, top strand). Figure II presents the results for (+)- and (-)-A, -AB, and -ABC. The gel obtained for (+)- and (-)-AB'C' is virtually identical with that of (+)- and (-)-CC-1065 in Figure I (unpublished results). Table III summarizes these data.

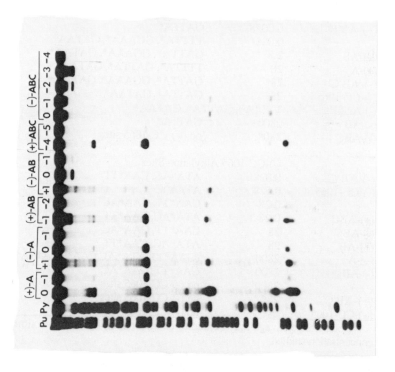

Figure II. Concentration dependence of alkylation by (+)- and (-)-A, (+)- and (-)-AB, and (+)- and (-)-ABC of the 117-bp MspI-BstNI fragment of M13mpl. Aliquots of 5'-labeled 117-bp DNA fragment were incubated with appropriate concentration of (+)- or (-)-A, (+)- or (-)-AB, or (+)- or (-)-ABC, as described in the Materials and Methods Section. +1, 0, -1, -2, -3, and -4 denote 2.8 mM and 280, 28, 2.8, 0.28, and 0.028 mM, respectively. Thermal cleavage conditions are described in the Materials and Methods Section. Pu, purine; Py, pyrimidine.

In confirmation of our earlier results,[22] Table III shows that the same set of sequences that reacts with (+)-CC-1065 also reacts with all of the (+)-CPI agents (although consensus studies were not carried out for each analogue, the consistent and exact correspondence of covalently reactive adenines strongly suggests that the alkylation sites themselves correspond as well).[22] While the 10-fold dilutions provide only approximate quantitation, the overall reaction rate at all of the sensitive sequences increases markedly as the molecular structure of the drug is lengthened from from (+)-A to (+)-ABC, since correspondingly lower concentrations of drug are needed to produce detectable cleavage. With (+)-ABC this effect reaches a plateau, and no further increase in alkylation reactivity results from further elaboration of the CC-1065 structure for these

Table III. Reactivity of (+)- and (-)-CC-1065 Alkylation Sites

agents	concn of earliest detection, μM	sequences alkylated at A* at earliest detection, 5' to 3'
	(+)-CC-1065 Alkylation Sites	
(+)-ABC	0.0028	TTTTA*, GATTA*, GATAA*
(+)-CC-1065	0.0028	GATTA*
	0.028	TTTTA*, GATAA*
(+)-AB'C'	0.0028	GATTA*
	0.028	TTTTA*, GGAAA*, GATAA*
(+)-AB	2.8	GATTA*, GGAAA*, GATAA*
(+)-A	28	TTTTA*, GATTA*, GATAA*
(-)-AB'C'	28[a]	GATTA*, GGAAA*, GATAA*
(-)-CC-1065	28[a]	GATTA*, GATAA*
(-)-A280		TTTTA*, GATTA*, GATAA*
(-)-AB	2800[a]	GATTA*
(-)-ABC	280[b]	no (+)-CC-1065 sites
	(-)-CC-1065 Alkylation Sites	
(-)-AB'C'	0.028	ATA*AC, CAA*TT
(-)-CC-1065	0.028	ATA*AC
	0.28	CAA*TT, GAA*AC
(+)-ABC	0.28	ATA*AC, CAA*TT
(-)-ABC	28	CAA*TT, GAA*AC
(+)-AB	28	ATA*AC, CAA*TT
(+)-A	2800	(GAA*TT)ATA*AC,CAA*TT, GAA*AC
(-)-AB	2800	CAA*TT, GAA*AC
(-)-A	2800[b]	no (-)-CC-1065 sites
(+)-AB'C'	28[b]	no (-)-CC-1065 sites
(+)-CC-1065	28[b]	no (-)-CC-1065 sites

[a]Detectable cleavage may be due to contamination by (+). [b]Maximum concentration studied.

sequences. At very high concentrations [1000-fold higher than required for detection of cleavage by their respective (+)-CPI enantiomers], (-)-CC-1065, (-)-AB'C', and (-)-AB produce detectable cleavage at these (+)-CC-1065 sites. This is most likely due to traces of enantiomeric contaminant in these synthetic agents, since as little as 0.1% (+)-enantiomer could produce the observed bands. A very different situation obtains for (-)-A, however, whose reactivity with the set of adenines in (+)-CC-1065 sites is within an order of magnitude of the reactivity of (+)-A. Since all of the (-)-CPI compounds are prepared from the same resolved intermediate, (-)-A cannot contain sufficient (+)-enantiomer contaminant to account for this result.[23]

Some interesting differences emerge for the (-)-CC-1065 alkylation sites. While these distinct alkylation sites are highly reactive toward (-)-CC-1065 and its close analogue (-)-AB'C', they are somewhat less reactive than the (+)-CC-1065 sites are toward the corresponding (+)-CPI agents, judging by the higher concentrations required to produce detectable cleavage. A second difference is that (-)-ABC is very poorly reactive with adenines at these (-)-CC-1065-sensitive sequences, requiring 100- to 1000-fold greater concentrations than the more structurally elaborated (-)-CC-1065 and (-)-AB'C'. This large rate discrepancy continues with the smaller

(–)-AB, and no adenine reactions corresponding to (–)-CC-1065 alkylation sites could be detected for (–)-A at the highest concentration tested. Indeed, (–)-A reacts at (+)-CC-1065 sites much more readily than it would appear to react at (–)-CC-1065 sites. The third intriguing observation from this part of Table III is that the less elaborate (+)-series analogues, (+)-ABC, (+)-AB, and (+)-A, react to a small extent at adenines associated with (–)-CC-1065 sites. To be sure, reaction at these adenines requires 10- to 1000-fold higher concentrations of each drug than reactive adenines in (+)-CC-1065 sites. However, this still allows these (+)-CPI analogues to be far more reactive than their respective (–)-CPI enantiomers with these same adenines. Indeed, the adenine (A*) in the 5'CAA*TT sequence, which is one of the few (–)-CC-1065 alkylated sequences not near to a (+)-CC-1065 bonding site, appears to react as readily with (+)-ABC as it does with (–)-CC-1065. A 5'GAA*TT sequence rather near the labeled end of this DNA fragment, which would be expected to have a good bonding site for (–)-CC-1065, also provides an alkylation site for (+)-A and (–)-AB at 2800 and 280 μM, respectively. [The observation that intense bands from heat-treated, 5'-end-labeled DNA are often accompanied by traces of slower running cleavage product presumed to result from β-elimination of the modified deoxyribose, as shown in Figure I, might make it somewhat difficult to unambiguously identify weak bands resulting from reaction at adenines in (–)-CC-1065 sites, which frequently occur immediately 5' to (+)-CC-1065 sites. Such a complication does not occur, however, for the 5'CAA*TT and GAA*TT sequences in this DNA fragment.] At the highest concentrations tested, (+)-AB'C' and (+)-CC-1065 did not produce detectable cleavage at adenines associated with (–)-CC-1065. Because of the lower reactivity of (–)-CC-1065 sites, traces of (–)-enantiomeric contaminant in synthetic (+)-AB'C' might not be detected at these concentrations.

In summary, those sequences in the M13mpl DNA strand described as (+)-CC-1065 alkylation sites appear to be intrinsically more reactive and more generally chirally discriminating than those sequences described as (–)-CC-1065 alkylation sites. Their relative reactivity is largely independent of the structural appendages attached to the (+)-CPI subunit, although binding enhancement by those appendages accelerates reaction at all susceptible sequences.[23] Only the smallest (–)-CPI agent, (–)-A, can react at these sequences, and this occurs with a 10-fold loss of rate relative to (+)-A. The sequences identified as (–)-CC-1065 alkylation sites are only chirally discriminatory to the structurally elaborated (+)-CC-1065 and (+)-AB'C'. They react, although more slowly than (+)-CC-1065 sites, with other (+)-CPI analogues [presumably with the bonded (+)-CPI lying to the 5'-side of the alkylated adenine, opposite to the 3'-orientation of (–)-CC-1065]. The (–)-CPI analogues in general appear to be intrinsically less reactive alkylating agents for DNA. The smaller analogues, (–)-A, (–)-AB, and (–)-ABC, are less reactive at any sequence than their (+)-enantiomers. This low reactivity of the (–)-CPI subunit is overcome, however, only by (–)-CC-1065 and (–)-AB'C', which contain the full carbon skeleton of (–)-CC-1065, and may react at adenines that react relatively poorly with smaller (+)-CPI analogues, and not at all with (+)-CC-1065 and (+)-AB'C'.

Biological Activity. Table IV summarizes some important biological characteristics of the (+)- and (–)-CPI compounds described above. These include the cytotoxicity against L1210 tumor cells in suspension culture, antitumor efficacy against P388 leukemia cells in mice and the dose required to achieve it, and the ability to cause delayed death in nontumored mice. In comparing the activities of enantiomers, three types of relationships are evident from Table IV: (1) (+)- and (–)-CC-1065, and also (+)- and (–)-AB'C', show comparable, high potency both *in vitro* and *in vivo*, as well as comparable efficacy against P388 leukemia in mice. However, the (+)- and (–)-compounds are strikingly differentiated in their biological action by the ability of only the (+)-agents to cause delayed death in nontumored mice. (2) (+)- and (–)-ABC, and also (+)- and (–)-AB, in sharp contrast to the more complex structures, show very disparate potencies between (+) and (–) counterparts. Since these potency differences approach two orders of magnitude or more, it becomes experimentally impossible to distinguish the contribution of a minor, potent contaminant from that of the bulk agent. Thus, the ID_{50} value for (-)-ABC might be due to the

Table IV. Physical Binding Interaction and Biological Activities of CC-1065 and Analogues.

	$\Delta T_m{}^a$	*in vitro* L1210 leukemia $ID_{50},^b$ nM	%LS	*in vivo* P388 leukemia[c] OD, μg/kg-day	delayed death[d]
(+)-CC-1065	29.8	0.03	60	50	yes
(-)-CC-1065	12.7	0.007	50	12	no
(+)-AB'C'	22.8	0.01	75	100	yes
(-)-AB'C'	12.4	0.01	50	50	no
(+)-ABC	19.1	0.004	(4/6)[f]	25	no
(-)-ABC	7.5	1.0	55	200	no[e]
(+)-AB	13.3	0.1	100	250	no[e]
(-)-AB	3.3	20	80	10000	no[e]
(+)-A	1.4	10	90	1200	no
(-)-A	0	100	75	50000[g]	nt

[a]Thermal melting of calf thymus DNA compared with or without the drug in 0.01 M phosphate buffer, pH 7.2, after the drug-DNA mixture, drug alone, or DNA alone was incubated at 25 °C for 24 h. The molar ratio of DNA (measured as nucleic acid phosphate) and drug was 13 with a drug concentration of 0.85 X 10^{-5} M. $\Delta T_m = T_m$ (drug-DNA mixture) - T_m (DNA alone). The T_m of calf thymus DNA is 65.6 °C. [b]Drug concentration causing 50% inhibition of cell growth with 3-day drug incubation. L1210 cells were at a concentration of 5 X 10^3 cells/mL at the start of the experiment. [c]Mice bearing ip-implanted P388 leukemia with ip drug treatment on days 1, 5, and 9; % ILS, percent increase in life span of treated compared to control mice. Most values are rounded averages of two or more experiments with repeat experiments showing good reproducibility. OD, optimum dose. [d]Groups of CD2F1 mice received a single iv dose over a range of dose levels including acutely lethal doses and deaths were recorded daily for 90 days. Deaths occurring later than 20 days were classified as "delayed." [e]Tested as racemic mixture of (+)- and (-)-enantiomers. [f]Number of day 30 survivors/total. [g]Highest dose tested.

possible presence of 0.5% (+)-ABC. Similarly, the 55% increased life span observed with 200 μg/kg (-)-ABC could in part be entirely due to contamination by (+)-ABC. Thus, the cytotoxicities and efficacies reported for (-)-ABC and (-)-AB must be interpreted with caution. What can be said with assurance is that these (-)-CPI agents are dramatically less active biologically than are their (+)-enantiomers. (3) (+)-A and (-)-A show yet another relationship. They differ in *in vitro* cytotoxic potency by only an order of magnitude. Since all of the various enantiomers shown in Chart I were prepared from the same diasteriomerically resolved intermediate, and since the observed potencies of (-)-ABC and (-)-AB are inconsistent with high (10%) levels of (+) contamination, it follows that the cytotoxic potency observed for (-)-A is real and does not reflect a significant contribution by (+)-A contamination.

Discussion

CC-1065 is also a natural product that binds strongly to the DNA minor groove in AT-rich regions, but unlike distamycin, it reacts with DNA to form a covalent adduct.[7-10] The irreversibility of this molecular lesion undoubtedly contributes to the six orders of magnitude greater cytotoxic potency of (+)-CC-1065[7b] relative to distamycin.[25] However, (+)-CC-1065 is not simply a natural product version of N-bromoacetyldistamycin, in which sequence- (AT-) selective minor groove binding constrains a reluctant alkylating center to react with nucleophilic atoms in proximity to the binding site.[2] If this were the case, it would be expected that truncated (+)-CPI analogues of (+)-CC-1065, whose properties, such as the ability to affect the thermal melting of DNA (Table IV), indicate much weaker binding, would be unable to alkylate DNA sequence selectively, if at all. In fact, while the overall rates of the alkylation reactions with DNA are lower for the less strongly binding (+)-CPI agents, it has been shown that the set of sequences in which adenines are alkylated is the same for (+)-CPI analogues regardless of binding ability.[22] We have interpreted this result as demonstrating that the sequence selectivity of (+)-CPI compounds is primarily determined, not by sequence-recognizing noncovalent binding, but by the sequence requirements inherent in the covalent-bonding step itself.[22] This might be postulated to arise from a sequence-dependent conformational flexibility allowing adenine in certain environments to approach the electrophilic center sufficiently closely for bond formation. It might also be postulated to involve sequence-dependent electrophilic activation of the (+)-CPI moiety simultaneously with adenine N3 nucleophilic attack, in a process similar to the bifunctional catalysis observed in many enzyme-substrate reactions.[26] From a purely structural perspective, such a process resembles suicide substrate-enzyme inactivation.

The strand breakage assay used in these studies can only detect covalent reaction sites of (+)-CC-1065.[27] Recent spectroscopic studies of (+)-CC-1065 with various synthetic oligomers reveal that the drug may bind strongly to certain sequences with which it does not react covalently.[28] The most dramatic example of this behavior is provided by the strong noncovalent complex between (+)-CC-1065 and the duplex of 5'CGCGAATTCGCG. This oligomer also provides an excellent binding site for netropsin, and indeed, the first crystal structure of a netropsin-oligomer complex was obtained with a 5-bromocytosine-modified duplex of this sequence.[29] Such evidence of reversible binding of (+)-CC-1065 to the same minor groove locations selected by netropsin or distamycin suggests that the purely noncovalent recognition mechanisms of (+)-CC-1065 are not greatly different from those of other minor groove binding agents. It also emphasizes the overlying powerful A vs T discriminating ability of the (+)-CPI covalent reaction.

$$\text{Drug} + \text{DNA} \underset{}{\overset{K_b}{\rightleftharpoons}} (\text{Drug-DNA})_{noncov} \xrightarrow{k_r} (\text{Drug-DNA})_{cov}$$

In terms of the generalized kinetic scheme in eq. 1, a drug may preferentially react with a particular sequence through either a very high binding affinity (K_b) at that sequence, or a covalent reaction whose activation energy is especially favorable for that sequence.[2] If k_r is not sequence-dependent, the sites of covalent reaction will be those having the highest on-rate of noncovalent binding (if k_r is fast relative to dissociation) or the highest K_b (if k_r is slow). (+)-CC-1065, although it has a very high binding affinity for AT-rich sequences, is also constrained by a sequence-dependent k_r. Simple (+)-CPI analogues with low binding affinity to DNA still react, though at a lower overall rate, at the same sequences that react with (+)-CC-1065, sequences that allow a favorable k_r for covalent reaction between adenine and the (+)-CPI moiety.

Implicit in the sequence recognition capability of the (+)-CPI alkylating segment is that alterations to the structure of that moiety could have significant effects on its absolute reactivity and relative reactivity (sequence selectivity) with DNA. The accessibility, through synthesis, of the unnatural enantiomer of the alkylating segment,[20a] and the methodology to prepare (-)-CC-1065 and other (-)-CPI analogues,[12] have provided an ideal system with which to explore these implications. The identical physical and chemical properties of the (-)-CPI and (+)-CPI agents focus comparison on the differences of their interactions with the chiral DNA helix. Indeed, we have found that (-)-CC-1065, while it superficially resembles (+)-CC-1065 in reacting covalently with certain adenines in the minor groove of DNA in AT-rich regions, shows subtle but significant differences. It lies in the opposite direction in the groove, it alkylates a nonidentical set of adenines, it reveals a different consensus requirement for the covalent adduct, and, finally, its simplified (-)-CPI analogues fail to react with DNA. The sole exception is (-)-A, which reacts at the (+)-CC-1065 sites rather than at the (-)-CC-1065 sites. How can these observations be rationalized in terms of structural models of drug-DNA bonding? The following discussion describes, not a rigorous modeling study, but a qualitative attempt to illustrate some of the possible interactions that might underlie the experimental observations described in this work.

Figure III presents stereo views of (+)-CPI (I) and (-)-CPI (II) adducts with the duplex oligomer of 5'CGTTA*ACG-3' (only the 5'GTTA*A-3' portion is shown for clarity). These structures were generated by the modeling program MOSAIC, an Upjohn-developed program based on Macromodel.[30] Using the AMBER set of force field parameters,[31,32] we performed energy minimizations starting from idealized DNA helices to which a ring-opened CPI moiety had been attached. These structures are intended to help illustrate possible interactions of these agents with recognition elements in the minor groove of DNA. The adduct is assumed to resemble the activated complex in the transition state for the alkylation reaction. The covalently modified adenine (A*) is in a good consensus bonding sequence for both (+)-CC-1065 (5'TTA*) and (-)-CC-1065 (5'TA*A). In structure I, the bonded (+)-CPI is oriented to place the acyl appendage [for simplicity the adduct of (+)-A was used for the modeling] in the minor groove, lying to the 5'-side of the covalently modified adenine. The energy-minimized structure places the phenolic hydroxyl group within hydrogen-bonding distance of the phosphate oxygen (O-O distance = 2.6 Å, I; Figure III) on the opposite strand, two base pairs down from the covalently bonded adenine. While the proximity of these groups in the energy-minimized model most likely reflects the magnification of such dipolar interactions in the gas-phase calculations employed, it also suggests that such an interaction is possible and thus might play a role in protic activation of the (+)-CPI moiety. This orientation also positions the pyrrolidine methylene carbon for a favorable van der Waals contact (C-C distance = 3.3 Å) with the C2 of the adenine one base pair down, on the opposite strand, and close to the O2 of the thymine of that base pair (C-O distance = 3.1 Å, I; Figure III). There is a favorable van der Waals distance between the carbon atom of the acyl appendage (methyl, in I) and the C2 of the adenine on the opposite strand, two base pairs in the 5'-direction from the bonded adenine (C-C distance = 3.8 Å, I; Figure III).

All of the observed contacts in I lie on the 5'-side of the adduct. Even the indole methyl substituent, while it faces the 3'-side of the adduct, lies far from the base pair to the 3'-side of the alkylated adenine. Thus, this model is consistent with the (+)-CC-1065 consensus data, which only show a general sequence requirement in the 5'-direction from the adduct, and with footprinting, as well as NMR data.[14] It further suggests that a favorable dipolar interaction between the oxo group of (+)-CPI and a phosphate group in the backbone of the helix is compatible with favorable van der Waals interactions of this alkylating moiety.

A different model (II, Figure III) is required to account for the observed reactivity of (-)-CC-1065 and (-)-AB'C' at the (-)-CC-1065 alkylation sites. [Recall that we are using the term "alkylation site" to refer not only to the specific adenine involved in bonding but also to the sequence context as summarized in the consensus sequence for the drug. In II, the acyl appendage lies in the minor groove to the 3'-side of the covalently modified adenine. This model accommodates the footprinting findings for (-)-CC-1065, the recovery of essentially only adenine N3 adducts, and the high degree of minor grooves overlap between (+)-CC-1065 sites and (-)-CC-1065 sites on the DNA fragments examined in this work. In the energy-minimized structure

I

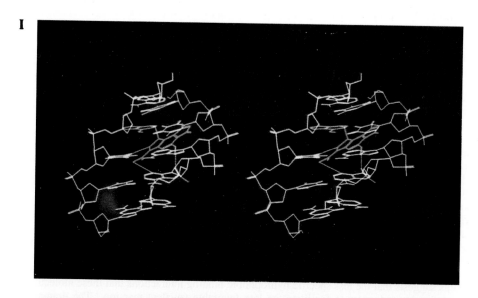

II

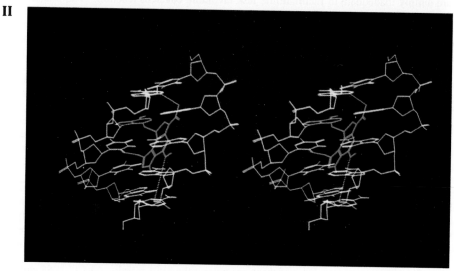

Figure III. Stereoscopic diagrams of the adducts of (+)-A (I, top) and (-)-A (II, bottom) with the duplex 5'CGTTA*ACG (showing only 5'GTTA*A) generated as described in the text. The drug moiety is shown in blue, and the covalently modified adenine (A*) and the phosphate group on the opposite strand, two base pairs away, are shown in magenta. White dotted lines depict some distances that are discussed in the text.

II, the acyl methyl group is brought into favorable contact with the adenine lying 3' to the adduct, particularly with its anomeric carbon (C-C distance = 3.6 Å) and its N3 atom (C-N distance = 3.8 Å, II; Figure III). The indole methyl group enjoys contact with the thymine O2 (C-O distance = 3.3 Å, II; Figure III) and the adenine C2 (C-C distance = 3.4 Å) and adenine N3 (C-N distance = 3.8 Å) in the base pair immediately 5' to the adduct. This 5' base pair recognition of II provides a modeling rationale for the observed 5' flanking AT requirement of (-)-CC-1065, even though the acyl appendage lies in the 3'-direction from the adduct. To achieve the favorable 3'-contacts of the adduct structure II, the phenol hydroxyl group interaction with the phosphate group (O-O distance = 5.3 Å, II; Figure III) has been sacrificed. If indeed the phosphate proximity in I is important to provide electrophilic activation of the cyclopropylcyclohexadienone moiety (nucleophilic ring opening of CPI compounds in solution is strongly acid catalyzed),[20b,30,33] then (-)-CPI compounds would be expected to suffer a sever loss of reactivity with DNA when oriented as in II.[34] The observations (Table III) that (-)-A, (-)-AB, and (-)-ABC are extremely unreactive at (-)-CC-1065 sites is consistent with this model. Particularly in the case of (+)- and (-)-A, where differential binding interactions are minimized, it is apparent that (-)-A reacts at (-)-CC-1065 sites (II) at least two orders of magnitude more slowly than (+)-A reacts at (+)-CC-1065 sites (I). Indeed, (-)-A appears better able to react at (+)-CC-1065 sites (I, *vide supra*). Clearly the (-)-CPI alkylating structure by itself, in contrast to (+)-CPI, does not enjoy uniquely facilitated alkylation.

Why then do (-)-CC-1065 and (-)-AB'C' readily alkylate DNA at the (-)-CC-1065 sites? The answer must lie with the interaction of the large bis(pyrroloindole) acyl appendage with the minor groove when the drug is oriented as depicted in II. The low reactivity of (-)-AB and (-)-ABC at these sites suggests that the less specific hydrophobic binding interactions that these molecules allow are simply inadequate to facilitate the less favorable covalent reaction. The dramatic alkylation rate enhancement resulting from attachment of the full carbon skeleton of the natural products to the (-)-CPI moiety strongly implies that specific AT recognizing interactions, such as favorable van der Waals contacts involving the rigidly held framework of pyrrolidine methylene groups, play a dominant role in both the absolute and relative reactivities or (-)-CC-1065 and (-)-AB'C' with DNA.

That the pyrrolidine methylene groups of the B' and C' segments can play a decisive role in minor groove binding is supported by the observation of an induced circular dichroism of this dimeric fragment [from the alkaline hydrolysis of (+)-CC-1065] in the presence of calf thymus DNA.[35] In contrast, the amide linked bis(indole-2-carboxylate) from the alkaline hydrolysis of (+)-ABC show no induced circular dichroism with DNA.[36] Important synthetic[37] and computational[38] studies by Boger have contributed to an appreciation of the strong minor groove affinity of B'-C' oligomers. Indeed, these studies are highly relevant to the DNA sequence recognition characteristics of (-)-CC-1065.

Interestingly, (-)-CC-1065 does not discriminate significantly among 5'AA*A, 5'TA*A, and 5'AA*T sequences (Table IV), as expected from an agent whose primary recognition mode fundamentally resembles that of noncovalent minor groove binders.[1c] There is, however, clearly a lower degree of reactivity at 5'TA*T sequences, and this may reflect a particular bonding limitation in II at that sequence. It is intriguing that the sequencing studies show that of the fourteen (-)-CC-1065 sites not overlapping (+)-CC-1065 alkylation sites, eight are within or adjacent to 5'AATT sequences. Four of these are among the most reactive (-)-CC-1065 sites (Table II). This sequence, as discussed above, is a highly favorable binding site for netropsin and related reversible binding agents and is an exceptionally stable, reversible (noncovalent) binding site for (+)-CC-1065. The implication of these observations is that AT-specific minor groove binding interaction attributable to the B' and C' segments appear to be the main determinant of the reactivity of (-)-CC-1065 at its target adenines. In terms of eq. 1, k_r for (-)-CC-1065 is not greatly facilitated by favorable transition-state stabilization at particular sequences. Rather, high binding affinity for certain sequences is necessary for significant alkylation to occur there.

The similar level of cytotoxic potency (ID_{50} values) of (+)- and (-)-CC-1065 against several cell lines observed by us (Table IV) and others[39] is somewhat fortuitous and misleading. Cytotoxicity, or course, may be the result of diverse lethal biochemical processes. While (-)-CC-1065 and (-)-AB'C' are, if anything, slightly more potent in the biological systems tested than are (+)-CC-1065 and (+)-AB'C' (Table IV), that are about an order of magnitude less reactive than their (+)-counterparts toward DNA under the reaction conditions used in the alkylation experiments (Table III). Thus, the rank order correlation of potency and reactivity observed for compounds reacting at (+)-CC-1065 sites must be considered separately from a similar correlation for compounds reacting at (-)-CC-1065 sites. This is expected if different biochemical mechanisms for cytotoxicity result from the different molecular lesions on DNA. Because it seems reasonable that the observed low level of reaction of the simpler (+)-CPI agents at adenines in (-)-CC-1065 sites (Table III) produces adducts lying to the 5'-side of the alkylated adenine, rather that the 3'-side, these (+)-CPI agents should not be included in correlations of reactivity and potency of (-)-CPI agents at (-)-CC-1065 sites. Further delineation of the biological and biochemical differences between (+)- and (-)-CC-1065 is in progress. One striking difference already noted (Table IV) is that (-)-CC-1065 does not cause delayed death in mice.

The full carbon structure right-hand appendage has a somewhat more subtle, but nevertheless, significant effect on the properties of (+)-CPI as well.[22] At some sequences, it appears that the specific noncovalent interactions of the two pyrrolidine rings with the minor groove can either promote an otherwise less favorable bonding reaction [operationally similar to (-)-CC-1065] or inhibit, perhaps through a competitive binding mode, a bonding reaction that is available to simpler (+)-CPI analogues. While these specific binding interactions are clearly not responsible for the large majority of (+)-CC-1065-reactive alkylation sites on DNA restriction fragments,[22] they appear to have profound biological significance. Only (+)-CC-1065 and (+)-AB'C' [and not the simpler (+)-CPI agents] cause delayed death in mice at therapeutic doses.[19,20,39] This difference could conceivably reflect adduct formation at unique sites with these agents. Alternatively, unique structural and functional consequences to DNA caused by these specific binding interaction, available only to (+)-CC-1065 and (+)-AB'C' covalent adducts, may be critical for the expression of this toxicity.

In earlier biosynthesis studies,[40] we had concluded that *S. zelensis* incorporated a single carbon atom from methionine to form the cyclopropyl group of (+)-CC-1065. Was it strictly fortuitous that the 7bR,8aS enantiomer was formed, or was this the result of selective pressure? Only the (+)-CPI moiety appears to recognize the minor groove through sequence-dependent covalent reactivity. The synthetically prepared unnatural enantiomer of the natural product, while it is also a potent cytotoxin, has a fundamentally different ultimate interaction with double-helical DNA and at least some different resulting biological properties. It is speculative but conceivable that these different biological properties would not have given the microorganism a particular survival advantage. Likewise, it is interesting to note that the B'C' pyrrolidine carbon atoms are not incorporated through tyrosine, as might, *a priori*, be supposed. Instead, two carbon atoms from serine are incorporated into each of the tyrosine-derived indoles of the B'C' segment.[40] It is this additional structural elaboration that leads to the subtle modulation of the (+)-CPI covalent reaction site or adduct structure, which in turn appears to underlie delayed toxicity in rodents. Conceivably, this structural elaboration might also result from selective pressure. An exciting and general implication of these considerations is that the very biosynthesis of natural product toxins may provide important clues about their essential reactivity features. If these often elaborate molecules have evolved through multiple levels of structural refinement to optimize key biological interactions, then a knowledge of the biosynthetic origins of the structural components could serve as a valuable inspiration to structure-activity investigations.

The synthetic toxin (-)-CC-1065 appears to have "pirated" an AT-specific minor groove binding ligand that was exquisitely designed through evolution to optimize the critical (to the producing organism) DNA interaction of (+)-CC-1065. It uses this structural key to gain entry to AT regions generally favorable to (+)-CC-1065 (and other minor groove binders). Once there, however, the rules governing the critical covalent reaction are quite different than for the natural product. One might also regard the synthetic (+)-CPI analogues, such as (+)-ABC, as having "pirated" the sequence-discriminating alkylating moiety of the natural product. Separated from the natural binding ligand, and efficiently delivered to the reactive DNA site by less specific binding ligands, these analogues are proving to be highly effective and potent antitumor agents in animal testing.[20]

It has been suggested that the predilection of natural product defense molecules for the minor groove is an evolutionary adaptation to the accessibility of this groove, which is not occupies by endogenous proteins.[1c] The extraordinary and highly evolved DNA binding and bonding interactions of (+)-CC-1065 invite even further intriguing speculations. The "recognition" by (+)-CC-1065 of its critical reactive sequences in DNA may involve both cognitive and response features; i.e., these DNA sequences may be pharmacological receptors.[41] They might, for example, be normal binding regions for regulatory proteins, and modifications of those sequences could affect protein binding, and ultimately the processes that those proteins control. Alternatively, unusual DNA local structures that are sequence dependent could be receptors for selective drug action. Our clearer understanding, resulting from the present work, of the structural basis for the binding and bonding selectivity of (+)-CC-1065 and related molecules will help to focus the study of the consequences of those interactions and the efforts to exploit them for therapeutic ends.

Acknowledgment. Supported in part by a grant from the U.S. Public Health Service (CA-49751) and the Robert A. Welch Foundation (F890) and the Burroughs Welcome Fund. We gratefully acknowledge Upjohn colleagues L. H. Li, for conducting *in vitro* and *in vivo* antitumor studies, and W. C. Krueger, for conducting the DNA melting temperature studies. Additional thanks go to our colleagues at the University of Texas: Chong-Soon Lee, for sequencing data, and David M. Bishop, for proofreading, editing, and preparing the manuscript.

References

(1) (a) Dickerson, R. E. *J. Mol. Biol.* **1983**, *166*, 419. (b) Calladine, C. R. *J. Mol. Biol.* **1982**, *161*, 343. (c) Dickerson, R. E.; Kopka, M. L.; Pjura, P. E. In *DNA-Ligand Interactions. From Drugs to Proteins*; Guschlbauer, W., Saenger, W., Eds.; Plenum Press: New York, 1987; pp. 45-66.

(2) Warpehoski, M. A.; Hurley, L. H. *Chem. Res. Toxicol.* **1988**, *1*, 315.

(3) (a) Kohn, K. W.; Harley, J. A.; Mattes, W. B. *Nucleic Acids Res.* **1987**, *15*, 10531. (b) Benasutti, M.; Ejade, S.; Whitlow, M. D.; Loechler, E. L. *Biochemistry* **1988**, *27*, 472.

(4) (a) Hertzberg, R. P.; Hecht, S. M.; Reynolds, V. L.; Molineux, I. J.; Hurley, L. H. *Biochemistry* **1986**, *25*, 1249. (b) Zakrzewska, K.; Pullman, B. *J. Biomol. Struct. Dyn.* **1986**, *27*, 472.

(5) Buckley, N. *J. Am. Chem. Soc.* **1987**, *109*, 7918.

(6) Baker, B. F.; Dervan, P. B. *J. Am. Chem. Soc.* **1989**, *111*, 2700.

(7) (a) Hanka, L. J.; Dietz, A.; Gerpheide, S. A.; Kuentzel, S. L.; Martin, D. G. *J. Antibiot.* **1978**, *31*, 1211. (b) Martin, D. G.; Chidester, C. G.; Duchamp, D. J.; Mizsak, S. A. *J. Antibiot.* **1980**, *33*, 902.

(8) Reviews: (a) Reynolds, V. L.; McGovren, J. P.; Hurley, L. H. *J. Antibiot.* **1986**, *39*, 319. (b) Hurley, L. H.; Needham-VanDevanter, D. R. *Acc. Chem. Res.* **1986**, *19*, 230.

(9) (a) Hurley, L. H.; Reynolds, V. L.; Swenson, D. H.; Perzoid, G. L.; Scahill, T. A. *Science* **1984**, *226*, 843. (b) Reynolds, V. L.; Molineux, I. J.; Kaplan, D. J.; Swenson, D. H.; Hurley, L. H. *Biochemistry* **1985**, *24*, 6228.

(10) Chidester, C. G.; Krueger, W. C.; Mizsak, S. A.; Duchamp, D. J.; Martin, D. G. *J. Am. Chem. Soc.* **1981**, *103*, 7629.

(11) Martin, D. G.; Kelly, R. C.; Watt, W.; Wicnienski, N.; Mizsak, S. A.; Nielsen. J. W.; Prairie, M. D. *J. Org. Chem.* **1988**, *53*, 4610.

(12) Kelly, R. C.; Gebhard, I.; Wicnienski, N.; Aristoff, P. A.; Johson, P. D.; Martin, D. G. *J. Am. Chem. Soc.* **1987**, *109*, 6837.

(13) (a) Swenson, D. H.; Li, L. H.; Hurley, L. H.; Rokem, J. S.; Petzold, G. L.; Dayton, B. D.; Wallace, T. L.; Lin, A. H.; Krueger, W. C. Cancer Res. 1982, 42, 2821. (b) Krueger, W. C.; Li, L. H.; Moscowitz, A.; Prairie, M. D.; Petzold, G. L.; Swenson, D. H. Biopolymers 1985, 24, 1549.

(14) Scahill, T. A.; Jensen, R. M.; Swenson, D. H.; Hatzenbuhler, N. T.; Petzold, G. L.; Brahme, N. M. *Biochemistry* **1990**, *24*, 2852.

(15) Needham-VanDevanter, D. R.; Hurley, L. H. *Biochemistry* **1986**, *25*, 8430.

(16) Hurley, L. H.; Needham-VanDevanter, D. R.; Lee, C.-S. *Proc. Natl. Acad. Sci. U.S.A.* **1987**, *84*, 6412.

(17) (a) Li, L. H.; Swenson, D. H.; Schpok, S.; Kuentzel, S. L.; Dayton, B. D.; Krueger, W. C. Cancer Res. 1982, 42, 999. (b) Bhuyan, B. K.; Newell, K. A.; Crampton, S. L.; Von Hoff, D. D. Cancer Res. 1982, 42, 3532.

(18) Martin, D. G.; Biles, C.; Gerpheide, S. A.; Hanka, L. J.; Krueger, W. C.; McGovren, J. P.; Mizsak, S. A.; Neil, G. J.; Stewart, J. C.; Visser, J. *J. Antibiot.* **1981**, *34*, 1119.

(19) McGovren, J. P.; Clarke, G. L.; Pratt, E. A.; DeKoning, T. F. *J. Antibiot.* **1984**, *37*, 63.

(20) (a) Warpehoski, M. A. *Tetrahedron Lett.* **1986**, *27*, 4103. (b) Warpehoski, M. A.; Gebhard, I.; Kelly, R. C.; Krueger, W. C.; Li, L. H.; McGovren, J. P.; Prairie, M. D.; Wicnienski, N.; Wierenga, W. *J. Med. Chem.* **1988**, *31*, 590. (c) DeKoning, T. R.; Kelly, R. C.; Wallace, T. L.; Li, L. H. *Proc. Am. Assoc. Cancer Res.* **1989**, *30*, 491.

(21) Swenson, D. H.; Petzold, G. L. Unpublished results.

(22) Hurley, L. H.; Lee, C.-S.; McGovren, J. P.; Warpehoski, M. A.; Mitchell, M. A.; Kelly, R. C.; Aristoff, P. A. *Biochemistry* **1988**, *27*, 3886.

(23) Hurley, L. H., Warpehoski, M. A., Lee, C.-S., McGovren, P. J., Scahill, T. A., Kelly, R. C., Mitchell, M. A., Wicnienski, N. A., Gebhard, I., Johnson, P. D., and Bradford, V. S. *J. Am. Chem. Soc.* **1990**, in press.

(24) The consensus sequence is a composite sequence derived by determining which of the nucleotide bases or base combinations occurs with the highest frequency at a given position relative to the alkylated adenine.

(25) Lown, J. W.; Krowicki, K.; Balzarini, J.; DeClercq, E. *J. Med. Chem.* **1986**, *29*, 1210.

(26) Fersht, A. In *Enzyme Structure and Function*, 2nd ed.; W. H. Freeman and Co.: New York, 1985.

(27) Footprinting studies with MPE-Fe(II) were initially thought to reveal only covalent bonding sites for (+)-CC-1065 (ref. 22); however, the footprint labeled B in Figure III of that reference may possibly reflect a strong noncovalent complex, particularly in view of the spectroscopic studies of reference [next reference number]. The attribution of that footprint to a covalent adduct on the opposite strand is an error; there is no evidence of a covalent adduct at that site on the opposite strand.

(28) Theriault, N. Y.; Krueger, W. C.; Prairie, M. D. *Chem.-Biol. Interact.* **1988**, *65*, 187.

(29) Kopka, M. L.; Yoon, C.; Goodsell, D.; Pjura, P.; Dickerson, R. E. *Proc. Natl. Acad. Sci. U.S.A.* **1985**, *82*, 1376.

(30) Still, C. W., Columbia University,

(31) Weiner, S. J.; Kollman, P. A.; Case, D. A.; Singh, U. C.; Ghio, C.; Alogona, G.; Profeta, S., Jr.; Weiner, P. *J. Am. Chem. Soc.* **1984**, *106*, 765.

(32) Weiner, S. J.; Kollman, P. A.; Nguyen, D. T.; Case, D. A. *J. Comput. Chem.* **1986**, *7*, 230.

(33) Warpehoski, M. A.; Harper, D. E. Unpublished results.

(34) Another feature distinguishing model I from the diastereomeric model II is the steric conformation around the newly formed N-C bond. Viewed along the axis of this bond, the adenine ring in structure I is almost fully staggered relative to the nearest carbon-carbon bond of the drug moiety. In structure II, in contrast, the adenine ring is nearly eclipsed by this carbon-carbon bond. An interaction such as this might significantly add to the instability of a transition state resembling II.

(35) Martin, D. G.; Mizsak, S. A.; Krueger, W. C. *J. Antibiot.* **1985**, *38*, 746.

(36) Warpehoski, M. A.; Krueger, W. C. Unpublished results.

(37) Boger, D. L.; Coleman, R. S.; Invergo, G. J. *J. Org. Chem.* **1987**, *52*, 1521.

(38) Coleman, R. S.; Boger, D. L. In *Studies in Natural Products Chemistry*; Atta-ur-Rahman, Ed.; Elsevier: Amsterdam, 1988; Vol. 2.

(39) Boger, D. L.; Coleman, R. S. *J. Am. Chem. Soc.* **1988**, *110*, 4796.

(40) Hurley, L. H.; Rokem, J. S. *J. Antibiot.* **1983**, *36*, 383.

(41) Hurley, L. H. *J. Med. Chem.* **1989**, *32*, 2027.

COURSE OF RECOGNITION AND COVALENT REACTIONS BETWEEN MITOMYCIN C AND DNA: SEQUENCE SELECTIVITY OF A CROSS-LINKING DRUG.

M. TOMASZ, H. BOROWY-BOROWSKI
Department of Chemistry, Hunter College,
City University of New York,
New York, NY 10021, USA

B.F. McGUINNESS
Department of Chemistry, Columbia University
New York, NY 10027, USA

ABSTRACT. Precovalent association of the reductively activated form of mitomycin C was mimicked by stable, structurally similar "active-form analogs". The binding constant K_b is relatively low (approx. 10^{-3} M^{-1} in 0.01 M NaCl/0.001 M sodium phosphate, pH 7.5 buffer). Binding is primarily electrostatic and is insensitive to DNA composition, sequence and denaturation. Mitomycin C itself shows no binding to DNA. Covalent cross-linking of poly(dG).poly(dC) by mitomycin C resulted in the isolation of a mitomycin-d(GpG) adduct, characterized as a mitomycin cross-link between adjacent guanines (intra-strand cross-link). A large preference by mitomycin C to form a cross-link at CG.CG over the GG.CC sequence was observed. The order of preferred sequences is now established as CG.CG>GG.CC>GC.GC≈0. It is shown that the orientation of the monoadduct precursors in the minor groove is fixed and unique and is such that it is optimal at CG.CG but prohibitive at GC.GC sequences for cross-link formation. A role for the pre-covalent binding in these processes is discussed.

Introduction

Mitomycin C (**1**; MC) is a potent antitumor antibiotic, used widely in clinical cancer chemotherapy. The main cellular target of its cytotoxic activity is, by all indications, DNA. The molecular basis of the action of MC resides in its ability to bind covalently to DNA and cross-link the complementary strands [1]. The chemical basis of these earlier observations was recently elucidated. Under physiological conditions MC is unreactive towards DNA. Reduction of the quinone function, however, triggers a series of spontaneous reactions within the MC molecule, giving rise to a reactive form, **2** [2]. This then alkylates DNA both monofunctionally and bifunctionally, the latter action resulting in cross-links between the two strands of DNA. Re-exposure to air oxidizes the bound activated hydroquinone form of MC and stable quinone-form MC-DNA adducts can be isolated (Scheme I). The structure of these adducts have been determined [3-5]. The structures (monoadducts **3** and **4** and bis-adduct **5**) revealed that the alkylation of DNA by MC is exquisitely specific: both of the alkylating functions of

551

B. Pullman and J. Jortner (eds.), Molecular Basis of Specificity in Nucleic Acid-Drug Interactions, 551–564.
© 1990 *Kluwer Academic Publishers. Printed in the Netherlands.*

MC, namely the aziridine at C-1 and the carbamate at C-10 react exclusively with N^2-positions of guanines. The same adduct pattern was observed in DNA from living cells treated with MC [5,6].

There is considerable indication that among these different types of MC-induced modifications of DNA the inter-strand cross-link is the most cytotoxic. It is also essential for the antitumor activity, as seen by comparison of monofunctional and bifunctional MC analogs [7]. It is of great interest, therefore, to determine what factors govern the formation of this cross-link in DNA. A fundamental, metabolic factor is the redox environment which determines not only the overall extent of drug activation but also the balance between the pathways of monofunctional and bifunctional activation of MC (Scheme I) [8,9].

Another critical factor is the sequence of DNA. Obviously, formation of a cross-link of DNA by small molecules requires two reactive nucleotides in close proximity within a DNA sequence. The sequence relationship between the two nucleotides varies with the cross-linking agent. Psoralen, for example forms inter-strand cross-links selectively between two Ts at the AT.AT sequence [10], the nitrogen mustard, mechlorethamine cross-links the two guanines at GNC.GNC [11,12], and the platinum drugs cis-DDP and trans-DDP form intra-strand cross-links preferentially between guanines at GG and GNG, respectively [13]. The molecular basis for these selectivities may be diverse. "Best match" of the distance between the two reactive atoms in the cross-linking agent to that between two reactive positions in static non-distorted DNA may not be the critical factor, per se, as exemplified dramatically by the case of the nitrogen mustard cross-link [11,12]. Here, cross-links were formed between two guanines more distal from one another than those in another sequence even though this cross-link induced more distortion of DNA in the final product. Thus, in addition to distance restrictions, other factors, such as sequence specific recognition in non-covalent drug-DNA association or rate acceleration of bond formation exerted by specific surrounding nucleotides may play decisive roles in the overall sequence specificity of cross-links .

We have explored the sequence specificity of the mitomycin C cross-link and its causes. In this presentation we review recent results, as well as report new findings, and propose a mechanism for the course of recognition and cross-linking of DNA by MC.

Materials and Methods.

"Active-form analogs" of MC **7-10** were synthesized as previously described [3]. Binding constants K_b for equilibrium binding of "active-form analogs" of MC to DNA were determined from data obtained by the absorption titration method [14]. This was based on the observed hypochromicity of the 310-315 nm absorption band of the mitosene chromophore upon addition of DNA. Drug and DNA were mixed in various buffers and

Scheme I:

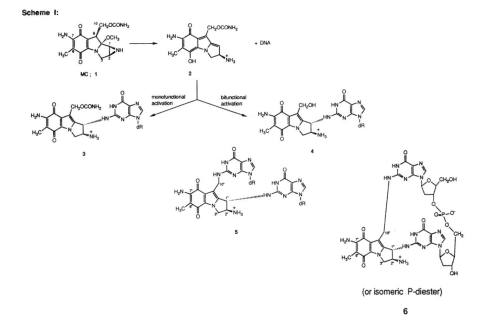

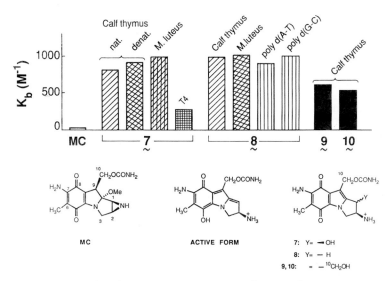

FIGURE 1: Binding constants K_b (in reciprocal mononucle-otide concentration M^{-1}) for binding of "active-form analogs" **7-10** to various DNAs as indicated in the figure. T4 is T4 phage DNA. The DNAs from natural sources were sonicated before use. Denatured calf thymus DNA was obtained by heating a solution of native DNA for 10 min at 100° then cooled in ice. Binding was measured in 0.01 M Tris buffer, pH 7.5.

absorbance of the solution at 310 nm was measured at 25° as function of added DNA concentration. K_b (intrinsic binding constant) was calculated from the data according to the general method [14]. Correlation coefficients for the linear fitted curve from experimental points were in the 0.95-0.99 range.

Cross-linking of synthetic oligo- and polynucleotides by MC in the presence of $Na_2S_2O_4$ as activating agent was done as described before [5, 18]. Quantitative analysis of the cross-link adducts was accomplished by digestion of the cross-linked polynucleotides to nucleosides by a mixture of pancreatic DNAse I, snake venom diesterase and alkaline phosphatase. The digest was separated into its components by reverse phase HPLC; the peaks in the elution pattern were detected by ultraviolet absorbance and quantitated by integration. All of this methodology was described in detail previously [4,5].

Results.

Non-covalent association of "active-form analogs" of MC with DNA and synthetic polynucleotides. Mitosene quinones 7 and 8 exhibited binding to calf thymus DNA (both native and denatured), M. luteus DNA, T4 phage DNA and the synthetic polynucleotides poly(dA-dT).poly(dA-dT) and poly(dG-dC).poly(dG-dC). The parent drug, MC (1) showed no binding affinity to any of the DNAs; this was confirmed by equilibrium dialysis experiments (data not shown). Binding by the 10-decarbamoyl mitosenes 9 and 10 was lower compared with that by 7 and 8 (Figure 1). The binding of 7 and 8 decreased with increasing pH, in the pH range of deprotonation of their $2"-NH_3(+)$ group (Figure 2). Increase of Tris buffer concentration resulted in lower binding (Figure 3). Similar results were obtained in solutions of increasing NaCl concentration containing 0.001 M sodium phosphate buffer, pH 6.0 (K_b = 2,601, 1,657 and 472 M^{-1}, at 0.001, 0.01 and 0.05 M NaCl respectively).

Isolation of the "intra-strand cross-link" adduct 6: Poly(dG-dC).poly(dG-dC) and poly(dG).poly(dC) were both submitted to cross-linking reactions with MC. After digestion of the resulting MC-polynucleotide complexes to nucleosides, the digests were analyzed by HPLC. The HPLC patterns were entirely different: Poly(dG-dC).poly(dG-dC) yielded the cross-link adduct 5, as reported previously [5] (Figure 4a). Poly(dG).poly(dC), however, produced predominantly a new adduct seen as a peak at 21 min (Figure 4b). The identity of this adduct was established as the "intra-strand cross-link" 6; this will be published elsewhere (B.F. McGuinness, M. Tomasz and K. Nakanishi, unpublished).

Relative yields of 5 and 6 from the same oligonucleotide: An equimolar mixture of 5'-d(ATGGGTAACGTAA) and 5'-d(TTAACGTT-ACCCAT) was reacted with MC under the cross-linking conditions (Methods). Quantitative analysis indicated the formation of both the inter-strand cross-link adduct 5 and intra-strand cross-link adduct 6, in 1:0.55 molar proportion, as determined by integration of the corresponding peak areas of the HPLC

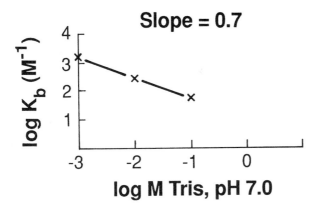

FIGURE 2: (a) Extent of binding of **7** and **8** to calf thymus DNA as function of pH.   The buffers used were 0.01 M MES, pH 5.5, 0.01 M Tris, pH 7.5 and 0.01 M Tris, pH 9.0.   (b) Spectrophotometric titration curve of **7**.   To an aqueous solution of **7** small increments of 0.1 N HCl or of 0.1 N NaoH were added.   After each addition the pH was measured by a pH-meter and the absorbance was determined at 285 nm.

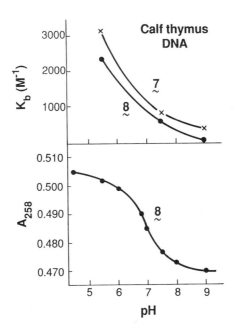

FIGURE 3:   Tris concentration dependence of the binding of **7**, to calf thymus DNA as a logarithmic plot.

pattern (Figure 4C).   From area measurements it was also
calculated that one of 6.25 molecules of duplex oligonucleotides had an inter-strand cross-link and one of 11.36 had one
intra-strand one.

Discussion.

It is now well established by diverse types of criteria, that
the various covalent modifications of DNA by reduced MC,
including the cross-link, are all relatively non-distortive;
see e.g. [15].   This fact allows one to narrow down the
potential cross-linkable sequences in DNA to GC.GC, CG.CG and
GG.CC.   (The last sequence would harbor an intra-strand cross-
link.)   Any Gs further apart would be cross-linkable only with
great distortion, since the distance between the mitomycin C-
1" and C-10" atoms as in 5 is only 3.36 A.   Indeed, both
cross-link adducts (5 and 6) originate from one of these
sequences [5,15,18].

Sequence Specificity of the Inter-Strand Cross-Link.   Experimental comparison for the preferences by MC among the three
sequences is now complete.   First, a virtually absolute
preference was found for CG.CG against GC.GC in inter-strand
cross-links, as shown by three laboratories independently [16-
18].   In each case, synthetic oligonucleotides were used as
sequence models.   In our own work a series of oligonucleotides
were submitted to the cross-linking reaction and the extent of
cross-linking was analyzed by HPLC (Figure 5).   The results
(Table I) showed that all oligonucleotides possessing a
central CG.CG gave good yields, while those having GC.GC
instead, exhibited complete resistance to cross-linking.
Other variations led to the same results: for example, in
group C the central sequences CGC.GCG and CGC.ICG both yielded
50% cross-links while CGC.GCI was resistant.   Thus, the
presence of the CG.CG sequence is required.
     What is the reason for this selectivity? Computation of the
relative binding energies of oligonucleotides cross-linked by
MC at CG.CG and GC.GC sequences revealed no significant
differences; the sequence preference could not be rationalized
by results of energy-minimization of the <u>final</u> products [16].
     Before proceeding further with the analysis of the causes of
this phenomenon, it is useful to draw the hypothetical scheme
of the steps involved in the cross-linking reaction of MC:

Scheme II:

$$M^* + DNA \xrightarrow{1} M^* \ldots DNA \xrightarrow{2} M^*\text{-}DNA \xrightarrow{3} M^*\!\!=\!\!DNA$$

$$\downarrow [O] \qquad\qquad \downarrow [O]$$

$$M\text{-}DNA \qquad\qquad M=DNA$$

(M^* is activated mitomycin, i.e. 2; $M^* \ldots$ DNA is non-covalent
complex; M^*-DNA is the monoadduct as in 3; M^* DNA is the

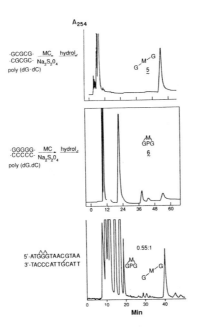

FIGURE 4: HPLC of nuclease digests of cross-linked oligo- and polynucleotides. (a) Digest of cross-linked poly(dG.dC).poly(dG-dC). (b) Digest of cross-linked poly(dG).-poly(dC). (c) Digest of the indicated oligonucleotide cross-linked by MC. The potential cross-links are indicated in the formula on the left of the figure.

TABLE I: Comparison of yields of cross-link at CG:CG and GC:GC sites in oligonucleotide duplexes.

Group	Sequence[a]	% Yield of Cross-Link	Comment
A	5'-TATAT [CG] ATATA 3'-ATATA [GC] TATAT	19	--
	5'-TATAT [GC] ATATA 3'-ATATA [CG] TATAT	0	--
B	5'-ATATA [CG] TATAT 3'-TATAT [GC] ATATA	47	--
	5'-ATATA [GC] ATATA 3'-TATAT [CG] TATAT	0	--
C	5'-ATATA [CGC] TATA 3'-TATAT [GCG] ATAT	47	5'-CGC- formed[b] 3'-GCG-
	5'-ATATA [CGC] TATA 3'-TATAT [GCI] ATAT	50	5'-CGC- formed 3'-GCI-
	5'-ATATA [CGC] TATA 3'-TATAT [ICG] ATAT	0	5'-CGC- not formed 3'-ICG-
D	5'-ATATA [CGCG] TATAT 3'-TATAT [GCGC] ATATA	80	Only one cross-link formed
	5'-ATATA [CG] TATAT 3'-TATAT [GC] ATATA	47	

cross-link adduct, as in **5**).   M is reoxidized form of the bound mitosene.

Evidence for step 1, the binding step, was recently shown by Crothers and coworkers [16], who observed competitive inhibition of the cross-linking of d(G-C)$_n$ in the presence of added d(A-T)$_n$ oligonucleotides and interpreted it as due to binding of M* to both AT-and GC- containing oligonucleotides with equal affinity (K$_b$ approx. 600 M^{-1}).   Since M* is a transient, unstable compound it is impossible to study its binding directly.   In the present work, we substituted M* (**2**) by structurally very similar, <u>stable</u> analogs and were able to characterize their binding to DNA.   The closest analog is **8** but **7** also showed identical binding properties.   As seen in Figure 1, they bind without any apparent discrimination in base composition, base sequence, or even secondary structure of DNA.   The observed intrinsic binding constants K$_b$ for the various DNAs (800 to 1000 M^{-1}) are in fair agreement with the estimated value of 600 M^{-1} by Teng et al. [16] for oligo(dG-dC) and oligo(dA-dT) under comparable conditions.   The decrease of K$_b$ with increasing ionic strength (Figure 3) indicates electrostatic binding primarily.   The somewhat weaker binding by **9** and **10** (Figure 1), demonstrates the contribution of H-bonding by the 10"-carbamate group in **7** and **8**, lacking in these derivatives.   The pH dependence of the binding follows the titration curve of the 2"-NH$_2$ of **7** or **9** (Figure 2), demonstrating clearly the predominance of the electrostatic component in the observed binding.

These results indicate that activated mitomycin gets concentrated along the DNA by weak, predominantly electrostatic binding (step 1 in Scheme II).   This should facilitate the first bonding step (step 2 in Scheme II).   It does not explain, however, the observed cross-link sequence specificity, as concluded also by Teng et al. [16].

We now show that the first bonding step (step 2) holds the key to the observed CG.CG specificity as follows.   It is immediately apparent from space-filling models that the monoadduct form of mitomycin may be situated inside the minor groove in two alternative orientations: A or B (Figure 6), differing by 180° rotation around the M-guanine bond.   In orientation A the C-10" position, bearing the carbamate points in the 3' direction; therefore it is poised for cross-link formation at CG.CG.   In B, the C-10" is oriented towards 5', therefore it could only form the GC.GC cross-link.   Since rotation of the large mitomycin (M) residue from A to B around the M1"-GN^2 bond is impossible in the tight minor grove, the orientation of the covalent monoadduct is fixed; consequently, the cross-link reflects this orientation.   Space-filling models of the monoadduct in DNA could only be fitted in the A orientation without distortion of the minor groove space [4]. More definitively, Remers, Kollmann and their collaborators [19,20] computed energy-minimized structures of MC monoadduct-oligonucleotide complexes, in which the flanking bases N and N' in the 5'-NGN' sequence were mutated systematically in

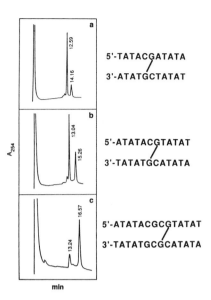

FIGURE 5:   HPLC separation of parent and cross-linked
oligonucleotides.   The later eluting component is the cross-
linked product.   From reference [18].

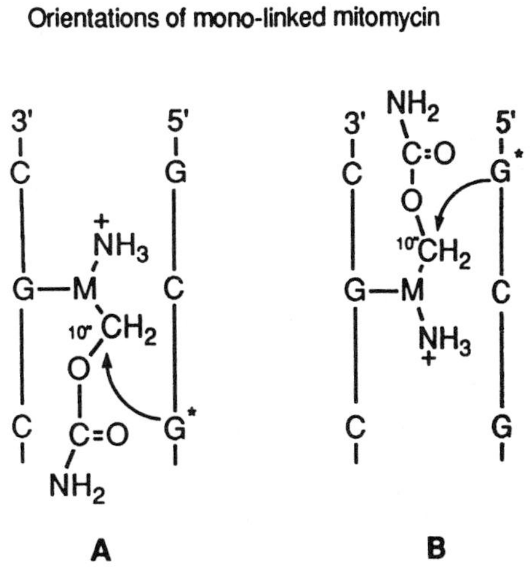

Orientations of mono-linked mitomycin

A   B

FIGURE 6: Opposite orientations of the mitomycin monoadduct.

order to include all four bases in both flanking positions. The striking conclusion of this work was that "there is one best way for mitomycin to bind in the minor groove of double helical DNA...", namely in orientation A according to our notation (Figure 6). According to their model this orientation allows for specific H-bonds between the M-7"-NH$_2$ group and a phosphate in the opposite strand, as well as between the protonated M-2"-NH$_3$(+) group and O^2 of the cytosine basepaired with the bonded G. Thus these modeling studies strongly support our proposed explanation of the CG.CG specificity of the cross-link.

We show experimental evidence that this orientation is indeed specific [18]: The single-stranded monoadducted oligonucleotide 11 (Figure 7) was hybridized with the parent strand and the monoadduct was reactivated in the hybrid form by Na$_2$S$_2$O$_4$. This resulted in >98% conversion to the cross-link 12. The monoadduct duplex assumed the thermodynamically most stable conformation(s) under the equilibrium conditions of the hybrid formation. From the yield of cross-linked product (>98%) it is apparent that >98% of the monoadduct had the A orientation in the equilibrated duplex originally; that is, the thermodynamically most stable conformation requires orientation A. In the control experiment using a monoadduct in the -GC- sequence (Figure 7, 13) no cross-link was observed. Thus, the unique orientation of the monoadduct as indicated by space-filling and theoretical models is dramatically demonstrated by this experiment and provides a satisfactory explanation for the observed absolute bias for CG.CG against GC.GC by the mitomycin inter-strand cross-link [21].

At what stage of the reaction does this orientation arise? It must be in place before the first bonding step (step 2 in Scheme II) because no change is possible afterwards. In order for bonding to occur, one has to assume that the observed non-specific overall binding (step 1) has a small component in equilibrium, in which the drug diffuses randomly into the minor groove, which may or may not be a stronger binding site than the "outside" binding sites or the major groove. It is, however, the only productive site with respect to bond formation in step 2, because the drug has to get close to the guanine bonding sites (N^2). This binding component also has the role of orienting the drug because without a proper orientation the drug simply cannot penetrate the minor groove, as discussed above. In support of these ideas Remers and Kollmann [19,20] presented computational evidence that activated mitomycins (2, as well as reduced 8) are best oriented non-covalently in the minor groove exactly as the bonded monoadduct. In summary, we believe that minor groove-specific binding which is masked by the overall non-specific binding is orientation-specific and this is responsible for the observed orientation of the bonded monoadduct.

Preference for Inter-Strand Over Intra-Strand Cross-Link Formation by MC. The fact taht the intra-strand cross-link adduct 6 retains the internucleotide phosphate group upon

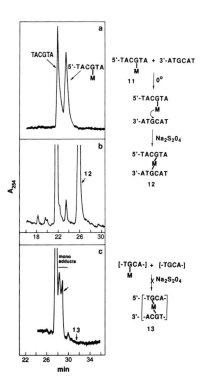

FIGURE 7: Sequence-dependent conversion of the mitomycin monoadduct to cross-link.   (a) HPLC pattern of TACGTA and its monoadduct derivative.   (b) Above compounds were mixed (2:1 ratio) and activated by $Na_2S_2O_4$: HPLC pattern of the reaction mixture.   (c) TATATGCATATA and its monoadduct mixture after activation by $Na_2S_2O_4$: HPLC of the reaction mixture.   (From ref. [18]).

nuclease digestion enables one to distringuish it from the inter-strand adduct **5**. The directly observed molar ratio of the two products from the oligonucleotide containing sites for both is 1:0.55 (Figure 4c). However, this value is obtained using an oligonucleotide which has one target site for **5** (CG.CG) but two for **6** (5'-GGG; see below). Simple correction of the relative yields by this statistical factor gives a 4-fold preference of inter-strand cross-links over intra-strand ones (1:0.27). Bulk DNA also yields **6**, in low proportion to **5** (unpublished observation). The cause of this bias is at present uncertain. Since formation of the new cross-link has not been reported before, no theoretical modeling studies have been carried out to our knowledge and therefore comparative net binding energies of the two cross-linked structures are

not known. Theoretical models for the respective <u>monoadduct</u>
precursors at -CG- and -GG- sequences were thoroughly compared
by Remers [20], however.   In these models the drug was
oriented in the usual way in both sequences (orientation A,
Figure 6).   The N^2-atom of the second guanine, poised for
attack to complete the cross-link (step 3 in Scheme II) is at
close binding distance to C-10" of the monoadduct <u>in both
models</u>.   Therefore, the orientation of the monoadduct is
equally favorable for cross-link formation at both sequences
and one has to search for an alternative explanation.   A
possibility is sequence-dependence of monoadduct formation
(step 2).   Again, Remers and coworkers studied the influence
of both the 5'-base and the 3'-base on monoadduct net binding
energy and found that 5'-CGT is energetically the most
favorable sequence [20]. This agrees qualitatively with our
observed   order   of   preference   for   the   two   cross-links.
Experimental determination of the sequence-selectivity of
mitomycin monoadduct formation should contribute critical
information about the causes of preference for DNA interstrand cross-links over intra-strand ones by mitomycin C.

Conclusions.

The hierarchy of the DNA base sequence specificity of cross-
link formation by mitomycin C is now established as

>   CG.CG>GG.CC>GC.GC≈O.

   The molecular basis for this finding was analyzed in terms
of a three-step reaction scheme, generally applicable for
cross-linking agents:

$$M^* + DNA \overset{1}{\leftrightharpoons} M^*...DNA \overset{2}{\longrightarrow} M^*-DNA \overset{3}{\longrightarrow} M^*{=}DNA$$

>   non-covalent   monoadduct   cross-link
>   complex   adduct

In step 1, a predominantly electrostatic DNA-binding affinity
leads to a concentration of activated mitomycin (M*) along the
DNA in a non-specific manner.   The subsequent steps 2 and 3
are restricted to occur at N^2 positions of guanines, in the
minor groove, as inherent in the covalent reactivity of
mitomycin.   For these covalent bondings to occur an M*
molecule has to diffuse into the minor groove.   M* fits only
in a certain orientation in the minor groove space, however,
governed by steric factors and favorable interactions between
DNA and drug functional groups.   This orientation allows the
drug molecule to form the first covalent bond (step 2) leading
to monoadduct.   The monoadduct retains the original orientation because rotation of the bound drug is kinetically

impossible. This orientation is also the thermodynamically stable one, as shown by our experiment. As a consequence, the last step (step 3), leading to the cross-link, can occur only at CG.CG and GG.CC sequences; cross-linking at GC.GC is prohibited.

The much lower rate of cross-linking at GG.CC than at CG.CG is likely to be based on an additional factor, since in this case monoadduct orientation allows for cross-liking at both sequences. This factor is probably the effect of the flanking base(s) on the rate of monoadduct formation. This hypothesis is open to experimental probing.

ACKNOWLEDGMENTS.

This research was supported by grant CA28681 from the National Cancer Institute and a "Research Centers for Minority Institutions" grant (RR03037) from the Division of Research Resources, NIH. We thank Dr. W.T. Bradner of Bristol Laboratories, Syracuse, NY, for a generous supply of mitomycin C. Professor Koji Nakanishi, Columbia University, is acknowledged for his advice and support.

REFERENCES AND FOOTNOTES.

1.   Iyer, V.N., & Szybalski, W. (1964) Science (Washington, D.C.) 145, 55-58.
2.   Moore, H.W. (1977) Science (Washington, D.C.) 197, 527-532; Danishefsky, S.J., & Ciufolini, M. (1984) J. Am. Chem. Soc., 106, 6424; Danishefsky, S.J., & Egbertson, M. (1986) J. Am. Chem. Soc., 108, 4648-4649; Tomasz, M., & Lipman, R. (1981) Biochemistry 20, 5056-5061.
3.   Tomasz, M., Jung, M., Verdine, G., and Nakanishi, K. (1984) J. Am. Chem. Soc., 106, 7367-7370.
4.   Tomasz, M., Chowdary, C., Lipman, R., Shimotakahara, S., Veiro, D., Walker, V., & Verdine, G.L. (1986) Proc. Natl. Acad. Sci. U.S.A. 83, 6702-6706.
5.   Tomasz, M., Lipman, R., Chowdary, C., Pawlak, J., Verdine, G.L., & Nakanishi, K. (1987) Science (Washington, D.C.) 235, 1204-1208.
6.   Chowdary, D., and Tomasz, M. (1987) Fed. Proc., Fed. Am. Soc. Exp. Biol. 46, 2037.
7.   Kojima, R., Driscoll, J., Mantel, N., and Goldin, A. (1972) Cancer Chemother. Rep. Part 2, 3, 121.
8.   Tomasz, M., Chawla, A.K., & Lipman, R. (1988) Biochemistry 27, 3182-3187.
9.   Sartorelli, A.C. (1986) Biochem. Pharmacol. 35, 67-69.
10. Gamper, H., Piette, J., & Hearst, J.E. (1984) Photochem. Photobiol. 40, 19-34.
11. Ojwang, J.O., Grenberg, D.A., and Loechler, E.L. (1989) Cancer Res. 49, 6529-6537.

12. Millard, J.T., Raucher, S., and Hopkins, P. B. (1990) J. Am. Chem. Soc. 112, 2459-2460.
13. Pinto, A.P., and Lippard, S.J. (1985) Proc. Natl. Acad. Sci. (USA) 82, 4616-4620.
14. Wolfe, A., Shimer, G.H., and Meeham, T., (1987) Biochem. 26, 6392-6396.
15. Norman, D., Live, D., Sastry, M., Lipman, R., Hingerty, B.E., Tomasz, M., Broyde, S., and Patel, D.J. (1990) Biochem. 29, 2861-2876.
16. Teng, S.P., Woodson, S.A., & Crothers, D.M. (1989) Biochemistry 28, 3901-3907.
17. Weidner, M.F., Millard, J.T., & Hopkins, P.B. (1989) J. Am. Chem. Soc. 111, 9270-9272.
18. Borowy-Borowski, H., Lipman, R. and Tomasz, (1990) Biochem. 29, 2999-3006.
19. Rao, N.S., Singh, C., & Kollman, P.A. (1986) J. Am. Chem Soc. 108, 2058-2068.
20. Remers, W.A., Rao, S.N., Wunz, T.P., & Kollman, P.A. (1988) J. Med. Chem. 31, 1612-1620.
21. Both space filling modeling and energy-minimized models indicate that this orientation was optimal independent of the nature of the flanking bases [20]. Nevertheless, it would be of interest to demonstrate it experimentally, e.g. by NMR, at a sequence other than 5'-CG.
22. It is conceivable that such a precovalent orientation has an accelerating effect on bond formation (step 2) [23]. Comparison of rates of monoadduct formation in duplex and single-stranded states of the same nucleic acid could throw light on this question.
23. Benasutti, M., Ejadi, S., Whitlow, M.D., and Loechler, E.L. (1988) Biochem. 27, 472-481.

TRIPLEX FORMING OLIGONUCLEOTIDE REAGENTS: RATIONALIZATION OF DNA SITE SELECTIVITY AND APPLICATION IN A PHARMACEUTICAL CONTEXT.

ROSS H. DURLAND
DONALD J. KESSLER
MADELEINE DUVIC* AND
MICHAEL HOGAN
Center for Biotechnology
Baylor College of Medicine
The Woodlands, TX 77381

* Department of Dermatology
University of Texas Health Science Center
Houston, TX

ABSTRACT. We show that triplex forming oligonucleotides (TFOs) can be designed against discrete target sites within the human c-myc and human epidermal growth factor receptor promotor. These TFOs bind with high sequence selectivity to functionally important sites in those promotors, with dissociation constants in the 10-9M to 10-7M range at physiological pH, temperature and divalent ion concentration. The distinguishing feature of this class of TFO is that it is stabilized by GGC and TAT triplets, bound in the antiparallel orientation relative to the polypurine-rich strand of the underlying duplex.

INTRODUCTION

The existence of triple helical DNA is an old observation and has now been well documented for synthetic polymers [1-3], for the binding of short SS oligonucleotides to cloned DNA fragments [4-9] and as the result of internal disproportionation of tandemly repeated, polypurine rich duplex DNA [9,10]. In particular, the elegant studies of Dervan [4,5] and Helene [6-9] have shown that by targeting a polypurine tracts within a naturally occurring DNA region, synthetic DNA oligonucleotides with the base composition (C_n, T_m) form stable triple helices at acidic pH. As expected from the polydT-polydA-polydT triplex described by Arnott [11] and the polydTC-polydGA-polydTC triplexes which have been well studied by Morgan and colleagues [12], formation of those localized triplexes is based upon formation TAT and C+GC triplets of the Hoogstein type [13], where C+ refers to protonated C. In triple helices with a polypyrimidine third strand, it had been shown that the third stand is bound parallel to the purine rich strand of the underlying duplex [4-9]. The parallel nature of those C+T triplexes is in good agreement with the parallel

B. Pullman and J. Jortner (eds.), Molecular Basis of Specificity in Nucleic Acid-Drug Interactions, 565–578.

structure inferred by Arnott and colleagues for the polydT-polydA-polydT triplex [11].

As has been well discussed [4-8], the requirement for protonation of C is a bothersome design problem since the pK of cytosine is near 3. As a first solution to that problem, Dervan has shown that the use of 5-methyl cytosine and 5 bromo uridine in the third strand appears to raise the apparent pK for triplex formation [5], thereby allowing for triplex formation in the pH 6 to pH 7 range (but not above). Clearly, a major goal for the field is to explore other nucleoside modifications or other nucleosides as TFO substituents, in order to allow for complex formation at physiological pH (near to 7.5).

Our interest in the TFO design problem is based on a program to determine the site selectivity of triplex formation on DNA under physiological conditions, using the c-myc control region as a model. Those preliminary observations [14] have shown that, at physiological pH, a discrete 27 base long oligonucleotide had the capacity to bind in a stable fashion and with high selectivity to a target site within the c-myc control region. The data suggested that the gross secondary structure of the complex might resemble the Arnott triplex model [11], but work below will show that the preferred structure of that type displays significant differences.

The c-myc target sequence in those preliminary studies was not a simple homopolymer in c-myc, but rather one of several DNA elements shown by the Bishop laboratory to be required in cis for initiation of mRNA synthesis from the c-myc promotor [15]. In that earlier work [14], the TFO of interest was designed so as to bind in the parallel orientation relative to the more purine rich strand of the duplex target, stabilized by TAT, AAT, and GGC base triplets. G-G hydrogen bonding of the Hoogstein type had been well known in polymeric triplexes [3], in four stranded DNA structures [16] and had been inferred for internal triplex formation at neutral pH [10] in H-DNA. Therefore the utility of GGC triplet as an element in TFO design is not unexpected. AAT triplets are also well conserved as tertiary interactions in tRNA [17] and had been shown in polymeric triple helices by Fresco [2].

Given that the Hoogstein base triplets which stabilize tRNA tertiary structure are antiparallel [17] and that the GGC triplets inferred by Kohwi et.al. within H-DNA are in an antiparallel structure [10], it is interesting to consider that as a general principal, antiparallel triple helices may be very stable. Likely H bonding schemes for such triplets are displayed in Fig. 1, which are presented as antiparellel triplets of the reverse Hoogstein type.

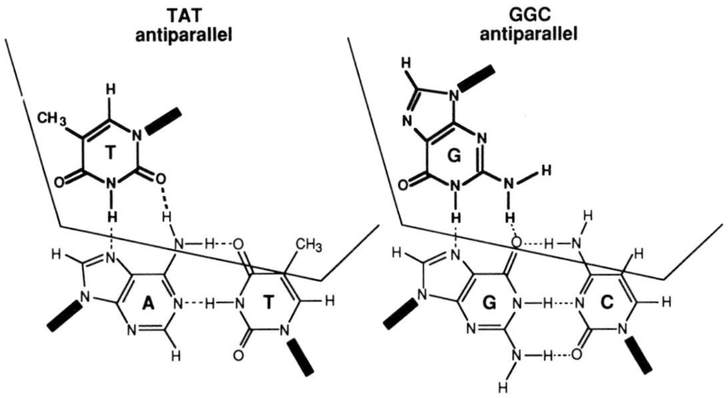

Figure 1. GGC and TAT triplets: An antiparallel solution. Triplet hydrogen bonding schemes are presented. They are of the reverse Hoogstein type for TAT and GGC triplets. They give rise to a net antiparallel orientation of a nucleotide element of a triplex forming oligonucleotide relative to the purine of the underlying duplex base pair. The details of the particular triplets presented here are reasonable, but are as yet unconfirmed by high resolution structure analysis. Consequently, they should be viewed as a structural hypothesis to explain binding thermodynamics and footprinting analysis. When bound in this way, TAT and GGC triplets are not perfectly isomorphous. However, more detailed modeling suggests that the resulting backbone discontinuity is not severe (not shown).

In this work, the refinement of TFOs targeted to the c-myc binding site is described, the goal being to determine the relative importance of TAT vs GGC triplets and the importance of antiparallel strand orientation for triple helices of this kind.

RESULTS AND DISCUSSION

TFO DESIGN PRINCIPLES: EXPERIMENTAL ANALYSIS WITH THE C-MYC MODEL.

We have investigated structure/stability relationships by monitoring TFO binding to a target in the human c-myc promotor region: sites -115 to -142 relative to the principle transcription origin (see Fig. 2A for a schematic).

```
A.                                    C-MYC PROMOTOR DOMAIN
                                          PUGT27 SERIES
                    -152                                    -115
                    5'-CTCCTCCCCACCTTCCCCACCCTCCCCACCCTCCCCA-3' C-MYC
                    3'-GAGGAGGGGTGGAAGGGGTGGGAGGGGTGGGAGGGGT-5' TARGET
                            3'-GGTTGGGGTGGGTGGGGTGGGTGGGGT-5' PUGT27p
                            5'-GGTTGGGGTGGGTGGGGTGGGTGGGGT-3' PUGT27a
                            3'-GGAAGGGGTGGGAGGGGTGGGAGGGGT-5' PU1p
                            3'-GGAAGGGGAGGGAGGGGAGGGAGGGGA-5' PU1Ap
                            5'-GGAAGGGGAGGGAGGGGAGGGAGGGGA-3' PU1Aa
                            3'-CCTTCCCCTCCCTCCCCTCCCTCCCCT-5' CT27p
                            3'-GGAACGGGAGGGAGCGGAGGGAGGCCGA-5' PU1ACp
                            5'-GGAACGGGAGGGAGCGGAGGGAGGCCGA-3' PU1ACa
                3'-GTGGTGGGGTGGTTGGGGTGGGTGGGGTGGGTGGGGT-5' PUGT37p
                5'-GTGGTGGGGTGGTTGGGGTGGGTGGGGTGGGTGGGGT-3' PUGT37a
                3'-GTGGTGGGGTGGTTGGGGTGGGTGGGGTGGGTGGGGTGG-5' PUGT34p
                            3'-GGTTGGGGTGGGTGGGGTGGGTGGGGT-A5'    5'A-PUGT27p
                            5'A-GGTTGGGGTGGGTGGGGTGGGTGGGGT-3'    5'A-PUGT27a
                            3'A-GGTTGGGGTGGGTGGGGTGGGTGGGGT-5'    3'A-PUGT27p
                            5'-GGTTGGGGTGGGTGGGGTGGGTGGGGT-A3'    3'A-PUGT27a
                3'-GTGGTGGGGTGGTTGGGGTGGGTGGGGTGGGTGGGGT-A5'    5'A-PUGT37p
                5'A-GTGGTGGGGTGGTTGGGGTGGGTGGGGTGGGTGGGGT-3'    5'A-PUGT37a
                3'A-GTGGTGGGGTGGTTGGGGTGGGTGGGGTGGGTGGGGT-5'    3'A-PUGT37p
                5'-GTGGTGGGGTGGTTGGGGTGGGTGGGGTGGGTGGGGT-A3'    3'A-PUGT37a
    3'A
     -O-CH2-CHOH-CH2-NH3+
    5'A
     +NH3-CH2-CH2-CH2-CH2-CH2-CH2-O-PO2-O-


B.                                    EGFR PROMOTOR DOMAIN
                                     EG36 AND EG31 BINDING SITES
    -354                                                                         -278
    5'-CATTCTCCTCCTCCTCTGCTCCTCCCGATCCCTCCTCCGCCGCCTGGTCCCTCCTCCTCCCGGCCCTGCCTCCCCGCG-3'
    3'-GTAAGAGGAGGAGGAGACGAGGAGGGCTAGGGAGGAGGCGGCGGACCAGGGAGGAGGAGGGCGGGACGGGAGGGGCGC-5'
           3'-TTGTGGTGGTGGTGTGTGGTGGTGGGGTTGGGTGGTGG-5'  3'-TGGGTGGTGGTGGGGGGGTGGGTGGGG-5'  parallel
           5'-TTGTGGTGGTGGTGTGTGGTGGTGGGGTTGGGTGGTGG-3'  5'-TGGGTGGTGGTGGGGGGGTGGGTGGGG-3'  antiparallel
                   EG36 FAMILY                                  EG31 FAMILY


    3'A
     -O-CH2-CHOH-CH2-NH3+
    5'A
     +NH3-CH2-CH2-CH2-CH2-CH2-CH2-O-PO2-O-
```

Figure 2. Duplex target sites for triplex formation. A. c-myc Sites. The functionally important domain −115 to −152 bases relative to the regulated c-myc P1 promotor has been presented, along with a series of TFOs designed to bind to this site. For each, the nomenclature (p,a) refers to the orientation the bound TFO would assume, if it were to bind so as to form GGC and TAT triplets with the third strand in a parallel or antiparallel orientation, respectively. In certain instances, the TFOs have been synthesized as the 3' or 5' amine derivative. Those derivatives are identified by a 3'A-, or 5'A- prefix.
B. EGFR Sites. The promotor domain of EGFR possesses a long purine rich domain of importance to promotor function (see text). The identity and proposed binding domain for two non overlapping target sites in this region have been identified, with the corresponding family of TFO.

As a result of that analysis, we have determined that by designing a 27 base TFO so as to form GGC and TAT triplets as an antiparallel colinear triplex (the triplex forming oligonucleotide TFO bound antiparallel with respect to the purine rich strand of the target) the resulting optimized TFO molecule (PUGT27a) displays an apparent dissociation constant for its target near to 5×10^{-10}M in a standard physiological binding buffer system: pH 7.8 in the presence of 20mM TRIS/HCL, 5mM MgCl2, 1mM spermine,

10% sucrose at 37C.

Binding data of that kind appear to be well described as a bimolecular equilibrium and can be obtained by titration of duplex DNA with a TFO, then observing the change of electrophoretic mobility of a radiolabelled 53bp target fragment due to triplex formation (Fig. 3A). As seen, the titration is consistent with a two state model in which the duplex (faster mobility) is converted to a discrete triplex species with reduced electrophoretic mobility.

Binding can also be assessed by titrating the radiolabelled TFO with an unlabelled plasmid DNA digest bearing the duplex DNA target site within one of several fragments. Briefly, we have digested a plasmid DNA bearing the c-myc gene [14] with AvaI and ScaI to yield 9 fragments in which a single 190bp fragment bears the -115 to -142 target site. That fragment array was then purified and used to titrate 5'32P-labelled PUGT27a in the standard binding buffer. In this analysis, stable binding should result in comigration of the TFO with the 190bp species during electrophoresis through an acrylamide matrix, the other 8 digestion fragments serving as controls.

As seen in Fig. 3B, such a titration yields detectable binding to only the 190bp fragment, with a midpoint near to 10-9 M of added duplex target. The midpoint is in good agreement with the preceding titration and confirms the high site selectivity for complex formation.

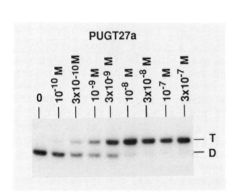

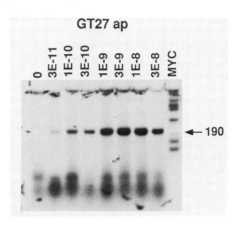

a   b

Figure 3. Low resolution analysis of triplex formation on the human c-myc gene promotor. A. Band shift analysis. Band shift analysis was performed on a cloned 53 bp DNA duplex fragment, spanning HhaI and HinfI sites at -111 to -154 relative to the human c-myc P1 promotor. The binding fragment was obtained by EcoRI and BamHI digestion of pRD1000, a c-myc subclone of the HhaI to HinfI domain inserted into the SacI and SmaI sites of pUC19. Band shift analysis was performed by incubation of the 32P-labelled 53 bp fragment with increasing concentration of PUGT27a (measured in strand equivalents) at 37C in the standard binding buffer: 20mM Tris/HCl, 5mM MgCl2, 1mM spermine 10% sucrose, pH 7.8. Subsequent to mixing, the solution was allowed to incubate for 60min at 37C, followed by electrophoresis at room temperature in an 8% acrylamide gel (19/1 acrylamide to bis) with 90mMTris/Borate pH7.8, 10mM MgC12 as the standard Mg+2 containing electrophoresis buffer. Duplex and triplex bands have been marked as D and T, respectively.   B. Co-migration analysis. The plasmid pMHX which contains the 5' end of the human c-myc gene [14] was digested with AvaI and ScaI, thereby liberating the -115 to -142 target site at the center of a 193bp fragment. In that digest, the remainder of the plasmid exists as a series of unrelated fragments with the following length in bp: 2521, 916, 784, 702, 540, 340, 107 & 63. Binding was performed by incubation of 32P labelled PUGT27a with increasing concentration of the AvaI, ScaI pMHX digest in the standard binding buffer, as described in Part A. Added plasmid concentration is listed in moles of binding site (plasmid molecule) equivalents per liter. Triplex formation was visualized by observing the co-migration of the 32P labelled TFO with the 193bp target duplex in a 5% acrylamide gel at room temperature, in the presence of the standard Mg+2 containing electrophoresis buffer. In this assay, specificity can be assessed from binding to fragments other than the 193 bp target. As seen, under conditions which approach staturation of the target band, the other 8 bands of the digest (visible in the lane marked "MYC") do not display detectible binding. For these measurements and for those below, the TFO of interest was prepared by the phosphoramidite method, followed by C18 HPLC (trityl on), detritylation and subsequent analysis by electrophoresis or analytical HPLC.

In order to verify that binding had occurred as hypothesized in the model of Fig. 2, we have employed DNase I footprinting, taking advantage of the generally slow DNaseI cleavage rate within a triple strand structure [14]. As for band shift and co-migration analysis, the footprinting was performed at 37C, pH 7.8 in the standard Tris/Mg+2 binding buffer (see above).
          Within experimental accuracy, the DNase I footprint for PUGT27a (Fig. 4) is as expected from the model in Fig. 2. We also find that if a 37b long antiparallel TFO is similarly designed (PUGT37a), so as to bind to the extended -115 to -152 domain (Fig. 1), the resulting triplex covers the full 37 base domain (Fig. 4). Bars to the side in 4 identifies the predicted span of "anti27" and "anti37" sites.
          In order to test the relative stability of the parallel vs. antiparallel triplex configuration, we have synthesized sequence isomers which could form GGC and TAT triplets as a parallel triple helix: PUGT27p and PUGT37p (see Fig. 2A). The DNase I footprints for the two isomers are quite different from that for the antiparallel series: that for PUGT27p is shifted 4 bases relative to GT27ap. That for GT37p is identical to that for GT27p, and consequently, distinctly different from the 37base binding site for the PUGT37a isomer. Bars "par27" and "par 37" identify the altered span of those sites.

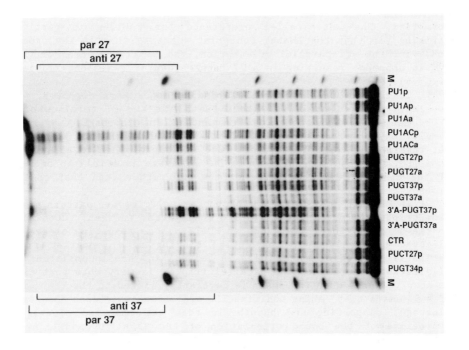

Figure 4. DNaseI footprinting. Triplex formation was assessed at high resolution, based upon reduction of DNaseI activity at the site of triplex formation [14]. DNaseI treatment was performed on the XmaI to SmaI fragment of pMHX (-99 to -193 relative to P1, labelled on the 3' terminus of the XmaI site) in the standard binding buffer at 20C, with 10U/ml of enzyme for 10 min. Following deprotination, samples were desalted by ethanol precipitation then analyzed by high resolution electrophoresis. The electrophoresis experiment described in Fig. 4 was performed at 5×10^{-7}M of added TFO, which is 100 times the measured dissociation constant for PUGT27a. The identity of TFO homologues at the top is as presented in Fig. 2A. Bars to the side are the DNaseI footprint predicted from the model in Part B. As seen, the span of the PUGT37a binding site is approximately 10 base pairs greater than that for PUGT27a. It is also interesting to note that TFOs which contain CGC triplets (CT27a, PUIACp, PUIACa) do not display detectible binding at this TFO concentration.

As outlined in Fig 5B., it is likely that differences in the binding of the two sets of isomers results from the fact that the antiparallel orientation for triplex formation is strongly favored. Because the c-myc binding site has a pseudo-two fold axis of symmetry, TFOs which were designed to bind parallel can reverse their orientation so as to bind as an antiparallel triplex. In the process, a number of bases at the 3' end of PUGT27p and PUGT37p are unable to participate in triple helix formation. The antiparallel preference for PUGT37p is particularly striking, in that the DNaseI footprint suggests that in the process of achieving the antiparallel orientation, the final ten bases of this TFO remain unbound, resulting in a 27 rather than 37 base long triple helix (FIG. 4).

To confirm the antiparallel binding model more directly, we have covalently linked eosin to either the 5' or the 3' terminus of these TFOs. Eosin is a photochemical reagent which cleaves DNA by means of singlet oxygen production [18]. It is negatively charged and thus does not itself bind to nucleic acids [18]. Therefore, when covalently linked to a TFO, the resulting pattern of DNA cleavage is likely to be a result of the binding characteristics of the TFO rather than that of the dye probe.

In Fig. 5A, we display the photochemical cleavage pattern resulting from triplex formation of eosin-TFO conjugates. As seen, upon addition to the 5' or to the 3' terminus of PUGT27p, the resulting cleavages are directed to the end of the duplex binding site predicted from the antiparallel binding model (Figs 2A and 5B) and are inconsistent with a significant fraction of the parallel triplex isomer. Note that cleavage is observed only when the eosin is covalently linked to the TFO. Addition of a 3' amine or 5' amine modified TFO plus free eosin does not result in cleavage, supporting the hupothesis that the observed cleavages are indicative of the bound orientation of the TFOs. In addition, eosin linked to CTRL PUGT37, an oligonucleotide with the same base composition as PUGT37a but randomized sequence, shows no cleavage, indicating that cleavage occurs only upon stable triplex formation. Figure 5B shows proposed complexes of eosin-TFOs with duplex DNA, superimposed upon the observed cleavage patterns (arrows).

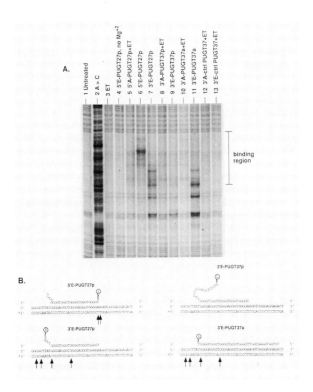

Figure 5. A. Eosin photochemical mapping. Eosin isothiocyanate has been affixed through an amine linker to TFOs (Fig. 2), employing standard aqueous thiourea coupling chemistry. Subsequent to coupling, the dye-oligonucleotide conjugates were purified by exclusion chromatography. Photochemistry was analyzed on the duplex DNA fragment described in Fig. 4. The eosin-TFO conjugates (5x10-7M final concentration) were added to labelled DNA fragment in the standard binding buffer at 20C. 5uL samples were then irradiated for 5min at 3W/cm2 with the green output of an Ar+ion laser, followed by heating to 80C for 15min in 10% piperidine to induce hydrolysis of photochemically oxidized bases [18]. Binding buffer, the eosin isothiocyanate-Tris complex (prepared as in ref 18) or uncomplexed TFO were added as controls at 5x10-7M, then similarly irradiated. Fragmentation patterns were analyzed by electrophoresis on a 7M urea, 5% acrylamide sequencing gel. Bands to the side refer to the size of the binding site domain, as predicted in Fig. 4B. Eosin limkage to the 5' or 3' ends of TFOs is indicated by the prefix 5'E or 3'E, respectively. In lane 4, MgC12 was omitted from the binding buffer to demonstrate the requirement for divalent ion dependent triplex formation. Lanes 12 and 13 show that a eosin modified TFO with a base composition identical to PUGT37a but with a randomized sequence does not lead to the observed DNA cleavages. B. Proposed structures of eosie-

linked TFO complexes. Eosin linked to the 5' or 3' ends of a TFO is
indicated by "E". Sites of observed cleavage are marked by arrows. The
labelled end of the duplex, in DNaseI and photochemical analysis is
marked with an asterisk.

Further examination of the eosin cleavage data indicate that there may be
two secondary bound antiparallel complexes, differing by 9 bases in the
positioning of their termini (see multiple bands in Fig 5A). It is likely
that this degeneracy is a result of the underlying 9 base repeat to the
binding domain, which results in a small number of secondary complexes
resulting from "slippage" (see Fig. 5B for a model). The absolute
cleavage efficiency at these various sites is probably not a reliable
indicator of the relative likelihood of the various alternative
complexes, since the details of triplex conformation as well as DNA
sequence is likely to affect the cleavage rate.
       Interestingly, no photochemical cleavage is observed for the 3'
eosin derivative of PUGT37p (Fig. 5). We suggest that this is because the
10 bases comprising the 3' end of PUGT37p do not make contact with the
duplex, as previously inferred from the truncated DNase 1 footprint for
this triplex (Fig. 4). As depicted in Fig. 5B, it is likely that since
eosin is affixed to this unstructured tail, the proximity of eosin to the
duplex is reduced, thereby reducing the likelihood the eosin-sensitized
cleavage.
       Such an unbound "tail" upon the PUGT37p complex might significantly
increase the hydrodynamic crossection of the resulting triplex, when
compared to the PUGT37ap complex (where the TFO appears to be uniformly
buried within the major helix groove (Fig. 4B). Consistent with that
expectation, we have found that, under conditions of high resolution
electrophoresis (Fig. 6) the PUGT37p triplex with the 53bp target
fragment migrates more slowly than when formed with the PUGT37a isomer

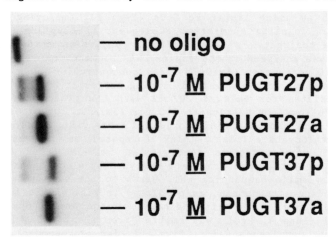

Figure 6. Band shifts with PUGT37. To additionally substantiate the
antiparallel binding model, TFOs were added at 1x10-7 to the 53bp c-myc
duplex target fragment. The resulting shift of electrophoretic mobility

was then assessed as in Fig. 3A. Binding of the 37b TFOs induces a larger reduction of mobility than for the 27b TFOs. As predicted from the model, the 27b triplexes are characterized by similar mobility. However, the PUGT37p complex displays reduced mobility relative to that for PUGT37ap. This is consistent the 3' end of PUGT37p being unbound as predicted by the antiparallel binding model in Fig. 5B, resulting in a larger hydrodynamic volume relative to the fully bound complex with PUGT37a and an associated reduction electrophoretic mobility.

RELATED TRIPLE STRAND STRUCTURES.

We have found that the characteristics of the c-myc triplex are not unique. To emphasize the generality of those observations, we have designed two families of TFO targeted against a purine-rich segment within the promotor domain of the human epidermal growth factor receptor (the EG36 and EG31 series, Fig 2B). As was the case for c-myc, this purine rich region has been shown to be crucial for promotor function by genetic criteria [19] and also serves as the binding site for cellular factors [20].

In Fig. 7, we display band shift (7A) and comigration analysis (7B) for the EG31 and EG36 series. As evidenced by both criteria, the antiparallel isomers EG31ap and EG36ap bind stably to their EGFR target domain (within a 232bp fragment) with undetectable nonspecific binding at other sites within the plasmid (7B). Under those standard binding conditions, the apparent dissociation constant for triplex formation appears to be near $5 \times 10^{-7}M$ for each, as assessed by either method. Binding of the corresponding parallel isomers cannot be detected by these methods, up to 10uM of added TFO (not shown).

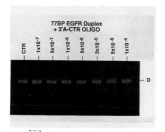
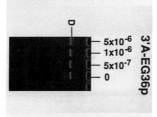
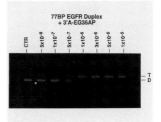

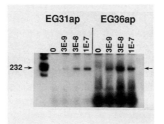

Figure 7. Binding analysis of EGFR specific TFOs.   A. Band Shift. Band shift analysis was performed at 37C in standard binding buffer as in Fig. 3, employing the synthetic 77bp. duplex displayed in Fig. 2B. Binding was analyzed at $1 \times 10^{-6}M$ of duplex, with increasing concentration of added TFO. Bands have been visualized by ethidium staining. As seen, binding is

associated with an apparent two state transition from duplex (D) to triplex (T) with a midpoint near to 5x10-7M. Importantly, addition of the random sequence isomer of 3'A-EG36ap (3'A-CTRL OLIGO) does not give rise to detectible binding over this concentration range. Similarly, binding of the parallel isomer, 3'A-EG36p, does not give rise to detectible binding. B. Co-migration analysis was performed as in Fig. 3B on the plasmid pGER9, originally described by Merlino and colleagues [20]. pGER9 was digested with HindIII and BglI thereby yielding a fragment profile of 1515, 1435, 232 and 118bp, with the target for both the EG31 and Eg36 family in the 232bp fragment. As seen, specific binding of the antiparallel isomers is detected in the 10-7 molar range, under conditions where binding cannot be detected to other potential sites in the plasmid.

DNaseI footprinting of TFO binding (Fig. 8) confirms that the binding site for EG36ap and EG31ap is as expected from the antiparallel binding model (Fig. 2B). Interestingly, DNaseI hypersensitivity is seen adjacent to the domain of DNaseI protection, which may be indicative of conformational change which is not localized to the immediate site of binding.

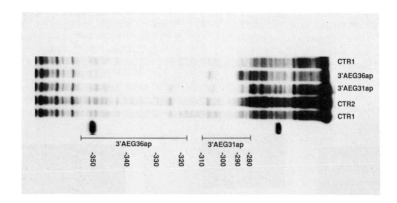

Figure 8. DNaseI footprinting of EGFR specific TFOs. DNaseI footprinting was performed as described in Fig. 4A at 5x10-6M of added TFO: 20C in the standard binding buffer. Analysis was performed on the HindIII to BglI fragment of the EGFR promotor (sites -388 to -156 in the numbering system of Fig 2B), which had been excised from pGER9 and purified by electrophoresis. Upon 3' labeling of the HindIII terminus, the footprints predicted from the binding model (Fig. 2B) would be positioned as identified by the bars to the side of the cleavage profile. As for the band shift analysis, addition of a randomized control TFO does not give rise to a measureable DNaseI footprint.

CONCLUSIONS

The hypothesis which guides much of the work of this symposium is that, if targeted to functionally important sites in the genome, reagents which bind with sequence specificity to DNA under physiological conditions might have useful pharmaceutical properties. The potential uses of such reagents would be as an agonist or antagonist of regulatory protein binding or as a vehicle for targeted DNA damage. Based upon the observation of site selective triplex formation in several laboratories, it is now clear that the time has arrived to begin to consider the possibility that, employing a triplex approach, gene-specific ligands can be designed and synthesized against important DNA domains, so as to display high affinity and DNA target selectivity at physiological temperature, pH and ion conditions. The molecules we have described in this symposium are one preliminary solution to that longer range drug development problem.

ACKNOWLEDGEMENTS

This work was supported by grants to M.H. from the Departmant of the Navy, The National Cancer Institute, and The Texas Advanced Technology Program and grants from The National Institutes of Health, to M.D.

REFERENCES

1.   Felsenfeld G. & Miles H.T (1967) Ann. Rev. Biochem. 36:407
2.   Broitman S.L., Im D.D. & Fresco J.R. (1987) 84:5120
3.   Letai A.G., Palladino M.A., Fromm E. & Fresco J.R. (1988) Biochem. 27:9108
4.   Povsic T.J. & Dervan (1989) JACS 111:3059
5.   Strobel S.A., Moser H.F. & Dervan P.B. (1988) JACS 110: 7927
6.   Helene C. & Toulne J.J. (1989) "Control of gene expression by oligodeoxnucleotides covalently linked to intercalating agents", in Oligodeoxnucleotides as Antisense inhibitors of Gene Expression. CRC Press Boca Raton Fl.
7.   Helene C. et.al. (1990) Nature
8.   Preaseuth D.L. et al (1988) PNAS 85:1349-1353
9.   Hanvey J.C., Klysik J. & Wells R.D. (1988) J. Biol. Chem 263:7397-7405
10. Kowhi Y. & Kohwi-Shigematsu T. (1988) PNAS 85:3781
11. Arnott S. & Selsing E. (1974) J. Mol. Biol. 88:509-521
12. Lee J.S., Johnson D.A. & Morgan R.A. (1979) Nucl. Acids Res. 6:3073-3091
13. Saenger W. Principles of Nucleic Acid Structure. Springer Verlag New York
14. M. Cooney, G. Czernuszewicz, E.H. Postel, S.J. Flint, and M.E. Hogan, Science 241, 456-459 (1988).
15. Hay N., Takimoto M. & Bishop M. (1989) Genes & Develop. 3:293
16. Sundquist W.I. & Klug A. (1989) Nature:825

17. Jack A., Ladner J. & Klug A. (1976) J.M.B. 108:619
18. Hogan M.E., Rooney T.E. & Austin R.H. (1987) Nature 328:554-557
19. Johnson A.C., Ishii S., Jinno Y., Pastan I & Merlino G.T. (1988) J. Biol. Chem. 263:5693-5699
20. Johnson A.C., Jinno Y., & Merlino G.T. (1988) Mol. & Cell. Biol. 8:4174-4184

EXPERIMENTAL PROOFS OF A DRUG'S DNA SPECIFICITY

W. LEUPIN
F. Hoffmann-La Roche Ltd.
Department of Pharmaceutical Research
CH-4002 Basel
Switzerland

ABSTRACT. In the first part of this account we briefly review biophysical and biochemical methods used for the evaluation of the binding affinity and specificity of small ligands interacting with double-stranded DNA. Based on their simplicity, low consumption of oligonucleotides, speed, and possible information content of the experiment, the inclusion of functional *in vitro* assays (e.g. restriction enzyme inhibition) in a series of tests for the evaluation of ligand/DNA interactions is heavily advocated. In the second part we demonstrate the large potential of one- and two-dimensional NMR spectroscopy applied to the characterization of ligand/DNA complexes, using either short (~35 base-pairs) random sequence DNA fragments, or synthetic DNA fragments of a given sequence. It follows from these considerations that the random sequence DNA should be preferentially used in screening type experiments, allowing us to distinguish modes of ligand binding (intercalation, groove binding, and nonspecific outside binding), to determine the kinetics of ligand binding (approximate lifetimes of the bound ligand) and, in favorable cases, to determine the specificity of the ligands for AT or GC base-pairs. The employment of a short DNA fragment of known sequence in NMR-studies allows the elucidation of detailed static and dynamic structures of a given ligand/DNA complex. Finally, isotopically labelled ligands used in heteronuclear two-dimensional NMR experiments of ligand/DNA interactions are shown to be an invaluable tool in such studies.

1. Introduction

Over the last few decades it has become obvious that DNA represents an important target for a variety of drugs. Given the central role of DNA in the regulation of biochemical processes, it becomes clear that such DNA binding ligands exhibit a wide spectrum of antibacterial, antiprotozoal, antiviral, and antitumor activity. Most of these ligands bind noncovalently to DNA by a combination of hydrophobic, electrostatic, hydrogen-bonding, and dipolar forces. At least three aspects of the binding of ligands to DNA could be expected to influence their biological activity: (i) the mode of binding to the DNA, (ii) the selectivity for AT or GC base-pairs, or specific sequences, and (iii) the kinetics of binding. The selectivity for a specific base-pair sequence in the DNA is of particular importance for the specific biological activity of a given drug, since depending upon their sequence specificity, ligands could selectively influence the activity of certain genes. Furthermore, it is known that certain drugs tend to disrupt chromatin structure (Doeneke, 1977; Paoletti et al., 1977; Erard et al., 1979; Portugal and Waring, 1987). The DNA conformation of e.g. a ligand binding site is dependent on its local base-pair sequence (Yoon et al., 1988; Wang et al., 1985) but can also be influenced by flanking base-

579

B. Pullman and J. Jortner (eds.), Molecular Basis of Specificity in Nucleic Acid-Drug Interactions, 579–603.
© 1990 *Kluwer Academic Publishers. Printed in the Netherlands.*

TABLE 1 List of assays for testing the DNA binding properties of a ligand

1st level tests:

- Photometric methods   Literature
  UV/VIS-absorption spectroscopy   [1]
    - fluorescence spectroscopy   [1]
    - circular dichroism   [1]
    - linear dichroism   [2]
    - ethidium bromide displacement
      measured by fluorescence spectroscopy   [3, 4]
- one-dimensional 1H-NMR studies of ligand/DNA
  complexes using
    - short, random sequence DNA   [5]
    - short fragments of polynucleotides   [6, 7]
- mobility shift assay   [8]
- inhibition of endonucleases   [9]
- nick-translation inhibition   [10]

2nd level tests:

- measurement of binding constants   Literature
    - by equilibrium dialysis   [1]
    - by spectrophotometric methods   [1, 11]
- footprinting studies (gels analyzed qualitatively)
  using:
    - DNAase I   [12, 13]
    - methidiumpropyl-EDTA · Fe (II)   [14]
    - hydroxy radicals   [15, 16]
- DNA unwinding assay
    - in solution   [17]
    - on gels   [18]
- alkylation interference   [19]

3rd level tests:

- Footprinting studies (footprinting gels analyzed   Literature
  quantitatively)   [11, 20]
- measurement of kinetic parameters using
    - surfactant-sequestration method   [21]
    - pressure-jump   [22]
    - temperature jump   [23, 24]
    - stopped flow   [25]
- measurements of thermodynamic properties
  using calorimetric methods   [26]
- two-dimensional NMR studies of complexes
  using short oligodeoxynucleotides   [27]
- X-ray studies of ligand/DNA complexes   [28,29]

Binding studies have most often been performed on bulk, random-sequence DNA from different sources, so sequence-dependent variabilities are averaged out. With the use of synthetic polynucleotides like poly(dA-dT), poly(dA).poly(dT) etc. the information content of the biophysical measurements has generally been improved. Measurements using short oligonucleotides of defined sequence have further increased the information content, even though they do not provide unambiguous data on ligand interaction at particular DNA sites, partially due to endeffects. However they can be very informative about differences in interactions of ligands between random, natural DNA, and defined oligonucleotides, respectively. Making such comparisons it should be kept in mind that the "naked" DNA used in most measurements is nothing but a model of the chromosomal DNA. Thus most results should be taken as experimental data characteristic for a given sequence and further generalizations like using the result as a measure for the interactions of a ligand with all possible DNA sequences may lead to misconceptions about the interactions of ligands with DNA.

We feel that the numerous tests can be divided up into different categories based on their simplicity and information content. This holds especially for the case, when many compounds should be screened for their DNA binding and selectivity, in contrast to the situation, when the full characterization of a given ligand/DNA complex with one ligand should be performed. That is why we differentiate the various tests into three levels, meaning that first a given ligand should be tested upon DNA binding only with tests of level 1. A further round of tests is then performed with only those compounds which show high DNA binding activity in the tests of level 1. Thus the time-consuming, more demanding tests of level 3 will only be performed with the few compounds, that show promising results in the tests of level 1 and 2. Table 1 shows a list of tests and their separation into the 3 levels, as has proved best in our hands. The various methods employed in characterization of ligand/DNA interactions have been extensively reviewed elsewhere (Gale et al., 1981; Dougherty and Pigram, 1982) and will not be discussed in detail.

2.1. LIGAND INDUCED INHIBITION OF RESTRICTION ENDONUCLEASES

The tests of level 1 can be divided up into functional and non-functional tests, meaning that in the functional tests the derived experimental data give some measure of the inhibition or attenuation of the action of enzymes, and , if the enzyme interacts with DNA in a sequence specific manner, the data can also probe the DNA sequence preference of the ligand.The inhibition of restriction endonucleases represents such a test (Polyanousky et al., 1978; Malcolm and Moffatt, 1981). Restriction endonucleases are enzymes that recognize DNA in a sequence-dependent manner. Enzymes of one subgroup, the type II restriction endonucleases, are very useful in probing the DNA structure and function, since they both recognize and cut a specific nucleotide sequence, the recognizing sequence being mostly 6 base-pairs in length. (For a recent review of the different restriction enzymes known see Roberts, 1990.) The principle behind the restriction enzyme inhibition assay is that a given restriction endounuclease shows a specific cutting pattern on a given piece of DNA. If a ligand is preincubated with the DNA, this cutting pattern may be altered if the ligand binds preferentially to the endonuclease cleavage site. Figure 1 shows the results of such a study, using the two restriction enzymes EcoRI and Sal I, the ligand Hoechst 33258, and the plasmid pBR322, linearized with Pvu II.

Due to the preference of the ligand Hoechst 33258 for a binding site containing at least 4 consecutive AT base-pairs (Pjura et al., 1987), it can be expected that the cleaving of the EcoR I enzyme (recognition sequence GAATTC) is much more strongly inhibited than the cleaving by the enzyme Sal I (recognition sequence GTCGAC). And indeed, this is experimentally observed, as follows from an inspection of the results shown in Figure 1.

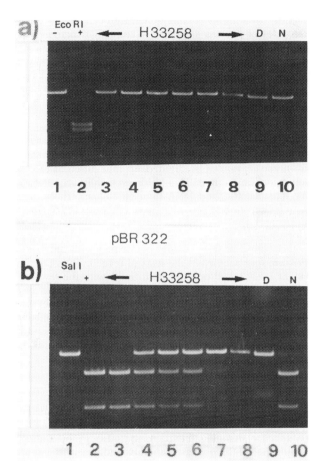

Figure 1   Inhibition of endonucleases. Electrophoresis in a 0.7 % agarose gel of 0.1 µg pBR322 DNA (linearized by Pvu II), following digestion with restriction endonucleases EcoR I (a) and Sal I (b) in the presence or absence of Hoechst 33258. Lane 1: pBR322 DNA (0.1 mM in base-pairs); lanes 2-8: increasing amounts of Hoechst 33258 (0; 6.7 µM; 0.067 mM; 0.13 mM; 0.33 mM; 0.67 mM; 1.33 mM); lanes 9,10: control, Distamycin (D) and Netropsin (N), 1.33mM).

It goes without saying, that for a detailed analysis of the influence of a ligand on the cleaving pattern of a DNA fragment more experimental parameters should be varied. In addition the use of isoschizomers of a given restriction enzyme has to be considered in order to check if the ligand induced enzyme inhibition is due to an interaction with the enzyme instead of the enzyme recognition site on the DNA.

In conclusion, this assay for detection of DNA binding is fast and efficient. Of course, the inhibition of a given restriction enzyme only results in information as to ligand binding to the corresponding specific restriction site. However, by the use of different restriction endonucleases and plasmids the number of screened specific DNA binding sites can be expanded.

2.2. UV/VIS-ABSORPTION SPECTROSCOPY

The UV/VIS-absorption spectroscopy represents one of the earliest spectroscopic methods applied to follow the effects of ligand binding to DNA (for a review see Dougherty and Pigram, 1982). The analysis of the UV/VIS-absorption spectra gained by titrating a ligand with DNA also allows the calculation of ligand/DNA dissociation constants and the excluded DNA binding site size of the ligand using Scatchard type analysis of the different spectra. This test is highly sensitive and can easily be applied to ligand/DNA systems where the first absorption band of the ligand is well separated (λ max > 310 nm) from that of the DNA. In cases where this spectral separation is small, a precise quantitative analysis of the drug/DNA complexation is not possible.

ABS.

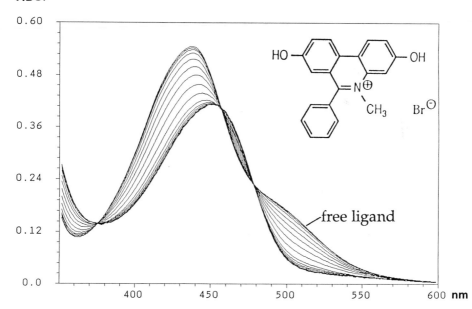

Figure 2   Titration of a 3,8-dihydroxy-phenanthridinium bromide with short (about 400 base-pairs) calf thymus DNA followed by UV/VIS-absorption spectroscopy (0.1 M NaCl; 50 mM phosphate; 0.1 mM NaN$_3$; pH = 7.0; T = 298 K). Concentration of ligand: 0.01 mM; DNA concentration varied from 2.90 µM up to 3.02 mM.

Figure 2 shows the titration of an intercalating 3,8-dihydroxy-phenanthridinium derivative with short (~400 base-pairs) calf thymus DNA as followed by UV/VIS absorption spectroscopy (Leupin et al., in preparation). The first absorption band does not exhibit the usual shift to longer wavelength and a diminishing extinction coefficient upon binding as expected for an intercalating chromophore. The observed opposite behaviour of the first absorption band upon titration is disturbing at first sight. However, a determination of the two deprotonation equilibria for the hydroxy substituents of free ligand yields values of pK = 8.40 and pK = 7.06 (whereby the mono- and di-deprotonated species exhibit their first absorption band at longer wavelength than the parent compound). This means that at pH = 7.0, the pH of the titration experiment, the free ligand is not fully protonated, thus providing an explanation for the series of spectra seen in Figure 1 under the assumption that the pK's of the two hydroxy substituents are different in the free and bound state, respectively. Indeed a titration of this ligand with the same DNA at pH 5.0 results in a series of UV/VIS-spectra expected for an intercalating ligand (Leupin et al., in preparation).

This example nicely demonstrates the importance of the knowledge of many physical properties of a ligand when performing even simple UV/VIS-absorption measurements of a ligand/DNA complex.

3.   Applications of NMR Studies in the Characterization of Ligand/DNA Complexes

The development of NMR spectroscopy and its successful application to the study of biopolymers in solution is reflected in the literature containing NMR spectroscopic characterizations of DNA fragments and their complexes with ligands: In the seventies

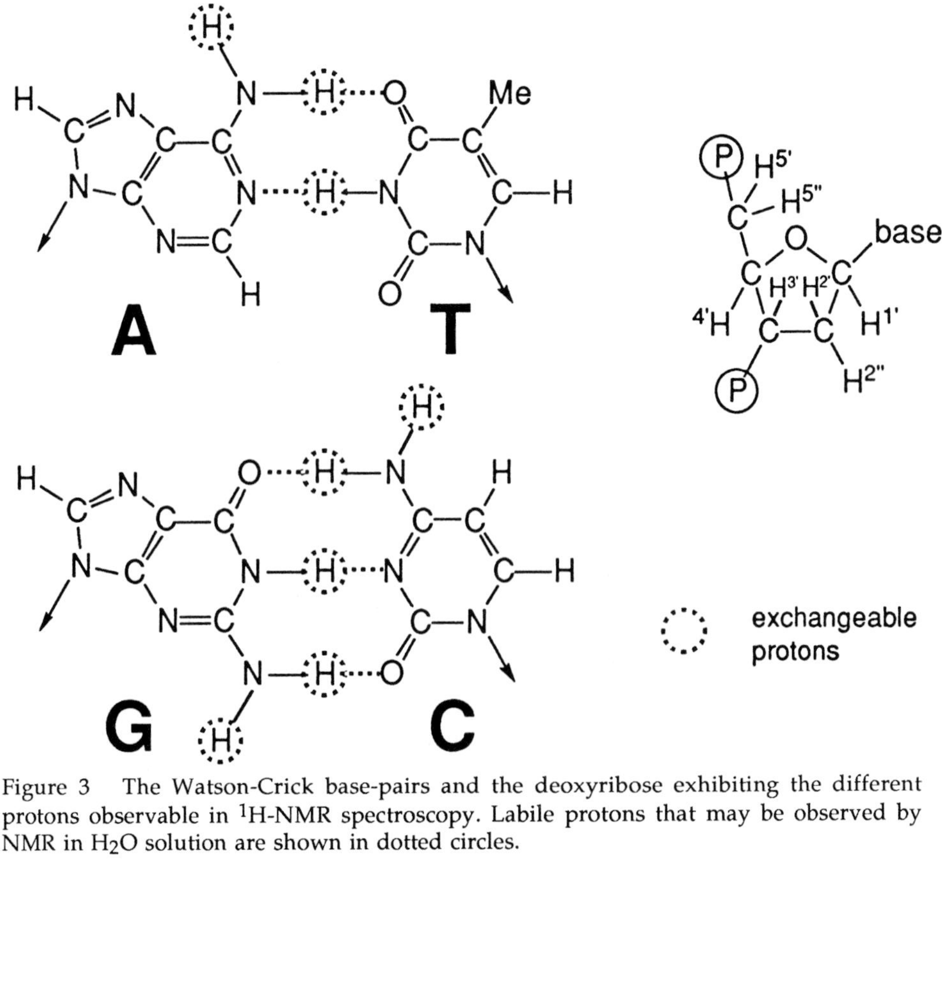

Figure 3   The Watson-Crick base-pairs and the deoxyribose exhibiting the different protons observable in ^{1}H-NMR spectroscopy. Labile protons that may be observed by NMR in H$_2$O solution are shown in dotted circles.

most papers in the field were concerned with one-dimensional ^{1}H-NMR studies of short oligonucleotides, short random sequence DNA, or tRNA molecules (for reviews see Kearns, 1977, 1984; Patel 1979). After the publication of the first two-dimensional NMR spectrum of a decadeoxynucleotide (Feigon et al., 1982a) many two-dimensional NMR-studies of oligonucleotides and their complexes with ligands appeared in the literature, applying the pioneering work of Wuthrich and coworkers in the field of proteins to the field of DNA fragments (Wuthrich, 1986; Patel et al., 1987; van de Ven and Hilbers, 1988).

The building blocks of DNA are depicted in Figure 3. From this Figure it follows that two different types of protons can be observed, labile and non-labile, giving rise to different ^{1}H-NMR spectra in H_2O and D_2O, respectively, with the labile protons only observable in H_2O solution. Figure 4 shows the ^{1}H-NMR spectrum in H_2O solution at low field of short (about 35 base-pairs) random sequence DNA fragments and of the decadeoxynucleotide duplex d-(GCATTAATGC)$_2$ under similar experimental conditions.

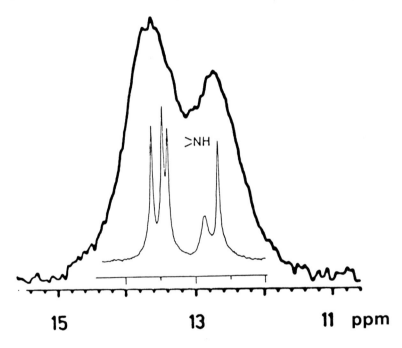

Figure 4 ^{1}H-NMR spectrum of the imino-protons of short (about 35 base-pairs), random sequence chicken erythrocyte DNA (heavy line) and of d(GCATTAATGC)$_2$ (thin line). Experimental conditions for random sequence DNA: 40 mM in base-pairs; 0.1 M NaCl; 10 mM Na cacodylate; 10 mM MgCl$_2$; pH = 7.0; T = 310 K. Experimental conditions for d(GCATTAATGC)$_2$: 2.8 mM in duplex; 0.1 M NaCl; 50 mM phosphate; 0.1 mM NaN$_3$; pH = 7.0; T = 288 K. Note that the integrated areas under the two spectra are not to scale.

The following conclusions can be drawn from these spectra: (1) The resonance lines of the NH-imino protons are well separated from all other protons, thus providing an ideal spectral window for observation of ligand induced changes affecting DNA protons, unperturbed by protons of the ligand. (2) The NMR spectrum of the random sequence DNA shows two broad envelopes arising from the AT-imino protons (centered at ~13.7

ppm) and the GC-imino protons (centered at ~12.7 ppm), but exhibits no resolved proton resonance. This is because each imino-proton of a base-pair gives rise to a resonance whose chemical shift is determined mainly by its intrinsic chemical shift (~14.6 ppm for AT and ~13.6 ppm for GC imino protons, see Patel, 1979) and by the ring-current shifts exerted on it by both its nearest-neighbour and, to a smaller extent, by its next-nearest-neighbour base-pairs. Thus at least 256 different resonance positions are expected for the imino-protons of each base-pair in random sequence DNA. These resonances overlap in such a way that only two broad envelopes of resonances are observed, and these can be clearly differentiated as arising from imino-protons of AT base-pairs and of GC base-pairs. The situation is much simpler for synthetic oligonucleotides like poly (dA-dT), poly (dG) · poly (dC) etc. because of the overlap of fewer imino proton resonances thus partially exhibiting one of the advantages when using short fragments of such DNA in ligand/DNA studies. (3) The imino protons of d-(GCATTAATGC)$_2$ manifest the twofold symmetry of the duplex structure formed by this self-complementary decanucleotide, since only 5 (rather than 10) separate imino proton resonances are observed, whereby each resonance corresponds to two (symmetrically equivalent) protons in the sequence, providing a much larger information content of this spectrum. With the application of two-dimensional NMR methods all proton resonances of the NMR spectrum of the decanucleotide can be assigned in a sequence-specific manner (Feigon et al., 1983; Hare et al., 1983; Scheek et al., 1983, 1984; Chazin et al., 1986; Otting et al., 1987). Such an assignment represents the basis for detailed characterization of ligand binding to oligonucleotides of specific sequence as will be discussed below. But first the possible data obtainable from NMR studies using random sequence DNA complexed with ligands will be discussed.

In this account we only present results of ^{1}H-NMR studies of oligonucleotides and their complexes with ligands. NMR studies of DNA-fragments and their complexes with ligands using ^{31}P-NMR techniques (e.g. see Gorenstein and Lai, 1989; Gorenstein et al., 1988; Gorenstein et al., 1989;Frey et al., 1985; and references therein), or ^{13}C-NMR studies (see Leupin et al., 1987; LaPlante et al., 1988 and references therein) will not be discussed here.

3.1. NMR STUDIES EMPLOYING SHORT, RANDOM SEQUENCE DNA

The interaction of small ligands with DNA has been followed by NMR using short, random sequence DNA (Feigon et al., 1984) or short fragments of poly(dG-dC) or poly(dA-dT) (Patel, 1979; Gupta et al., 1984; Wilson et al., 1990) whereby most information was gained by observing the ^{1}H resonances of the NH-imino protons of the Watson-Crick base-pairs (see Figures 3 and 4). Such binding studies give important qualitative information about the properties of the ligand/DNA complexes. Thus it has become possible by such measurements e.g. to estimate an averaged lifetime of ethidium bromide in its intercalation site of about 10 milliseconds (Feigon et al, 1984; Leupin et al., 1985), which agrees well with other lifetime determinations of ethidium/DNA complexes. Employing ethidium bromide, des-3-amino-ethidium, and des-8-amino-ethidium in these studies, we have found that removal of the 3-NH$_2$ group of the ethidium shortens the lifetime of the ligand/DNA complex, whereas removal of the 8-NH$_2$ group did not change it. Thus it is likely that the 3-NH$_2$ group is involved in stabilizing the ethidium/DNA intercalation complex. This simple assay allows the elucidation of substitution patterns in a given chromophore to yield more stable ligand/DNA complexes.

	R$_3$	R$_8$	
	NH$_2$	NH$_2$	Ethidium
	H	NH$_2$	des-3-amino-ethidium
	NH$_2$	H	des-8-amino-ethidium

In an extensive study (Feigon et al., 1984) covering over 70 clinical and experimental antitumor drugs, it was shown that the kinetic binding parameters determined by these NMR measurements, at least for the intercalating compounds, do correlate the cytotoxicity and useful antitumor activity of the ligands with long ligand/DNA residence times. This provides a reason for the previously puzzling observation that while many of the most useful antitumor drugs are DNA intercalators, there exist many ligands that bind DNA with high affinity by intercalation, yet do not possess cytotoxic activity. It should be pointed out that such estimations of ligand residence times on DNA can be performed very fast, with about 400 microliters of buffer solution, containing a DNA concentration of about 20 to 40 mM (in base-pairs). The measurement has to be performed with at least one given ligand to DNA ratio at two different temperatures, finally analysing the data by simple mathematical treatments (see Feigon et al., 1982b, 1984; Leupin et al., 1985). As a main result of such NMR studies with random DNA the mode of binding of most ligands can be immediately delineated from the behavior of the ligand induced chemical shift changes of the imino protons. As an empirical rule it was found (Feigon et al., 1984) that intercalators shift the AT and GC imino protons to higher field, whereas minor groove binders shift these resonances to lower field. This very clear spectral difference upon titrating DNA with a ligand allows an easy determination of the binding mode. If the binding mode of a ligand has previously been determined by other means, the binding mode determination by NMR agrees in all cases checked so far.

In addition to kinetics and mode of binding, the NMR method allows also the determination of specificity for binding to either GC or AT binding sites, at least for the intercalators. However, the reliability of this method is limited by the fact that (for fast exchange) the assumption of equal drug induced shifts for the AT- and GC-imino proton resonances for the protons next to the bound drug has to be made. Nevertheless, the conclusions about relative base-pair specificity derived from the NMR data were consistent with other data in the literature.

In summary one can conclude that NMR studies with random sequence DNA/ligand complexes give important qualitative information on ligand binding. In order to obtain detailed structures of ligands bound to specific DNA sequences it is necessary to employ short DNA fragments of defined sequence (see below). However the use of short random sequence DNA (or short fragments of synthetic polynucleotides) in NMR measurements of ligand/DNA complexes exhibits distinct advantages: (1) it provides an inexpensive and quick screen of mode of binding, kinetics, and base-pair specificity; (2) an averaged effect of binding to all possible sites rather than to one or more specific sites is observed, and this may be more biologically relevant; and (3) the ligand/DNA complexes can be studied over a wide range of temperatures. Finally this system provides an excellent method for selecting specific drugs for study with DNA duplexes of known sequence.

3.2. NMR STUDIES EMPLOYING OLIGONUCLEOTIDES

The information content of a ^{1}H-NMR spectrum of an oligonucleotide is very large. The spectrum can be assigned in a sequence specific manner by the application of two-dimensional NMR techniques using standard methods (Feigon et al., 1982a, 1983; Scheek et al., 1983, 1984; Hare et al., 1983; Chazin et al., 1986). These assignments provide the starting point for determination of structural changes in the DNA introduced by the binding of a ligand and for investigation of detailed static and dynamic aspects of the intermolecular interactions in the ligand/DNA complex (Reid, 1987). Note that for such detailed studies the spectra of the free DNA, the free ligand, the complexed DNA, and the bound ligand must be individually assigned.

Many NMR studies of ligand/oligonucleotide complexes have been performed and published in the literature, in which the ligands have been e.g. minor groove binders [netropsin (Patel and Shapiro, 1986); distamycin A (Klevit et al., 1986; Pelton and Wemmer 1989, 1990); chromomycin A (Gao and Patel, 1989); or the bis-quaternary ammonium heterocycle SN 6999 (Leupin et al., 1986)], intercalators [daunomycin (Phillips and Roberts, 1980); actinomycin D (Scott et al., 1988; Brown et al., 1984)], bisintercalators [echinomycin (Gao and Patel, 1988); bis-9-aminoacridines (Assa-Munt et al., 1985a,b)], and covalently attached ligands [anthramycin (Boyd et al., 1990)]. Below we present some experimental findings of such studies from our own work just to demonstrate the power of two-dimensional NMR spectroscopy applied to the study of ligand/oligonucleotide complexes.

3.2.1. ^{1}H-NMR STUDIES OF COMPLEXES BETWEEN BIS-9-AMINOACRIDINES AND d(ATATATATAT)$_2$

Employing the decanucleotide d(ATATATATAT)$_2$ we studied the mode of binding of a series of bis-9-aminoacridine derivatives connected by different types of linker chains (Table 2; Assa-Munt et al., 1985a). The length and character (ionic, aliphatic, rigid, flexible) of the linker chains are found to have a profound effect on the binding of these ligands to DNA. Bis(acridine) derivatives with linker chains shorter than 0.9 nm (e.g. containing six methylene groups) monointercalate under the experimental conditions employed in the NMR experiment, whereas derivatives with linker chains longer than 0.98 nm bisintercalate. All bisintercalating ligands in this series do not violate the so-called neighbor-exclusion principle. A qualitative analysis of the respective NMR spectra reveals that different types of linking chains of the acridine chromophores result in different NMR spectra, which can be explained by formation of different ligand/DNA structures. This emphasizes the important effect of the linker chain on the structure of the intercalation complex of bisintercalators. This conclusion has also been drawn from the comparison of the three bisintercalating quinoxaline antibiotics echinomycin, triostin A, and TANDEM, differing slightly in the linker chains connecting the two quinoxaline moieties (Ughetto et al., 1985) whereby the linker seems to be responsible for the GC specificity of echinomycin (Van Dyke and Dervan, 1984) and the AT specificity of TANDEM (Low et al., 1984).

In a further study employing bis(acridines) and d(ATATATATAT)$_2$, the lifetimes of the intercalating chromophores in a specific binding site have been determined (Assa-Munt et al., 1985b) by measuring ^{1}H NMR spectra and relaxation rates at different temperatures. The surprising result was that the lifetime of an individual intercalation complex for most bis(acridines) has the same order of magnitude as that of the free 9-aminoacridine, despite large differences in the binding constants. Thus an important conclusion from this study is that certain bisintercalators rapidly migrate along the DNA helix despite exhibiting large binding constants. This might explain the unexpected low biological activity of many bisintercalators, because the bisintercalator dissociates slowly off the DNA but its individual chromophores have a short residence time at a given binding site.

TABLE 2   Structure of some bis(acridine) compounds studied

linker X	chain length (nm) (N to N distance)	bis-intercalation
—(CH$_2$)$_4$—	0.63	no
—(CH$_2$)$_6$—	0.88	no
—(CH$_2$)$_7$—	1.00	yes
—(CH$_2$)$_8$—	1.13	yes
—(CH$_2$)$_{10}$—	1.38	yes
—(CH$_2$-NC-CH$_2$CH$_2$)— (O, H)	0.86	no
—(CH$_2$CH$_2$N-C-CH$_2$CH$_2$CH$_2$)— (O, H)	0.98	yes
—(CH$_2$)$_3$-N-(CH$_2$)$_4$— (H)	1.12	yes
—(CH$_2$)$_2$N C—C C—CN(CH$_2$)$_2$— (O, N—N, CH$_3$, C, O, H)	1.37	yes

3.2.2. HOMO- AND HETERONUCLEAR NMR-STUDIES OF THE 1 : 1 COMPLEX OF d(GCATTAATGC)₂ WITH ¹³C-LABELED AND NON-LABELED SN 6999

A more detailed NMR study was performed (Leupin et al., 1986) using the bisquaternary ammonium heterocycle SN 6999 and the decadeoxynucleotide d(GCATTAATGC)2 (see Figure 5).

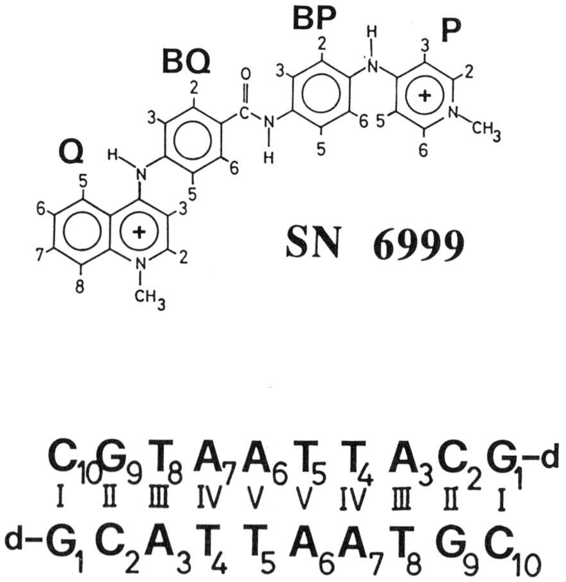

$$C_{10} G_9 T_8 A_7 A_6 T_5 T_4 A_3 C_2 G_1\text{-d}$$
$$\text{I} \quad \text{II} \quad \text{III} \quad \text{IV} \quad \text{V} \quad \text{V} \quad \text{IV} \quad \text{III} \quad \text{II} \quad \text{I}$$
$$\text{d-}G_1 C_2 A_3 T_4 T_5 A_6 A_7 T_8 G_9 C_{10}$$

Figure 5   Structures of 4-[p[p-(4-quinoylamino)benzoamido]-anilino]pyridine (SN 6999) and the decadeoxyribonucleoside nonaphosphate d(GCATTAATGC)₂. In the latter the numeration of the base-pairs with roman numerals reflects the two-fold symmetry observed for the solution conformation of the free duplex. For the ligand an unambiguous identification of individual protons includes the letter for the ring and the ring position, e.g. P2.

About 35 intermolecular nuclear Overhauser effects (NOEs) between the ligand and the DNA have been assigned. At that stage no quantitative distance measurements were attempted, and the experiments focused on observing the absence or presence of intermolecular NOEs. The 2H protons of all adenines, which are the only non-labile base-protons pointing into the minor groove, showed NOEs with many ligand protons, indicating that the ligand binds in the minor grove of the DNA, interacting with the central AT base-pairs. Inspection of DNA models and the models resulting from the X-ray study of the netropsin/d(CGCGAATTCGCG)₂ complex (Kopka et al., 1985) indicate that only N3 of adenine and the 2-keto group of thymine can act as hydrogen-bond acceptors in the minor groove. The occurrence of such hydrogen bonds cannot be directly determined by ¹H NMR but may be inferred from observation of ¹H-¹H NOEs involving DNA protons near these hydrogen-bond acceptors. Thus the intermolecular NOEs between the amide proton of SN 6999 and $T_5 1'H$, $A_6 1'H$, and $A_7 2H$ make it most likely that this amide proton is bound to the 2-keto group of T5. It further appears that the amino proton between BQ and Q could be hydrogen bonded to the 2-keto group of

T4. Similar hydrogen-bonding interactions were reported for the complexes of netropsin with d(CGCGAATTCGCG)$_2$ (Patel, 1982), d(GGAATTCC)$_2$ or d(GGTATACC)$_2$ (Patel and Shapiro, 1986). These studies also revealed that it is a disadvantage to use self-complementary DNA fragments in NMR studies of ligand/DNA complexes. In such a DNA fragment each potential ligand binding site contains a symmetrical equivalent second binding site (see structures A and B in Figure 6). In such isolated complexes A or B with the asymmetric drug bound, the magnetic equivalence of symmetrically disposed DNA protons would be lost, for example A$_6$2H on the (+) strand would not be equivalent to A$_6$2H on the (-) strand. However for many ligands an equilibrium between symmetrically equivalent ligand binding sites on the DNA exists, drastically lowering the usable information content of the NMR spectra of such a ligand/DNA complex. In addition these studies reveal that the application of NMR spectroscopy for the elucidation of ligand/DNA complexes allows the study of several different oligonucleotides complexed to the same ligand, thus giving the opportunity to study the different effects of a given ligand exerted on different DNA sequences.

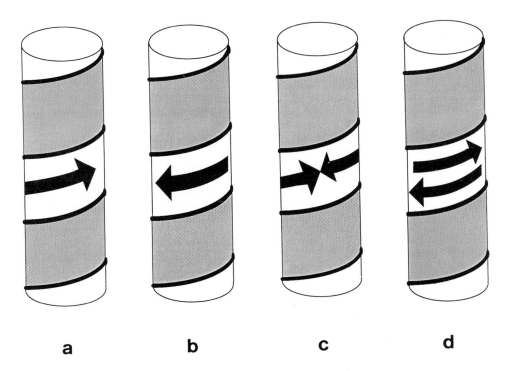

a   b   c   d

Figure 6   Schematic representations of the different complexes experimentally observed between DNA and ligands binding in the minor groove. Shaded area: major groove; arrow: ligand. a) and b): e.g. SN 6999, netropsin; c) chromomycin; d) distamycin A. See text for further explanations.

Figure 6 summarizes the results obtained by two-dimensional NMR studies of minor groove binding ligands in a schematic fashion. Thus, it has become possible to determine in which direction of a given DNA strand a ligand is aligned (Figures 6A, 6B) as determined for netropsin (Patel, 1982; Patel and Shapiro, 1986) or SN 6999 (Leupin et al., 1986). It has also been possible to observe a dimerization of the ligand chromomycin A when bound to DNA, bridged by a magnesium ion (Gao and Patel, 1989) and doubling the size of the specific binding site of the ligand, thus increasing the sequence specificity of the ligand (see Figure 6C). A dimerization upon DNA binding has also been observed for distamycin A (Pelton and Wemmer, 1989, 1990). However, the result of this dimerization is the binding of two distamycin A ligands side by side in the minor groove of the DNA, expanding the size of the minor groove (see Figure 6D). From all these studies it follows that such NMR studies increase the basic knowledge about ligand/DNA interactions, providing a crucial increase in experimental data as a base for the rational design of new ligands binding to DNA with high sequence selectivity.

Table 3 contains a collection of advantages, disadvantages, and possible future developments of NMR spectroscopy applied to the study of ligand/DNA interactions. A very important development represents the ready availability of isotopically labeled (^{13}C, ^{15}N, ^{2}H) DNA ligands as well as ligand receptors (for reviews in the field of proteins see Otting and Wuthrich, 1990; McIntosh and Dahlquist, 1990; LeMaster, 1990) in connection with improved heteronuclear two-dimensional and three-dimensional NMR techniques (see e.g. Otting and Wuthrich, 1990; Fesik and Zuiderweg, 1990).

TABLE 3:   General Considerations about NMR Measurements of Drug/DNA Complexes

a) **Advantages of NMR**
   The application of NMR methods allows the elucidation of
   - simultaneous observation of DNA and ligand resonances possible
   - kinetics of binding process
   - structure determination in solution ("X-ray in solution")

b) **Disadvantages of NMR**
   - requires large amounts (5 - 10 milligrams) of DNA fragments of high purity
   - required concentration > 1 mM in duplex
   (Note: many DNA/ligand systems are insoluble at such concentration)
   - limitation by molecular weight (≤ 25 000 Daltons)

c) **Future development**
   - specific isotope labeling (e.g. ^{2}H, ^{13}C, ^{15}N) of DNA and/or ligand
   - editing of NMR spectra; development of new pulse sequences for homo- and heteronuclear NMR
   - development of new NMR hardware
   - development of programs for computer aided sequence specific resonance assignment

In a recent study (Leupin et al., 1990) we have demonstrated the great potential of combining specific isotope labeling of a ligand with heteronuclear two-dimensional NMR techniques by studying the 1:1 complex formed between d(GCATTAATGC)$_2$ and the ligand SN 6999 with ^{13}C labeled P-CH$_3$ and Q-CH$_3$ groups (see Figure 5 for numbering), and employing [^{1}H,^{1}H]-NOESY with a ^{13}C(ω_1)-half filter (Otting and Wuthrich, 1990). This NMR technique makes use of the fact that all one-bond coupling constants $^1J(^1H,^{13}C)$ are large compared to all other homonuclear and heteronuclear coupling constants with only minor variations in the different ^{13}CH fragments. Thus it becomes possible to tune the ^{13}C-half filters for separating the diagonal peaks of ^{13}C-bound protons and the cross-peaks with ^{13}C-bound protons from all other peaks. In

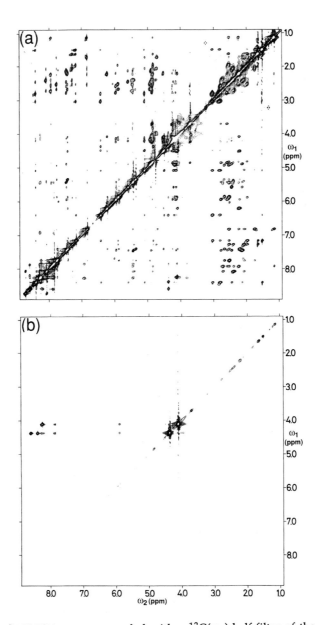

Figure 7 [^{1}H,^{1}H]-NOESY spectra recorded with a ^{13}C(ω_1)-half-filter of the 1:1 complex formed between d-(GCATTAATGC)$_2$ and the ligand SN 6999 with ^{13}C-labeled P-CH$_3$ and Q-CH$_3$ groups (Figure 5). DNA duplex concentration 3 mM, solvent ^{2}H$_2$O, 0.05 M phosphate buffer, 0.1 M NaCl, 0.1 mM NaN$_3$, pH = 7.0, T = 301 K, ^{1}H frequency 500 MHz. (a) Sum spectrum. (b) difference spectrum. Spectrum taken from Leupin et al., 1990.

studies of intermolecular complexes by delineation of the contacts between the interacting molecules with ^{1}H-1H NOEs, limitations arise whenever the individual components contain hydrogen atoms with identical chemical shifts. In such situations it may be impossible to distinguish between intramolecular and intermolecular NOEs. In the system SN 6999/d(GCATTAATGC)$_2$ this difficulty arises for the ring methyl protons P-CH$_3$ and Q-CH$_3$ in SN 6999, for which the chemical shifts coincide closely with those of several deoxyribose protons 4'H, 5'H, and 5"H of the DNA (Leupin et al., 1986), and there is the additional difficulty that most of the 4', 5', and 5" protons have not been individually assigned. In contrast, the chemical shifts of the aromatic protons of SN 6999 are unique in the complex with d(GCATTAATGC)$_2$, so that numerous ligand/DNA contact sites could be identified using conventional [^{1}H,^{1}H]-NOESY (Leupin et al., 1986). Overall, this system is thus already well characterized and the intermolecular ^{1}H-^{1}H NOEs with the N-methyl groups of the ligand are therefore particularly suitable for a demonstration of the potentialities of heteronuclear filters for studies of intermolecular interactions (Fesik et al., 1988; Senn et al., 1987). Figure 7 illustrates the simplification of the two-dimensional [^{1}H,^{1}H]-NOESY spectrum for the ligand/DNA complex upon isotope labeling and application of the ^{13}C(ω_1) half filter technique.

In the [^{1}H,^{1}H]-NOESY experiment with ^{13}C(ω_1)-half-filter the difference spectrum (Fig. 7b) contains the diagonal peaks from the two ^{13}C-labeled methyl groups and from protons bound to ^{13}C in natural abundance, and all the cross-peaks between ^{13}C-bound protons in ω_1 and all other protons in ω_2. The sum spectrum (Fig. 7a) contains all the other diagonal peaks and cross-peaks, which would be present also in the conventional [^{1}H,^{1}H]-NOESY experiment.

Using this heteronuclear NMR technique it has become possible to assign the nuclear Overhauser effects between the methyl protons of the ligand and protons of the DNA in a spectral region which is very crowded in the conventional [^{1}H,^{1}H]-NOESY spectrum, thus drastically decreasing the possibility of wrongly assigning intermolecular proton-proton contacts. The intermolecular proton-proton contacts between P-^{13}CH$_3$ and Q-^{13}CH$_3$ of SN 6999 and the decanucleotide observed in the difference spectrum from the [^{1}H,^{1}H]-NOESY experiment with ^{13}C(ω_1)- half-filter corroborate the following conclusions drawn from [^{1}H,^{1}H]-NOESY experiments with the unlabeled system (Leupin et al., 1986): (i) the methyl protons P-^{13}CH$_3$ and Q-^{13}CH$_3$ are close to protons accessible in the minor groove of the DNA; (ii) all intermolecular contacts found here are consistent with the previous proposal that SN 6999 binds only within the AT-stretch of d-(GCATTAATGC)$_2$. The different number of intermolecular [^{1}H,^{1}H]-NOESY cross-peaks with P-^{13}CH$_3$ and Q-^{13}CH$_3$ protons identified in the spectra and in earlier work (Leupin et al., 1986), using data from conventional [^{1}H,^{1}H]-NOESY spectra recorded under otherwise identical conditions is due to the limitations of earlier experiments imposed by the chemical shift degeneracies between P-^{13}CH$_3$ and Q-^{13}CH$_3$ protons of SN 6999 and 4'H, 5'H and 5"H of the DNA. Thus the present study emphasizes the usefulness of [^{1}H,^{1}H]-NOESY spectra recorded with X-half-filters (Otting and Wüthrich, 1990) as an aid in identifying intermolecular proton-proton contacts in DNA/drug complexes. Clearly, the use of this approach can be extended to DNA complexes with other classes of ligands, eventually also including other macromolecules such as proteins.

3.3 STRUCTURE DETERMINATION OF BIOMOLECULES BY NMR AND BY DIFFRACTION TECHNIQUES

It is important to note that X-ray diffraction and NMR can both be used independently to determine the complete three-dimensional structure of proteins or nucleic acids and their complexes with ligands. The two techniques can yield complementary information, whereby the complementarity results from the fact that the time scales of the two measurements are different, and that single crystals of the biomolecules are needed for the X-ray method, in contrast to the NMR measurements where only some solubility (about 1 millimolar concentration) in aqueous solution is needed. Thus it becomes possible to compare corresponding structures of biomolecules in the crystalline and noncrystalline state. Such a comparison can be relevant, since the solution conditions of the NMR measurement can be varied drastically, giving the possibility of observing any structural changes upon changing the solution conditions. In addition the NMR method can be employed in cases where no single crystals can be grown. Finally kinetic parameters for internal and intermolecular dynamics for biomolecules and their complexes with ligands can be quantitatively determined by NMR methods.

Direct comparisons of corresponding three-dimensional structures in the crystalline and noncrystalline state have mainly been determined for various proteins. Whereas the structures for the α-amylase-inhibitor tendamistat are very close in the crystalline state (Pflugrath et al., 1986) and in solution (Kline et al., 1986), this does not hold for the metallothionein-2 from rat liver, for which different global structures were reported in aqueous solution (Schultze et al., 1988) and in the crystal (Furey et al., 1986). However, very recently a revised structure for the crystalline metallothionein-2 was reported (Stout et al., 1989) which agrees well with the structure in solution determined by NMR. Corresponding comparisons are very rare in the field of ligand/DNA complexes, an example being e.g. the observation of Hoogsteen pairing for the two base-pairs flanking the binding site of the bis-intercalating Echinomycin, both in the crystalline state (Ughetto et al., 1985) and in solution (Gao and Patel, 1988). However, in all these studies the ligand was binding at or near the end of a short DNA duplex. Thus the question arises if the formation of Hoogsteen base-pairs is not due to end effects as might be concluded from footprinting studies of echinomycin with longer DNA fragments (McLean and Waring, 1988). Further NMR and X-ray studies should clarify this point.

For the complexes of the dye Hoechst 33258 complexed with the dodecanucleotide d(CGCGAATTCGCG)$_2$ there are two X-ray studies published in the literature (Pjura et al., 1987; Teng et al., 1988) which reveal somewhat similar but nonetheless distinctly different binding modes: the ligand binds in the minor groove and either straddles the four AATT residues in the center of the DNA duplex symmetrically or is displaced by one base-pair so that the piperazine ring is located in the segment in which the minor groove has broadened out because of the presence of the first GC base-pair (For a discussion of the two structures see also Carrondo et al., 1989). In order to study the complex of the dye Hoechst 33258 with d(CGCGAATTCGCG)$_2$ in solution, a NMR study was undertaken (Parkinson et al., 1989). However from their preliminary data the authors were not able to give a clear indication as to which structure would preferentially occur in solution. We are presently studying the complex between Hoechst 33258 and the dodecanucleotide d(GTGGAATTCCAC)$_2$ by NMR (Fede et al., to be published). An analysis of our preliminary data allows us to conclude that the residence time of the ligand on the DNA is long relative to the NMR time scale, as judged by the doubling of the proton resonances of the DNA in the 1 : 1 complex, concomitant with the loss of the two-fold symmetry of the self-complementary DNA duplex upon binding of the asymmetric ligand. All central AT-imino protons shift to lower field upon complexation of Hoechst 33258, indicative of minor groove binding around the center of the dodecanucleotide. However, the positioning of the ligand on the DNA duplex has to be determined by analysis of [^{1}H,^{1}H]-NOESY spectra. This analysis in the search for intermolecular proton-proton contacts is in progress and the full results will be published elsewhere.

4.   Conclusions

The characterization of a given ligand/DNA complex can be undertaken from the results of many biochemical and biophysical test systems. We advocate to use as many assays as possible in order to circumvent the possible pitfalls and artefacts arising from the results of one assay for a given ligand/DNA complex. For screening purposes we propose the application of one-dimensional ^{1}H-NMR for the measurement of ligand-induced changes in chemical shifts of the NH-iminoprotons of the Watson-Crick base-pairs in short fragments of random-sequence DNA or synthetic polynucleotides as a fast check for mode of binding, kinetics of binding, and AT- or GC-base-pair specificity of a given ligand. In addition we advocate functional test systems, like the ligand induced inhibition of restriction enzymes, which are fast and reliable methods, giving rise to no "false negative" results.

Despite its drawbacks (e.g. minimal solubility required, limit in molecular weight) the two-dimensional ^{1}H-NMR method represents the method of choice for full characterization of a given ligand/DNA complex (using short oligonucleotides of known sequence). The ongoing developments in the design of new NMR hardware, the development in new NMR techniques, and the concomitant developments in easier and faster isotope labeling techniques will make this method indispensable in the future and an excellent complement to X-ray spectroscopy.

5.   Acknowledgements

I thank my colleagues mentioned in the references for the fruitful and enjoyable collaboration and Mrs. E. Hiss and Miss A. Perrin for the careful processing of the manuscript.

6.   References

Assa-Munt, N., Denny, W.A., Leupin, W., and Kearns, D.R. (1985a) "^{1}H NMR study of the binding of bis(acridines) to d(AT)5.d(AT)5. 1. Mode of binding", Biochemistry 24, 1441-1449

Assa-Munt, N., Leupin, W., Denny, W.A., and Kearns, D.R. (1985b) "^{1}H NMR study of the binding of bis(acridines) to d(AT)5.d(AT)5. 2. Dynamic aspects", Biochemistry 24, 1449-1460

Boyd, F.L., Cheatham, S.F., Remers, W., Hill, G.C., and Hurley, L.H. (1990) "Characterization of the structure of the Anthramycin-d(ATGCAT)$_2$ adduct by NMR and molecular modeling studies. Determination of the stereochemistry of the covalent linkage site, orientation in the minor groove of DNA, and effect on local DNA structure", Biochemistry 112, 3279-3289

Brown, S.C., Mullis, K., Levenson, C., and Shafer, R.H. (1984) "Aqueous solution structure of an intercalated Actinomycin D-dATGCAT complex by two-dimensional and one-dimensional proton NMR", Biochemistry 23, 403-408

Carrondo, M.A.A.F., Coll, M., Aymami, J., Wang, A.H.-J., van der Marel, G.A., van Boom, J.H., and Rich, A. (1989) "Binding of a Hoechst Dye to d(CGCGATATCGCG) and its influence on the conformation of the DNA fragment", Biochemistry 28, 7849-7859

Chazin, W.J., Wuthrich, K., Hyberts, S., Rance, M., Denny, W.A., and Leupin, W. (1986) "[1]H Nuclear magnetic resonance assignments for d-(GCATTAATGC)[2] using experimental refinements of established procedures", J. Mol. Biol. 190, 439-453

Dickerson, R.E. (1983) "Base sequence and helix structure variability in B- and A-DNA", J. Mol. Biol. 166, 419-441

Doeneke, D. (1977) "Ethidium bromide (EB) binding to nucleosomal DNA", Exp. Cell Res. 109, 309-315

Dougherty, G. and Pigram, W.J. (1982) "Spectroscopic analysis of drug-nucleic acid interactions", CRC Crit. Rev. Biochem. 12, 103

Erard, M., Das, G.C., de Murcia, G., Mazen, A., Pouyet, J., Champagne, M., and Daune, M. (1979) "Ethidium bromide binding to core particle: Comparison with native chromatin", Nucleic Acids Res. 6, 3231-3253

Feigon, J., Wright, J.M., Leupin, W., Denny, W.A., and Kearns, D.R. (1982a) "Use of two-dimensional NMR in the study of a double-stranded DNA decamer", J. Am. Chem. Soc. 104, 5540-5541

Feigon, J., Leupin, W., Denny, W.A., and Kearns, D.R. (1982b) "Binding of ethidium derivatives to natural DNA: a 300 MHz ^{1}H NMR study", Nucleic Acids Res. 10, 749-762

Feigon, J., Leupin, W., Denny, W.A., and Kearns, D.R. (1983) "Two-dimensional proton nuclear magnetic resonance investigation of the synthetic deoxyribonucleic acid decamer d(ATATCGATAT)", Biochemistry 22, 5943-5951

Feigon, J., Denny, W.A., Leupin, W., and Kearns, D.R. (1984) "Interactions of antitumor drugs with natural DNA: ^{1}H NMR study of binding mode and kinetics", J. Med. Chem. 27, 450-465

Fesik, S.W. (1989) in "Computer Aided Drug Design. Methods and Applications" (Perun, T.J. and Propst, C.L., editors) M. Dekker Inc., New York, pages 133-184

Fesik, S.W. and Zuiderweg, E.R.P. (1990) "Heteronuclear three-dimensional NMR spectroscopy of isotopically labelled biological macromolecules", Quart. Rev. Biophysics 23, 97-131

Fesik, S.W., Luly, J.R., Erickson, J.W., and Abad-Zapatero, C. (1988) "Isotope-edited proton NMR study on the structure of a pepsin/inhibitor complex", Biochemistry 27, 8297-830

Frey, M.H., Leupin, W., Sorensen, O.W., Denny, W.A., Ernst, R.R., and Wuthrich, K. (1985) "Sequence-specific assignment of the backbone ^{1}H- and ^{31}P-NMR lines in a short DNA duplex with homo- and heteronuclear correlated spectroscopy", Biopolymers 24, 2371-2380

Frederick, C.A., Williams, L.D., Ughetto, G., van der Marel, G.A., van Boom, J.H., Rich, A., and Wang, A.H.-J. (1990) " Structural comparison of anticancer drug-DNA complexes: Adriamycin and Daunomycin", Biochemistry 29, 2538-2549

Furey, W.F., Robbins, A.H., Clancy, L.L., Winge, D.R., Wang, B.C., and Stout, C.D. (1986) "Crystal structure of Cd,Zn metallothionein", Science 231, 704-710

Gale, E.F., Cundliffe, E., Reynolds, P.E., Richmond, M.H., and Waring, M.J. (1981) "The Molecular Basis of Antibiotic Action", second edition, John Wiley & Sons, London

Gao, X. and Patel, D.J. (1988) "NMR studies of Echinomycin bisintercalation complexes with d(A1-C2-G3-T4) and d(T1-C2-G3-A4) duplexes in aqueous solution: Sequence-dependent formation of Hoogsteen A1.T4 and Watson-Crick T1.A4 base pairs flanking the bisintercalation site", Biochemistry 27, 1744-1751

Gao, X. and Patel, D.J. (1989) "Solution structure of the Chromomycin-DNA complex", Biochemistry 28, 751-762

Gorenstein, D.G. and Lai, K. (1989) "31P NMR spectra of Ethidium, Quinacrine, and Daunomycin complexes with poly(adenylic acid).poly(uridylic acid) RNA duplex and calf thymus DNA", Biochemistry 28, 2804-2812

Gorenstein, D.G., Schroeder, S.G., Fu, J.M., Metz, J.T., Roongta, V., and Jones, C.R. (1988) "Assignments of ^{31}P NMR resonances in oligodeoxynucleotides: Origin of sequence-specific variations in the deoxyribose phosphate backbone conformation and the ^{31}P chemical shifts of double-helical nucleic acids", Biochemistry 27, 7223-7237

Gupta, G., Sarma, M.H., and Sarma, R.H. (1984) "Structure and Dynamics of Netropsin.poly(dA-dT).poly(dA-dT) complex: 500 MHz ^{1}H NMR Studies", J. Biomol. Struct. Dyn. 1, 1457-1472

Hansch, C., Li, R.-L., Blaney, J.M., and Langridge, R. (1982) "Comparison of the inhibition of Escherichia coli and Lactobacillus casei dihydrofolate reductase by 2,4-diamino-5-(substituted-benzyl)pyrimidines: Quantitative Structure-Activity Relationships, X-ray crystallography, and computer graphics in structure-activity analysis", J. Med. Chem., 25, 777-784

Hare, D.R., Wemmer, D.E., Chou, S.-H., Drobny, G., and Reid, B.R. (1983) "Assignment of the non-exchangeable proton resonance of d(CGCGAATTCGCG) using two-dimensional nuclear magnetic resonance methods", J. Mol. Biol. 171, 319-336

Kearns, D.R. (1977) "High-resolution nuclear magnetic resonance studies of double helical polynucleotides", Annu. Rev. Biophys. Bioeng. 6, 477-523

Klevit, R.E., Wemmer, D.E., and Reid, B.R. (1986) "1H NMR studies on the interaction between Distamycin A and a symmetrical DNA dodecamer", Biochemistry 25, 3296-3303

Kline, A.D., Braun, W., and Wuthrich, K. (1986) "Studies by ^{1}H nuclear magnetic resonance and distance geometry of the solution conformation of the α-amylase inhibitor Tendamistat", J. Mol. Biol. 189, 377-382

Kopka, M.L., Yoon, C., Goodsell, D., Pjura, P., and Dickerson, R.E. (1985) "The binding of an antitumor drug to DNA: Netropsin and d(CGCGAATT^{Br}CGCG)", J. Mol. Biol. 183, 553-563

LaPlante, S.R., Boudreau, E.A., Zanatta, N., Levy, G.C., Borer, P.N., Ashcroft, J., and Cowburn, D. (1988) "^{13}C NMR of the bases of three DNA oligonucleotide duplexes: Assignment methods and structural features", Biochemistry 27, 7902-7909

LeMaster, D.M. (1990) "Deuterium labelling in NMR structural analysis of larger proteins", Quart. Rev. Biophysics 23, 133-174

Leupin, W., Chazin, W.J., Hyberts, S., Denny, W.A., and Wuthrich, K. (1986) "NMR studies of the complex between the decadeoxynucleotide d-(GCATTAATGC)$_2$ and a minor-groove-binding drug", Biochemistry 25, 5902-5910

Leupin, W., Wagner, G., Denny, W.A., and Wuthrich, K. (1987) "Assignment of the ^{13}C nuclear magnetic resonance spectrum of a short DNA-duplex with ^{1}H-detected two-dimensional heteronuclear correlation spectroscopy", Nucleic Acids Res. 15, 267-275

Leupin, W., Otting, G., Amacker, H., and Wuthrich, K. (1990) "Application of ^{13}C(ω_1)-half-filtered [^{1}H,^{1}H]-NOESY for studies of a complex formed between DNA and a ^{13}C-labelled minor-groove-binding drug", FEBS Letters 263, 313-316

Leupin, W., Feigon, J., Denny, W.A., and Kearns, D.R. (1985) "Substituent effects on the binding of ethidium and its derivatives to natural DNA", Biophys. Chem. 22, 299-305

Low, C.M.L., Olsen, R.E., and Waring, M.J. (1984) "Sequence preferences in the binding to DNA of Triostin A and TANDEM as reported by DNase I footprinting", FEBS Letters 176, 414-420

Malcolm, A.D.B. and Moffat, J.R. (1981) "Differential reactivities at restriction enzyme sites", Biochim. Biophys. Acta 655, 128-135

McIntosh, L.P. and Dahlquist, F.W. (1990) "Biosynthetic incorporation of ^{15}N and ^{13}C for assignment and interpretation of nuclear magnetic resonance spectra of proteins", Quart. Rev. Biophysics 23, 1-38

McLean, M.J. and Waring, M.J. (1988) "Chemical probes reveal no evidence of Hoogsteen base pairing in complexes formed between Echinomycin and DNA in solution", J. Mol. Recognition 1, 138-151

Mirau, P.A. and Kearns, D.R. (1983) "The effect of intercalating drugs on the kinetics of the B to Z transition of poly(dG-dC)", Nucleic Acids Res. 11, 1931-1941

Neidle, S. and Abraham, Z. (1984) "Structural and sequence-dependent aspects of drug intercalation into nucleic acids", CRC Crit. Rev. Biochem. 17, 73-121

Otting, G. and Wuthrich, K. (1990) "Heteronuclear filters in two-dimensional [^{1}H,^{1}H]-NMR spectroscopy: combined use with isotope labelling for studies of macromolecular conformation and intermolecular interactions", Quart Rev. Biophysics 23, 39-96

Otting, G., Grutter, R., Leupin, W., Minganti, C., Ganesh, K.N., Sproat, B.S., Gait, M.J., and Wuthrich, K. (1987) "Sequential NMR assignments of labile protons in DNA using two-dimensional nuclear-Overhauser-enhancement spectroscopy with three jump-and-return pulse sequences", Eur. J. Biochem. 166, 215-220

Paoletti, J., Magee, B.B., and Magee, P.T. (1977) "The structure of chromatin: Interaction of Ethidium bromide with native and denatured chromatin", Biochemistry 16, 351-357

Parkinson, J.A., Barber, J., Douglas, K.T., Rosamund, J., and Sharples, D. (1989) "Nuclear magnetic resonance probing of binding interactions used by minor groove binding, DNA-directed ligands: Assigment of the binding site of Hoechst 33258 on the self-complementary oligonucleotide, d(CGCGAATTCGCG)" J. Chem. Soc. Chem. Commun., 1023-1025

Patel, D.J. (1979) "NMR studies of drug-nucleic acid interactions at the synthetic DNA level in solution", Acc. Chem. Res. 12, 118-125

Patel, D.J. (1982) "Antibiotic-DNA interactions: Intermolecular nuclear Overhauser effects in the Netropsin-d(CGCGAATTCGCG) complex in solution", Proc. Natl. Acad. Sci. USA 79, 6424-6428

Patel, D.J. and Shapiro, L. (1986) "Sequence-dependent recognition of DNA duplexes. Netropsin complexation to the AATT site of the d(GGAATTCC) duplex in aqueous solution", J. Biol. Chem., 261, 1230-1240

Patel, D.J., Shapiro, L., and Hare, D. (1987) DNA and RNA: NMR studies of conformations and dynamics in solution", Quart. Rev. Biophysics 20, 35-112

Pelton, J.G. and Wemmer, D.E. (1989) "Structural characterization of a 2:1 Distamycin A.d(CGCAAATTGGC) complex by two-dimensional NMR", Proc. Natl Acad. Sci. USA 86, 5723-5727

Pelton, J.G. and Wemmer, D.E. (1990) "Binding modes of Distamycin A with d(CGCAAATTTGCG)$_2$ determined by two-dimensional NMR" J. Am. Chem. Soc. 112, 1393-1399

Pflugrath, J.W., Wiegand, G., Huber, R., and Vertesy, L. (1986) "Crystal structure determination, refinement and the molecular model of the α-amylase inhibitor Hoe-467 A", J. Mol. Biol. 189, 383-386

Phillips, D.R. and Roberts, G.C.K. (1980) "Proton nuclear magnetic resonance study of the self-complementary hexanucleotide d(pTpA)$_3$ and its interaction with Daunomycin", Biochemistry 19, 4795-4801

Pjura, P.E., Grzeskowiak, K., and Dickerson, R.E. (1987) "Binding of Hoechst 33258 to the minor groove of B-DNA", J. Mol. Biol. 197, 257-271

Pohl, F.M., Jovin, T.M., Baehr, W., and Holbrook, J.J. (1972) "Ethidium bromide as a cooperative effector of a DNA structure", Proc. Natl. Acad. Sci. USA 69, 3805-3809

Polyanousky, O.L., Nosikov, V.V., Zhuze, A.L., Braga, E.A., and Karyshev, A.V. (1978) "Regulation of restriction endonuclease activity with antibiotics", Adv. Enzyme Regul. 17, 307

Portugal, J. and Waring, M.J. (1987) "Analysis of the effects of antibiotics on the structure of nucleosome core particles determined by DNAase I cleavage", Biochimie 69, 825-840

Powers, R., Olsen, R.K., and Gorenstein, D.G. (1989) "Two-dimensional ^{1}H and ^{31}P NMR spectra of a decamer oligodeoxyribonucleotide duplex and a quinoxaline ([meCys3,meCys7]TANDEM) drug duplex complex", J. Biomol. Struct. Dyn. 7, 515-556

Pullman, B. (1989) "Molecular mechanisms of specificity in DNA-antitumor drug interactions", Advances in Drug Res. 18, 1-113

Quigley, G.J., Ughetto, G., van der Marel, G.A., van Boom, J.H., Wang, A.H.-J., and Rich, A. (1986) "Non-Watson-Crick GC and AT base pairs in a DNA-antibiotic complex", Science 232, 1255-1258

Reid, B.R. (1987) "Sequence specific assignments and their use in NMR studies of DNA structure", Quart. Rev. Biophysics 20, 1-34

Roberts, R.J. (1990) "Restriction enzymes and their isoschizomers", Nucleic Acids Res. 18, 2331-2365

Scott, E.V., Zon, G., Marzilli, L.G., and Wilson, W.D. (1988) "2D NMR investigation of the binding of the anticancer drug Actinomycin D to duplexed d(ATGCGCAT); Conformational features of the unique 2:1 adduct", Biochemistry 27, 7940-7951

Senn, H., Otting, G., and Wuthrich, K. (1987) "Protein structure and interactions by combined use of sequential NMR assignments and isotope labelling", J. Am. Chem. Soc. 109, 1090-1092

Stout, C.D., McRee, D.E., Robbins, A.H., Collett, S.A., Williamson, M., and Xuong, X.H. (1989) Abstracts of Internatl. Chem. Congress of Pacific Bassin Societies, Honolulu, Hawaii, USA, Dec. 17-22, Volume I, Page 04-57

Scheek, R.M., Russo, N., Boelens, R., Kaptein, R., and van Boom, J.H. (1983) "Sequential resonance assignments in DNA ^{1}H NMR spectra by two-dimensional NOE spectroscopy", J. Am. Chem. Soc. 105, 2914-2916

Scheek, R.M., Boelens, R., Russo, N., van Boom, J.H., and Kaptein, R. (1984) "Sequential resonance assignments in ^{1}H NMR spectra of oligonucleotides by two-dimensional NMR spectroscopy", Biochemistry 23, 1371-1376

Schultze, P., Worgotter, E., Braun, W., Wagner, G., Vasak, M., Kagi, J.H.R., and Wuthrich, K. (1988) "Conformation of [Cd$_7$]-metallothionein-2 from rat liver in aqueous solution determined by nuclear magnetic resonance spectroscopy", J. Mol. Biol. 203, 251-268

Teng, M.-K., Usman, N., Frederick, C.A., and Wang, A.H.-J. (1988) "The molecular structure of the commplex of Hoechst 33258 and the DNA dodecamer d(CGCGAATTCGCG)", Nucleic Acids Res. 16, 2671-2690

Ughetto, G., Wang, A.H.-J., Quigley, G.J., van der Marel, G., van Boom, J.H., and Rich A., (1985) "A comparison of the structure of Echinomycin and Triostin, A complexed to a DNA fragment", Nucleic Acids Res. 13, 2305-2323

Van de Ven, F.J.M. and Hilbers, C.W. (1988) "Nucleic acids and nuclear magnetic resonance", Eur. J. Biochem. 178, 1-38

Van Dyke, M.W. and Dervan, P.B. (1984) "Echinomycin binding sites on DNA", Science 225, 1122-1127

Wakelin, L.P.G., Atwell, G.J., Rewcastle, G.W., and Denny, W.A. (1987) "Relationships between DNA binding kinetics and biological activity for the 9-aminoacridine-4-carboxamide class of antitumor agents", J. Med. Chem. 30, 855-861

Wang, A.H.-J., Gessner, R.V., van der Marel, G.A., van Boom, J.H., and Rich, A. (1985) "Crystal structure of Z-DNA without an alternating purine-pyrimidine sequence", Proc. Natl. Acad. Sci. USA 82, 3611-3615

Wang, A.H.-J., Ughetto, G., Quigley, G.J., and Rich, A. (1987) "Interactions between an anthracycline antibiotic and DNA: Molecular structure of Daunomycin complexed to d(DpGpTpApCpG) at 1.2 A resolution", Biochemistry 26, 1152-1163

Wilson, W.D., Tanious, F.A., Barton, H.J., Wydra, R.L., Jones, R.L., Boykin, D.W., and Strekowski, L. (1990) "The interaction of unfused polyaromatic heterocycles with DNA: intercalation, groove-binding and bleomycin amplification", Anti-Cancer Drug Des. 5, 31-42

Wuthrich, K. (1986) "NMR of Proteins and Nucleic Acids", Wiley, New York

Yoon, C., Prive, G.G., Goodsell, D.S., and Dickerson, R.E. (1988) "Structure of an alternating B-DNA helix and its relationship to A-tract DNA", Proc. Natl. Acad. Sci. USA 85, 6332-6336

Zimmer, C., and Wahnert, U. (1986) "Nonintercalating DNA-binding ligands: Specificity of the interactions and their use as tools in biophysical, biochemical and biological investigations of the genetic material", Prog. Biophys. Mol. Biol. 47, 31-112